中 外 物 理 学 精 品 书 系

本 书 出 版 得 到 “ 国 家 出 版 基 金 ” 资 助

国家出版基金项目
NATIONAL PUBLICATION FOUNDATION

中 外 物 理 学 精 品 书 系

引 进 系 列 · 47

Topology and Geometry for Physics

物理学中的拓扑与几何

（影印版）

〔德〕埃施里格（H．Eschrig） 著

北京大学出版社
PEKING UNIVERSITY PRESS

著作权合同登记号　图字:01-2014-3697
图书在版编目(CIP)数据

物理学中的拓扑与几何 = Topology and geometry for physics:英文/(德)埃施里格(Eschrig,H.)著.—影印本.—北京:北京大学出版社,2014.10
(中外物理学精品书系)
ISBN 978-7-301-24830-0

Ⅰ. ①物… Ⅱ. ①埃… Ⅲ. ①拓扑—应用—物理学—英文 ②几何—应用—物理学—英文 Ⅳ. ①O4

中国版本图书馆 CIP 数据核字(2014)第 216802 号

书　　　名:Topology and Geometry for Physics(物理学中的拓扑与几何)(影印版)
著作责任者:〔德〕埃施里格(H. Eschrig)　著
责 任 编 辑:刘　啸
标 准 书 号:ISBN 978-7-301-24830-0/O・1007
出 版 发 行:北京大学出版社
地　　　址:北京市海淀区成府路 205 号　100871
网　　　址:http://www.pup.cn
新 浪 微 博:@北京大学出版社
电 子 信 箱:zpup@pup.cn
电　　　话:邮购部 62752015　发行部 62750672　编辑部 62752038　出版部 62754962
印　刷　者:北京中科印刷有限公司
经　销　者:新华书店
730 毫米×980 毫米　16 开本　25.5 印张　486 千字
2014 年 10 月第 1 版　2014 年 10 月第 1 次印刷
定　　　价:69.00 元

序　言

物理学是研究物质、能量以及它们之间相互作用的科学。她不仅是化学、生命、材料、信息、能源和环境等相关学科的基础，同时还是许多新兴学科和交叉学科的前沿。在科技发展日新月异和国际竞争日趋激烈的今天，物理学不仅囿于基础科学和技术应用研究的范畴，而且在社会发展与人类进步的历史进程中发挥着越来越关键的作用。

我们欣喜地看到，改革开放三十多年来，随着中国政治、经济、教育、文化等领域各项事业的持续稳定发展，我国物理学取得了跨越式的进步，做出了很多为世界瞩目的研究成果。今日的中国物理正在经历一个历史上少有的黄金时代。

在我国物理学科快速发展的背景下，近年来物理学相关书籍也呈现百花齐放的良好态势，在知识传承、学术交流、人才培养等方面发挥着无可替代的作用。从另一方面看，尽管国内各出版社相继推出了一些质量很高的物理教材和图书，但系统总结物理学各门类知识和发展，深入浅出地介绍其与现代科学技术之间的渊源，并针对不同层次的读者提供有价值的教材和研究参考，仍是我国科学传播与出版界面临的一个极富挑战性的课题。

为有力推动我国物理学研究、加快相关学科的建设与发展，特别是展现近年来中国物理学者的研究水平和成果，北京大学出版社在国家出版基金的支持下推出了“中外物理学精品书系”，试图对以上难题进行大胆的尝试和探索。该书系编委会集结了数十位来自内地和香港顶尖高校及科研院所的知名专家学者。他们都是目前该领域十分活跃的专家，确保了整套丛书的权威性和前瞻性。

这套书系内容丰富，涵盖面广，可读性强，其中既有对我国传统物理学发展的梳理和总结，也有对正在蓬勃发展的物理学前沿的全面展示；既引进和介绍了世界物理学研究的发展动态，也面向国际主流领域传播中国物理的优秀专著。可以说，“中外物理学精品书系”力图完整呈现近现代世界和中国物理科学发展的全貌，是一部目前国内为数不多的兼具学术价值和阅读乐趣的经典物理丛书。

“中外物理学精品书系”另一个突出特点是，在把西方物理的精华要义“请进来”的同时，也将我国近现代物理的优秀成果“送出去”。物理学科在世界范围内的重要性不言而喻，引进和翻译世界物理的经典著作和前沿动态，可以满足当前国内物理教学和科研工作的迫切需求。另一方面，改革开放几十年来，我国的物理学研究取得了长足发展，一大批具有较高学术价值的著作相继问世。这套丛书首次将一些中国物理学者的优秀论著以英文版的形式直接推向国际相关研究的主流领域，使世界对中国物理学的过去和现状有更多的深入了解，不仅充分展示出中国物理学研究和积累的“硬实力”，也向世界主动传播我国科技文化领域不断创新的“软实力”，对全面提升中国科学、教育和文化领域的国际形象起到重要的促进作用。

值得一提的是，“中外物理学精品书系”还对中国近现代物理学科的经典著作进行了全面收录。20 世纪以来，中国物理界诞生了很多经典作品，但当时大都分散出版，如今很多代表性的作品已经淹没在浩瀚的图书海洋中，读者们对这些论著也都是“只闻其声，未见其真”。该书系的编者们在这方面下了很大工夫，对中国物理学科不同时期、不同分支的经典著作进行了系统的整理和收录。这项工作具有非常重要的学术意义和社会价值，不仅可以很好地保护和传承我国物理学的经典文献，充分发挥其应有的传世育人的作用，更能使广大物理学人和青年学子切身体会我国物理学研究的发展脉络和优良传统，真正领悟到老一辈科学家严谨求实、追求卓越、博大精深的治学之美。

温家宝总理在 2006 年中国科学技术大会上指出，“加强基础研究是提升国家创新能力、积累智力资本的重要途径，是我国跻身世界科技强国的必要条件”。中国的发展在于创新，而基础研究正是一切创新的根本和源泉。我相信，这套“中外物理学精品书系”的出版，不仅可以使所有热爱和研究物理学的人们从中获取思维的启迪、智力的挑战和阅读的乐趣，也将进一步推动其他相关基础科学更好更快地发展，为我国今后的科技创新和社会进步做出应有的贡献。

“中外物理学精品书系”编委会　主任
中国科学院院士，北京大学教授
王恩哥
2010 年 5 月于燕园

Helmut Eschrig

Topology and Geometry for Physics

Springer

Preface

The real revolution in mathematical physics in the second half of twentieth century (and in pure mathematics itself) was algebraic topology and algebraic geometry. Meanwhile there is the Course in Mathematical Physics by W. Thirring, a large body of monographs and textbooks for mathematicians and of monographs for physicists on the subject, and field theorists in high-energy and particle physics are among the experts in the field, notably E. Witten. Nevertheless, I feel it still not to be easy for the average theoretical physicist to penetrate into the field in an effective manner. Textbooks and monographs for mathematicians are nowadays not easily accessible for physicists because of their purely deductive style of presentation and often also because of their level of abstraction, and they do not really introduce into physics applications even if they mention a number of them. Special texts addressed to physicists, written both by mathematicians or physicists in most cases lack a systematic introduction into the mathematical tools and rather present them as a patchwork of recipes. This text tries an intermediate approach. Written by a physicist, it still tries a rather systematic but more inductive introduction into the mathematics by avoiding the minimalistic deductive style of a sequence of theorems and proofs without much of commentary or even motivating text. Although theorems are highlighted by using italics, the text in between is considered equally important, while proofs are sketched to be spelled out as exercises in this branch of mathematics. The text also mainly addresses students in solid state and statistical physics rather than particle physicists by the focusses and the choice of examples of application.

Classical analysis was largely physics driven, and mathematical physics of the nineteens century was essentially the classical theory of ordinary and partial differential equations. Variational calculus, since the very beginning of theoretical mechanics a standard tool of physicists, was seen with great reservation by mathematicians until D. Hilbert initiated its rigorous foundation by pushing forward functional analysis. This marked the transition into the first half of twentieth century, where under the influence of quantum mechanics and relativity mathematical physics turned mainly into functional analysis (as for instance witnessed by the textbooks of M. Reed and B. Simon), complemented by the theory of Lie groups and by tensor analysis. Physicists, nowadays more or less familiar with these branches, still are on average mainly analytically and very little algebraically educated, to say nothing of topology. So it could happen that for nearly sixty years it was overlooked that not every quantum mechanical observable may be represented by an operator in Hilbert space, and only in the middle of the eighties of last century with Berry's phase, which is such an observable, it was realized how polarization in an infinitely extended crystal is correctly described and that textbooks even by most renowned authors contained meaningless statements about this question.

This author feels that all branches of theoretical physics still can expect the strongest impacts from use of the unprecedented wealth of results of algebraic topology and algebraic geometry of the second half of twentieth century, and to introduce theoretical physics students into its basics is the purpose of this text. It is still basically a text in mathematics, physics applications are included for illustration and are chosen mainly from the fields the author is familiar with. There are many important examples of application in physics left out of course. Also the cited literature is chosen just to give some sources for further study both in mathematics and physics. Unfortunately, this author did not find an English translation of the marvelous Analyse Mathématique by L. Schwartz,[1] which he considers (from the Russian edition) as one of the best textbooks of modern analysis. A rather encyclopedic text addressed to physicists is that by Choquet-Bruhat et al.,[2] however, a compromise between the wide scope and limitations in space made it in places somewhat sketchy.

The order of the material in the present text is chosen such that physics applications could be treated as early as possible without doing too much violence to the inner logic of the mathematical building. As already said, central results are highlighted in italics but purposely avoiding the structure of a sequence of theorems. Sketches of proofs are given, if they help understanding the matter. They are understood as exercises for the reader to spell them out in more detail. Purely technical proofs are omitted even if they prove central issues of the theory. A compendium is appended to the basic text for reference also of some concepts (for instance of general algebra) used in the text but not treated. This appendix is meant as an expanded glossary and, apart form very few exceptions, not covered by the index.

Finally, I would like to acknowledge many suggestions for improvement and corrections by people from the Springer-Verlag.

Dresden, May 2010 Helmut Eschrig

---

[1] Schwartz, L.: Analyse Mathématique. Hermann, Paris (1967).

[2] Choquet-Bruhat, Y., de Witt-Morette, C., Dillard-Bleick, M.: Analysis, Manifolds and Physics, Elsevier, Amsterdam, vol. I (1982), vol. II (1989).

Contents

Basic notations

Sets $A, B, \ldots, X, Y, \ldots$ are subjects of the axioms of set theory. $A = \{x \mid P(x)\}$ denotes the family of elements x having the property P; if the elements x are members of a set X, $x \in X$, then the above family is a set, a subset (part) of the set X: $A \subset X$. X is a superset of $A, X \supset A$. $\subset, \supset$ will always be used to allow equality. A proper subset (superset) would be denoted by $A \subsetneq X (X \supsetneq A)$. Union, intersection and complement of A relative to X have their usual meaning. The product of n sets is in the usual manner the set of ordered n-tuples of elements, one of each factor.

Set and space as well as subset and part are used synonymously. Depending on context the elements of a space may be called points, n-tuples, vectors, functions, operators, or something else. **Mapping** and **function** are also used synonymously. A function f from the set A into the set B is denoted $f : A \to B : x \mapsto y$. It maps *each* point $x \in A$ *uniquely* to some point $y = f(x) \in B$. A is the domain of f and $f(A) = \{f(x) \mid x \in A\} \subset B$ is the range of f; if $U \subset A$, then $f(U) = \{f(x) \mid x \in U\}$is the image of U under f. The inverse image or preimage $U = f^{-1}(V) \subset A$ of $V \subset B$ under f is the set $f(U) = \{x \mid f(x) \in V\}$. V need not be a subset of the range $f(A)$; f^{-1} (V) may be empty. Depending on context, f may be called real, complex, vector-valued, function-valued, operator-valued, ...

The function $f : A \to B$ is called **surjective** or onto, if $f(A) = B$. It is called **injective** or one-one, if for each $y \in f(A)$, $f^{-1}(\{y\}) = f^{-1}(y)$ consists of a single point of A. In this case the inverse function $f^{-1} : f(A) \to A$ exists. A surjective and injective function is **bijective** or onto and one-one. If a bijection between A and B exists then the two sets have the same cardinality. A set is **countable** if it has the cardinality of the set of natural numbers or of one of its subsets.

The **identity mapping** $f : A \to A : x \mapsto x$ is denoted by Id_A. Extensions and restrictions of f are defined in the usual manner by extensions or restrictions of the domain. The restriction of $f : A \to B$ to $A' \subset A$ is denoted by $f|_{A'}$. If $f : A \to B$ and $g : B \to C$, then the **composite mapping** is denoted by $g \circ f : A \to C : x \mapsto g(f(x))$.

The monoid of natural numbers (non-negative integers, 0 included) is denoted by $\mathbb{N}$. The ring of integers is denoted by $\mathbb{Z}$, sometimes the notation $\mathbb{N} = \mathbb{Z}_+$ is used. The field of rational numbers is denoted by $\mathbb{Q}$, that of real numbers is denoted by $\mathbb{R}$ and that of complex numbers by $\mathbb{C}$. $\mathbb{R}_+$ is the non-negative ray of $\mathbb{R}$.

The symbol $\Rightarrow$ means 'implies', and $\Leftrightarrow$ means 'is equivalent to'. 'Iff' abbreviates 'if and only if' (that is, $\Leftrightarrow$), and $\Box$ denotes the end of a proof.

Chapter 1
Introduction

Topology and continuity on the one hand and geometry or metric and distance on the other hand are intimately connected pairs of concepts of central relevance both in analysis and physics. A totally non-trivial concept in this connection is parallelism.

As an example, consider a mapping f from some two-dimensional area into the real line as in Fig. 1.1a. Think of a temperature distribution on that area. We say that f is continuous at point $\boldsymbol{x}$, if for any neighborhood V of $y = f(\boldsymbol{x})$ there exists a neighborhood U of $\boldsymbol{x}$ (for instance U_1 in Fig. 1.1a for V indicated there) which is mapped into V by f. It is clear that the concept of neighborhood is central in the definition of continuity.

As another example, consider the mapping g of Fig. 1.1b. The curve segment W_1 is mapped into V, but the segment W_2 is not: its part above the point $\boldsymbol{x}$ is mapped into an interval above $y = g(\boldsymbol{x})$ and its part below $\boldsymbol{x}$ is mapped disruptly into a lower interval. Hence, there is no segment of the curve W_2 which contains $\boldsymbol{x}$ as an inner point and which is mapped into V by g. The map g is continuous on the curve W_1 but is discontinuous at $\boldsymbol{x}$ on the curve W_2. (The function value makes a jump at $\boldsymbol{x}$.) Hence, it cannot be continuous at $\boldsymbol{x}$ as a function on the two-dimensional area. To avoid conflict with the above definition of continuity, the curve W_1 must not be considered a neighborhood of $\boldsymbol{x}$ in the two-dimensional area.

If f is a mapping from a metric space (a space in which the distance $d(\boldsymbol{x},\boldsymbol{x}')$ between any two points $\boldsymbol{x}$ and $\boldsymbol{x}'$ is defined) into another metric space, then it suffices to consider open balls $B_\varepsilon(\boldsymbol{x}) = \{\boldsymbol{x}'|d(\boldsymbol{x},\boldsymbol{x}') < \varepsilon\}$ of radius ε as neighborhoods of $\boldsymbol{x}$. The metric of the n-dimensional Euclidean space $\mathbb{R}^n$ is given by $d(\boldsymbol{x},\boldsymbol{x}') = (\sum_{i=1}^n (x^i - x'^i)^2)^{1/2}$ where the x^i are the Cartesian coordinates of $\boldsymbol{x}$. It also defines the usual topology of the $\mathbb{R}^n$. (The open balls form a base of that topology; no two-dimensional open ball is contained in the set W_1 above.)

Later on in Chap. 2 the topology of a space will be precisely defined. Intuitively any open interval containing the point x may be considered a neighborhood of x on the real line $\mathbb{R}$ (open intervals form again a base of the usual topology on $\mathbb{R}$). Recall that the product $X \times Y$ of two sets X and Y is the set of ordered pairs (x, y),

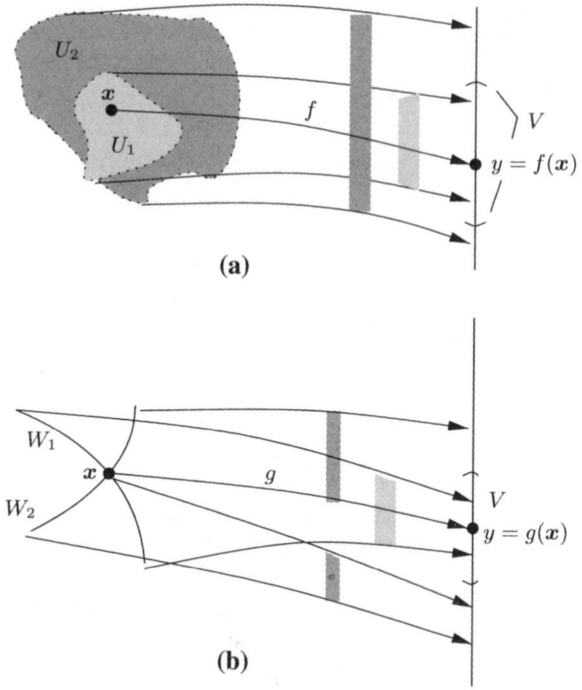

Fig. 1.1 Mappings from a two-dimensional area into the real line. **a** mapping f continuous at $\boldsymbol{x}$, **b** mapping g discontinuous at $\boldsymbol{x}$. The arrows and shaded bars indicate the range of the mapping of the sets U_1, U_2, W_1 and parts of W_2, respectively

$x \in X, y \in Y$. If X and Y are topological spaces, this leads naturally to the product topology in $X \times Y$ with a base of sets $\{(x, y)|x \in U, y \in V\}$ where U and V are in the base of the topology of X and Y, respectively. If this way the Cartesian plane is considered as the topological product of two real lines, $\mathbb{R}^2 = \mathbb{R} \times \mathbb{R}$, then the corresponding base is the set of all open rectangles. (This base defines the same topology in $\mathbb{R}^2$ as the base of open balls.) Note that neither distances nor angles need be defined so far in $\mathbb{R} \times \mathbb{R}$: topology is insensitive to stretchings or skew distortions as long as they are continuous.

Consider next the unit circle, 'the one-dimensional unit sphere' S^1, as a topological space with all open segments as base of topology, and the open unit interval $I =]0, 1[$ on the real line, with open subintervals as base of topology. Then, the topological product $S^1 \times I$ is the unit cylinder with its natural topology. Cut the cylinder on a line 'above one point of S^1', turn one cut edge around by 180° and glue the edges together again. A **Möbius band** is obtained (Fig. 1.2). This rises the question, can a Möbius band be considered as a topological product similar to the case of the unit cylinder? (Try it!) The true answer is no.

There are two important conclusions from that situation: (i) besides the local properties of a topology intuitively inferred from its base there are obviously important global properties of a topology, and (ii) a generalization of topological product is needed where gluings play a key role.

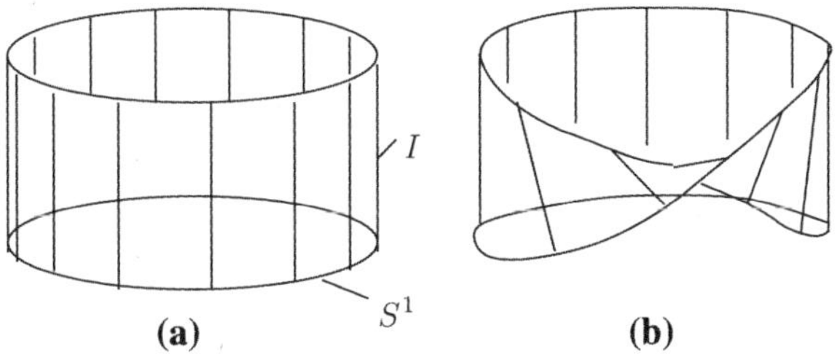

Fig. 1.2 **a** The unit cylinder and **b** the Möbius band

This latter generalization is precisely what a (topological) manifold is. The unit cylinder cut through in the above described way may be unfolded into an open rectangle of the plane $\mathbb{R}^2$. Locally, the topology of the unit cylinder and of the Möbius band and of $\mathbb{R}^2$ are the same. Globally they are all different. (The neighborhoods at the left and right edge of the rectangle are independent while on the unit cylinder they are connected.) Another example is the ordinary sphere S^2 embedded in the $\mathbb{R}^3$. Although its topology is locally the same as that of $\mathbb{R}^2$, globally it is different from any part of the $\mathbb{R}^2$. (From the stereographic projection which is a continuous one-one mapping it is known that the global topology of the sphere S^2 is the same as that of the completed or better compactified plane $\overline{\mathbb{R}^2}$ with the 'infinite point' and its neighborhoods added.) The S^2-problem was maybe first considered by Merkator (1512–1569) as the problem to project the surface of the earth onto planar charts. The key to describe manifolds are atlases of charts.

Topological space is a vast category, topological product is a construction of new topological spaces from simpler ones. Manifold is yet another construction to a similar goal. An m-dimensional manifold is a topological space the local topology of which is the same as that of $\mathbb{R}^m$. Not every topological space is a manifold. Since a manifold is a topological space, a topological product of manifolds is just a special case of topological product of spaces. A simple example is the two-dimensional torus $\mathbb{T}^2 = S^1 \times S^1$ of Fig. 1.3.

More special cases of topological spaces with richer structure are obtained by assigning to them additional algebraic and analytic structures. Algebraically, the

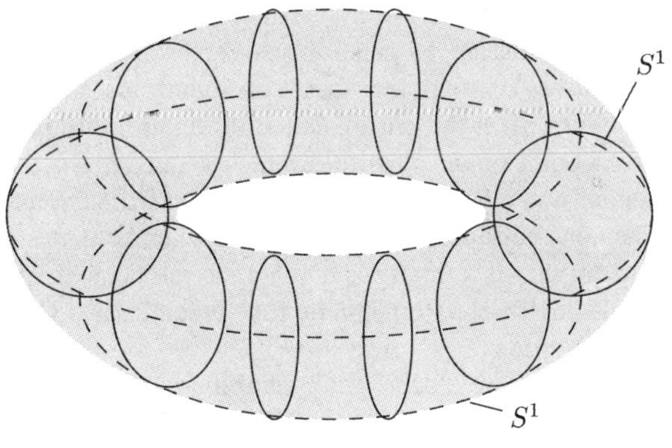

Fig. 1.3 The two-dimensional torus $\mathbb{T}^2 = S^1 \times S^1$

$\mathbb{R}^n$ is usually considered as a vector space (see Compendium at the end of this book) over the scalar field of real numbers, that is, a linear space. It may be attached with the usual topology which is such that multiplication of vectors by scalars, $(\lambda, \boldsymbol{x}) \mapsto \lambda \boldsymbol{x}$, and addition of vectors, $(\boldsymbol{x}, \boldsymbol{y}) \mapsto \boldsymbol{x} + \boldsymbol{y}$, are continuous functions from $\mathbb{R} \times \mathbb{R}^n$ to $\mathbb{R}^n$ and from $\mathbb{R}^n \times \mathbb{R}^n$ to $\mathbb{R}^n$, respectively. As was already mentioned, this topology can likewise be derived as a product topology from n factors $\mathbb{R}$ or from the Euclidean metric related to the usual Euclidean scalar product of vectors. The latter defines lengths and angles. For good reasons a metric will be used only on a much later stage as it is too restrictive for many considerations. So far, linear operations are defined and continuous, for instance linear dependence is defined, but angles and orthogonality remain undefined. If $\boldsymbol{e}_i, i = 1 \ldots n$ are n linearly independent vectors of $\mathbb{R}^n$, then any vector $\boldsymbol{x} \in \mathbb{R}^n$ can be written as $\boldsymbol{x} = \sum_i x^i \boldsymbol{e}_i$ with uniquely defined components x^i in the basis $\{\boldsymbol{e}_i\}$.

If X and Y are two topological vector spaces, then their algebraic direct sum $Z = X \oplus Y$ with the product topology is again a topological vector space. Any vector $\boldsymbol{z} \in Z$ is uniquely decomposed into $\boldsymbol{z} = \boldsymbol{x} + \boldsymbol{y}$, $\boldsymbol{x} \in X$, $\boldsymbol{y} \in Y$, and the canonical projections $\boldsymbol{pr}_1$ and $\boldsymbol{pr}_2$, $\boldsymbol{pr}_1(\boldsymbol{z}) = \boldsymbol{x}$, $\boldsymbol{pr}_2(\boldsymbol{z}) = \boldsymbol{y}$ are continuous. (Orthogonality of $\boldsymbol{x}$ and $\boldsymbol{y}$ again is not an issue here.)

Analysis is readily introduced in topological vector spaces. Let $\boldsymbol{f} : \mathbb{R}^n \to \mathbb{R}^m$ be any function, $\boldsymbol{f}(\boldsymbol{x}) = \boldsymbol{y}$ or more explicitly with respect to bases, $\boldsymbol{f}(x^1, \ldots, x^n) = (y^1, \ldots, y^m)$, that is, $f^i(\boldsymbol{x}) = y^i$. If the limits

$$\left.\frac{\partial f^i}{\partial x^k}\right|_{\boldsymbol{x}} = \lim_{t \to 0} \frac{f^i(\boldsymbol{x} + t\boldsymbol{e}_k) - f^i(\boldsymbol{x})}{t} \tag{1.1}$$

exist *and are continuous in* $\boldsymbol{x}$, then the vector function $\boldsymbol{f}$ is differentiable with derivative

$$\frac{\partial \boldsymbol{f}}{\partial \boldsymbol{x}} = \left(\frac{\partial f^1}{\partial \boldsymbol{x}}, \ldots, \frac{\partial f^m}{\partial \boldsymbol{x}}\right) = \left(\frac{\partial f^i}{\partial x^k}\right). \tag{1.2}$$

For $n = 1$ think of a velocity vector as the derivative of $\boldsymbol{x}(t)$, for $m = n = 4$ think of the electromagnetic field tensor as twice the antisymmetric part of the derivative of the four-potential $A^\mu(x^\nu)$. Higher derivatives are likewise obtained.

Manifolds are in general not vector spaces (cf. Figs. 1.2, 1.3) and therefore derivatives of mappings between manifolds cannot be defined in a direct way. However, if m-dimensional manifolds are sufficiently smooth, one may at any given point of the manifold attach a tangent vector space to it and project in a certain way a neighborhood of that point from the manifold into this tangent space. Then one considers derivatives in those tangent spaces. If a point moves in time on a manifold, its velocity is a vector in the tangent space. If space–time is a curved manifold, the electromagnetic four-potential is a vector and the field a tensor in the tangent space.

The derivative of a vector field meets however a new difficulty: the numerator of Eq. 1.1 is the difference of vectors at different points of the manifold which lie in different tangent spaces. Such differences cannot be considered before the introduction of affine connections between tangent spaces in Chap. 7. However, there are two types of derivative which may be introduced more directly and which are considered in Chap. 4: Lie derivatives and exterior derivatives. They yield also the basis for the study of Pfaff systems of differential forms playing a key role for instance in Hamilton mechanics and in thermodynamics. In any case, analysis leads to an important new construct of a manifold with a tangent space attached to each of its points, the tangent bundle.

As an example, the circle S^1 as a one-dimensional manifold is shown in the upper part of Fig. 1.4 together with its tangent spaces $T_x(S^1)$ at points x of S^1. All those tangent spaces together with the base manifold S^1 form again a manifold: If all tangent spaces are turned around by 90° as in the lower part of Fig. 1.4, a neighborhood of the tangent vector indicated in the upper part is obviously smoothly deformed only. Hence it is natural to introduce a topology in the whole construct which is locally equivalent to the product topology of $V \times \mathbb{R}$ where V is an open set of S^1 and hence in the whole this topology is equivalent to that of an infinite cylinder, the vertically infinitely extended version of Fig. 1.2a. (Note that the tangent vector spaces to different points of a manifold are considered disjoint by definition. In the upper panel of Fig. 1.4 the lines in clockwise direction from S^1 must therefore be considered on a sheet of paper different from that for the lines in counterclockwise direction in order to avoid common points.) In this topology, the canonical projection π from the tangent spaces to their base points in S^1 is

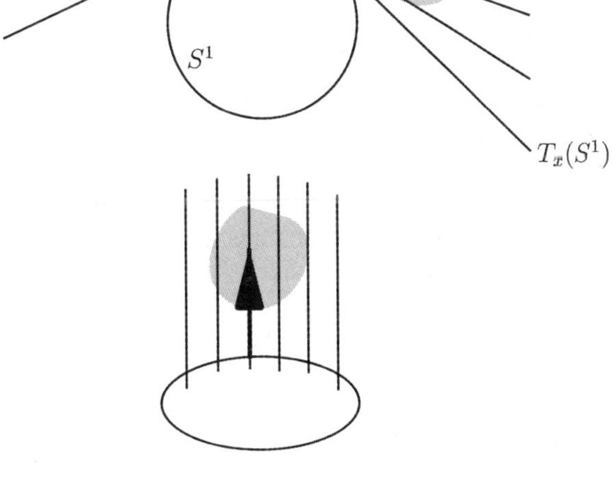

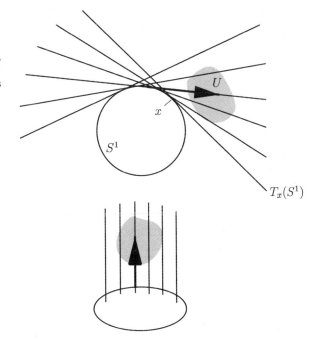

Fig. 1.4 The circle S^1 attached with a bundle of one-dimensional tangent spaces $T_x(S^1)$ (*upper part*). A neighborhood U of a tangent vector marked by an *arrow* is indicated. If the tangent spaces are turned around as shown in the *lower part*, the neighborhood U is just smoothly deformed

continuous. Such a rather special construct of a manifold is called a bundle, in the considered case a tangent bundle $T(S^1)$ which is a special case of a vector bundle.

All tangent spaces to a manifold are isomorphic to each other, they are isomorphic to $\mathbb{R}^m$ if the manifold M has a given (constant) dimension m (its local topology is that of $\mathbb{R}^m$). Such a bundle of isomorphic structures is in general called a fiber bundle, in the considered case the tangent bundle $T(M)$ with base M and typical fiber $\mathbb{R}^m$ (tangent space). Fiber bundles are somehow manifolds obtained by gluings along fibers. The complete definition of bundles given in Chap. 7 includes additionally transformation groups of fibers. The characteristic fiber of a fiber bundle need not be a vector space, it can again be a manifold. As already stated, a fiber bundle is again a new special type of manifold. Hence, one may construct fiber bundles with other fiber bundles as base...

Given tangent and cotangent spaces in every point of a manifold, the latter as the duals to tangent spaces, a tensor algebra may be introduced on each of those dual pairs of spaces. This leads to the concept of tensor fields and the corresponding tensor analysis. Totally antisymmetric tensors are called forms and play a particularly important role because E. Cartan's exterior calculus and the integration of forms leading to de Rham's cohomology provide the basis for the deepest interrelations between topology, analysis and algebra. In particular field theories like Maxwell's theory are most elegantly cast into cases of exterior calculus. Tensor fields and forms as well as their Lie derivatives along a vector field and the exterior derivative of forms are treated in Chap. 4. Besides the tensor notation related to coordinates which is familiar in physics, the modern coordinate invariant notation is introduced which is more flexible in generalizations to manifolds, in particular in the exterior calculus.

On the real line $\mathbb{R}$, differentiation and integration are in a certain sense inverse to each other due to the Fundamental Theorem of Calculus

$$\int_a^x f'(y)dy = f(x) - f(a). \tag{1.3}$$

In general, however, while differentiation needs only an affine structure, integration needs the definition of a measure. However, it turns out that the integration of an exterior differential n-form on an n-dimensional manifold is independent of the actual local coordinates of charts. It is treated in Chap. 5. This implies the classical integral theorems of vector analysis and is the basis of de Rham's cohomology theory which connects local and global properties of manifolds.

There are two classical roots of modern algebraic topology and homology, of which two textbooks which have many times been reprinted still maintain actuality not only for historical reasons. These are that of Herbert Seifert and William Threlfall, Dresden [1], and that of Pawel Alexandroff and Heinz Hopf, Göttingen/Moscow [2]. Seifert was the person who coined the name fiber space, then in a meaning slightly different from what is called fiber bundle nowadays.

On the basis of integration of simplicial chains, Chap. 5 provides cohomology theory in some detail as the purely algebraic skeleton of the theory of integration of forms with its astonishingly far reaching generalizations for any type of graded algebras or modules. Cohomology theory is intimately related to the general continuation problems in mathematics and physics: given a certain quantity defined on a domain U of a space X, can it continuously, smoothly, analytically, ... be continued to a quantity defined on a larger domain. Cohomology theory forms nowadays the most powerful core of algebraic topology and led to a wealth of results not only in mathematical physics but also in nearly every branch of pure mathematics itself. Here, the focus nevertheless is on topological invariants. Besides, as another example of application of (co)homology theory in mathematics with physical relevance Morse's theory of critical points of real functions on manifolds is presented.

Physicists are well acquainted with the duality between alternating tensors of rank r and alternating tensors of rank $d-r$ in dimensions $d=3$ and $d=4$, provided by the Levi-Civita pseudo-tensor (alternating d-form). Its general basis is Hodge's star operator, which is treated in the last section of Chap. 5 in connection with Maxwell's electrodynamics as a case of application of the exterior calculus. As another application of homology and homotopy theory, the dynamics of electrons in a perfect crystal lattice as a case of topological classification of embedding one- and two-dimensional manifolds into the 3-torus of a Brillouin zone is considered in some detail.

The most general type of cohomology is sheaf cohomology, and sheaf theory is nowadays used to prove de Rham's theorem. Since sheaf theory is essentially a technique to prove isomorphisms between various cohomologies and is quite abstract for a physicist, it is not included here, and de Rham's theorem is not proved although it is amply used. The interested reader is referred to cited mathematical literature.

Let $\boldsymbol{X}$ be a tangent vector field on a manifold M. In a neighborhood $U(x)$ of each point $x \in M$ it generates a flow $\varphi_t : U(x) \to M, -\varepsilon<t<\varepsilon$ of local transformations with a group structure $\varphi_t\varphi_{t'} = \varphi_{t+t'}$, $\varphi_0 = \mathrm{Id}_{U(x)}$ (identical transformation), $\varphi_t^{-1} = \varphi_{-t}$ so that one may formally write $\varphi_t = \exp(t\boldsymbol{X})$.

If the points of a manifold themselves form a group and $M \times M \to M$: $(x,y) \mapsto xy$, and $M \to M : x \mapsto x^{-1}$ are smooth mappings, then M is a Lie group. The tangent fiber bundle $T(M)$ based on the Lie group M has the Lie algebra $\mathfrak{m}$ of M as its typical fiber.

Besides being themselves manifolds, Lie groups play a central role as transformation groups of other manifolds. The theory of Lie groups and of Lie algebras forms a huge field with relevance in physics by itself. In this text, the focus is on two aspects, most relevant in the present context: covering groups, the most prominent example of which in physics is the interrelation of spin and angular momentum, and the classical groups and some of their descendants. Two amply used links between Lie groups and their Lie algebras are the exponential mapping and the adjoint representations. All these parts of the theory of topological groups are considered in Chap. 6. The Compendium at the end of the volume contains in addition a sketch of the representation theory of the finite dimensional simple Lie algebras, part of which is well known in physics in the theory of angular momenta and in the treatment of unitary symmetry in quantum field theory.

The simplest fiber bundles, the so called principal fiber bundles have Lie groups as characteristic fiber. Their investigation lays the ground for moving elements of one fiber into another with the help of a connection form.

Given a linear base of a vector space which sets linear coordinates, a tensor is represented by an ordered set of numbers, the tensor components. Physicists are taught early on, however, that a tensor describes a physical reality independent of its representation in a coordinate system. It is an equivalence class of doubles of linear bases in the vector space and representations of the tensor in that base, the transformations of both being linked together. Tensor fields on a manifold M live in the tangent spaces of that manifold (more precisely in tensor products of copies of tangent and cotangent spaces). All admissible linear bases of the tangent space at $x \in M$ form the frame bundle as a special principal fiber bundle with the transformation group of transformations of bases into each other as the characteristic fiber. The tensor bundle, the fibers of which are formed by tensors relative to the tangent spaces at all points $x \in M$, is now a general fiber bundle associated with the frame bundle, and the interrelation between both is precisely describing the above mentioned equivalence classes, making up tensors. Connection forms on frame bundles allow to transport tensors from one point $x \in M$ to another point $x' \in M$ *on paths through* M, the result of the transport depending on the path, if M is not flat. Only after so much work, the directional derivatives of tensor fields on manifolds can be treated in Chap. 7. Now, also the curvature form and the torsion form as local characteristics of a manifold as well as the corresponding torsion and curvature tensors living in tensor bundles over manifolds are provided.

With the help of parallel transport, deep results on global properties of manifolds are obtained in Chap. 8: surprising interrelations between the holonomy and homotopy groups of the manifold. In order to provide some inside into the flavor of these mathematical constructs, the exact homotopy sequence and the homotopy of sections are treated in some detail, although not so much directly used in physics. The exact homotopy sequence is quite helpful in calculating homotopy groups of various manifolds, some of which are also used at other places in the text. The homotopy of sections in fiber bundles provides the general basis of understanding characteristic classes, the latter topological invariants becoming more and more used in physics. These interrelations are presented in direct connection with very topical applications in physics: gauge field theories and the quantum physics of geometrical phases called Berry's phases. They are also in the core of modern treatments of molecular physics beyond the simplest Born-Oppenheimer adiabatic approximation.

By introducing an everywhere non-degenerate symmetric covariant rank 2 tensor field, the Levi-Civita connection is obtained as the uniquely defined metric-compatible torsion-free connection form. This leads to the particular case of Riemannian geometry, which is considered in Chap. 9, having in particular the theory of gravitation in mind the basic features of which are discussed. The text concludes with an outlook on complex generalizations of manifolds and a short introduction to Hermitian and Kählerian manifolds. Besides providing the basis of modern treatment of analytic complex functions of many variables, a tool present everywhere in physics, the Kählerian manifolds as torsion-free Hermitian manifolds form in a certain sense the complex generalization of Riemannian manifolds.

References

1. Seifert, K.J.H., Threlfall, W.R.M.H.: Lehrbuch der Topologie (Teubner, Leipzig, 1934). Chelsea, New York (reprint 1980)
2. Alexandroff, P.S., Hopf, H.: Topologie (Springer, Berlin, 1935). Chelsea, New York (reprint 1972)

Chapter 2
Topology

The first four sections of this chapter contain a brief summary of results of analysis most theoretical physicists are more or less familiar with.

2.1 Basic Definitions

A **topological space** is a double $(X, \mathcal{T})$ of a set X and a family $\mathcal{T}$ of subsets of X specified as the **open sets** of X with the following properties:

1. $\emptyset \in \mathcal{T},\ X \in \mathcal{T}\quad (\emptyset$ is the empty set),
2. $(\mathcal{U} \subset \mathcal{T}) \Rightarrow \left(\bigcup_{U\in\mathcal{U}} U \in \mathcal{T}\right),$
3. $(U_n \in \mathcal{T}$ for $1 \le n \le N \in \mathbb{N}) \Rightarrow \left(\bigcap_{n=1}^{N} U_n \in \mathcal{T}\right),$

that is, $\mathcal{T}$ is closed under unions and under finite intersections. If there is no doubt about the family $\mathcal{T}$, the topological space is simply denoted by X instead of $(X, \mathcal{T})$.

Two **topologies** $\mathcal{T}_1$ and $\mathcal{T}_2$ on X may be compared, if one is a subset of the other; if $\mathcal{T}_1 \subset \mathcal{T}_2$, then $\mathcal{T}_1$ is **coarser** than $\mathcal{T}_2$ and $\mathcal{T}_2$ is **finer** that $\mathcal{T}_1$. The coarsest topology is the **trivial topology** $\mathcal{T}_0 = \{\emptyset, X\}$, the finest topology is the **discrete topology** consisting of all subsets of X.

A **neighborhood** of a point $x \in X$ (of a set $A \subset X$) is an open[1] set $U \in \mathcal{T}$ containing x as a point (A as a subset). The complements $C = X \setminus U$ of open sets $U \in \mathcal{T}$ are the **closed sets** of the topological space X. If $A \in X$ is any set, then the **closure** $\overline{A}$ of A is the smallest closed set containing A, and the **interior** $\mathring{A}$ of A is the largest open set contained in A; $\overline{A}$ and $\mathring{A}$ always exist by Zorn's lemma. $\mathring{A}$ is

[1] In this text neighborhoods are assumed open; more generally a neighborhood is any set containing an open neighborhood.

the set of **inner points** of A. $\overline{A}$ is the set of **points of closure** of A; points every neighborhood of which contains at least one point of A. (The complement of $\overline{A}$ is the largest open set not intersecting A.) The **boundary** ∂A of A is the set $\overline{A} \setminus \mathring{A}$. A is **dense** in X, if $\overline{A} = X$. A is **nowhere dense** in X, if the interior of $\overline{A}$ is empty: $(\overline{A})^\circ = \emptyset$. X is **separable** if $X = \overline{A}$ for some *countable* set A.

(One might wonder about the asymmetry of axioms 2 and 3. However, if closure under all intersections would be demanded, no useful theory would result. For instance, a point of the real line $\mathbb{R}$ can be obtained as the intersection of an infinite series of open intervals. Hence, with the considered modification of axiom 3, points and all subsets of $\mathbb{R}$ would be open and closed and the topology would be discrete as soon as all open intervals are open sets.)

The **relative topology** $\mathcal{T}_A$ on a subset A of a topological space $(X, \mathcal{T})$ is $\mathcal{T}_A = \{A \cap T | T \in \mathcal{T}\}$, that is, its open sets are the intersections of A with open sets of X. Consider the closed interval $[0, 1]$ on the real line $\mathbb{R}$ with the usual topology of unions of open intervals on $\mathbb{R}$. The half-open interval $]x, 1], 0 < x < 1$, of $\mathbb{R}$ is an open set in the relative topology on $[0, 1] \subset \mathbb{R}$!

Most of the interesting topological spaces are **Hausdorff**: any two distinct points have disjoint neighborhoods. (A non-empty space of at least two points and with the trivial topology is not Hausdorff.) In a Hausdorff space single point sets $\{x\}$ are closed. (Exercise, take neighborhoods of all points distinct from x.)

Sequences are not an essential subject in this book. Just to be mentioned, a sequence of points in a topological space X converges to a point x, if every neighborhood of x contains all but finitely many points of the sequence. A partially ordered set I is directed, if every pair a, b of elements of I has an upper bound $c \in I, c \geq a, c \geq b$. A set of points of X is a net, if it is indexed by a directed index set I. A net converges to a point x, if for every neighborhood U of x there is an index b so that $x_a \in U$ for all $a \geq b$. In Hausdorff spaces points of convergence are unique if they exist.

The central issue of topology is continuity. A function (mapping) f from a topological space X into a topological space Y (maybe the same space X) is **continuous** at $x \in X$, if given any (in particular small) neighborhood V of $f(x) \subset Y$ there is a neighborhood U of x such that $f(U) \subset V$ (compare Fig. 1.1 of Chap. 1). The function f is continuous if it is continuous at every point of its domain. In this case, the inverse image $f^{-1}(V)$ of any open set V of the target space Y of f is an open set of X. (It may be empty.) The coarser the topology of Y or the finer the topology of X the more functions from X into Y are continuous. Observe that, *if X is provided with the discrete topology, then every function $f : X \to Y$ is continuous, no matter what the topology of Y is.* If $f : X \to Y$ and $g : Y \to Z$ are continuous functions, then their composition $g \circ f : X \to Z$ is obviously again a continuous function.

Consider functions $f(x) = y : [0, 1] \to \mathbb{R}$. What means continuity at $x = 1$ if the relative topology of $[0, 1] \subset \mathbb{R}$ is taken?

f is continuous iff it maps convergent nets to convergent nets; in metric spaces sequences suffice instead of nets.

A **homeomorphism** is a bicontinuous bijection f (f and f^{-1} are continuous functions *onto*); it maps open sets to open sets and closed sets to closed sets. A homeomorphism from a topological space X to a topological space Y provides a one–one mapping of points and a one–one mapping of open sets, hence it provides an equivalence relation between topological spaces; X and Y are called **homeomorphic**, $X \sim Y$, if a homeomorphism from X to Y exists. There exists always the identical homeomorphism Id_X from X to X, and a composition of homeomorphisms is a homeomorphism. The topological spaces form a category the morphisms of which are the continuous functions and the isomorphisms are the homeomorphisms (see Compendium C.1 at the end of the book).

A **topological invariant** is a property of topological spaces which is preserved under homeomorphisms.

2.2 Base of Topology, Metric, Norm

If topological problems are to be solved, it is in most cases of great help that not the whole family $\mathcal{T}$ of a topological space $(X, \mathcal{T})$ need be considered.

A subfamily $\mathcal{B}$ of $\mathcal{T}$ is called a **base of the topology** $\mathcal{T}$ if every $U \in \mathcal{T}$ can be formed as $U = \cup_\beta B_\beta,\ B_\beta \in \mathcal{B}$. A family $\mathcal{B}(x)$ is called a **neighborhood base** at x if each $B \in \mathcal{B}(x)$ is a neighborhood of x and given any neighborhood U of x there is a B with $U \supset B \in \mathcal{B}(x)$. A topological space is called **first countable** if each of its points has a countable neighborhood base, it is called **second countable** if it has a countable base.

The **product topology** on the product $X \times Y$ of topological spaces X and Y is defined by the base consisting of sets

$$\{(x,y) \,|\, x \in B_X,\ y \in B_Y\}, \quad B_X \in \mathcal{B}_X, \quad B_Y \in \mathcal{B}_Y, \tag{2.1}$$

where $\mathcal{B}_X$ and $\mathcal{B}_Y$ are bases of topology of X and Y, respectively. It is the coarsest topology for which the canonical projection mappings $(x,y) \mapsto x$ and $(x,y) \mapsto y$ are continuous (exercise). The $\mathbb{R}^n$ with its usual topology is the topological product $\mathbb{R} \times \cdots \times \mathbb{R}, n$ times.

A very frequent special case of topological space is a metric space. A set X is a **metric space** if a non-negative real valued function, the **distance function** $d : X \times X \to \mathbb{R}_+$ is given with the following properties:

1. $d(x,y) = 0, \quad \text{iff}\, x = y,$
2. $d(x,y) = d(y,x),$
3. $d(x,z) \leq d(x,y) + d(y,z)$ (triangle inequality).

An **open ball** of radius r with its center at point $x \in X$ is defined as $B_r(x) = \{x' \,|\, d(x,x') < r\}$. The class of all open balls forms a base of a topology of X, the **metric topology**. It is Hausdorff and first countable; a neighborhood base of point x is for instance the sequence $B_{1/n}(x),\ n = 1, 2, \ldots$

The metric topology is uniquely defined by the metric as any topology is uniquely defined by a base. There are, however, in general many different metrics defining the same topology. For instance, in $\mathbb{R}^2 \ni x = (x^1, x^2)$ the metrics

$$d_1(\boldsymbol{x},\boldsymbol{y}) = ((x^1 - y^1)^2 + (x^2 - y^2)^2)^{1/2} \quad \text{Euclidean metric,}$$
$$d_2(\boldsymbol{x},\boldsymbol{y}) = \max\{|x^1 - y^1|, |x^2 - y^2|\},$$
$$d_3(\boldsymbol{x},\boldsymbol{y}) = |x^1 - y^1| + |x^2 - y^2| \quad \text{Manhattan metric}$$

define the same topology (exercise).

A sequence $\{x_n\}$ in a metric space is **Cauchy** if

$$\lim_{m,n\to\infty} d(x_m, x_n) = 0. \tag{2.2}$$

A metric space X is **complete** if every Cauchy sequence converges in X (in the metric topology). The rational line $\mathbb{Q}$ is not complete, the real line $\mathbb{R}$ is, it is an isometric completion of $\mathbb{Q}$. An **isometric completion** $\tilde{X}$ of a metric space X always exists in the sense that $\tilde{X} \supset X$ is complete, $\tilde{X} = \overline{X}$ (closure of X in $\tilde{X}$), and the distance function $d(x,x')$ is extended to $\tilde{X}$ by continuity. $\tilde{X}$ is unique up to isometries (distance preserving transformations) which leave the points of X on place. A complete metric space is a **Baire space**, that is, it is not a countable union of nowhere dense subsets. The relevance of this statement lies in the fact that if a complete metric space is a countable union $X = \cup_n U_n$, then some of the U_n must have a non-empty interior [1, Section III.5].

A metric space X is complete, iff every sequence $C_1 \supset C_2 \supset \dots$ of closed balls with radii $r_1, r_2, \dots \to 0$ has a non-empty intersection.

Proof Necessity: Let X be complete. The centers x_n of the balls C_n obviously form a Cauchy sequence which converges to some point x, and $x \in \cap_n C_n$. Sufficiency: Let x_n be Cauchy. Pick n_1 so that $d(x_n, x_{n_1}) < 1/2$ for all $n \geq n_1$ and take x_{n_1} as the center of a ball C_1 of radius $r_1 = 1$. Pick $n_2 \geq n_1$ so that $d(x_n, x_{n_2}) < 1/2^2$ for all $n \geq n_2$ and take x_{n_2} as the center of a ball C_2 of radius $r_2 = 1/2\dots$ The sequence $C_1 \supset C_2 \supset \dots$ has a non-empty intersection containing some point x. It is easily seen that $x = \lim x_n$. □

Let X be a metric space and let $F : X \to X : x \mapsto Fx$ be a **strict contraction**, that is a mapping of X into itself with the property

$$d(Fx, Fx') \leq \lambda d(x, x'), \quad \lambda < 1. \tag{2.3}$$

(A contraction is a mapping which obeys the weaker condition $d(Fx, Fx') \leq d(x, x')$; every contraction is obviously continuous since the preimage of any open ball $B_r(Fx)$ contains the open ball $B_r(x)$. Exercise.) A vast variety of physical problems implies **fixed point equations**, equations of the type $x = Fx$. Banach's contraction mapping principle says that *a strict contraction F on a complete metric space X has a unique fixed point.*

Proof Uniqueness: Let $x = Fx$ and $y = Fy$, then $d(x,y) = d(Fx,Fy) \leq \lambda d(x,y)$, $\lambda < 1$. Hence, $d(x,y) = 0$ that is $x = y$. Existence: Pick x_0 and let $x_n = F^n x_0$. Then, $d(x_{n+1}, x_n) = d(Fx_n, Fx_{n-1}) \leq \lambda d(x_n, x_{n-1}) \leq \dots \leq \lambda^n d(x_1, x_0)$. Thus, if $n > m$, by the triangle inequality and by the sum of a geometrical series,

$d(x_n, x_m) \le \sum_{l=m+1}^{n} d(x_l, x_l - 1) \le \lambda^m (1-\lambda)^{-1} d(x_1, x_0) \to 0$ for $m, n \to \infty$ implying that $\{x_n\} = \{Fx_{n-1}\}$ is Cauchy and converges towards an $x \in X$. By continuity of $F, x = Fx$. □

Equation systems, systems of differential equations, integral equations or more complex equations may be cast into the form of a fixed point equation. A simple case is the equation $x = f(x)$ for a function $f : [a, b] \to [a, b]$, $[a, b] \subset \mathbb{R}$, obeying the Lipschitz condition

$$|f(x) - f(x')| \le \lambda |x - x'|, \quad \lambda < 1, \quad x, x' \in [a, b].$$

If for instance $|f'(x)| \le \lambda < 1$ for $x \in [a, b]$, the Lipschitz condition is fulfilled. From Fig. 2.1 it is clearly seen how the solution process $x_n = f(x_{n-1})$ converges. The convergence is fast if $|f'(x)| \ll 1$. Consider this process for $|f'(x)| > 1$. Next consider $a = -\infty$; why is a simple contraction not sufficient and a strict contraction needed to guarantee the existence of a solution?

There are always many ways to cast a problem into a fixed point equation. If $x = Fx$ has a solution x_0, it is easily seen that $x = \tilde{F}x$ with $\tilde{F}x = x + p(Fx - x)$ has the same solution x_0. If F is not a strict contraction, $\tilde{F}$ with a properly chosen p sometimes is, although possibly with a very slow convergence of the solution process. Sophisticated constructions have been developed to enforce convergence of the solution process of a fixed point equation.

Another frequent special case of topological space is a **topological vector space** X over a field K. (In most cases $K = \mathbb{R}$ or $K = \mathbb{C}$.) It is also a vector space (see Compendium) and its topology is such that the mappings

$$K \times X \to X : (\lambda, x) \mapsto \lambda x,$$
$$X \times X \to X : (x, x') \mapsto x + x'$$

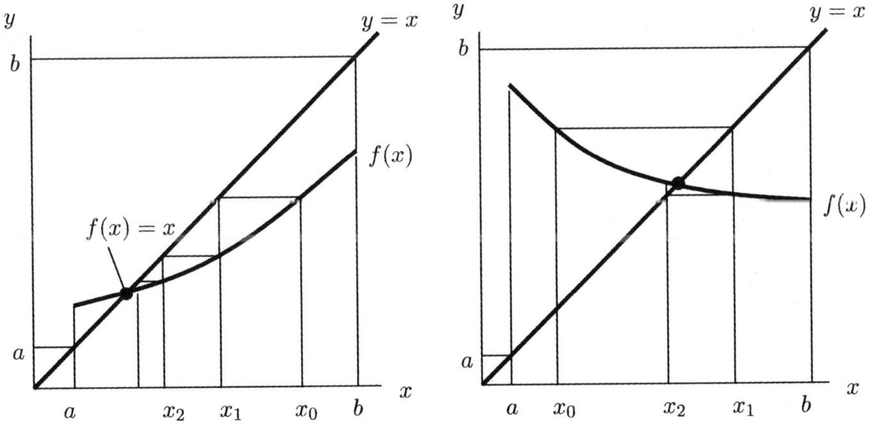

Fig. 2.1 Illustration of the fixed point equation $x = f(x)$ for $f'(x) > 0$ (*left*) and $f'(x) < 0$ (*right*)

are continuous, where K is taken with its usual metric topology and $K \times X$ and $X \times X$ are taken with the product topology. If $\mathcal{B}(0)$ is a neighborhood base at the origin of the vector space X, then the set $\mathcal{B}(x)$ of all open sets $B_\beta(x) = x + B_\beta(0) = \{x + x' | x' \in B_\beta(0)\}$ with $B_\beta(0) \in \mathcal{B}(0)$ is a neighborhood base at x. For any open (closed) set A, $x + A$ is open (closed). For two sets $A \subset X, B \subset X$ the vector sum is defined as $A + B = \{x + x' | x \in A, x' \in B\}$.

Linear independence of a set $E \subset X$ means that if $\sum_{n=1}^{N} \lambda^n x_n = 0$ (upper index at λ^n, not power of λ) holds for any finite set of N *distinct* vectors $x_n \in E$, then $\lambda^n = 0$ for all $n = 1, \dots, N$. Linear independence (as well as its opposite, linear dependence) is a property of the algebraic structure of the vector space, not of its topology. A **base** E in a topological vector space is a linearly independent subset the **span** of which (the set of all linear combinations over K of finitely many vectors out of E) is dense in $X : \overline{\mathrm{span}_K E} = X$. It is a base of vector space, not a base of topology. It may, however, depend on the topology of X. The maximal number of linearly independent vectors in E is the **dimension** of the topological vector space X; it is a finite integer n or infinity, countable or not. If the dimension of a topological vector space X is $n < \infty$, then X is homeomorphic to K^n. If it is infinite, the dimension is to be distinguished from the algebraic dimension of the vector space (see Compendium). It can be shown that a topological vector space X is **separable** if it admits a countable base. Any vector x of $\mathrm{span}_K E$ has a unique representation $x = \sum_{n=1}^{N} \lambda^n x_n$, $x_n \in E$ with some finite N. Hence, if X is Hausdorff, then every vector $x \in X$ has a unique representation by a converging series $x = \sum_{n=1}^{\infty} \lambda^n x_n$, $x_n \in E$ (exercise).

Two subspaces (see Compendium) M and N of a vector space X are called **algebraically complementary**, if $M \cap N = \{0\}$ and $M + N = X$. X is then said to be the **direct sum** $M \oplus N$ of the vector spaces M and N. Consider all possible sets $x + M$, $x \in X$. They either are disjoint or identical (exercise). Let $\tilde{x}$ be the equivalence class of the set $x + M$. By an obvious canonical transfer of the linear structure of X into the set of classes $\tilde{x}$ these classes form a vector space; it is called the **quotient space** X/M of X by M (Fig. 2.2). Let the topology of X be such that the one point set $\{0\}$ is closed. Then, for any $x \in X$, $M_x = \{\lambda x | \lambda \in K\}$ is a closed subspace of X (exercise).

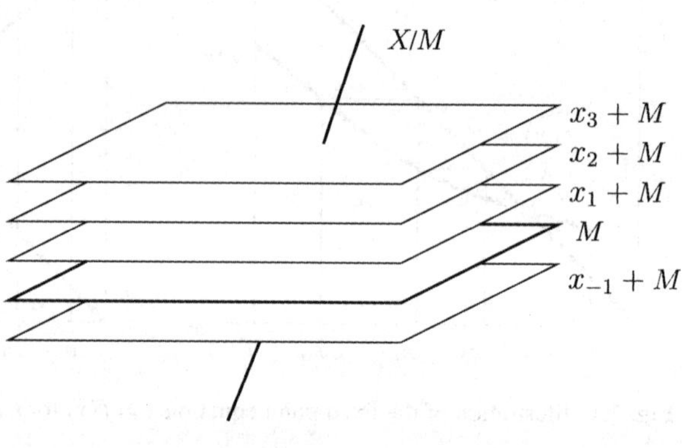

Fig. 2.2 A subspace M of a vector space X and cosets $x_i + M$ with x_i linearly independent of M. Note that an angle between X/M and M has no meaning so far

It is just by custom that the cosets $x_i + M$ were drawn as parallel planes in Fig. 2.2, and that X/M was drawn as a straight line. Angles, curvature and all that is not defined as long as X is considered as a topological vector space only. Any continuous deformation of Fig. 2.2 is admitted. Even if a metric is defined on a one-dimensional vector space, say, it would not make a difference if it would be drawn as a straight line or a spiral provided it is consistently declared how to relate the point λx to the point x. These remarks are essential in later considerations.

A topological vector space X is said to be **metrizable** if its topology can be deduced from a metric that is translational invariant: $d(x, x') = d(x + a, x' + a)$ for all $a \in X$. Many topological vector spaces, in particular all metrizable vector spaces, are **locally convex**: they admit a base of topology made of convex sets. (A set of a vector space is convex if it contains the 'chord' between any two of its points, that is, if x and x' are two points of the set then all points $\lambda x + (1 - \lambda)x'$, $0 < \lambda < 1$ belong to the set.)

In most cases a metrizable topological vector space is metrized either by a family of seminorms or by a norm. A **norm** is a real function $x \mapsto ||x||$ with the properties

1. $||x + x'|| \leq ||x|| + ||x'||$,
2. $||\lambda x|| = |\lambda|\,||x||$,
3. $||x|| = 0, \;\text{iff}\, x = 0$.

From the first two properties the non-negativity of a norm follows; if the last property is abandoned one speaks of a **seminorm**.The metric of a norm is given by $d(x, x') = ||x - x'||$. A complete metrizable vector space is a **Fréchet space**, a complete normed vector space is a **Banach space**. Fréchet spaces whose metric does not come from a single norm are used in the theory of generalized functions (distributions).

A **linear function (operator)** $L : X \to Y$ from a vector space X into a vector space Y over the same field K is a function with the property

$$L(\lambda x + \lambda' x') = \lambda L(x) + \lambda' L(x'), \quad \lambda, \lambda' \in K. \tag{2.4}$$

A function from a vector space X into its field of scalars K is called a **functional**, if it is linear it is called a **linear functional**. A linear function from a topological vector space into a topological vector space is continuous, iff it is continuous at the origin $x = 0$ (exercise). A linear function from a normed vector space X into a normed vector space Y (for instance the one-dimensional vector space K) is **bounded** if

$$||L|| = \sup_{0 \neq x \in X} \frac{||L(x)||_Y}{||x||_X} < \infty. \tag{2.5}$$

The operator notation Lx is often used instead of $L(x)$. A linear function from a normed vector space into a normed vector space is bounded, iff it is continuous (exercise). With the norm (2.5) (prove that it is indeed a norm), the set $\mathcal{L}(X, Y)$ of all bounded linear operators with linear operations among them defined in the natural way is again a normed vector space; *it is Banach if Y is Banach.*

Proof Let $\{L_n\}$ be Cauchy. Since $|||L_m|| - ||L_n||| \leq ||L_m - L_n|| \to 0$, $\{||L_n||\}$ is a Cauchy sequence of real numbers converging to some real number C. For each $x \in X$, $\{L_n x\}$ is a Cauchy sequence in Y. Since Y is complete, $L_n x$ converges to some point $y \in Y$. Define L by $Lx = y$. Then, $||Lx|| = \lim_{n\to\infty} ||L_n x|| \leq \lim_{n\to\infty} ||L_n||\,||x|| = C||x||$, where (2.5) was used. Hence, L is a bounded operator. Moreover, $||(L - L_n)x|| = \lim_{m\to\infty} ||(L_m - L_n)x|| \leq \lim_{m\to\infty} ||(L_m - L_n)||\,||x||$ and therefore $\lim_{n\to\infty} ||L - L_n|| = \lim_{n\to\infty} \sup_{x\neq 0} ||(L - L_n)x||/||x|| \leq \lim_{m,n\to\infty} ||L_m - L_n|| = 0$. Hence, L_n converges to L in the operator norm. □

The **topological dual** X^* of a topological vector space X is the set of all continuous **linear functionals**

$$f : X \to K : x \mapsto \langle f, x\rangle \in K, \quad \langle f, \lambda x + \lambda' x'\rangle = \lambda\langle f, x\rangle + \lambda'\langle f, x'\rangle, \tag{2.6}$$

from X into K provided with the natural linear structure $\langle \lambda f + \lambda' f', x\rangle = \lambda\langle f, x\rangle + \lambda'\langle f', x\rangle$. It is again a normed vector space with the norm $||f||$ given by (2.5) with f instead of L, $||f|| = \sup_{0\neq x\in X} |\langle f, x\rangle|/||x||_X$. As there are the less continuous functions the coarser the topology of the domain space is, the question arises, what is the coarsest topology of X for which all bounded linear functionals are continuous. This topology of X is called the **weak topology**. A neighborhood base of the origin for this weak topology is given by all intersections of finitely many open sets $\{x|\,|\langle f, x\rangle| < 1/k\}$, $k = 1, 2\ldots$ for all $f \in E^*$, a base of the vector space X^*. For instance, if $X = \mathbb{R}^n$, these open sets comprise all infinite 'hyperplates' of thickness $2/k$ sandwiching the origin and normal in turn to one of the n base vectors $\boldsymbol{f}^i$ of $X^* = \mathbb{R}^n$ (Fig. 2.3). Taken for every k, the intersections of n such 'hyperplates' containing $\{0\} \in X$ form a neighborhood base of the origin of $\mathbb{R} \times \cdots \times \mathbb{R}, n$ factors, in the product topology which in this case is equivalent to the standard norm topology of $\mathbb{R}^n$. Hence, the $\mathbb{R}^n$ with both the weak and the norm topologies are homeomorphic to each other and can be identified with each other. This does not hold true for an infinite dimensional space X.

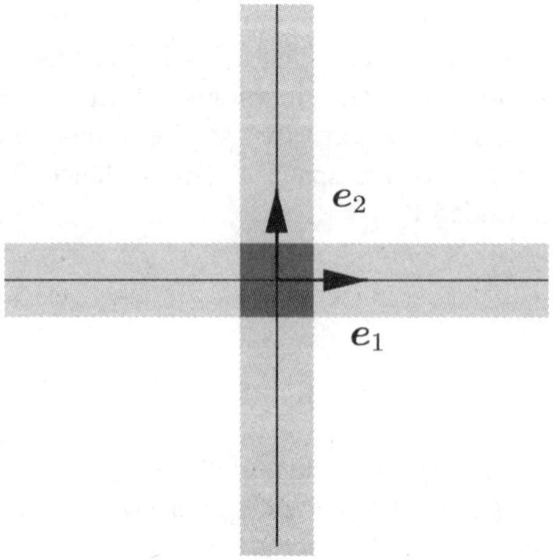

Fig. 2.3 Open sets of a neighborhood base of the origin of the $\mathbb{R}^2$ in the weak topology and their intersection

The topological dual of a normed vector space X is $X^* = \mathcal{L}(X, K)$ (with the norm $||f||$ as above); if K is complete (as $\mathbb{R}$ or $\mathbb{C}$) then X^* is a Banach space (no matter whether X is complete or not). The second dual of X is the dual $X^{**} = (X^*)^*$ of X^*. Let $J : X \to X^{**} : x \mapsto \tilde{x}$ where $\langle \tilde{x}, f \rangle = \langle f, x \rangle$ for all $f \in X^*$.

If X is a Banach space then the above mapping J is an isometric isomorphism of X onto a subspace of X^{**}*, hence, one may consider* $X \subset X^{**}$.

The proof of this statement makes use of the famous Hahn–Banach theorem which provides the existence of ample sets of continuous linear functionals [1, Section III.2.3]. X is said to be **reflexive**, if the above mapping J is *onto* X^{**}. In this case one may consider $X = X^{**}$.

An **inner product** (or scalar product) in a complex vector space X is a **sesquilinear function** $X \times X \to \mathbb{C} : (x, y) \mapsto (x|y)$ with the properties

1. $(x|y) = \overline{(y|x)}$,
2. $(x|y_1 + y_2) = (x|y_1) + (x|y_2)$,
3. $(x|\lambda y) = \lambda(x|y)$ (convention in physics),
4. $(x|x) > 0$ for $x \neq 0$.

(In mathematics literature, the convention $(\lambda x|y) = \lambda(x|y)$ is used instead of 3.) An inner product in a real vector space X is the corresponding bilinear function $X \times X \to \mathbb{R}$ with the same properties 1 through 4. ($\bar{\lambda}$ is the complex conjugate of λ, in $\mathbb{R}$ of course $\bar{\lambda} = \lambda$.) If an inner product is given,

$$||x|| = (x|x)^{1/2} \tag{2.7}$$

has all properties of a norm (exercise, use the Schwarz inequality given below). A normed vector space with a norm of an inner product is called an **inner product space** or a pre-Hibert space. A complete inner product space is called a **Hilbert space**. Some authors call it a Hilbert space only if it is infinite-dimensional; a finite-dimensional inner product space is also called a **unitary space** in the complex case and a **Euclidean space** in the real case. Two Hilbert spaces X and X' are said to be **isomorphic** or unitarily equivalent, $X \approx X'$, if there exists a **unitary operator** $U : X \to X'$, that is, a surjective linear operator for which $(Ux|Uy) = (x|y)$ holds for all $x, y \in X$ (actually it is bijective, exercise).

In an inner product space the **Schwarz inequality**

$$|(x|y)| \leq ||x||\,||y|| \tag{2.8}$$

holds, and in a *real* inner product space the **angle** between vectors x and y is defined as

$$\cos(\measuredangle(x, y)) = \frac{(x|y)}{||x||\,||y||}. \tag{2.9}$$

Proof of the Schwarz inequality Let $\hat{y} = y/||y||$ and $x_1 = (\hat{y}|x)\hat{y}$, $x_2 = x - x_1$ implying $(x_1|x_2) = 0$, $x = x_1 + x_2$. Then, $||x||^2 = (x_1 + x_2|x_1 + x_2) = ||x_1||^2 + ||x_2||^2 \geq ||x_1||^2 = |(x|y)|^2/||y||^2$. □

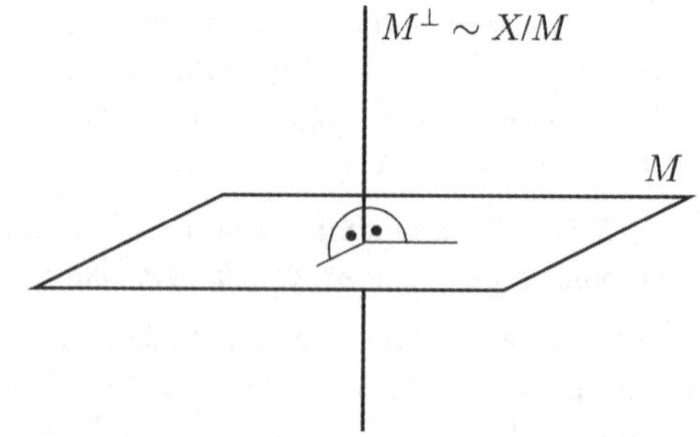

Fig. 2.4 Orthogonal complement $M^\perp$ to a closed subspace M of an inner product space

Even in a complex inner product space, orthogonality is defined: two vectors x and y are **orthogonal** to each other, if $(x|y) = 0$. An **orthonormalized base** in an inner product space is a base E (of topological vector space, see p. 16) with $||e|| = 1$ and $(e|e') = 0$, $e \neq e'$ for all $e, e' \in E$. Let $\{e_n\}_{n=1}^N \subset E$. A slight generalization of the proof of the Schwarz inequality proves **Bessel's inequality**: $||x||^2 \geq \sum_{n=1}^N |(e_n|x)|^2$. If M is a closed subspace of an inner product space X, then the set of all vectors of X which are orthogonal to all vectors of M forms the **orthogonal complement** $M^\perp$ of M in X (Fig. 2.4, compare to Fig. 2.2). Every vector $x \in X$ has a unique decomposition $x = x_1 + x_2$, $x_1 \in M$, $x_2 \in M^\perp$ (exercise), that is, $X = M + M^\perp$.

If X and X' are two Hilbert spaces over the same field K then their **direct sum** $X \oplus X'$ is defined as the set of all ordered pairs $(x, x'), x \in X, x' \in X'$ with the scalar product $((x, x')|(y, y')) = (x|y)_X + (x'|y')_{X'}$. (Hence, in the above case also $X = M \oplus M^\perp$ holds.) The direct sum of more that two, possibly infinitely many Hilbert spaces is defined accordingly. (The vectors of the latter case are the sequences $\{x^i\}$ for which the sum of squares of norms converges.)

The **tensor product** $X \otimes X'$ of Hilbert spaces X and X' is defined in the following way: Consider pairs $(x, x') \in X \times X'$ and define for each pair a bilinear function $x \otimes x'$ on the product vector space $X \times X'$ by $x \otimes x'(y, y') = (x|y)(x'|y')$. Consider the linear space of all finite linear combinations $\varphi = \sum_{n=1}^N c_n x_n \otimes x'_n$ and define an inner product $(\varphi|\psi)$ by linear extension of $(x \otimes x'|y \otimes y') = (x|y)(x'|y')$. The completion of this space is $X \otimes X'$. (Exercise: show that $(\varphi|\psi) = 0$ if $\varphi = \sum_{n=1}^N c_n x_n \otimes x'_n = 0$ and that $(\varphi|\psi)$ has the four properties of a scalar product.)

Finally, let X be a Hilbert space and let $y \in X$. Then, $f_y(x) = (y|x)$ is a continuous linear function $f_y : X \to K : x \mapsto (y|x)$, hence $f_y \in X^*$. The Riesz lemma says that there is a *conjugate* linear bijection $y \mapsto f_y$ between X and its dual X^* [1].

We close the section with a number of examples of vector spaces from physics:

$\mathbb{R}^n = \mathbb{R}^{n*}$, the set of real n-tuples $\boldsymbol{a} = \{a^1, a^2, \ldots, a^n\}$, is used as a mere topological vector space with the product topology of $\mathbb{R} \times \mathbb{R} \times \cdots \times \mathbb{R}$ (n factors) or as a Euclidean space (real finite-dimensional Hilbert space, $(a|b) = \boldsymbol{a} \cdot \boldsymbol{b} = \sum a^i b^i$ implying the same topology) in the sequel, depending on context (cf. the discussion in connection with Figs. 2.2 and 2.4). Both concepts play a central role in the theory of real manifolds. As a mere topological vector space it is the **configuration space** of a many-particle system, as an Euclidean space the position space or the momentum space of physics. For instance in the **physics of vibrations**, $\mathbb{C}^n \approx \mathbb{R}^{2n}$ by the isomorphism $z^j = x^j + iy^j \mapsto (a^{2j-1}, a^{2j}) = (x^j, y^j)$ is used, where only the x^j describe actual amplitudes. In the sequel, vectors of the space K^n ($K = \mathbb{R}$ or $\mathbb{C}$) are denoted by bold-face letters and the inner product is denoted by a dot.

l^p as sequence spaces the points of which are complex or real number sequences $a = \{a^i\}_{i=1}^{\infty}$ are defined for $1 \le p < \infty$ with the norm ($a \in l^p$, iff $||a||_p < \infty$)

$$l^p: \ ||a||_p = \left(\sum_{i=1}^{\infty} |a^i|^p\right)^{1/p}, \quad 1 \le p < \infty. \tag{2.10}$$

Young's inequality says $|a^i b^i| \le |a^i|^p/p + |b^i|^q/q$ for $1/p + 1/q = 1$. (It suffices to take real positive a^i, b^i to prove it. Determine the maximum of the function $f_{b^i}(a^i) = b^i a^i - |a^i|^p/p$.) Therefore, if $1 < p < \infty, 1/p + 1/q = 1$, $||a||_p < \infty$, $||b||_q < \infty$ then $|\langle b, a\rangle| = |\sum b^i a^i| < \infty$, that is, $b \in l^q$ is a continuous linear functional on $l^p \ni a$, $l^q \subset l^{p*}$. It can be proved that $l^q = l^{p*}$ [2, Section IV.9]. Since X^* is always a Banach space, l^p, $1 < p < \infty$ is a Banach space. Additionally, the normed sequence spaces $l^\infty \supset c_0 \supset f$, all with norm

$$l^\infty: \ ||a||_\infty = \sup_i |a^i|, \tag{2.11}$$

$$c_0 \subset l^\infty: \ \lim_{i\to\infty} a^i = 0,$$

$$f \subset l^\infty: \ a^i = 0 \ \text{ for all but finitely many } i$$

are considered. It can be shown that l^∞ and c_0 are Banach spaces and $l^{1*} = l^\infty$ and $c_0^* = l^1$. Hence, l^1 is also a Banach space. It is easily seen that f has a countable base as a vector space. Moreover, it is dense in l^p, $1 \le p < \infty$ (in the topology of the norm $||\cdot||_p$) and in c_0 (in the topology of the norm $||\cdot||_\infty$). Hence, those spaces have a countable base and are separable. Finally, l^2 with the inner product $(a|b) = \sum_i \overline{a^i} b^i$ is the Hilbert space of **Heisenberg's quantum mechanics**. Every infinite-dimensional separable Hilbert space is isomorphic to l^2 [1, Section II.3].

$L^p(M, d\mu)$ [1]: Let $(M, d\mu)$ be a measure space, for instance $\mathbb{R}^n$ or a part of it with Lebesgue measure $d^n x$. Denote by f the class of complex or real functions on M which differ from each other at most on a set of measure zero. Clearly, linear combinations respect classes. $L^p(M, d\mu)$ is the functional linear space of classes f for which

$$||f||_p = \left(\int_M |f|^p d\mu\right)^{1/p} < \infty. \tag{2.12}$$

For $p=\infty$, $||f||_\infty = \text{ess sup}\,|f|$, that is the smallest real number c so that $|f|>c$ at most on a set of zero measure. For $1\leq p\leq\infty$, $||f||_p$ is a norm, and $L^p(M,d\mu)$ is complete. $L^p(M,d\mu)^* = L^q(M,d\mu)$, $1/p+1/q=1$, $1\leq p<\infty$ with $\langle g,f\rangle = \int_M gfd\mu$. If $M=\mathbb{R}_+$ and $d\mu = \sum_{n=1}^{\infty}\delta(x-n)dx$, then $L^p(M,d\mu)=l^p$. If $\mu(M)<\infty$, then $L^p(M,d\mu)\subset L^{p'}(M,d\mu)$ for $p\geq p'$. The Hilbert space of Schrödinger's **quantum states** of a spinless particle is $L^2(\mathbb{R}^3,d^3x)$, for a spin-S particle is $L^2(\mathbb{R}^3,d^3x)\otimes\mathbb{C}^{2S+1}$, where $\mathbb{C}^{2S+1}$ is the $(2S+1)$-dimensional state space of spin. The L^p-spaces are for instance used in **density functional theories**.

Fock space: Let $\mathcal{H}$ be a Hilbert space of single-particle quantum states, and let $\mathcal{H}^0=K$ (field of scalars) and $\mathcal{H}^n=\mathcal{H}\otimes\mathcal{H}\otimes\cdots\otimes\mathcal{H}$ (n factors). For any vector $\psi_{k_1}\otimes\psi_{k_2}\otimes\cdots\otimes\psi_{k_n}\in\mathcal{H}^n$, let $S_n\psi_{k_1}\otimes\psi_{k_2}\otimes\cdots\otimes\psi_{k_n}=\sum_{\mathcal{P}}\psi_{k_{\mathcal{P}(1)}}\otimes\psi_{k_{\mathcal{P}(2)}}\otimes\cdots\otimes\psi_{k_{\mathcal{P}(n)}}$ and $A_n\psi_{k_1}\otimes\psi_{k_2}\otimes\cdots\otimes\psi_{k_n}=\sum_{\mathcal{P}}(-1)^{|\mathcal{P}|}\psi_{k_{\mathcal{P}(1)}}\otimes\psi_{k_{\mathcal{P}(2)}}\otimes\cdots\otimes\psi_{k_{\mathcal{P}(n)}}$, where the summation is over all permutations $\mathcal{P}$ of the numbers $1,2,\ldots,n$ and $|\mathcal{P}|$ is its order. Let $S_0=A_0=\text{Id}_{\mathcal{H}^0}$. Then,

$$\mathcal{F}_B(\mathcal{H})=\oplus_{n=0}^{\infty}S_n\mathcal{H}^n$$

is the bosonic Fock space, and

$$\mathcal{F}_F(\mathcal{H})=\oplus_{n=0}^{\infty}A_n\mathcal{H}^n$$

is the fermionic Fock space. An orthonormal base in both cases may be introduced as the set of occupation number eigenstates for a fixed orthonormal basis $\{\psi_k\}$ in $\mathcal{H}$

$$|\rangle,\ |n_1,n_2,\ldots,n_N\rangle,\ N=1,2,\ldots,\ n_k=0,1,2,\ldots\text{(bosons) and } n_k=0,1\text{ (fermions)}.$$

The state with vector $|\rangle\in\mathcal{H}^0$ is called the vacuum state. The Fock space is the closure (in the topology of the direct sum of tensor products of $\mathcal{H}$) of the span of all occupation number eigenstates.

2.3 Derivatives

Let $F:\Omega\rightarrow Y$ be a mapping (vector-valued function) from an open set Ω of a normed vector space X into a topological vector space Y. If the limes

$$D_xF(x_0)=\frac{d}{dt}F(x_0+tx)|_{t=0}=\lim_{\substack{t\neq 0,t\to 0\\ x_0+tx\in\Omega}}\frac{F(x_0+tx)-F(x_0)}{t} \tag{2.13}$$

exists it is called a partial derivative or (for $||x||=1$) **directional derivative** in the direction of x of the function F at x_0. $D_xF(x_0)$ is a vector of the space Y. $D_xF(x_0)$ is of course defined for any value of norm of x; by replacing in the above definition t by λt it is readily seen that $D_{\lambda x}F(x_0)=\lambda D_xF(x_0)$. (However, $D_xF(x_0)$ as a function of x need not be linear; for instance it may exist for some x and not for others.) If the directional derivative (for fixed x) exists for all $x_0 \in \Omega$ then $D_xF(x_0)$ is another function (of the variable x_0) from Ω into Y (which need not be continuous), and the second directional derivative $D_{x'}D_xF(x_0)$ may be considered if it exists for some x', and so on. If, given x_0, the directional derivative $D_xF(x_0)$ exists for all x as a continuous linear function from X into Y, then it is called the Gâteaux derivative.

Caution: The existence of all directional derivatives is not sufficient for the chain rule of differentiation to be valid; see example below.

Let Y also be a normed vector space. If there is a continuous linear function $DF(x_0) \in \mathcal{L}(X, Y)$ so that

$$F(x_0 + x) - F(x_0) = DF(x_0)x + R(x)||x||, \quad \lim_{x \to 0} R(x) = 0, \tag{2.14}$$

then $DF(x_0)$ is called the **total derivative** or the Fréchet derivative of F at x_0. $R(x)$ is supposed continuous at $x = 0$ with respect to the norm topologies of X and Y, and $R(0) = 0$. (For $x \neq 0$, $R(x)$ is uniquely defined to be $[F(x_0 + x) - F(x_0) - DF(x_0)x]/||x||$.) Given x (and x_0), $DF(x_0)x$ is again a vector in Y, that is, for given x_0, $DF(x_0)$ is a continuous linear function from X into Y. If $DF(x_0)$ exists for all $x_0 \in \Omega$, then DF is a mapping from Ω into $\mathcal{L}(X, Y)$ and DFx (x fixed) is a mapping from Ω into Y. Hence, the second derivative $D(DFx)(x_0)x' = D^2F(x_0)xx'$ may be considered, and so on. For instance, D^2F is a mapping from Ω into $\mathcal{L}(X, \mathcal{L}(X, Y))$, the space of continuous bilinear functions from $X \times X$ into Y and, given x and x', D^2Fxx' is a mapping from Ω into Y.

The total derivative may not exist even if all directional derivatives do exist. As an example [3, §10.1], consider $X = \mathbb{R}^2$, $Y = \mathbb{R}$ and the real function of two real variables x^1 and x^2

$$F(x^1, x^2) = \begin{cases} \dfrac{2(x^1)^3 x^2}{(x^1)^4 + (x^2)^2} & \text{for } (x^1, x^2) \neq (0, 0), \\ 0 & \text{for } (x^1, x^2) = (0, 0). \end{cases}$$

Let $0 = (0, 0)$ and $\boldsymbol{x} = (x^1, x^2) \neq 0$. Then, $(F(0 + t\boldsymbol{x}) - F(0))/t = (2t^3(x^1)^3 x^2)/(t^4(x^1)^4 + t^2(x^2)^2)$. For $x^2 = 0$ this is 0, and for $x^2 \neq 0$ it is of order $O(t)$, hence, $D_{\boldsymbol{x}}F(0) = 0$ for all $\boldsymbol{x}$. Nevertheless, $F(x^1, (x^1)^2) = x^1$: the slope of the graph of F on the curve $x^2 = (x^1)^2$ is unity. This means that $DF(0)$, which should be zero according to the directional derivatives, in fact does not exist: $R(\boldsymbol{x}) \to 0$ does not hold for $\boldsymbol{x} = (x^1, (x^1)^2)$. (Exercise: Show that $D_{\boldsymbol{x}}F(\boldsymbol{x}_0)$ is discontinuous at $\boldsymbol{x}_0 = 0$.)

If $D_xF(x'_0)$ exists for all x and for all x'_0 in a neighborhood U of x_0 and is continuous as a function of x'_0 at x_0, then $DF(x_0)$ exists and $DF(x_0)x = D_xF(x_0)$.

Proof For small enough x so that $x_0 + x \in U$, consider the function $r(x_0, x) = F(x_0 + x) - F(x_0) - D_xF(x_0)$ with values in Y. Take any vector f of the dual space Y^* of Y and consider the scalar function $f(t) = \langle f, F(x_0 + tx)\rangle$ of the real variable $t, 0 \leq t \leq 1$. This function has a derivative

$$\frac{df}{dt} = \lim_{\Delta t \to 0} \left\langle f, \frac{F(x_0 + tx + \Delta tx) - F(x_0 + tx)}{\Delta t} \right\rangle = \langle f, D_x F(x_0 + tx) \rangle$$

and hence $f(1) - f(0) = \langle f, D_x F(x_0 + \tau x) \rangle$ for some $\tau, 0 \leq \tau \leq 1$. Therefore, $\langle f, r(x_0, x) \rangle = \langle f, D_x F(x_0 + \tau x) - D_x F(x_0) \rangle$. Choose f with $||f|| = 1$ for which

$$|\langle f, r(x_0, x) \rangle| \geq \frac{1}{2} ||f|| \, ||r(x_0, x)|| = \frac{1}{2} ||r(x_0, x)||$$

holds. (It exists by the Hahn–Banach theorem.) It follows that $||r(x_0, x)|| \leq 2 |\langle f, D_x F(x_0 + \tau x) - D_x F(x_0) \rangle| \leq 2 ||D_x F(x_0 + \tau x) - D_x F(x_0)||$. Finally, put $x = ||x|| \hat{x}$ and get $||r(x_0, x)|| \leq 2 ||D_{\hat{x}}(x_0 + \tau x) - D_{\hat{x}}(x_0)|| \, ||x||$. Hence, in view of the continuity of $D_{\hat{x}}(x'_0)$ at $x'_0 = x_0$ it follows that $r(x_0, x) = R(x) ||x||$ with $\lim_{x \to 0} R(x) = 0$. □

In the special case $Y = K$, the scalar field of X, the mapping $F : X \to K$ is a functional, and $DF(x_0) \in \mathcal{L}(X, K) = X^*$ is a continuous linear functional and hence an element of the dual space X^*, if it exists. For instance, if $X = K^n$ then $DF(x_0) = \boldsymbol{y} \in K^n$ (gradient). If X is a functional space, $DF(x_0)$ is called the **functional derivative** of F at x_0. If $X = L^p(K^n, d^n z) \ni f(z)$ then $DF(f_0) = g(z) \in L^q(K^n, d^n z), 1/p + 1/q = 1$. The functional derivative in the functional space L^p is a function (more precisely class of functions) of the functional space L^q. A trivial example which nevertheless is frequently met in physics is $F(f) = (g|f)$ with $D(g|f)(f) = g$ (derivative of a linear function).

If $X = K^n \ni \boldsymbol{x} = x^1 \boldsymbol{e}_1 + x^2 \boldsymbol{e}_2 + \cdots + x^n \boldsymbol{e}_n$ and $Y = K^m \ni \boldsymbol{y} = y^1 \boldsymbol{e}'_1 + y^2 \boldsymbol{e}'_2 + \cdots + y^m \boldsymbol{e}'_m$, then $\boldsymbol{F}(\boldsymbol{x}) = F^1(\boldsymbol{x}) \boldsymbol{e}'_1 + F^2(\boldsymbol{x}) \boldsymbol{e}'_2 + \cdots + F^m(\boldsymbol{x}) \boldsymbol{e}'_m$ and $\langle \boldsymbol{f}^i, \boldsymbol{DF}(\boldsymbol{x}_0) \boldsymbol{e}_k \rangle = \partial F^i(\boldsymbol{x}_0) / \partial x^k$, $\langle \boldsymbol{f}^i, \boldsymbol{e}'_k \rangle = \delta^i_k$. In this case,

$$\boldsymbol{DF}(\boldsymbol{x}_0) = \begin{pmatrix} \frac{\partial F^1(\boldsymbol{x}_0)}{\partial x^1} & \frac{\partial F^1(\boldsymbol{x}_0)}{\partial x^2} & \cdots & \frac{\partial F^1(\boldsymbol{x}_0)}{\partial x^n} \\ \frac{\partial F^2(\boldsymbol{x}_0)}{\partial x^1} & \frac{\partial F^2(\boldsymbol{x}_0)}{\partial x^2} & \cdots & \frac{\partial F^2(\boldsymbol{x}_0)}{\partial x^n} \\ \vdots & \vdots & & \vdots \\ \frac{\partial F^m(\boldsymbol{x}_0)}{\partial x^1} & \frac{\partial F^m(\boldsymbol{x}_0)}{\partial x^2} & \cdots & \frac{\partial F^m(\boldsymbol{x}_0)}{\partial x^n} \end{pmatrix} \tag{2.15}$$

is the **Jacobian matrix** of the function $\boldsymbol{F} : K^n \to K^m$. For any $\boldsymbol{y}^* \in Y^*$, $\langle \boldsymbol{y}^*, \boldsymbol{DF}(\boldsymbol{x}_0) \boldsymbol{x} \rangle = \boldsymbol{y}^* \cdot \boldsymbol{DF}(\boldsymbol{x}_0) \cdot \boldsymbol{x}$, where the dot $\cdot$ marks the inner product in the spaces K^n and K^m. For $m = n$ the determinant

$$\frac{D(y^1, \ldots, y^n)}{D(x^1, \ldots, x^n)} = \det \left(\frac{\partial F^i(\boldsymbol{x}_0)}{\partial x^j} \right) \tag{2.16}$$

is the **Jacobian**.

Employing higher derivatives, the **Taylor expansion** of a function F from the normed linear space X into a normed linear space Y reads

$$F(x_0 + x) = F(x_0) + DF(x_0)x + \frac{1}{2!} D^2 F(x_0) xx + \cdots + \frac{1}{k!} D^k F(x_0) \underbrace{xx \cdots x}_{(k \text{ factors})} + \cdots, \tag{2.17}$$

provided x_0 and x_0+x belong to a convex domain $\Omega \subset X$ on which F is defined and has total derivatives to all orders, which are continuous functions of x_0 in Ω and provided this Taylor series converges in the norm topology of Y. As explained after (2.14), $D^kF(x_0) \in \mathcal{L}(X, \mathcal{L}(X, \cdots, \mathcal{L}(X,Y)\cdots))$ is a k-linear function from $X \times X \times \cdots \times X$ (k factors) into Y. For instance, in the case $X = K^n, Y = K^m$ this means

$$\langle \boldsymbol{y}^*, D^k\boldsymbol{F}(\boldsymbol{x}_0)\boldsymbol{x}\cdots\boldsymbol{x}\rangle = \sum_{i,i_1,\dots,i_k} y_i^* \frac{\partial F^i(\boldsymbol{x}_0)}{\partial x^{i_1}\cdots\partial x^{i_k}} x^{i_1}\cdots x^{i_k}. \tag{2.18}$$

Proofs of this Taylor expansion theorem and the following generalizations from standard analysis can be found in textbooks, for instance [4].

Recall that $\mathcal{L}(X,Y)$ is a normed vector space with the norm (2.5) which is Banach if Y is Banach. Hence, $\mathcal{L}(X,\mathcal{L}(X,Y))$ is again a normed vector space which is Banach if Y is Banach. If $L_2 : X \to \mathcal{L}(X,Y) : x, x' \mapsto L_2(x,x') =: L_2xx'$ is a bilinear function from $X \times X$ into Y, its $\mathcal{L}(X,\mathcal{L}(X,Y))$-norm is (cf. (2.5))

$$\begin{aligned}||L_2||_{\mathcal{L}(X,\mathcal{L}(X,Y))} &= \sup_{x\in X}\frac{||L_2xx'||_{\mathcal{L}(X,Y)}}{||x||_X} = \sup_{x\in X}\frac{\sup_{x'\in X}||L_2xx'||_Y/||x'||_X}{||x||_X}\\ &= \sup_{x,x'\in X}\frac{||L_2xx'||_Y}{||x||_X||x'||_X}.\end{aligned}$$

By continuing this process, $\mathcal{L}(X,\mathcal{L}(X,\cdots,\mathcal{L}(X,Y)\cdots))$ (depth k) is a normed vector space which is Banach if Y is Banach, and the norm of a k-linear function $L_kx^{(1)}x^{(2)}\cdots x^{(k)}$ is

$$||L_k||_{\underbrace{\mathcal{L}(X,\mathcal{L}(X,\dots,\mathcal{L}(X,Y)\dots))}_{\text{depth } k}} = \sup_{x^{(1)}\dots x^{(k)}\in X}\frac{||L_kx^{(1)}\cdots x^{(k)}||_Y}{||x^{(1)}||_X\cdots||x^{(k)}||_X}. \tag{2.19}$$

A general **chain rule** holds for the case *if $F : X \supset \Omega \to Y, F(\Omega) \subset \Omega', G : Y \supset \Omega' \to Z$ and $H = G \circ F : X \supset \Omega \to Z$. Then,*

$$DH(x_0) = DG(F(x_0)) \circ DF(x_0) \tag{2.20}$$

if the right hand side derivatives exist. In this case, $DF(x_0) \in \mathcal{L}(X,Y)$ and $DG(F(x_0)) \in \mathcal{L}(Y,Z)$ and hence $DH(x_0) \in \mathcal{L}(X,Z)$. Moreover, if $DF : \Omega \to \mathcal{L}(X,Y)$ is continuous at $x_0 \in \Omega$ and $DG : \Omega' \to \mathcal{L}(Y,Z)$ is continuous at $F(x_0) \in \Omega'$, then $DH : \Omega \to \mathcal{L}(X,Z)$ is continuous at $x_0 \in \Omega$.

Coming back to the warning on p. 23, take the function $F : \mathbb{R} \to \mathbb{R}^2 : t \mapsto (t,t^2)$, and for $G : \mathbb{R}^2 \to \mathbb{R}$ take the function of the example on p. 23. Then, $H(t) = (G \circ F)(t) = t$ and hence $DH(0) = 1$. Would one from $D_{(x^1,x^2)}G(0,0) = 0$ for all (x^1,x^2) infer that $DG(0,0) = 0$, then one would get erroneously $DH(0) = DG(0,0) \circ DF(0) = 0$. In more familiar notation for this case,

$$\left.\frac{dH}{dt}\right|_{t=0} \neq \left.\frac{\partial G}{\partial x^1}\right|_{(0,0)} \left.\frac{dx^1}{dt}\right|_0 + \left.\frac{\partial G}{\partial x^2}\right|_{(0,0)} \left.\frac{dx^2}{dt}\right|_0 = 0.$$

The chain rule does not hold because the *total derivative* of G does not exist at $(0,0)$; $\partial G/\partial x^2$ is discontinuous there.

If X, Y and Z are the finite-dimensional vector spaces K^n, K^m and K^l with general (not necessarily orthonormal) bases fixed, then the $l \times n$ Jacobian matrix of $DH(x_0)$ is just the matrix product of the $l \times m$ and $m \times n$ Jacobian matrices (2.15) of $DG(F(x_0))$ and $DF(x_0)$. It follows that in the case $l = m = n$ the Jacobian of H is the product of the Jacobians of G and F:

$$\frac{D(z^1,\dots,z^n)}{D(x^1,\dots,x^n)} = \frac{D(z^1,\dots,z^n)}{D(y^1,\dots,y^n)}\frac{D(y^1,\dots,y^n)}{D(x^1,\dots,x^n)}.$$

Just this is suggested by the notation (2.16) of a Jacobian.

If $F : X \supset \Omega \to \Omega' \subset Y$ is a bijection and $DF(x_0)$ and $DF^{-1}(F(x_0))$ both exist, then

$$(DF(x_0))^{-1} = DF^{-1}(F(x_0)). \tag{2.21}$$

This follows from the chain rule in view of $F^{-1} \circ F = \mathrm{Id}_\Omega$ and $D\mathrm{Id}(x_0) = \mathrm{Id}$. (From the definition (2.14) it follows for a linear function $F \in \mathcal{L}(X,Y)$ that $DF(x_0) = F$ independent of $x_0 \in X$.) The case $X = Y = K^n$ now implies

$$\frac{D(x^1,\dots,x^n)}{D(y^1,\dots,y^n)} = \left(\frac{D(y^1,\dots,y^n)}{D(x^1,\dots,x^n)}\right)^{-1}$$

for the Jacobian. For $n = 1$ this is the rule $dx/dy = (dy/dx)^{-1}$.

A function F from an open domain Ω of a normed space X into a normed space Y is called a **class $C^n(\Omega, Y)$ function** if it has continuous derivatives $D^kF(x_0)$ up to order $k = n$ (continuous as functions of $x_0 \in \Omega$). If the domain Ω and the target space Y are clear from context, one speaks in short on a class C^n function (or even shorter of a C^n function). A C^0 function means just a continuous function. A C^∞ function is also called **smooth**. A smooth function still need not have a Taylor expansion. For instance the real function

$$f_\varepsilon(x) = \begin{cases} \exp(-\varepsilon^2/(\varepsilon^2 - x^2)) & \text{for } |x| < \varepsilon \\ 0 & \text{for } |x| \geq \varepsilon \end{cases}$$

is C^∞ on the whole real line, but has no Taylor expansion at the points $x = \pm\varepsilon$ although all its derivatives are equal to zero and continuous there. (Up to the normalization factor it is a δ_ε-function.) A function which has a Taylor expansion converging in the whole domain Ω is called a class $C^\omega(\Omega, Y)$ function or an **analytic function**. A complex-valued function of complex variables is analytic, iff it is C^1 and its derivatives obey the Cauchy–Riemann equations.

A C^n (C^∞, C^ω) **diffeomorphism** is a bijective mapping from $\Omega \subset X$ onto $\Omega' \subset Y$ which, along with its inverse, is $C^n, n > 0$, (C^∞, C^ω).

With pointwise linear combinations of functions with constant coefficients, $(\lambda F + \lambda' G)(x) = \lambda F(x) + \lambda' G(x)$, the class C^n (C^∞, C^ω) is made into a vector space. The vector spaces C^n, C^∞ include normed subspaces C^n_b, C^∞_b (of all functions with finite norm) by introducing the norm

$$||F||_{C_b^{n/\infty}} = \sup_{\substack{x_0 \in \Omega \\ k \leq n/\infty}} ||D^k F(x_0)||, \quad ||F||_{C_b^0} = \sup_{x_0 \in \Omega} ||F(x_0)||, \tag{2.22}$$

with the norms (2.19) on the right hand side of the first expression. These spaces are again Banach if Y is Banach. Convergence of a sequence of functions in these norms means uniform convergence on Ω, of the sequence of functions and of the sequences of all derivatives up to order n, or of unlimited order. (Besides, every space $C^n_b, m \leq n \leq \infty$, is dense in the normed space C^m_b.)

The mapping $D : C^1_b(\Omega, Y) \to C^0_b(\Omega, \mathcal{L}(X,Y)) : F \mapsto DF$ *is a continuous linear mapping with norm not exceeding unity.*

Proof As a bounded linear mapping, $D \in \mathcal{L}(C^1_b(\Omega, Y), C^0_b(\Omega, \mathcal{L}(X,Y)))$, the norm of D is $||D|| = \sup_F ||DF||_{C^0_b(\Omega, \mathcal{L}(X,Y))}/||F||_{C^1_b(\Omega,Y)}$. From (2.22) it is directly seen that the numerator of this quotient cannot exceed the denominator, hence $||D|| \leq 1$ and D is indeed bounded and hence continuous. □

If the normed vector space Y in addition is an algebra with unity I (see Compendium) and the norm has the additional properties

4. $||I|| = 1$,
5. $||yy'|| \leq ||y||\,||y'||$,

then it is called a **normed algebra**. If it is complete as a normed vector space, it is called a **Banach algebra**. If Y is a normed algebra, then with pointwise multiplication, $(FG)(x) = F(x)G(x)$, the class C^n_b (C^∞_b) with the norm (2.22) is made into a normed algebra. (Show that FG is C^n_b if F and G both are C^n_b.)

The **derivative of a product** in the algebra $C^n, n \geq 1$ is obtained by the **Leibniz rule**

$$D(FG) = (DF)G + F(DG). \tag{2.23}$$

(Exercise: Consider $\Phi(x) = (F(x), G(x))$, $\Psi(u,v) = uv$ and $H(x) = (\Psi \circ \Phi)(x)$ and apply the chain rule to obtain (2.23).)

An **implicit function** is defined in general in the following manner: *Let X be a topological space, let Y be a Banach space and let Z be a normed vector space. Let $F : X \times Y \supset \Omega \to Z$ be a continuous function and consider the equation*

$$F(x,y) = c, \quad c \in Z \text{ fixed}. \tag{2.24}$$

Assume that $D_yF(x_0, y_0) \in \mathcal{L}(Y,Z)$ *exists for all* $y \in Y$ *and is continuous on* Ω *(as a function of* x_0, y_0*), that* $F(a,b) = c$ *and that* $Q = D_yF(a,b)$ *is a linear bijection from* Y *onto* Z*, so that* $Q^{-1} \in \mathcal{L}(Z,Y)$*. Then, there are open sets* $A \ni a$ *and* $B \ni b$ *in* X *and* Y*, so that for every* $x \in A$ Eq. (2.24) *has a unique solution* $y \in B$ *which implicitly by* Eq. (2.24) *defines a continuous function* $G : A \to Y : x \mapsto y = G(x)$.

The proofs of this theorem and of the related theorems below are found in textbooks, for instance [4]. It is essential, that Y is Banach.

Let X *be also a normed vector space and assume* $F \in C^1(\Omega, Z)$*. Then the above function* G *has a continuous total derivative at* $x = a$*, and*

$$D_xG(a) = -(D_yF(a,b))^{-1} \circ D_xF(a,b), \quad b = G(a). \tag{2.25}$$

Formally, one may differentiate (2.24) by applying the chain rule,

$$D_xF(a,b)dx + D_yF(a,b)dy = 0, \quad x \in X,\ y \in Y,$$

where $dx = D\mathrm{Id}_X = 1$ and $dy = D_xG(a)$, and solve this relation for dy/dx.

In order to prove that $DG(a)$ of (2.25) is a continuous function of a, that is, that $G \in C^1(A,Y)$, the continuity of $(D_yF(a,b))^{-1}$ as function of a and b must be stated. Since $D_yF(a,b) \in \mathcal{L}(Y,Z)$, this implies the derivative of the inverse of a linear function with respect to a parameter which is of interest on its own:

Let X *and* Y *be Banach spaces and let* $\mathcal{U}$ *and* $\mathcal{U}^{-1}$ *be the sets of invertible continuous linear mappings out of* $\mathcal{L}(X,Y)$ *and* $\mathcal{L}(Y,X)$*. Then, both* $\mathcal{U}$ *and* $\mathcal{U}^{-1}$ *are open sets.*

Proof for $\mathcal{U}$*; for* $\mathcal{U}^{-1}$ *interchange* X *and* Y Let $U_0 \in \mathcal{U}$ and $U \in \mathcal{L}(X,Y)$ such that $||\mathrm{Id}_X - U_0^{-1} \circ U||_{\mathcal{L}(X,X)} < 1$. Then,

$$(U_0^{-1} \circ U)^{-1} = \mathrm{Id}_X + (\mathrm{Id}_X - U_0^{-1} \circ U) + (\mathrm{Id}_X - U_0^{-1} \circ U)^2 + \cdots$$

converges and hence $U = U_0 \circ (U_0^{-1} \circ U) \in \mathcal{U}$ $(U^{-1} = (U_0^{-1} \circ U)^{-1} \circ U_0^{-1})$. Every $U_0 \in \mathcal{U}$ has a neighborhood, $||U_0 - U|| < 1/||U_0||$, where this is realized. □

Let X *and* Y *be Banach spaces and* $\mathcal{U}$ *as above. Let* $\Phi : X \supset A \to \mathcal{U} \subset \mathcal{L}(X,Y) : x_0 \mapsto U(x_0)$ *be* C^1*. Then* $\tilde{\Phi} : x_0 \mapsto (U(x_0))^{-1}$ *is* C^1*, and its derivative is given by*

$$D(\tilde{\Phi})(x_0)x = -\tilde{\Phi}(x_0) \circ D\Phi(x_0)x \circ \tilde{\Phi}(x_0) \in \mathcal{L}(Y,X), \quad x \in X. \tag{2.26}$$

The proof of continuity of the left hand side with respect to the x_0-dependence consists of an investigation of the relation $\Phi(x) \circ \tilde{\Phi}(x) = \mathrm{Id}_Y$. It is left to the reader (see textbooks of analysis). Differentiating this equation with respect to x at point x_0 yields $\Phi(x_0) \circ D\tilde{\Phi}(x_0)x + D\Phi(x_0)x \circ \tilde{\Phi}(x_0) = 0$. Composing with $(\Phi(x_0))^{-1} = \tilde{\Phi}(x_0)$ from the left results in the above relation. If $X = K^n$, then $\mathcal{U}$ can only be non-empty if also $Y = K^n$. After introducing bases $U(\boldsymbol{x})$ is represented by a regular $n \times n$ matrix $\boldsymbol{M}(\boldsymbol{x})$. One obtains the familiar result $(\boldsymbol{x} \cdot \partial/\partial \boldsymbol{x})\boldsymbol{M}^{-1}|_{x_0} = \boldsymbol{M}^{-1} \cdot (\boldsymbol{x} \cdot \partial/\partial \boldsymbol{x})\boldsymbol{M}|_{x_0} \cdot \boldsymbol{M}^{-1}$. Along a straight line $\boldsymbol{x} = t\boldsymbol{e}$, or for a one parameter dependent matrix this reduces to $d\boldsymbol{M}^{-1}/dt = \boldsymbol{M}^{-1} \cdot (d\boldsymbol{M}/dt) \cdot \boldsymbol{M}^{-1}$.

2.4 Compactness

Compactness is the abstraction from closed bounded subsets of $\mathbb{R}^n$. Before introducing this concept, a few important properties of n-dimensional closed bounded sets are reviewed.

The **Bolzano–Weierstrass theorem** says that in an n-dimensional closed bounded set every sequence has a convergent subsequence. An equivalent formulation is that every infinite set of points of an n-dimensional closed bounded set has a cluster point.

A **cluster point** of a subset A of a topological space X is a point $x \in X$ every neighborhood of which contains at least one point of *A distinct from x*. (Compare the definition of a point of closure on p. 12. A cluster point is a point of closure, but the reverse is not true in general.)

Weierstrass theorem: A continuous function takes on its maximum and minimum values on an n-dimensional closed bounded set.

Brouwer's fixed point theorem: On a *convex* n-dimensional closed bounded set B the fixed point equation $x = F(x), F : B \to B$ continuous, has a solution.

These theorems do not necessarily hold in infinite dimensional spaces. Consider for example the closed unit ball (e.g. centered at the origin) in an infinite dimensional real Hilbert space. Clearly the sequence of distinct orthonormal unit vectors does not converge in the norm topology: the distance between any pair of orthogonal unit vectors is $||e_i - e_j|| = (e_i - e_j|e_i - e_j)^{1/2} = \sqrt{2}$. It is easily seen that open balls of radius $1/(2\sqrt{2})$ centered halfway on these unit vectors do not intersect. The unit ball is too roomy for the Bolzano–Weierstrass theorem to hold; it accommodates an infinite number of non-overlapping balls of a fixed non-zero radius. This consideration yields the key to compactness.

A set C of a topological space is called a **compact set**, if every open cover $\{U\}$, a family of open sets with $\cup U \supset C$, contains a finite subcover, $\cup_{i=1}^n U_i \supset C$. A compact set in a Hausdorff space (the only case of interest in this volume) is called a **compactum**.

Compactness is a topological property, *the image C' of a compact set C under a continuous mapping F is obviously a compact set*: Take any open cover of C'. Since the preimage $F^{-1}(U')$ of an open set U' is an open set $U \subset C$, these preimages form an open cover of C. A selection of a finite subcover of these preimages also selects a finite subcover of C'.

A compactum is closed.

Proof Let x be a point of closure of a compactum C, that is, every neighborhood of x contains at least one point $c \in C$. Let $x \notin C$. Since C is Hausdorff, for every $c \in C$ there are disjoint open sets $U_c \ni c$ and $V_{x,c} \ni x$. Since the sets U_c obviously form an open cover of C, a finite subcover $U_{c_i}, i = 1, \ldots, n$ may be selected. Then, $V = \cap_i V_{x,c_i}$ is a neighborhood of x not intersecting C, which contradicts the preposition. Hence, C contains all its points of closure. □

It easily follows that *the inverse of a continuous bijection f of a compactum C onto a compactum C′ is continuous, that is, the bijection is a homeomorphism.*

Proof Indeed, any closed subset A of the compactum C is a compactum; any open cover of A together with the complement of A forms an open cover of C and hence there is a finite subcover which is also a subcover of A. Now, since f is continuous, $f(A)$ is also a compactum and hence a closed subset of C'. Consequently f maps closed sets to closed sets, and because it is a bijection, it also maps open sets to open sets. □

Now, the Bolzano–Weierstrass theorem is extended:

Every infinite set of points of a compact set C has a cluster point.

Proof Assume that the infinite set $A \subset C$ has no cluster point. A set having no cluster point is closed. Indeed, if a is a point of closure of A, then $a \in A$ or a is a cluster point of A. Select any infinite sequence $\{a_i\} \subset A$ of distinct points a_i. The sets $\{a_i\}_{i=n}^{\infty}$ are closed for $n = 1, 2, \ldots$ and the intersection of any finite number of them is not empty. Their complements U_n in C form an open cover of C, for which hence there exists no finite subcover. C is not a compact set. □

As a consequence, an unbounded set of a metric space cannot be compact. Hence, the simple Heine–Borel theorem, that a closed bounded subset of $\mathbb{R}^n$, $n<\infty$ is compact, has a reversal: A compact subset of $\mathbb{R}^n$ is closed and bounded. (Recall that a metric space is Hausdorff.) This immediately also extents the Weierstrass theorem:

A continuous real-valued function on a compact set takes on its maximum and minimum values.

It maps the compact domain onto a compact set of the real line, which is closed and bounded and hence contains its minimum and maximum. However, a much more general statement on the existence of extrema will be made later on.

A closed subset of a compact set is a compact set.

Proof Take any open cover of the closed subset C' of the compact set C. Together with the set $C \backslash C'$, open in C, it also forms an open cover of C. A finite subcover of C also yields a finite subcover of C'. □

A set of a topological space is called **relatively compact** if its closure is compact. A topological space is called **locally compact** if every point has a relatively compact neighborhood. A function from a domain in a metric space X into a metric space Y is called a **compact function** or compact operator if it is continuous and maps bounded sets to relatively compact sets.

Brouwer's fixed point theorem has now two important generalizations which are given without proof (see textbooks of functional analysis):

Tychonoff's fixed point theorem: *A continuous mapping $F : C \to C$ in a compact convex set C of a locally convex vector space has a fixed point.*

Schauder's fixed point theorem: *A compact mapping $F : C \to C$ in a closed bounded convex set C of a Banach space has a fixed point.*

Both theorems release the precondition of Banach's fixed point theorem on F to be a strict contraction (p. 14). As a price, uniqueness is not guaranteed any more. Tychonoff's theorem also releases the precondition of completeness of the space.

Every locally compact space has a ***one point compactification*** *that is, a compact space* $X^c = X \cup \{x_\infty\}$ *and a homeomorphism* $P : X \to X^c \setminus \{x_\infty\}$.

Proof Let $x_\infty \notin X$ and let $\{U\}_\infty$ be the class of open sets of X for which $X \backslash U$ is compact in X. (X itself belongs to this class since $\varnothing$ is compact.) Take the open sets of X^c to be the open sets of X and all sets containing x_∞ and having their intersections with X in $\{U\}_\infty$. This establishes a topology in X^c and the homeomorphism. Let now $\{V\}$ be an open cover of X^c. It contains at least one set $V_\infty = U \cup \{x_\infty\}$, and $X^c \backslash V_\infty$ is compact in X. Hence, $\{V\}$ has a finite subcover. □

The compactified real line (circle) $\overline{\mathbb{R}}$ and the compactified complex plane (Riemann sphere) $\overline{\mathbb{C}}$ are well known examples of one point compactifications.

To get more general results for the existence of extrema, the concept of semicontinuity is needed. A function F from a domain of a topological space X into $\overline{\mathbb{R}}$ is called **lower (upper) semicontinuous** at the point $x_0 \in X$, if either $F(x_0) = -\infty$ ($F(x_0) = +\infty$) or for every $\varepsilon > 0$ there is a neighborhood of x_0 in which $F(x) > F(x_0) - \varepsilon$ ($F(x) < F(x_0) + \varepsilon$).

A lower semicontinuous function need not be continuous, its function value even may jump from $-\infty$ to ∞ at points of discontinuity. However, at every point of discontinuity it takes on the lowest limes of values. (For every net converging towards $x_0 \in X$ the function value at x_0 is equal to the lowest cluster point of function values on the net.) A lower semicontinuous function is **finite from below**, if $F(x) > -\infty$ for all x. Analogous statements hold for an upper semicontinuous function.

If F is a finite from below and lower semicontinuous function from a non-empty compactum A into $\mathbb{R}$, then F is even bounded below and the minimum problem $\min_{x \in A} F(x) = \alpha$ *has a solution* $x_0 \in A, \alpha = F(x_0)$.

An analogous theorem holds for a maximum problem. The proof of these statements is simple: Consider the infimum of F on A, pick a sequence for which $F(x_n) \leq \inf F(x) + 1/n$ and select a cluster point x_0 and a subnet converging to x_0. Hence, $\inf F(x) = F(x_0) > -\infty$ since F is finite from below.

Extremum problems are ubiquitous in physics. Many physical principles are directly variational. Extremum problems are also in the heart of duality theory which in physics mainly appears as theory of Legendre transforms. Moreover, since every system of partial differential equations is equivalent to a variational problem, extremum problems are also central in (particularly non-linear) analysis, again with central relevance for physics.

It has become evident above that compactness of the domain plays a decisive role in extremum problems. On the other hand, bounded sets in infinite-dimensional normed spaces are not compact in the norm topology, while many variational problems, in particular in physics, are based on infinite-dimensional functional spaces. (David Hilbert introduced the concept of functional inner product space to bring forward the variational calculus.) Rephrased, those functional spaces are not locally compact in the norm topology. The question arises, can one introduce a more cooperative topology in those spaces. The coarser a topology, the less open sets exist, and the more chances appear for a set to be compact. On p. 18, the weak topology was introduced as the coarsest topology in the vector space X, for which all bounded linear functionals are continuous. In a finite-dimensional space it was shown to be equivalent to the norm topology. In an infinite-dimensional space it is indeed coarser than the norm topology, but sometimes not coarse enough to our goal.

Let X be a Banach space and X^* its dual. In general, X^{**}, the space of all bounded linear functionals on X^*, may be larger than X. The weak topology of X^* is the coarsest topology in which all bounded linear functionals, that is all $f \in X^{**}$ are continuous. The **weak* topology** is the coarsest topology of X^* in which all bounded functionals $f \in X$ are continuous. Since these are in general less functionals, in general the weak* topology is coarser than the weak topology. If X is reflexive, then $X^{**} = X$ (and $X^{***} = X^*$), and the weak and weak* topologies of X^* (and also of $X^{**} = X$) are equivalent. (A Banach space is in general not any more first countable in the weak and weak* topologies; this is why instead of sequences nets are needed.)

The Banach–Alaoglu theorem states *that the unit ball of the dual X^* of a Banach space X is compact in the weak* topology*. As a corollary, *the unit ball of a reflexive Banach space is compact in the weak topology*.

A proof which uses Tichonoff's non-trivial theorem on topological products may be found in textbooks on functional analysis. Now, the way is paved for applications of the existence theorems of extrema. The price is that in the weak* topology there are much less semicontinuous functions than in the norm topology. Nevertheless, for instance the theory of functional Legendre transforms, relevant in **density functional theories** is pushed far ahead [5, and citations therein].

A few applications of the concept of compactness in functional analysis are finally mentioned which are related to the material of this volume. They use the facts that every compactum X is a **regular topological space**, that is, every non-empty open set contains the closure of another non-empty open set, and every compactum is a **normal topological space**, which means that every single point set $\{x\}$ is closed and every pair of disjoint closed sets is each contained in one of a pair of disjoint open sets.

Proof For each pair (x, y) of points in a pair of disjoint closed sets (C_1, C_2), $C_1 \cap C_2 = \emptyset, x \in C_1, y \in C_2$, there is a pair of disjoint open sets $(U_{x,y}, U_{y,x})$, $x \in U_{x,y}$, $y \in U_{y,x}$, since a compactum is Hausdorff. C_2 as a closed subset of a compactum is compact, and hence has a finite open cover $\{U_{y_1,x}, U_{y_2,x}, \ldots, U_{y_n,x}\}$, $y_i \in C_2$. Put $U_x = \cup_i U_{y_i,x}, U^x = \cap_i U_{x,y_i}$, and $U_1 = \cup_j U_{x_j}, U_2 = \cap_j U^{x_j}$, $x_j \in C_1$.

Regularity: Let U be an open set. Put $C_1 = X \backslash U$ and $C_2 = \{y\}$, $y \in U$ (C_2 is closed since X is Hausdorff.). Take $U_2 \ni y$ constructed above. $\overline{U_2} \subset U$.

Normality: For the above constructions, obviously $C_1 \subset U_1$, $C_2 \subset U_2$, $U_1 \cap U_2 = \emptyset$. □

In a regular topological space every point has a *closed neighborhood base*.

For the proofs of the following theorems see textbooks of functional analysis.

Urysohn's theorem: *For every pair* (C_0, C_1) *of disjoint closed sets of a normal space* X *there is a real-valued continuous function,* $F \in C^0(X, \mathbb{R})$, *with the properties* $0 \leq F(x) \leq 1, F(x) = 0$ *for* $x \in C_0$, $F(x) = 1$ *for* $x \in C_1$.

Tietze's extension theorem: *Let* X *be a compactum and* $C \subset X$ *closed. Then every* $C^0(C, \mathbb{R})$*-function has a* $C^0(X, \mathbb{R})$*-extension.*

A function F defined on a locally compact topological space X with values in a normed vector space Y is said to be a **function of compact support**, if it vanishes outside of some compact set (in general depending on F). The **support of a function** F, $\operatorname{supp} F$ is the smallest closed set outside of which $F(x) = 0$. If X is a locally compact normed vector space, then corresponding to the classes $C^n, 0 \leq n \leq \infty$ (p. 27) there are **classes** C_0^n of continuous or n times continuously differentiable functions of compact support. Like the classes C^n, the classes C_0^n are vector spaces or in the case of an algebra Y algebras with respect to pointwise operations on functions.

In the context of this volume, particularly $C_0^\infty(K^n, Y)$ functions, $K = \mathbb{R}$ or $\mathbb{C}$, are of importance. One could normalize the vector space $C_0^n, 0 \leq n \leq \infty$ with the C_b^n-norm (2.22), however, if X itself is not compact, C_{0b}^n would not be complete in this norm topology even if Y would be Banach. For instance, the function sequence

$$F_n(x) = \sum_{k=1}^{n} \frac{1}{2^k} \Phi(x-k), \quad \Phi \in C_0^n(\mathbb{R}, \mathbb{R}), \quad n = 1, 2, \ldots,$$

is Cauchy in the C_b^n norm, but its limit does not have compact support. The completion of the $C_0^0(X, Y) = C_{0b}^0(X, Y)$ space of continuous functions of compact support in the C_b^0-norm is the space $C_\infty(X, Y)$ of continuous functions vanishing for $||x||_X \to \infty$, that is, for every $\varepsilon > 0$ there is a compact $C_\varepsilon \subset X$ outside of which $||F(x)||_Y < \varepsilon$. (Hence, $C_0^0(X, Y)$ is dense in $C_\infty(X, Y)$ in the C_b^0-norm; moreover, all $C_0^n(X, Y)$, $0 \leq n \leq \infty$ are dense in $C_\infty(X, Y)$ in the C_b^0-norm. If X is not compact, of course non of those classes is dense in any C_b^n in the C_b^n-norm: Let for instance $||F_1(x)||_Y = 1$ for all $x \in X$, then $||F - F_1||_{C_b^n} = 1$ for all $F \in C_0^m$. These are simple statements on uniform approximations of functions by more well behaved functions.)

Functions of compact support are very helpful in analysis, geometry and physics. They are fairly wieldy since their study is much the same as that of functions on a closed bounded subset of $\mathbb{R}^n$. The tool of continuation of structures from this rather simple situation to much more complex spaces, that is to connect local with global structures, is called partition of unity. It works for all locally compact spaces which are countable unions of compacta. (Caution: Not every countable union of compact sets is locally compact.) However, the most general class of spaces where it works are the paracompact spaces.

A **paracompact space** is a Hausdorff topological space for which every open cover, $X \subset \cup_{\alpha \in A} U_\alpha$, has a **locally finite** refinement, that is an open cover $\cup_{\beta \in B} V_\beta$ for which every V_β is a subset of some U_α and every point $x \in X$ has a neighborhood W_x which intersects with a finite number of sets V_β only.

A **partition of unity** on a topological space X is a family $\{\varphi_\alpha | \alpha \in A\}$ of $C_0^\infty(X, \mathbb{R})$-functions such that

1. there is a locally finite open cover, $X \subset \cup_{\beta \in B} U_\beta$,
2. the support of each φ_α is in some U_β, $\{\operatorname{supp} \varphi_\alpha | \alpha \in A\}$ is locally finite,
3. $0 \leq \varphi_\alpha(x) \leq 1$ on X for every α,
4. $\sum_{\alpha \in A} \varphi_\alpha(x) = 1$ on X.

The last sum is well defined since, given x, only a finite number of items are non-zero due to the locally finite cover governing the partition. The partition of unity is called **subordinate** to the cover $\cup_{\beta \in B} U_\beta$.

A paracompact space could also be characterized as a space which permits a partition of unity. It can be shown that every second countable locally compact Hausdorff space is paracompact. This includes locally compact Hausdorff spaces which are countable unions of compact sets, in particular it includes $\mathbb{R}^n$ for finite n. However, *any* (not necessarily countable) *disconnected* union (see next section) of paracompact spaces is also paracompact.

The function f_ε on p. 26 is an example of a real C_0^∞-function on $\mathbb{R}$. A simple example of functions φ_α on $\mathbb{R}^n$ is obtained by starting with the C^∞-function

$$f(t) = \begin{cases} e^{-1/t} & \text{for } t > 0 \\ 0 & \text{for } t \leq 0 \end{cases}$$

and putting $g(t) = f(t)/(f(t) + f(1-t))$, which is C^∞, $0 \leq g(t) \leq 1, g(t) = 0$ for $t \leq 0, g(t) = 1$ for $t \geq 1$. Then, $h(t) = g(t+2)g(2-t)$ is C_0^∞, $0 \leq h(t) \leq 1, h(t) = 0$ for $|t| \geq 2, h(t) = 1$ for $|t| \leq 1$. Now, with a dual base $\{\boldsymbol{f}^i\}$ in $\mathbb{R}^n$, the $C_0^\infty(\mathbb{R}^n, \mathbb{R})$-function

$$\psi(\boldsymbol{x}) = h(x^1)h(x^2)\cdots h(x^n), \quad x^i = \boldsymbol{f}^i \cdot \boldsymbol{x}, \tag{2.27}$$

has the properties $0 \leq \psi(\boldsymbol{x}) \leq 1, \psi(\boldsymbol{x}) = 0$ outside the compact n-cube $|x^i| \leq 2$ with edge length 4 which is contained in an open n-cube with edge length $4 + \varepsilon$, $\varepsilon > 0$, and $\psi(\boldsymbol{x}) = 1$ inside the n-cube with edge length 2, all centered at the origin of $\mathbb{R}^n$. The total $\mathbb{R}^n$ may be covered with open n-cubes of edge length $4 + \varepsilon$ centered at points $\boldsymbol{m} = (x^1 = 3m_1, x^2 = 3m_2, \ldots, x^n = 3m_n)$, m_i integer. Then,

$$\varphi_{\boldsymbol{m}}(\boldsymbol{x}) = \frac{\psi(\boldsymbol{x} - \boldsymbol{m})}{\sum_{\boldsymbol{m}'} \psi(\boldsymbol{x} - \boldsymbol{m}')}, \quad \sum_{\boldsymbol{m}} \varphi_{\boldsymbol{m}}(\boldsymbol{x}) = 1, \tag{2.28}$$

is a partition of unity on $\mathbb{R}^n$ (Fig. 2.5).

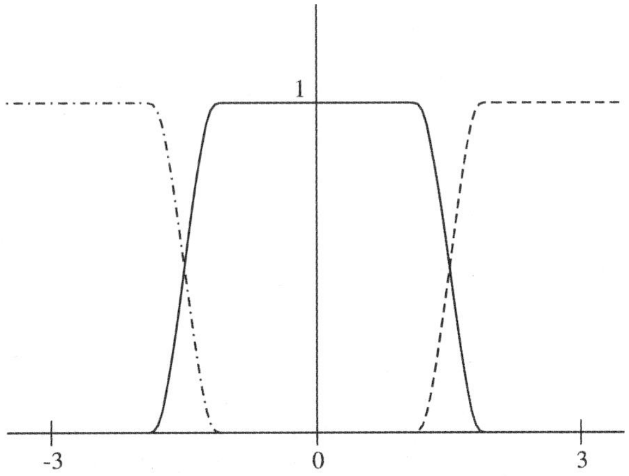

Fig. 2.5 Partition of unity on $\mathbb{R}$ with functions (2.28)

Besides applications in the theory of generalized functions and in the theory of manifolds, the partition of unity has direct applications in physics. For instance in **molecular orbital theory** of molecular or solid state physics the single particle quantum state (molecular orbital) is expanded into local basis orbitals centered at atom positions. For convenience of calculations one would like to have the density and self-consistent potential also as a site expansion of local contributions, hopefully to be left with a small number of multi-center integrals. This is however not automatically provided since the density is bilinear in the molecular orbitals, and the self-consistent potential is non-linear in the total density. If $v(\boldsymbol{x})$ is the self-consistent potential in the whole space $\mathbb{R}^3$ and $\sum_{\boldsymbol{R}} \varphi_{\boldsymbol{R}}(\boldsymbol{x})$ is a partition of unity on $\mathbb{R}^3$ with functions centered at the atom positions $\boldsymbol{R}$, then

$$v(\boldsymbol{x}) = \sum_{\boldsymbol{R}} (v(\boldsymbol{x})\varphi_{\boldsymbol{R}}(\boldsymbol{x})) = \sum_{\boldsymbol{R}} v_{\boldsymbol{R}}(\boldsymbol{x})$$

is the wanted expansion with potential contributions $v_{\boldsymbol{R}}$ of compact support. Thus, the number of multi-center integrals can be made finite in a very controlled way.

Finally, distributions (generalized functions) with compact support are shortly considered which comprise Dirac's δ-function and its derivatives.

Consider the whole vector space $C^\infty(\mathbb{R}^n, \mathbb{R})$ and instead of (2.22) for every compact $C \subset \mathbb{R}^n$ introduce the seminorm

$$p_{C,m}(F) = \sup_{\substack{x \in C \\ |l| \le m}} |D^l F(\boldsymbol{x})|, \quad D^0 F = F, \quad l = (l_1, \ldots, l_n), \ l_j \ge 0,$$
$$D^l F(\boldsymbol{x}) = \frac{\partial^{l_1 + l_2 + \cdots + l_n}}{(\partial x^1)^{l_1} (\partial x^2)^{l_2} \cdots (\partial x^n)^{l_n}} F(x^1, x^2, \ldots, x^n), \quad l_1 + l_2 + \cdots + l_n = |l|. \tag{2.29}$$

It is a seminorm because it may be $p_{C,m}(F) = 0, F \neq 0$ (if $\operatorname{supp} F \cap C = \emptyset$). In the topology of the family of seminorms for all $C \subset \mathbb{R}^n$ and all m, convergence of a sequence of functions means uniform convergence of the functions and of all their derivatives on every compactum. This topology is a metric topology, and the vector space $C^\infty(\mathbb{R}^n, \mathbb{R})$ topologized in this way is also denoted $\mathcal{E}(\mathbb{R}^n, \mathbb{R})$ or in short $\mathcal{E}$.

Indeed, *consider a sequence of compacta* $C_1 \subset C_2 \subset \cdots$ *with* $\cup_{i=1}^\infty C_i = \mathbb{R}^n$ *(for instance closed balls with a diverging sequence of radii). Then, the function*

$$d(F,G) = \sum_{i=1}^{\infty} 2^{-i} \frac{d_{C_i}(F,G)}{1 + d_{C_i}(F,G)}, \quad d_C(F,G) = \sum_{m=0}^{\infty} 2^{-m} \frac{p_{C,m}(F-G)}{1 + p_{C,m}(F-G)} \tag{2.30}$$

is a distance function.

Proof Clearly, $d(F,G) \neq 0$, if $F(\boldsymbol{x}) \neq G(\boldsymbol{x})$ for some $\boldsymbol{x}$ since the C_i cover $\mathbb{R}^n$. To prove the triangle inequality, consider the obvious inequality $(\alpha+\beta)/(1+\alpha+\beta) \leq \alpha/(1+\alpha) + \beta/(1+\beta)$ for any pair α, β of non-negative real numbers. In view of $|\alpha-\beta| \leq |\alpha-\gamma| + |\gamma-\beta|$ for any three real numbers α, β, γ it follows $|\alpha-\beta|/(1+|\alpha-\beta|) \leq |\alpha-\gamma|/(1+|\alpha-\gamma|) + |\gamma-\beta|/(1+|\gamma-\beta|)$. This yields the triangle inequality for each fraction on the right hand side of the second equation (2.30). Since each of these fractions is ≤ 1, the series converges to a finite number also obeying the inequality for d_C. For d it is obtained along the same line. □

Any topological vector space the topology of which is given by a *countable, separating* family of seminorms, which means that the difference of two distinct vectors has at least one non-zero seminorm, can be metrized in the above manner.

$\mathcal{E}(\mathbb{R}^n, \mathbb{R})$ *is a Fréchet space.*

Proof Completeness has to be proved. In $\mathcal{E}$, $\lim_{i,j\to\infty} d(F_i, F_j) = 0$ means that on every compactum $C \subset \mathbb{R}^n$ the sequence F_i together with the sequences of all derivatives converge uniformly. Hence, on every C and consequently on $\mathbb{R}^n$ the limit exists and is a C^∞-function F. □

The elements f of the dual space $\mathcal{E}^*$ of $\mathcal{E}$, that is the bounded linear functionals on $\mathcal{E}$, are called **distributions** or **generalized functions.** $\mathcal{E}$ is called the **base space** of the distributions $f \in \mathcal{E}^*$. Formally, the writing

$$\langle f, F \rangle = \int d^n x f(\boldsymbol{x}) F(\boldsymbol{x}), \quad F \in \mathcal{E}, \tag{2.31}$$

is used based on the linearity in F of integration. However, $f(\boldsymbol{x})$ has a definite meaning only in connection with this integral. Every ordinary L^1-function f with compact support defines via the integral (2.31) in the Lebesgue sense a bounded linear functional on $\mathcal{E}$, hence these functions (more precisely, equivalence classes of functions forming the elements of L^1) are special $\mathcal{E}^*$-distributions. Derivatives of distributions are defined via the derivatives of functions $F \in \mathcal{E}$ by formally integrating by parts. Hence, per definition distributions have derivatives to all orders. This holds also for L^1-functions (with compact support) considered as distributions. Derivatives of discontinuous functions as distributions comprise **Dirac's δ-function**

$$\langle \delta_{\boldsymbol{x}_0}, F \rangle = \int d^n x\, \delta(\boldsymbol{x} - \boldsymbol{x}_0) F(\boldsymbol{x}) = F(\boldsymbol{x}_0).$$

Elements of $\mathcal{E}^*$ are not the most general distributions. In the spirit of formula (2.31), more general distributions are obtained by narrowing the base space. In physics, **densities and spectral densities** are in general distributions, if they comprise point masses or point charges or point spectra (that is, eigenvalues).

Let $U \subset \mathbb{R}^n$ be open and consider all $F \in \mathcal{E}$ with $\operatorname{supp} F \subset U$. If $\langle f, F \rangle = 0$ for all those F, then the distribution f is said to be zero on U, $f(\boldsymbol{x}) = 0$ on U. The **support of a distribution** f is the smallest closed set in $\mathbb{R}^n$ outside of which f is zero. Since for a bounded functional f on $\mathcal{E}$ the value (2.31) must be finite for all $F \in \mathcal{E}$, *$\mathcal{E}^*$ is the space of distributions with compact support.* (Dirac's δ-function and its derivatives have one-point support.)

Another most important case in physics regards **Fourier transforms of distributions**. Consider the subspace $\mathcal{S}$ of **rapidly decaying functions** of the class $C^\infty(\mathbb{R}^n, \mathbb{C})$ for which for every k and $\boldsymbol{m}$

$$\sup_{\boldsymbol{x}} |\boldsymbol{x}^{\boldsymbol{m}} D^k F(\boldsymbol{x})| < \infty, \quad \boldsymbol{x}^{\boldsymbol{m}} = \prod_{i=1}^{n} (x^i)^{m_i}, \quad D^k F \text{ like in (2.29).}$$

It is a topological vector space with the family of seminorms

$$p_{k,P}(F) = \sup_{\boldsymbol{x}} |P(\boldsymbol{x}) D^k F(\boldsymbol{x})|, \quad P: \text{polynomial in } \boldsymbol{x}. \tag{2.32}$$

Clearly, $\mathcal{S}$ is closed with respect to the operation with differential operators with polynomial coefficients. Since obviously $\mathcal{S} \subset C^\infty(\mathbb{R}^n, \mathbb{C}) \cap C_\infty(\mathbb{R}^n, \mathbb{C})$ (p. 33), $C_0^\infty(\mathbb{R}^n, \mathbb{C})$ is dense in $\mathcal{S}$ in the topology (2.32) of $\mathcal{S}$. In fact, *$\mathcal{S}$ is a complete (in the topology of $\mathcal{S}$) subspace of $\mathcal{E}(\mathbb{R}^n, \mathbb{C})$; it is again a Fréchet space.* The Fourier transform of a function of $\mathcal{S}$ is

$$\begin{aligned} (\mathcal{F}F)(\boldsymbol{k}) &= \frac{1}{(2\pi)^{n/2}} \int d^n x\, e^{-i(\boldsymbol{k}\cdot\boldsymbol{x})} F(\boldsymbol{x}), \\ F(\boldsymbol{x}) &= (\bar{\mathcal{F}}(\mathcal{F}F))(\boldsymbol{x}) = \frac{1}{(2\pi)^{n/2}} \int d^n x\, e^{i(\boldsymbol{x}\cdot\boldsymbol{k})} (\mathcal{F}F)(\boldsymbol{k}). \end{aligned} \tag{2.33}$$

Depending on context, the prefactor may be defined differently. It can be shown that $\mathcal{F}: \mathcal{S} \to \mathcal{S}$ is an isomorphism and $\mathcal{F}\bar{\mathcal{F}} = \mathrm{Id}_{\mathcal{S}}$, that is $\mathcal{F}^{-1} = \bar{\mathcal{F}}$.

The dual $\mathcal{S}^*$ of $\mathcal{S}$ is the space of **tempered distributions**, $\mathcal{S}^* \supset \mathcal{E}^*$. It is a module on the ring of polynomials (see Compendium), and is closed under differentiation. The Fourier transform in $\mathcal{S}^*$ is defined through the Fourier transform in $\mathcal{S}$ as

$$\langle \mathcal{F}f, F\rangle = \langle f, \mathcal{F}F\rangle. \tag{2.34}$$

Again, $\mathcal{F} : \mathcal{S}^* \to \mathcal{S}^*$ is an isomorphism, $\mathcal{F}\overline{\mathcal{F}} = \mathrm{Id}_{\mathcal{S}^*}$. If $f(\boldsymbol{k}) \equiv 1 \in \mathcal{S}^*$ is considered as a tempered distribution, then $\mathcal{F}f = (2\pi)^{n/2}\delta_0$.

A simple result relevant in the theory of Green's functions is the **Paley–Wiener theorem**: *The Fourier transform of a distribution with compact support on* $\mathbb{R}^n$ *can be extended into an analytic function on* $\mathbb{C}^n$.

Proofs of the above and more details can be found in textbooks of functional analysis, for instance [2]. (Closely related is also the theory of generalized solutions of partial differential equations, which are elements of Sobolev spaces.)

2.5 Connectedness, Homotopy

So far, the focus was mainly on the local topological structure which can be expressed in terms of neighborhood bases of points, although the concepts of vector space and of compactness and in particular of partition of unity provide a link to global topological properties. Connectedness has the focus on global properties, though with now and then local aspects. Intuitively, connectedness seems to be quite simple. In fact, it is quite touchy, and one has to distinguish several concepts.

A topological space is called **connected**, if it is not a union of two disjoint non-empty open sets; otherwise it is called **disconnected** (Fig. 2.6). Connectedness is equivalent to the condition that it is not a union of two disjoint non-empty closed sets, and also to the condition that the only open-closed sets are the empty set and the space itself. A subset of X is connected, if it is connected as the topological subspace with the relative topology; it need neither be open nor closed in the topology of X (cf. the definition of the relative topology). *If A is connected then every A' with $A \subset A' \subset \overline{A}$ is connected* (exercise).

Caution: Two disjoint sets which are not both open or both closed may have common boundary points being points of one of the sets and hence their union may be connected. The union of disjoint sets need not be disconnected.

The **connected component** of a point x of a topological space X is the largest connected set in X containing x. The relation $R(x, y)$: (y belongs to the connected

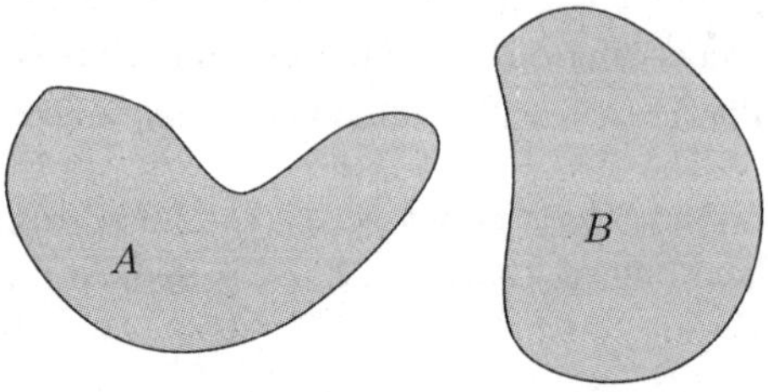

Fig. 2.6 Two connected sets A and B the union of which is disconnected

component of x) is an equivalence relation. The elements of the quotient space X/R are the connected components of X.

A topological space is called **totally disconnected**, if its connected components are all its one point sets $\{x\}$. Let $\pi : X \to X/R$ be the canonical projection onto the above quotient space X/R. The **quotient topology** of X/R is the finest topology in which π is continuous. Its open (closed) sets are the sets B for which $\pi^{-1}(B)$ is open (closed) in X. X/R is totally disconnected in the quotient topology.

Every set X is connected in its trivial topology and totally disconnected in its discrete topology. The rational line $\mathbb{Q}$ in the relative metric topology as a subset of $\mathbb{R}$ is totally disconnected. Indeed, let $\alpha < \beta$ be two rational numbers and let $\gamma, \alpha < \gamma < \beta$ be an irrational number. Then, $]-\infty, \gamma[$ and $]\gamma, +\infty[$ are two disjoint open intervals of $\mathbb{Q}$ the union of which is $\mathbb{Q}$. Hence, no two rational numbers belong to the same connected component of $\mathbb{Q}$. This example shows that the topology in which a space is totally disconnected need not be the discrete topology. In $\mathbb{Q}$, every one point set is closed (since $\mathbb{Q}$ as a metric space is Hausdorff) but not open. Open sets of $\mathbb{Q}$ are the rational parts of open sets of $\mathbb{R}$.

The image $F(A)$ of a connected set A in a continuous mapping is a connected set. Indeed, if $F(A)$ would consist of disjoint open sets then their preimages would be disjoint open sets constituting A. On the other hand, the preimage $F^{-1}(B)$ of a connected set B need not be connected (construct a counterexample). However, as connectedness is a topological property, a homeomorphism translates connected sets into connected sets in both directions. Check that, *if X is connected and Y is totally disconnected, for example if Y is provided with the discrete topology, then the only continuous functions $F : X \to Y$ are the constant functions on X.*

Let R be any equivalence relation in the topological space X. Since the canonical projection $\pi : X \to X/R$ is continuous in the quotient topology, it follows easily that *if the topological quotient space X/R is connected and every equivalence class in X with respect to R is connected, then X is connected.*

A topological space X is disconnected, iff there exists a continuous surjection onto a discrete two point space. (The target space may be $\{0, 1\}$ with the discrete topology; then, some of the connected components are mapped onto $\{0\}$ and some onto $\{1\}$.)

The topological product of non-empty spaces is connected, iff every factor is connected.

Proof Although the theorem holds for any number of factors, possibly uncountably many in Tichonoff's product, here only the case of finitely many factors is considered. (Though the proof works in the general case, only Tichonoff's product was not introduced in our context.) Let X_i be the factors of the product space X and $\pi_i : X \to X_i$ the canonical projections. Since these are continuous in the topological product, if X is connected, then every X_i as the image of X in a continuous mapping is connected. Now, assume that all X_i are connected but X is not. Then, there is a continuous surjection F of X onto $\{0, 1\}$. Let for some $\bar{x} = (\bar{x}_1, \ldots, \bar{x}_n)$, $\bar{x}_i \in X_i$, $F(\bar{x}) = 0$. Consider the subset $(x_1, \bar{x}_2, \ldots, \bar{x}_n)$, where x_1 runs through X_1, and the restriction of F on this subset. This restriction is a continuous function on X_1 and hence is $\equiv 0$ since X_1 is connected. Starting from every point of this subset, let now x_2 run through X_2 to obtain again $F \equiv 0$ for the restriction of F. After n steps, $F \equiv 0$ on X in contradiction to the assumption that F is surjective. □

A concept seemingly related to connectedness but in fact independent is local connectedness. A topological space is called **locally connected**, if every point has a neighborhood base of connected neighborhoods. (Not just one neighborhood, *all* neighborhoods of the base must be connected.)

A connected space need not be locally connected. For instance, consider the subspace of $\mathbb{R}^2$ consisting of a horizontal axis and vertical lines through all rational points on the horizontal axis, in the relative topology deduced from the usual topology of the $\mathbb{R}^2$. It is connected, but no point off the horizontal axis has a neighborhood base of only connected sets. (Compare the above statement on $\mathbb{Q}$.) On the other hand, every discrete space with more than one point, although it is totally disconnected, is locally connected! Indeed, since every one point set is open and connected in this case, it forms a connected neighborhood base of the point. (Check it.) This seems all odd, nevertheless local connectedness is an important concept.

A topological space is locally connected, iff every connected component of an open set is an open set. This is not the case in the above example with the vertical lines through rational points of a horizontal axis, since the connected components of open sets off the horizontal axis are not open.

Proof of the statement Pick any point x and any neighborhood of it and consider the connected component of x in it. Since it is open, it is a neighborhood of x. Hence, x has a neighborhood base of connected sets, and the condition of the theorem is sufficient. Reversely, let A be an open set in a locally connected space, A' one of its connected components and x any point of A'. Let U be a neighborhood of x in A. It contains a connected neighborhood of x which thus is in A'. Hence, x is an inner point of A' and, since x was chosen arbitrarily, A' is open. □

As a consequence, a locally connected space is a collection of its connected components *which are all open-closed.*

A topological quotient space of a locally connected space is locally connected.

Proof Let X be locally connected and let $\pi : X \to X/R$ be the canonical projection. Let $U \subset X/R$ be an open set and U' one of its connected components. Let $x \in \pi^{-1}(U')$, and let A be the connected component of x in $\pi^{-1}(U)$. Then, $\pi(A)$ is connected (since π is continuous) and contains $\pi(x)$. Hence, $\pi(A) \subset U'$ and $A \subset \pi^{-1}(U')$. Since X is locally connected and $\pi^{-1}(U)$ is open (again because π is continuous), $\pi^{-1}(U')$ is also open due to the previous theorem. Now, by the definition of the quotient topology, U' is also open, and the previous theorem in the opposite direction says that X/R is locally connected. □

The subsequently discussed further concepts of connectedness are based on **homotopy**. Let $I = [0, 1]$ be the closed real unit interval. Two continuous functions

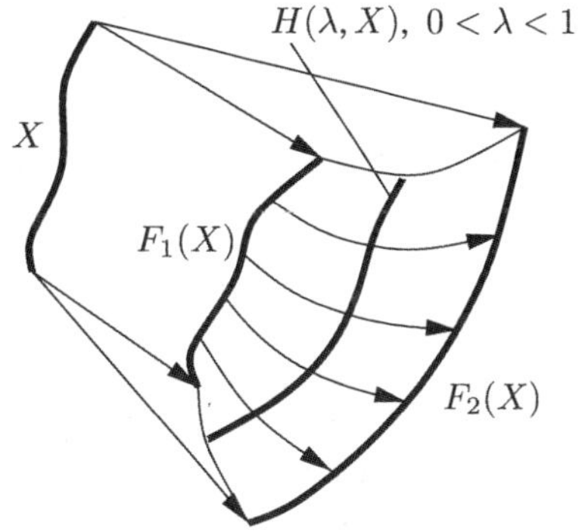

Fig. 2.7 Homotopic functions F_1 and F_2

F_1 and F_2 from the topological space X into the topological space Y are called **homotopic**, $F_1 \cong F_2$, if there exists a continuous function $H : I \times X \to Y$: $H(0,\cdot) = F_1$, $H(1,\cdot) = F_2$. H is called the homotopy translating F_1 into F_2 (Fig. 2.7). Since its definition is only based on the existence of continuous functions, homotopy is a purely topological concept.

The F_i may be considered as points in the functional space $C^0(X,Y)$. Then, $H(\lambda,\cdot)$, $0 \le \lambda \le 1$ is a path in $C^0(X,Y)$ from F_1 to F_2. If X and Y are normed vector spaces or manifolds, sometimes, in a narrower sense, the functions F_i, H are considered to be C^n-functions, $0 \le n \le \infty$. One then speaks of a C^n-homotopy. Of course, every C^n-homotopy is also a C^m-homotopy for $m \le n$. Homotopy is the C^0-homotopy. In the following statements homotopy may be replaced by C^n-homotopy with slight modifications in the construction of products H_2H_1 (see for instance [4, $VI.8]).

The product $H_2\ H_1$ of two homotopies, H_1 translating F_1 into F_2 and H_2 translating F_2 into F_3, may be introduced as a homotopy translating F_1 into F_3 in the following natural way by concatenating the two translations:

$$(H_2H_1)(\lambda,x) = \begin{cases} H_1(2\lambda,x) & \text{for } 0 \le \lambda \le 1/2 \\ H_2(2\lambda-1,x) & \text{for } 1/2 \le \lambda \le 1. \end{cases}$$

Hence, if $F_1 \cong F_2$ and $F_2 \cong F_3$, then also $F_1 \cong F_3$. This means that homotopy is an equivalence relation among continuous functions. The corresponding equivalence classes $[F]$ of functions F are called **homotopy classes**. If a homeomorphism P of X onto itself is homotopic to the identity mapping $P \cong \mathrm{Id}_X$, then $F \circ P \cong F$ (exercise).

Two topological spaces X and Y are called **homotopy equivalent**, if there exist continuous functions $F : X \to Y$ and $G : Y \to X$ so that $G \circ F \cong \mathrm{Id}_X$ and $F \circ G \cong \mathrm{Id}_Y$. Two homeomorphic spaces are also homotopy equivalent, the inverse is, however, in general not true. A topological space is called **contractible**, if it is homotopy equivalent to a one point space. For instance, every topological vector space is contractible. The homotopy class of a constant function mapping X to a single point is called the **null-homotopy class**.

Of particular interest are the homotopy classes of functions from n-dimensional unit spheres S^n into topological spaces X possibly with a topological group structure. The latter means that the points of X form a group (with unit element $e \in X$) and the group operations are continuous. The unit sphere S^n may be considered as the set of points $\boldsymbol{s} \in \mathbb{R}^{n+1}$ with $\sum_{i=1}^{n+1}(s^i)^2 = 1$. S^0 is the two point set $S^0 = \{-1, 1\}, S^1$ is the circle, S^2 is the ordinary sphere, and so on. For $-1 < s^1 < 1$, the points $(s^2, \ldots, s^{n+1})$ with coordinates on $S^n, n > 0$, form an $(n-1)$-dimensional sphere (of radius r depending on s^1).

The case $n = 0$ is special and is treated separately. A topological space X is called **pathwise connected** (also called arcwise connected), if for every pair (x, x') of points of X there is a continuous function $H : I \to X, H(0) = x, H(1) = x'$. For a general topological space X, pathwise connectedness of pairs of points is an equivalence relation, and the equivalence classes are the **pathwise connected components** of X. If X is pathwise connected, then it is connected (exercise). The inverse is not in general true. Let X be the union of the sets of points $(x, y) \in \mathbb{R}^2$ with $y = \sin(1/x)$ and $(0, y)$, $y \in \mathbb{R}$ in the relative topology as a subset of $\mathbb{R}^2$. It is connected, but points with $x = 0$ and $x \neq 0$ are not pathwise connected. (Points $(0, y)$ with $|y| \leq 1$ are also not locally connected.) X is **locally pathwise connected**, if every point has a neighborhood base of pathwise connected sets. If X is locally pathwise connected, then it is locally connected, but again the inverse is not in general true.

For the following, $n \geq 1$, and *until otherwise stated, X is considered pathwise connected.* A homeomorphism between the sphere S^n, $n \geq 1$ and the n-dimensional unit cube with a particular topology is needed. Consider the open unit cube $I^n = \{\boldsymbol{x} | -1/2 < x^i < 1/2\}$ with its usual topology and its one point compactification $\overline{I^n}$, obtained by identifying the surface ∂I^n of I^n with the additional point $\boldsymbol{x}_\infty$ of $\overline{I^n}$. $\overline{I^n}$ is obviously homeomorphic to the one point compactification $\overline{\mathbb{R}^n}$ of $\mathbb{R}^n$, but it is also homeomorphic to S^n where a homeomorphism may be considered which maps $\boldsymbol{x}_\infty \in \overline{I^n}$ and $\boldsymbol{s}_0 = (1, 0, \ldots, 0) \in S^n$ onto each other. For $n = 1$ a homeomorphism between the unit circle and $\overline{\mathbb{R}}$ is obvious, for $n = 2$ it is a stereographic projection of the unit sphere S^2 onto the one-point compactified plane $\overline{\mathbb{R}^2}$. A similar mapping for $n > 2$ is easily found (exercise). The homeomorphism between S^n and $\overline{I^n}$ which maps $\boldsymbol{x}_\infty \in \overline{I^n}$ and $\boldsymbol{s}_0 = (1, 0, \ldots, 0) \in S^n$ onto each other is denoted by $\boldsymbol{P}$.

A word on notation her: x, x_0 denote points of X not having themselves coordinates since X in general is not a vector space; $\boldsymbol{x}$, $\boldsymbol{x}_\infty$ denote points in $\overline{I^n} \subset \mathbb{R}^n$ having coordinates $x^1, x^2, \ldots, x^n$ (not unique for $\boldsymbol{x}_\infty$); $\boldsymbol{s}, \boldsymbol{s}_0$ denote points on $S^n \subset \mathbb{R}^{n+1}$ having coordinates $s^1, s^2, \ldots, s^{n+1}$, $\sum_i (s^i)^2 = 1$.

Now, fix x_0 in the topological space X and consider the class $C_n(x_0)$ of continuous functions $F : S^n \to X$ with $F(\boldsymbol{s}_0) = x_0$ fixed. Denote the homotopy classes of functions $F \in C_n(x_0)$ by $[F]$. It is not the whole homotopy class of F in X, because for the group construction below it is necessary that the mapping of $\boldsymbol{s}_0 \mapsto x_0$ is fixed in every function F. The mapping F can be composed of two steps

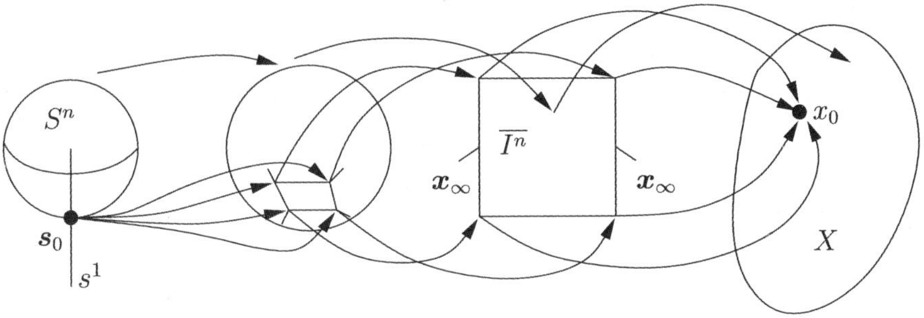

Fig. 2.8 Mapping of S^n onto $\overline{I^n}$ and $\overline{I^n}$ into X. It is visualized how the point s_0 is expanded into the square $\boldsymbol{x}_\infty$ which frames the image I^n of $S^n \setminus \{s_0\}$, and then $\boldsymbol{x}_\infty$ is mapped to x_0

(Fig. 2.8): first map S^n homeomorphically onto $\overline{I^n}$ by $\boldsymbol{P}$, implying $s_0 \mapsto \boldsymbol{x}_\infty$, and then map $\overline{I^n}$ into X by the continuous function $\tilde{F}$ with $\boldsymbol{x}_\infty \mapsto x_0$. Because $\boldsymbol{P}$ is a bijection, there is also a bijection between $\tilde{F}$ and $F = \tilde{F} \circ \boldsymbol{P}$, and $F(s_0) = x_0$.

This composition allows to explicitly define a group structure in the set of homotopy classes $[F]$ in the following way: For any two $C_n(x_0)$-functions F_1 and F_2 define a product $F_2F_1 \in C_n(x_0)$ by

$$(\tilde{F}_2\tilde{F}_1)(\boldsymbol{x}) = \begin{cases} \tilde{F}_1(2x^1 + 1/2, x^2, \ldots, x^n) & -1/2 \le x^1 \le 0 \\ \tilde{F}_2(2x^1 - 1/2, x^2, \ldots, x^n) & 0 \le x^1 \le 1/2, \end{cases}$$
$$F_2F_1 = (\tilde{F}_2\tilde{F}_1) \circ \boldsymbol{P}. \tag{2.35}$$

$(\tilde{F}_2\tilde{F}_1)$ is continuous, since the two functions $\tilde{F}_1$ and $\tilde{F}_2$ are glued together where $\tilde{F}_1(1/2, \ldots) = \tilde{F}_1(\boldsymbol{x}_\infty) = x_0 = \tilde{F}_2(\boldsymbol{x}_\infty) = \tilde{F}_2(-1/2, \ldots)$. Note that $\tilde{F}$ is supposed continuous with respect to the topology of $\overline{I^n}$ in which the surface ∂I^n is contracted into one point $\boldsymbol{x}_\infty$. Moreover, for $x^1 = -1/2$ or $x^1 = 1/2$, that is $\boldsymbol{x} = \boldsymbol{x}_\infty$, (2.35) yields $(\tilde{F}_2\tilde{F}_1)(\boldsymbol{x}_\infty) = x_0$, hence $F_2F_1 \in C_n(x_0)$. True, also $(\tilde{F}_2\tilde{F}_1)(0, \ldots) = x_0$ which for $|x_i| < 1/2, i = 2, \ldots, n$ is not demanded in the class $C_n(x_0)$. The construct (2.35) effectively pinches the section $x^1 = 0$ of $\overline{I^n}$ for $n > 1$ into one point. Via $\boldsymbol{P}$, this section corresponds to a meridian S^{n-1} of S^n containing the pole s_0. By moving from F_2F_1 to the homotopy class $[F_2F_1]$, this additional restriction (the pinch) is released.

In particular for $n = 1$, $\overline{I^1}$ is the line of length 1 with its endpoints identified (loop); hence it can again be considered as a circle. The mapping $\boldsymbol{P}$ which maps the pole s_0 to the connected endpoints of the second circle is trivial in this case. The point $\boldsymbol{x} = 0$ corresponds to the diametrically opposed point of the circle. In a product (2.35) of two mappings, this point is also mapped to x_0 making the product into a double loop (Fig. 2.9). The final correct product definition in the set of homotopy classes $[F]$ of functions with base point $F(s_0) = x_0$ is

$$[F_2][F_1] = [F_2F_1]. \tag{2.36}$$

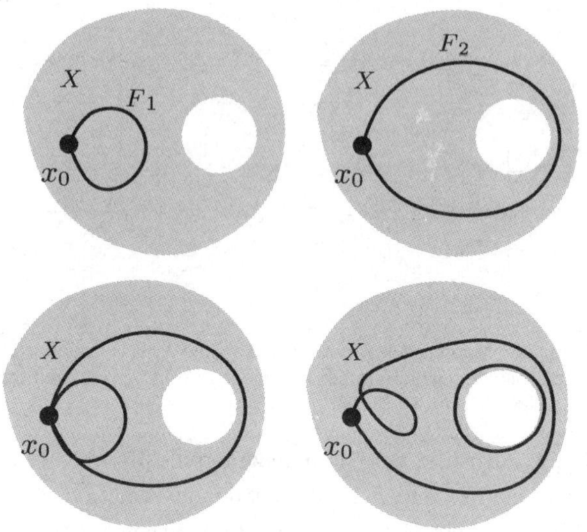

Fig. 2.9 Two loops $F_1, F_2 \in C_1(x_0)$ of the topological space X (shadowed area) and their product (2.35) (lower left panel). Also, another representative of $[F_2F_1]$ and a loop homotopic to F_2F_1 in X is shown (lower right panel). Since $[F_1] \cong E$ in this case, $[F_2F_1] \cong [F_1F_2] \cong [F_2]$

Next, having defined the product in (2.35, 2.36), it must be shown to be associative. Consider first

$$(\tilde{F}_3(\tilde{F}_2\tilde{F}_1))(\boldsymbol{x}) = \begin{cases} \tilde{F}_1(4x^1+3/2,\dots) & -1/2 \le x^1 \le -1/4 \\ \tilde{F}_2(4x^1+1/2,\dots) & -1/4 \le x^1 \le 0 \\ \tilde{F}_3(2x^1-1/2,\dots) & 0 \le x^1 \le 1/2 \end{cases}$$

and

$$((\tilde{F}_3\tilde{F}_2)\tilde{F}_1)(\boldsymbol{x}) = \begin{cases} \tilde{F}_1(2x^1+1/2,\dots) & -1/2 \le x^1 \le 0 \\ \tilde{F}_2(4x^1-1/2,\dots) & 0 \le x^1 \le 1/4 \\ \tilde{F}_3(4x^1-3/2,\dots) & 1/4 \le x^1 \le 1/2. \end{cases}$$

These two results differ only in a quite simple homeomorphism (piecewise linear in x^1, identity in the other coordinates) of $\overline{I^n}$ onto itself which is homotopic to $\mathrm{Id}_{\overline{I^n}}$. Hence, they are homotopic to each other (see p. 41). They also both map $\boldsymbol{x}_\infty$ to x_0. Thus, $[F_3]([F_2][F_1]) = [F_3(F_2F_1)] = [(F_3F_2)F_1] = ([F_3][F_2])[F_1]$.

If $\tilde{E}$ is the constant mapping $\tilde{E}(\boldsymbol{x}) \equiv x_0$ then obviously $e = [E]$ is a unity: $e[F] = [F] = [F]e$ for all $[F]$. Moreover, for $\tilde{F}_-(\boldsymbol{x}) = \tilde{F}(-x^1, x^2, \dots)$ (2.35) yields $[F_-][F] = e = [F][F_-]$. Indeed, $(\tilde{F}_-\tilde{F})(x^1, x^2, \dots) = (\tilde{F}_-\tilde{F})(-x^1, x^2, \dots)$: The image $(\tilde{F}_-\tilde{F})(\overline{I}^n)$ is a double layer in X. By symmetrically contracting the interval $-1/2 \le x^1 \le 1/2$ into $x^1 = 0$ with $x^2, \dots$ left constant $(\tilde{F}_-\tilde{F})(\overline{I}^n)$ shrinks continuously on itself into $x_0 = \tilde{E}(\overline{I}^n)$ by successive 'annihilation' of parts of the double layer. In total a group $\pi_n(X, x_0) = \{[F] | F \in C_n(x_0)\}$ is obtained with the group multiplication (2.36).

Now, consider *any* point x of the pathwise connected space X and a continuous path $H : I \to X$ with $H(0) = x_0$, $H(1) = x$. Given $F \in C_0(x_0)$, a function $F' \in C_0(x)$ may be constructed in the following manner:

$$\tilde{F}'(\boldsymbol{x}) = \begin{cases} \tilde{F}(2\boldsymbol{x}) & |x^i| \le 1/4,\ i = 1, \ldots, n, \\ H(t) & (2-t)\boldsymbol{x} \in \partial I^n,\ 0 \le t \le 1. \end{cases}$$

The base point x_0 of F is dragged along the path H to x. Apart from this path, the sets $\tilde{F}(\overline{I}^n)$ and $\tilde{F}'(\overline{I}^n)$ are the same which hence is also true for $F(\overline{I}^n)$ and $F'(\overline{I}^n)$. Moving F through its homotopy class $[F]$ with base point x_0 obviously also moves F' through its homotopy class $[F']$ with base point x. Moreover, it is easily seen (exercise) that $\tilde{F}_2\tilde{F}_1$ via H induces $\tilde{F}_2'\tilde{F}_1'$ for which $[F_2'F_1'] = [F_2'][F_1']$. Hence, the mapping $\tilde{H} : [F] \mapsto [F']$ is a homomorphism of groups. Two concatenated paths H_1 and H_2 obviously induce a composition of homomorphisms $\tilde{H}_2 \circ \tilde{H}_1$. Concatenate now the path H with its reversed $H_-(t) = H(1-t)$. Then H_-H provides the identity map $\mathrm{Id}_{C_0(x_0)}$ while HH_- provides $\mathrm{Id}_{C_0(x)}$. $\tilde{H}$ and $\tilde{H}_-$ are thus inverse to each other, and the homomorphism $\tilde{H}$ is in fact an isomorphism. The groups $\pi_n(X,x)$ and $\pi_n(X,x_0)$ are isomorphic, or, in other words, $\pi_n(X,x_0) \approx \pi_n(X)$ does not depend on x_0. The group $\pi_n(X)$ is called the **nth homotopy group** of the pathwise connected topological space X.

Since the case $n = 1$ is of particular interest in the theory of integration on manifolds (see Chap. 5), $\pi_1(X)$ is called the **fundamental group** of X.

Formally, a '0-dimensional open cube' can be considered as a one point set $I^0 = \{\boldsymbol{x}\}$, and its one point 'compactification' (I^0 is of course also compact) as the discrete two point set $\overline{I^0} = \{\boldsymbol{x}, \boldsymbol{x}_\infty\}$. The homeomorphism $\boldsymbol{P}$ between $S^0 = \{-1, 1\}$ and $\overline{I^0}$ maps -1 to $\boldsymbol{x}$ and 1 to $\boldsymbol{x}_\infty$. Now, $F : S^0 \to X$ is a two point mapping, and $F \in C_0(x_0)$ means that $F(-1) = x$ where x is *any* point of X, and $F(1) = x_0$. The classes $[F]$ thus map -1 into the pathwise connected components of X, and x_0 does not play any role. *For a pathwise connected topological space X, $\pi_0(X) = \{e\}$ is trivial.*

By inspection of (2.35) it is seen that interchanging the factors in the multiplication amounts to interchanging the halves $x^1 \le 0$ and $x^1 \ge 0$ in $\overline{I^n}$. For $n > 1$, the positioning of these two halves relative to each other does not play a role because of the pinch of the section $x^1 = 0$ involved in (2.35). Therefore, the interchanging of the two halves can be provided by a homeomorphism of I^n onto itself which is also homotopic to the identity mapping: note that $\overline{I^n}$ is homeomorphic to a cylinder with axis perpendicular to the x^1-axis. Rotate it by 180° to transform continuously from the identity to the interchanging of the above two halves. *The groups $\pi_n(X)$, $n > 2$ of a pathwise connected topological space X are commutative.* For that reason, in the literature the group operation of homotopy groups is often denoted as addition instead of multiplication.

In the case $n = 1$ the interchanging may still be provided by a homeomorphism, however, the argument of deformation into a cylinder does not work any more, and the interchanging is not any more homotopic to the identity mapping. *The fundamental group $\pi_1(x)$ need not be commutative.* Consider for instance a two-dimensional space X with two holes and a loop first orbiting clockwise around the first hole and then counterclockwise around the second. Check that this loop is not homotopic to the loop with the sequence of orbiting interchanged.

If X itself has a group structure, that is, X *is a topological group* with multiplication denoted by a dot (to distinguish it from the multiplication (2.36)), and $x_0 = e$, then another product of $C_n(e)$-functions and the inverse of a $C_n(e)$-function may alternatively be defined by pointwise application of the group operations. The $C_n(e)$-unity is the constant mapping on e. Let $F_1 \cong F_1'$ and $F_2 \cong F_2'$ and consider the homotopies H_i translating F_i into F_i', $(H_i(0,\cdot) = F_i,\ H_i(1,\cdot) = F_i')$. Then $H_1 \cdot H_2$ is a homotopy translating $F_1 \cdot F_2$ into $F_1' \cdot F_2'$, hence $[F_1 \cdot F_2] = [F_1' \cdot F_2']$: the group multiplication in X is compatible with the homotopy class structure of $C_n(e)$ and the multiplication $[F_1] \cdot [F_2]$ is properly defined. Clearly, $e = [E]$ is also the unity for the dot multiplication. Moreover, with (2.35), $\tilde{F}_1\tilde{F}_2 = (\tilde{F}_1\tilde{E}) \cdot (\tilde{E}\tilde{F}_2)$ is easily verified (check it). The conclusion is $[F_1][F_2] = [F_1] \cdot [F_2]$: the dot-multiplication yields again the same homotopy group $\pi_n(X, e)$ *of the pathwise connected component of e in X* as previously. Since the multiplication (from left or right) with any element x of the component X^e of e in X yields a translation of that component which is also a homeomorphism of that component X^e onto itself, $\pi_n(X, e) \approx \pi_n(X, x) \approx \pi_n(X^e)$ for any x of the component of e in X.

However, if the topological group X *is not pathwise connected*, in a wider sense the homotopy group $\pi_n(X)$ with the dot-multiplication can still be constructed. In this case, $\pi_0(X)$ is non-trivial, and the elements of $\pi_0(X)$ are in a one–one correspondence with the pathwise connected components of X. Let $x \notin X^e$ be a group element not in the pathwise connected component of e, and let $x_0 \in X^e$, that is, there is a continuous path connecting x_0 with e. Since in a topological group the group operations are continuous, it follows that there is a continuous path from $x \cdot x_0$ to $x \cdot e = x$; $x \cdot x_0 \in X^x$, and likewise $x_0 \cdot x \in X^x$. It is easily seen that all pathwise connected components of a group X are homeomorphic to each other (exercise). It follows further that there is a continuous path connecting $x \cdot x_0 \cdot x^{-1}$ with $x \cdot e \cdot x^{-1} = e$. Hence, $x \cdot x_0 \cdot x^{-1} \in X^e$ for every $x \in X$ and every $x_0 \in X^e : X^e$ is an invariant subgroup of X. It is easily seen that $X/X^e \approx \pi_0(X)$. On the other hand, $x \cdot x_1 \cdot x^{-1} \mapsto x_1', x_1 \in X, x \in X$ is an automorphism of X for any fixed x which, as was seen, transforms pathwise connected components of X into themselves.

Consider $C_n(x)$-functions F from the S^n-sphere into X with *any* base point x, not necessarily in X^e. The homotopy classes $[F]$ in $C_n(x)$ form a larger group $\pi_n(X)$ which now is only defined with the group multiplication $[F_1] \cdot [F_2]$. The above considered automorphism of X yields in a canonical way an automorphism of $\pi_n(X)$. Denote the elements of $\pi_0(X)$ by $[H]$; then the anticipated automorphism is given by

$$[F]' = [H] \cdot [F] \cdot [H]^{-1}, \quad [F] \in \pi_n(X),\ [H] \in \pi_0(X). \tag{2.37}$$

If $[F] \subset C_n(e)$ then $[F]' \subset C_n(e)$, hence, $\pi_n(X^e)$ is an invariant subgroup of $\pi_n(X)$, and $\pi_0(X) \approx \pi_n(X)/\pi_n(X^e)$. Because of the above discussed structure of the pathwise connected classes of X, obviously also $\pi_n(X^e) = \pi_n(X)/\pi_0(X)$ and hence

$$\pi_n(X) = \pi_0(X) \times \pi_n(X^e), \quad n > 0 \tag{2.38}$$

for the **homotopy groups of a topological group** X. They can be quite different from $\pi_n(X^e)$ (and need not be commutative for any $n \geq 0$ since $\pi_0(X)$ need not be commutative any more).

A topological space X is called n-**connected** (sometimes called n-simple), if every continuous image in X of the n-dimensional sphere S^n is contractible. A topological group X is n-connected, if $\pi_n(X) \approx \pi_0(X)$. An n-connected space need not be connected. A 0-connected space is pathwise connected, a 1-connected space is called **simply connected.**[2] Although n-connectedness is very similarly defined for different n, these properties are largely unrelated (except for the role of π_0). Some authors apply n-connectedness only to pathwise connected spaces X. However, for many applications this is an unnecessary restriction.

Some examples are given without proof. Some of them are intuitively clear. (1) A convex open subspace of a topological vector space is n-connected for *any* $n \geq 0$. (2) The sphere S^n or the complement to the origin in $\mathbb{R}^{n+1}$ is k-connected for $0 \leq k \leq n-1$; for $n > 1$ it is simply connected. (3) $\pi_n(S^n) = \mathbb{Z}$ (as an additively written Abelian group). For an integer $m \in \mathbb{Z} = \pi_n(S^n)$, $|m|$ is the cardinality of $F^{-1}(x)$ for any $x \in S^n$. It is called the **degree of** the **mapping** F. (4) $\pi_n(S^m)$, $n > m$ is a largely unsolved problem although many special cases have meanwhile been compiled; $\pi_3(S^2) = \mathbb{Z}$ is a theorem by Hopf, and $\pi_2(S^1) = 0$ is easily understood. (5) For the torus $\mathbb{T}^2$ (see Fig. 1.3), $\pi_1(\mathbb{T}^2) = \mathbb{Z} \times \mathbb{Z}$. One integer of $(m_1, m_2) \in \mathbb{Z} \times \mathbb{Z}$ counts the oriented windings around the circumference of the tire, and the other those around its cross section.

These concepts are further exploited in Chaps. 5 and 8. Although the physical relevance of homotopy was anticipated already by Poincaré, it turned out to be one of the most difficult and unsolved tasks of topology to calculate the homotopy groups of certain manifolds and to exploit them for classification. It was already known to Poincaré that every compact simply connected two-dimensional manifold without boundary is homeomorphic to the sphere S^2. His conjecture that the same is true in three dimensions and every compact simply connected three-dimensional manifold without boundary is homeomorphic to the 3-sphere S^3 withstood hard attempts by able mathematicians for hundred years to prove it and was eventually proved only quite recently by G. Perelman.

---

[2] There is a more general definition of simple connectedness and fundamental group in terms of covering space. For pathwise connected locally pathwise connected spaces X it is equivalent to the definition given here [6].

2.6 Topological Charges in Physics

In quantum physics, thermodynamic phases are characterized by order parameters: the particle densities of various particles, atom displacements of crystalline solids, the magnetization density vector, the anomalous Green function of the superconducting or superfluid state and so on. In an inhomogeneous, in particular defective state those order parameters are functions of space (and maybe time). The various defects can often be classified by discrete **topological charges**, and then those classes turn out to be stable: because of the discrete nature of the charges there is no continuous transformation of one class into another. The topological charges are often generating elements of homotopy groups.

Consider as a simple example a superconducting state in three dimensions penetrated by a vortex line. The space X of the superconducting state is $\mathbb{R}^3$ with the vortex line cut out. It is homotopy equivalent to a circle S^1 around the vortex line. The order parameter $\Delta = |\Delta|e^{i2\pi\phi}$ of a conventional superconducting state (spin singlet s wave) is a complex number having a phase ϕ the gradient of which is proportional to the supercurrent while the absolute value $|\Delta|$ is the gap which is fixed for a given material and for given temperature and pressure. A constant phase factor is irrelevant, the state is degenerate with respect to an arbitrary complex phase factor. The loop S^1 in the complex plane of all phase factors is the order parameter space Γ of degenerate states in that case. With a defect present in X, the order parameter in general will be position dependent with values out of Γ. This position dependence defines a mapping $F : X \to \Gamma$. Since Δ is a well defined function on X, the gradient $\partial\phi/\partial\boldsymbol{x}$ of the phase must integrate along any closed loop to an integer, $\oint ds \cdot (\partial\phi/\partial\boldsymbol{x}) =$ integer, and this integer must be the same for all homotopy equivalent loops. On a loop not encircling the vortex line this integer must be zero, since the loop may be continuously contracted within X to a point, and a non-zero integer cannot continuously be changed to zero. On a loop once encircling the vortex line the integral of the gradient of the phase ϕ may be any integer N characterizing the vortex line. For a loop m times winding around the vortex line it then is Nm. N is the number of magnetic flux quanta in the vortex line. It generates a group of elements Nm with $m \in \mathbb{Z}$. This group is obviously isomorphic with the group $\mathbb{Z}$, which in this case is the fundamental group $\pi_1(\Gamma = S^1)$ of homotopy classes of mappings from S^1 which is homotopy equivalent to X into $\Gamma = S^1$.

On a discrete lattice, the sum of unit lattice periods along a loop is similar to a phase and must be an integer number of lattice vectors along the loop. For a loop enclosing a defect free region of the crystal this sum is zero. For a loop around a displacement line this is the Burgers vector of the displacement. Here the space X of the crystalline phase is again the same as above and is again homotopy equivalent to the circle S^1, this time around the displacement line. Any loop yields m times the Burgers vector.

Such situations will in more generality and more detail be considered in Chap. 8. Here, some principal remarks are in due place. The Hamiltonian of a macroscopic system has in general a number of symmetries, it is invariant with respect to transformations of a symmetry group G, translational, rotational invariance, gauge symmetries and others. Some of the symmetries may be approximate, but obeyed to a sufficient level of accuracy. For instance in a rare gas liquid the coupling of the nuclear spin with the rotational motion is so weak that invariance with respect to spatial rotation and spin rotation may be considered separately. At sufficiently high temperature, the state of the system is completely disordered, so that its thermodynamic (macroscopic) variables are invariant under the symmetry transformations of G. The thermodynamic state γ fulfills the relation $\gamma = g\gamma$ for all $g \in G$ and is thus uniquely determined. In the course of lowering the temperature, phase transitions may take place with developing non-zero order parameters so that now γ is not any more invariant with respect to all symmetry transformations g of G, but may still be invariant with respect to a subgroup H of G. Then, γ generates an orbit $\{g\gamma | g \in G\}$ which is isomorphic to the quotient space $\Gamma = G/H$ of left cosets of H in G. It is this quotient space which figures as the order parameter space Γ in the above considerations.

In the above example of a line defect in $\mathbb{R}^3$ it was essential only that the defect free space X was homotopically equivalent to a circle S^1. The number of topological 'charges' of the defect is then equal to the number of generators of the homotopy group $\pi_1(\Gamma)$ (one in the above cases). The same would be true for a point defect in $\mathbb{R}^2$ or a line defect propagating in time (defect world sheet) in four-dimensional space–time. For a point defect in $\mathbb{R}^3$, X is homotopy equivalent to a sphere S^2 enclosing the defect, and hence the number of its topological charges is equal to the number of generators of $\pi_2(\Gamma)$.

In general, the number of topological charges of a defect of codimension d in a state with order parameter space Γ present in an n-dimensional position space (i.e., the dimension of the defect is $n - d$) is equal to the number of generators of the homotopy group $\pi_{d-1}(\Gamma)$.

In order to develop a non-zero topological quantum number (non-trivial topological charge), a defect of codimension d in a state with order parameter space Γ must have a non-trivial homotopy group $\pi_{d-1}(\Gamma)$. Consider as an example an isotropic magnetically polarizable material. The Hamiltonian does not prefer any direction in space, besides translational invariance which need not be considered here (it assures that a magnetization vector smoothly depending on position has low energy) the continuous symmetry group is $SO(3)$ (cf. Chap. 6). At sufficiently high temperature, above the magnetic order temperature, the magnetic polarization is disordered on an atomic scale and the state γ is invariant: $\gamma = g\gamma$ for all $g \in G = SO(3)$. Below the ordering temperature the magnetization density vector is non-zero. Its absolute value is determined by the material, temperature and pressure. Its direction may be arbitrary, and all directions are energetically degenerate. Smooth long wavelength changes of direction have low excitation energy. If the non-zero magnetization points in a certain direction, the state is still invariant with respect to rotations of the group $H = SO(2)$ around the axis of polarization. The order parameter space is

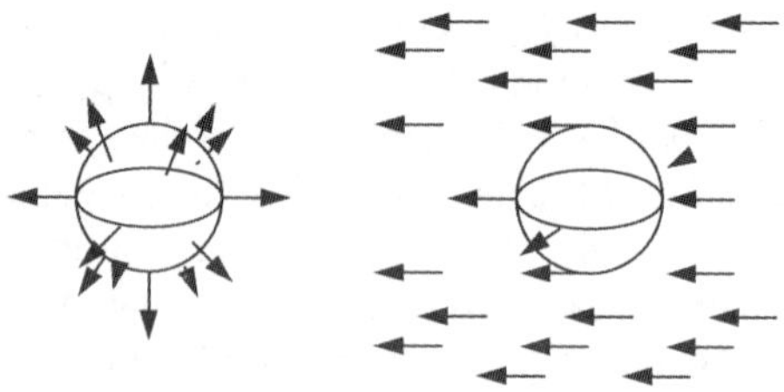

Fig. 2.10 Point defect of (a) an isotropic magnetic material, so-called hedgehog, and (b) of an easy plane anisotropic magnetic material with no non-trivial topological charge possible

$SO(3)/SO(2)$ and consists of all vectors of a given length pointing in all possible spatial directions. Topologically this group is homeomorphic to the sphere S^2. Hence, $\Gamma = S^2$. For a point defect in 3-space (codimension 3), $\pi_2(S^2) = \mathbb{Z}$ (see end of last section). Hence, the point defect may have a non-trivial topological charge in this case.

A point defect is a small spot where the magnetization density vanishes. Outside of a sphere of a small radius it is again fully developed, but may for instance everywhere point in radial direction (Fig. 2.10a). The change of direction outside of this sphere is everywhere smooth, but there is no smooth transition into a homogeneously magnetized state with constant magnetization direction. This 'hedgehog' point defect has non-trivial topological charge and is stable: the defect cannot be resolved by smooth magnetization changes.

Consider now an anisotropic magnetic material of the type easy plane. Again the magnitude of the magnetization density vector is fixed at given temperature and pressure, but can only point in the directions within a plane, $\Gamma = SO(2) \sim S^1$. Now $\pi_2(S^1) = 0$: the sphere S^2 is simply connected and cannot be continuously wound around a circle. Hence no non-trivial topological charge of a point defect is possible in this case. From Fig. 2.10b it is easily inferred that no hedgehog-like structure is possible without singularity lines outside a sphere around the defect of the magnetization vector field of constant magnitude. From the singularity lines the magnetization density vector would point into all planar radial directions. If this is a linear defect, it is governed by $\pi_1(S^1) = \mathbb{Z}$, and a topological charge can exist on the linear defect in an easy plane magnet.

A point defect of codimension 4 in four-dimensional space–time would be capable of carrying a topological charge, if $\pi_3(\Gamma)$ is non-trivial. Just to mention it, the Belavin–Polyakov instanton of a Yang–Mills field is such a case even without a defect (Chap. 8).

Structures with topological charges may intrinsically exist without a material defect. Consider the plane $\mathbb{R}^2$ with a non-zero magnetization density which approaches a homogeneous magnetization density vector of a fixed direction at infinite distance from the origin of $\mathbb{R}^2$. This state may be considered as a state in the compactified plane $\overline{\mathbb{R}^2} \sim S^2$ which is homeomorphic to a sphere via the stereographic projection. Since the order parameter space Γ of an $SO(3)$ spin is

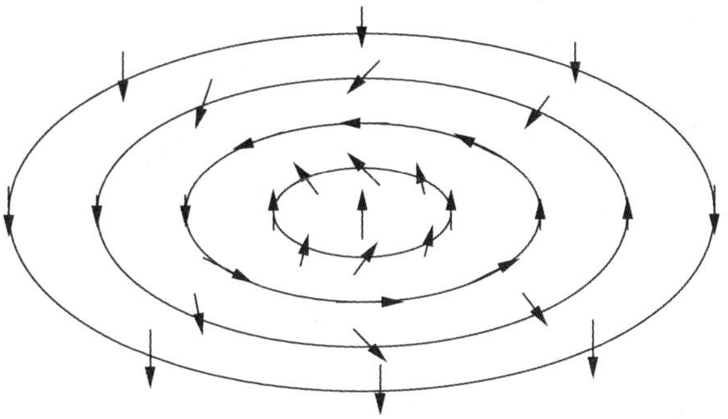

Fig. 2.11 Baby skyrmion on a planar magnet with magnetization density vector up in the center and down at infinity by a spiral rotation around the radial direction

also S^2, one has $\pi_2(\Gamma) = \pi_2(S^2) = \mathbb{Z}$, and hence there exists a topological charge. The corresponding magnetic state is called a 'baby **skyrmion**' and is the only skyrmion structure for which a picture can be drawn. It is shown in Fig. 2.11.

This state has a three-dimensional analogue since $\overline{\mathbb{R}^3} \sim S^3$ and $\pi_3(S^2) = \mathbb{Z}$ is the famous Hopf theorem. The corresponding Hopf mapping of S^3 onto S^2 is however not easy to draw. In general, skyrmions are special solitons in n dimensions corresponding to non-trivial homotopy groups $\pi_n(\Gamma)$. Originally, T. H. R. Skyrme proposed a subgroup of the product of the left and right chiral copies of $SU(N)$ as the order parameter space Γ to obtain local field structures as candidates of baryons in Yang–Mills field theories. For a more detailed discussion of the Hopf mapping and citations for further reading see [7].

More examples of topological charges can be found in [8].

The section is closed by a consideration of the topological stability of the **Fermi surface** of a Fermi liquid. (A more detailed discussion of Fermi surfaces is given in Sect. 5.9.) Again, first the two-dimensional case is considered which can easily be visualized. For a non-interacting isotropic Fermi gas, the single-particle Green function at imaginary frequency $\omega = ip_0$ is

$$G(ip_0,\boldsymbol{p}) = \frac{1}{ip_0 - v_F(p - p_F)}, \tag{2.39}$$

where $\boldsymbol{p}$ is the momentum vector, $p = |\boldsymbol{p}|$, p_F is the Fermi momentum, and v_F is the Fermi velocity. The energy dispersion close to the Fermi surface $p = p_F$ is $\varepsilon = v_F(p - p_F)$. States with $p < p_F$ have negative energies (measured from the chemical potential $\varepsilon = \varepsilon_F$) and are occupied, while states with $p > p_F$ have positive energies and are unoccupied. The Fermi surface $p = p_F$ in two-dimensional momentum space is a circle (Fig. 2.12) separating the occupied momentum region from the unoccupied one.

The Green function $G(ip_0,\boldsymbol{p})$ has a singularity line $p_0 = 0, p = p_F$ forming the Fermi surface and is otherwise a complex analytic function for imaginary frequencies. If one maps the contour C in the $(p_0,\boldsymbol{p})$-space onto the complex plane of G^{-1} with Re $G^{-1} = -v_F(p - p_F)$, Im $G^{-1} = p_0$, it maps the circle C onto the

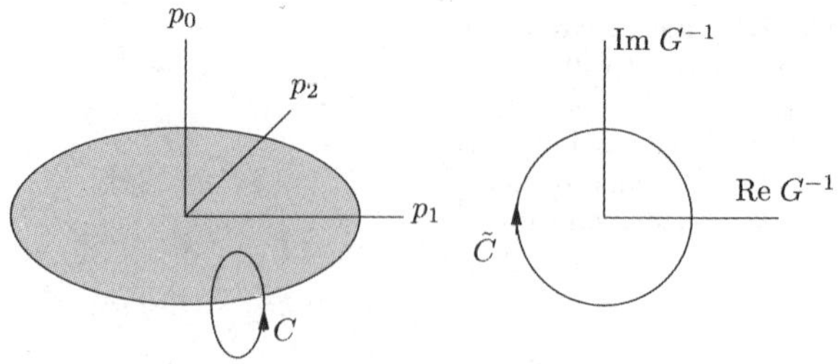

Fig. 2.12 *Left*: Fermi surface in two-dimensional momentum space and imaginary frequency axis with a loop C around the Fermi surface. *Right*: the corresponding loop $\tilde{C}$ of the complex function $G^{-1}(ip_0,\boldsymbol{p})$

circle $\tilde{C}$. Writing $G^{-1} = |G|^{-1}e^{-i\phi}$ it is seen that the phase of G increases by 2π when running around C, while for any loop not encircling the Fermi surface it returns to the start value. (This is like the phase of the superconducting order parameter when running around a vertex line.) The Fermi surface is like a defect line in momentum space.

If now the interaction between the particles is continuously switched on, the Green function changes smoothly. It cannot smoothly get rid of its denominator because of this topological charge on the Fermi surface, hence it must have the form

$$G(ip_0,\boldsymbol{p}) = \frac{Z}{ip_0 - v'_{\mathrm{F}}(p - p_{\mathrm{F}})}, \tag{2.40}$$

where Z is the spectral amplitude renormalization factor, and the Fermi velocity may change. (That p_{F} does not change is an independent result, the Luttinger theorem.) Hence the Fermi surface is topologically stabilized and can only disappear when Z becomes zero (which is only possible in a non-analytic way).

The only change for the case of three spatial dimensions is that now $\boldsymbol{p}$ is a 3-vector in the three-dimensional hyperplane of the four-dimensional frequency-momentum space of points $(p_0,\boldsymbol{p})$ for $p_0 = 0$, which contains the only singularities of (2.40) on the Fermi surface being now a 2-sphere. For every planar section in the three-dimensional momentum space through its origin, Fig. 2.12 visualizes further on the situation, and the Fermi surface is topologically stable.

A more general situation is present for electrons as spin 1/2 fermions in a crystalline solid instead of 'spinless fermions' in an isotropic medium which was considered so far. Here, the Green function is a complex valued matrix quantity indexed with band and spin indices. The change of its phase, normalized to 2π, as a complex number when going around a loop (contour integral of the gradient of the phase as considered in the case of a superconductor with a vertex line) is now to be replaced by the quantity

$$N = \operatorname{tr} \oint \frac{ds_p}{2i} \cdot G(ip_0,\boldsymbol{p}) \frac{\partial}{\partial p} G^{-1}(ip_0,\boldsymbol{p})$$

where the trace of the matrix product is to be formed, the contour integral is along the previous contour C, and $\partial/\partial p$ is the four-dimensional gradient in the frequency-momentum space. The dot means the scalar product of the line element vector with this gradient. This is the general structure of a homotopy invariant.[3]

Now, several sheets of Fermi surface may coexist of arbitrary shape. The shape may change when the interaction is tuned up and individual sheets may appear or disappear on the cost of other sheets. (If a Fermi radius shrinks to zero, in most cases the Fermi velocity also approaches zero, and the singularity disappears. Exceptions are so-called Dirac quasi-particles where the Fermi velocity remains non-zero in the Fermi points.) Nevertheless, between such changes the Fermi surface is topologically stable, and the only additional reason for its change is the vanishing of the spectral amplitude renormalization function $Z(p_0, \boldsymbol{p})$ on some part of the Fermi surface.

A much deeper analysis can be found in [9].

References

1. Reed M., Simon B.: Methods of Modern Mathematical Physics, Vol I. Functional Analysis Academic Press, New York (1973)
2. Yosida K.: Functional Analysis. Springer-Verlag, Berlin (1965)
3. Kolmogorov A., Fomin S.: Introductory Real Analysis. Prentice Hall, Englewood Cliffs NJ (1970)
4. Schwartz L.: Analyse Mathématique. Hermann, Paris (1967)
5. Eschrig H.: The Fundamentals of Density Functional Theory 1. Edition am Gutenbergplatz, Leipzig, p 226 (2003)
6. Choquet-Bruhat Y., de Witt-Morette C., Dillard-Bleick M.: Analysis, Manifolds and Physics, vol I: Basics. Elsevier, Amsterdam (1982)
7. Protogenov AP.: Knots and links in order parameter distributions of strongly correlated systems. Physics–Uspekhi **49**, 667–691 (2006)
8. Thouless DJ.: Toplogical Quantum Numbers in Nonrelativistic Physics. World Scientific, Singapore (1998)
9. Volovik GE.: The Universe in a Helium Droplet. Clarendon Press, Oxford (2003)

[3] The general expression for a topological charge enclosed by an n-sphere S^n (generator of the homotopy group $\pi_n(\Gamma)$) is $N_n = (n!|S^n|)^{-1} \int_{S^n} d\phi \wedge \cdots \wedge d\phi$, where $|S^n| = 2\pi^{(n+1)/2}/\Gamma((n+1)/2)$ is the volume of the n-sphere, $d\phi = \sum_{i=1}^n dx^i \partial\phi/\partial x^i$ is the 1-form of the gradient of the phase ϕ, and the $\wedge$-product has n factors, see later in Sect. 5.1.

Chapter 3
Manifolds

Vector space is already a large category of topological spaces. However, due to its linear structure, it is already too narrow for many applications in physics. Indeed, the topological and analytic structure is uniquely defined from a neighborhood of the origin alone. Manifold, on the one hand, is a generalization of metrizable vector space, maintaining only the local structure of the latter. On the other hand, every manifold can be considered as a (in general non-linear) subset of some vector space.

Both aspects are used to approach the theory of manifolds. In Algebraic Geometry one usually starts from the definition of manifolds in some vector space by means of a set of algebraic equations for a coordinate system in the vector space [1]. In physics, one rather knows local properties of manifolds and then asks for possibilities of continuation into the large. This is the standard approach in Differential Geometry [2], a rather complete classic; and [3], a well readable for physicists. This approach is taken in this text also.

With respect to the analytic structure, manifolds may be continuous, C^n, smooth or analytic. In this text the most important smooth case is treated, and for the sake of an effective terminology, **manifold** means **smooth manifold** throughout this text.

Since dimension of a vector space is a locally defined property, a manifold has a dimension. Although infinite dimensional manifolds have relevance in physics too, this text confines itself to n-dimensional manifolds, $n < \infty$, for basis manifolds of bundle spaces (which latter often form special infinite dimensional manifolds).

3.1 Charts and Atlases

An atlas of a manifold is a collection of charts projecting pieces of the manifold on open sets of an n-dimensional Euclidean space $\mathbb{R}^n$. *In all what follows* $\mathbb{R}^n$ *is taken to be a topological vector space homeomorphic to the Euclidean space* while the Euclidean metric given by the inner product structure is not used (cf. p. 13). The most familiar case is an atlas of the surface of the earth as a two-dimensional manifold. It is important to identify points of different charts of an atlas which are projections of the same point from overlapping domains of the manifold. Throughout this volume, points of an n-dimensional Euclidean space will be denoted by bold-faced letters as it was already done. Sets of the $\mathbb{R}^n$ will from now on also be denoted by (capital) bold-faced letters.

A **pseudo-group** S of class C^m, $m = 0, 1, \ldots, \infty, \omega$ consists of

1. a family S of homeomorphisms $\psi_{\beta\alpha} : \boldsymbol{U}_\alpha \to \boldsymbol{U}_\beta$ of class $C^m(\boldsymbol{U}_\alpha, \boldsymbol{U}_\beta)$, where $\boldsymbol{U}_\alpha$, $\boldsymbol{U}_\beta$ are open sets of $\mathbb{R}^n$,
2. for every $\psi_{\beta\alpha} \in S$, its restriction to any open subset of $\boldsymbol{U}_\alpha$ also belongs to the family S,
3. Conversely, if $\psi : \boldsymbol{U} \to \boldsymbol{U}'$ is a homeomorphism, $\boldsymbol{U} = \cup_{\alpha\in A}\boldsymbol{U}_\alpha$, and $\psi|_{\boldsymbol{U}_\alpha} \in S$ for all $\alpha \in A$, then $\psi \in S$,
4. $\mathrm{Id}_{\boldsymbol{U}}$ belongs to the family S for every open set $\boldsymbol{U} \in \mathbb{R}^n$,
5. for every $\psi_{\beta\alpha} \in S$, $(\psi_{\beta\alpha})^{-1} = (\psi^{-1})_{\alpha\beta} \in S$,
6. if $\psi_{\beta\alpha} \in S$ and $\psi_{\gamma\beta} \in S$, then $\psi_{\gamma\beta} \circ \psi_{\beta\alpha} = \psi_{\gamma\alpha} : \boldsymbol{U}_\alpha \to \boldsymbol{U}_\gamma \in S$.

A **complete atlas** $\mathcal{A}$ of **charts** $(U_\alpha, \varphi_\alpha)$, $\alpha \in A$, of a topological space M which is compatible with the pseudo-group S of class C^m consists of:

1. an open cover $M = \cup_{\alpha\in A} U_\alpha$ of M,
2. every φ_α is a homeomorphism from the open set $U_\alpha \in M$ to an open set $\boldsymbol{U}_\alpha \in \mathbb{R}^n$,
3. if $U_\alpha \cap U_\beta \neq \emptyset$, then $S \ni \psi_{\beta\alpha} = \varphi_\beta \circ \varphi_\alpha^{-1} : \varphi_\alpha(U_\alpha \cap U_\beta) \to \varphi_\beta(U_\alpha \cap U_\beta)$,
4. the complete atlas is not a proper subset of any other atlas compatible with S.

A complete atlas compatible with a pseudo-group S of class C^m, $m > 0$, is also called a **differentiable structure** on M. Figure 3.1 shows the interrelations of open sets U_α of M and open sets $\boldsymbol{U}_\alpha$ of $\mathbb{R}^n$ as well as the interrelation between homeomorphisms φ_α and $\psi_{\beta\alpha}$. The $\psi_{\beta\alpha}$ map images of the same point of M in different charts onto each other. A collection of charts not obeying the condition 4

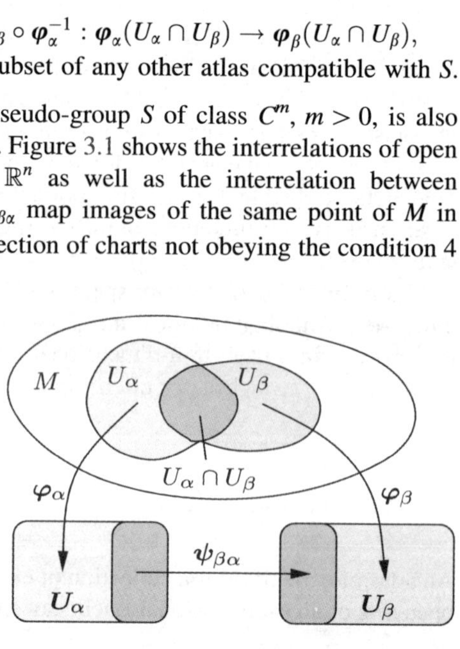

Fig. 3.1 Charts and homeomorphisms of an atlas

for complete atlases is simply called an **atlas**. It is not difficult to show that, given a pseudo-group S of class C^m, every atlas is subset of a complete atlas and that a complete atlas of a topological space M is uniquely generated by an atlas.

In all what follows either the family of all C^m-homeomorphisms of open sets of the $\mathbb{R}^n$ will be taken as the pseudo-group S, or (for $m > 0$ and to enforce orientation of manifolds, see end of next section) those homeomorphisms with positive Jacobian will be taken as the pseudo-group S_0. It is easily seen that these families fulfil all conditions 1–6 of a pseudo-group of class C^m.

With this convention, in both cases a complete atlas of a topological space M is uniquely defined by the space M itself and an atlas of M. The latter generates a complete atlas compatible with S or with S_0. An **admissible chart** of an atlas is a chart belonging to the corresponding complete atlas.

So far, only topological concepts (open sets and homeomorphisms) were used, and with respect to the topological space M these will be the only concepts to apply. The aim of mapping parts of M onto parts of $\mathbb{R}^n$ is to use the much richer structure of $\mathbb{R}^n$, its metric and linear structure as a vector space, in order to bring real numbers and analysis into the game. This is achieved by specifying a coordinate origin 0 in $\mathbb{R}^n$ and fixing a base $\{\boldsymbol{e}_1, \ldots, \boldsymbol{e}_n\}$ of vector space. Every point $\boldsymbol{x} \in \boldsymbol{U}_\alpha$ is then given by coordinates, $\boldsymbol{x} = \sum_{i=1}^n x^i \boldsymbol{e}_i$ for which $\boldsymbol{x} = (x^1, \ldots, x^n)$ will be written. The homeomorphism $\boldsymbol{\varphi}_\alpha$ means now an ordered set of n real-valued functions on $U_\alpha \in M : \boldsymbol{\varphi}_\alpha(x) = (\varphi^1_\alpha(x), \ldots, \varphi^n_\alpha(x)),\ x \in U_\alpha$. One may also write $\varphi^i_\alpha = \pi^i \circ \boldsymbol{\varphi}_\alpha$, where $\pi^i(\boldsymbol{x}) = x^i$ is the projection on $\boldsymbol{e}_i$ in $\mathbb{R}^n$. This all is not a big step ahead since the points $x \in M$ arc still not given by numbers. However, instead of moving through M one now can move through its charts; only once in a while one has to transit from one chart to another one by means of the **transition functions** $\boldsymbol{\psi}_{\beta\alpha}(\boldsymbol{x}) = (\psi^1_{\beta\alpha}(\boldsymbol{x}), \ldots, \psi^n_{\beta\alpha}(\boldsymbol{x})),\ \boldsymbol{x} = (x^1, \ldots, x^n) \in \boldsymbol{U}_\alpha \subset \mathbb{R}^n$. This is now already an ordered set of n real-valued functions of n real variables. *It was only these transition functions of which class C^m could be required.*

The set $U_\alpha \in M$ is now called a **coordinate neighborhood** and the set of functions $\varphi^i_\alpha(x)$ is called a **local coordinate system** on $U_\alpha \in M$. Since in the Euclidean space $\mathbb{R}^n$ the origin $\boldsymbol{x} = 0$ may be chosen arbitrarily by using affine-linearity, for every α separately it can always be chosen to be in $\boldsymbol{U}_\alpha$. If $\boldsymbol{U}_\alpha = \{\boldsymbol{x} \mid |x^i| < a\}$, then this is called a **coordinate cube** centered at $(\boldsymbol{\varphi}_\alpha)^{-1}(0) = x_0 \in M$.

Finally, the commutative diagram (see Compendium C.1)

$$\begin{array}{ccc} & U_\alpha \cap U_\beta \in M & \\ \boldsymbol{\varphi}_\alpha|_{U_\alpha \cap U_\beta} \swarrow & & \searrow \boldsymbol{\varphi}_\beta|_{U_\alpha \cap U_\beta} \\ \boldsymbol{U}_\alpha \supset \boldsymbol{\varphi}_\alpha(U_\alpha \cap U_\beta) & \xrightarrow{\boldsymbol{\psi}_{\beta\alpha}|_{\boldsymbol{\varphi}_\alpha(U_\alpha \cap U_\beta)}} & \boldsymbol{\varphi}_\beta(U_\alpha \cap U_\beta) \subset \boldsymbol{U}_\beta \end{array} \tag{3.1}$$

is mentioned.

3.2 Smooth Manifolds

An n-dimensional C^m**-manifold** is a paracompact topological space with a complete atlas compatible with the structure S of all C^m-homeomorphisms of open sets of $\mathbb{R}^n$ or with the structure S_0 of homeomorphisms with positive Jacobian.

The local topology of an n-dimensional manifold is very simple: it is the same as that of $\mathbb{R}^n$. In particular *a manifold is Hausdorff* (by our definition of a paracompact space). Like $\mathbb{R}^n$ it is also *normal* (disjoint closed sets are contained in disjoint open sets), *first countable and locally compact, locally simply connected and locally pathwise simply connected.* Since manifolds can be obtained by gluing together an arbitrary number of pieces in a most general way, they can be quite monstrous and their global topology may get out of control. A standard tool of studying global properties is by getting them as locally finite sums of local properties, in particular via a partition of unity. For that reason, it is demanded that M be paracompact. Alternatively, many authors demand that M be second countable; it can be shown that in combination with the local topology paracompactness then follows.

The geometry, on the contrary, is in general already locally involved. It will be studied from Chap. 7 on.

In all what follows, if not otherwise explicitly mentioned, **manifold** *means a finite-dimensional C^∞-manifold, that is, a* **smooth manifold***.*

Examples

$\mathbb{R}^n$: It is itself an n-dimensional manifold with the standard smooth pseudo-group S (see Sect. 3.1) and the complete atlas containing (generated by) the chart $(\mathbb{R}^n, \mathrm{Id}_{\mathbb{R}^n})$.

n-dimensional topological vector space X: (not necessarily provided with a geometry by an inner product). If $\{\boldsymbol{e}_i\}$ is an arbitrarily chosen base of X and $\{\boldsymbol{f}^j\}$ is the corresponding dual base, $\langle \boldsymbol{f}^j, \boldsymbol{e}_i\rangle = \delta^j_i$, then the projections $\pi^j(\boldsymbol{x}) = \langle \boldsymbol{f}^j, \boldsymbol{x}\rangle = x^j$ define a local coordinate system which is also a global one. A change of the basis $\{\boldsymbol{e}_i\}$ is a smooth homeomorphism of $\mathbb{R}^n$ to $\mathbb{R}^n$, and those changes in open sets of X provide a simple atlas compatible either with the standard pseudo-group S_0, if transformations with positive Jacobian are taken only, or with the standard pseudo-group S in the general case. (Further on the adjective standard is omitted.) There exist many more charts in a complete atlas, e.g. with curved coordinate systems.

Sphere $S^n \subset \mathbb{R}^{n+1}$: $\{\boldsymbol{x} | \sum_{i=1}^{n+1}(x^i)^2 = 1\}$. Let the 'south pole' be $\boldsymbol{s} = (1, 0, \ldots, 0)$ and the 'north pole' $\boldsymbol{n} = (-1, 0, \ldots, 0)$. A complete atlas is generated by the two charts $(S^n \setminus \{\boldsymbol{s}\}, p_{\boldsymbol{s}})$ and $(S^n \setminus \{\boldsymbol{n}\}, p_{\boldsymbol{n}})$, where $p_{\boldsymbol{s}}$ and $p_{\boldsymbol{n}}$ are the stereographic projections from the south pole and from the north pole, respectively (Fig. 3.2).

n-dimensional projective space P^n: Define an equivalence relation $\doteqdot$ in $\mathbb{R}^{n+1} \setminus \{0\}$ as $\boldsymbol{x} \doteqdot \boldsymbol{y}$, iff $\boldsymbol{x} = a\boldsymbol{y}$, $\mathbb{R} \ni a \neq 0$. Denote the equivalence classes (straight lines through the origin) by $x = [\boldsymbol{x}]$. Then, $P^n = ((\mathbb{R}^{n+1} \setminus \{0\})/\doteqdot) = \{x = [\boldsymbol{x}] | \boldsymbol{x} \in (\mathbb{R}^{n+1} \setminus \{0\})\}$. $(x^1, \ldots, x^{n+1})$ are the **homogeneous coordinates** of x; they are determined up to the factor $a \neq 0$. The $n+1$ open sets $U_i = \{x \,|\, x^i \neq 0\}$,

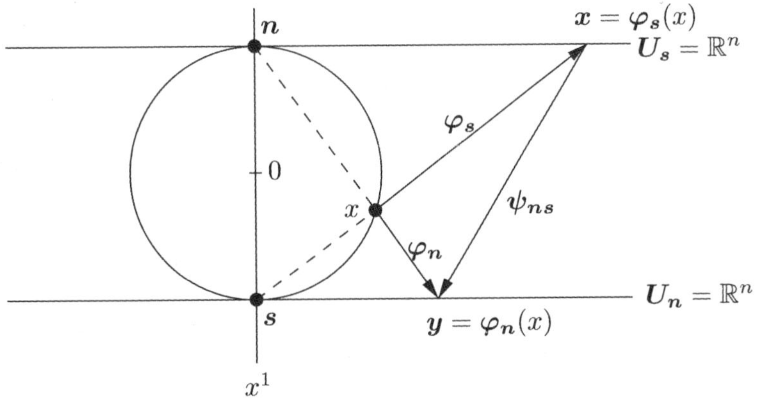

Fig. 3.2 The manifold S^1 with two stereographic projections

$i = 1, \ldots, n+1$ form an open cover of P^n. Unique coordinates in U_i may be chosen $\xi_i^j = x^j/x^i$, $j \neq i$, $\boldsymbol{\varphi}_i(x) = \sum_{j(\neq i)} \xi_i^j \boldsymbol{e}_j$. It follows $\boldsymbol{\psi}_{ki}(\boldsymbol{\xi}_i) = \sum_{j(\neq k)} \xi_k^j \boldsymbol{e}_j$, $\xi_k^j = \xi_i^j(x^i/x^k) = \xi_i^j/\xi_i^k$ for $j \neq i, k$ and $\xi_k^i = 1/\xi_i^k$. These $\boldsymbol{\psi}_{ki}$ are smooth functions on $\boldsymbol{U}_i \cap \boldsymbol{U}_k \subset \mathbb{R}^n$. The global topology of P^n is more involved than that of S^n.

The projective space P^1 is depicted in Fig. 3.3. There are two homogeneous coordinates x^1, x^2 forming a plane with removed origin. The open sets U_1, U_2 consist of the plane with removed x^2- and x^1-axis, respectively. There is only one coordinate $\xi_1 = \xi_1^2$ and $\xi_2 = \xi_2^1$, respectively, related by $\psi_{21}(\xi_1) = 1/\xi_1 = \xi_2$. Hence, the Jacobian of ψ_{12} reduces to the derivative $d\xi_2/d\xi_1 = -1/(\xi_1)^2 < 0$. Note that a 180° rotation of the (x^1, x^2)-plane is the identity mapping Id_{P^1} of the projective space P^1.

Möbius band (Fig. 1.2b, p. 3 in Chap. 1): Take the rectangle $M = \{(x, y) \in \mathbb{R}^2 | -\pi \leq x \leq \pi, -1 < y < 1\}$ and glue the two edges $x = \pm\pi$ in such a way together that the points $(-\pi, y)$ and $(\pi, -y)$ are identified with each other. Replace x by the polar angle ϕ along the circumference of the glued together tape. Every open set $U_\xi = \{(\phi, y) | \xi < \phi < \xi + 2\pi, -1 < y < 1\}$ with ordinary planar coordinates of the original rectangle is a coordinate neighborhood on the Möbius band

Fig. 3.3 The plane of homogeneous coordinates x^1, x^2 of the projective space P^1. On the *left panel*, the set U_1, $x^1 \neq 0$, is shadowed, on the *right panel* U_2

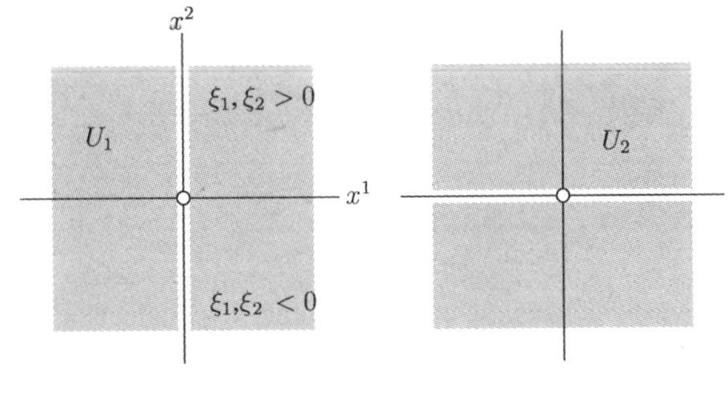

which is a two-dimensional manifold. However, the mappings between the overlap sets of neighborhoods U_ξ for different ξ have unavoidably partially positive and partially negative Jacobians. A complete atlas with the structure S_0 is not possible in this case.

A manifold for which a complete atlas compatible with S_0 exists is called an **orientable manifold**. For $n > 1$, all presented examples except the Möbius band and P^n for n even are orientable manifolds. The Möbius band as well as P^n, n even, are not orientable. An orientable manifold may have two orientations. If $(U_\alpha, \varphi_\alpha)$ are the charts of an oriented atlas of an orientable manifold, then another atlas with charts $(U_{\tilde\alpha}, \varphi_{\tilde\alpha})$ with $\psi_{\alpha\tilde\alpha}(x^1, x^2, \dots, x^n) = (-x^1, x^2, \dots, x^n)$ as transition functions between these charts has the opposite orientation. (Show that the inversion $\boldsymbol{x} \mapsto -\boldsymbol{x}$ of the $\mathbb{R}^{n+1}, n$ even, inverts orientation; since $\boldsymbol{x}$ and $-\boldsymbol{x}$ represent the same point of P^n in homogeneous coordinates, P^n, n even, cannot be orientable.)

Any open subset $M_1 \subset M$ of a manifold M is again a manifold with the charts $(U_\alpha \cap M_1, \varphi_\alpha|_{U_\alpha \cap M_1})$, if $(U_\alpha, \varphi_\alpha)$ are the charts of M. M_1 is called an **open submanifold** of M. (A detailed discussion follows in Sect. 3.5.)

The **product manifold** of two manifolds $(M_1, \mathcal{A}_1)$ and $(M_2, \mathcal{A}_2)$ with complete atlases $\mathcal{A}_1$ and $\mathcal{A}_2$ is the product $M_1 \times M_2$ of the topological spaces M_1 and M_2 with the product topology. Its complete atlas is created by the charts $(U^1_\alpha \times U^2_\beta, \varphi^1_\alpha \times \varphi^2_\beta)$ with evident notation. The dimension of the product manifold is $\dim M_1 + \dim M_2$. For instance the two-dimensional torus is the product manifold $T^2 = S^1 \times S^1$.

A **smooth mapping** F from a manifold $(M, \mathcal{A}_M)$ into a manifold $(N, \mathcal{A}_N)$ is a mapping $F : M \to N$ so that for every pair of charts $(U, \varphi_U) \in \mathcal{A}_M$, $(V, \varphi_V) \in \mathcal{A}_N$ the mapping $\varphi_V \circ F \circ (\varphi_U)^{-1} : \varphi_U(U) \to \varphi_V(V)$ is C^∞. ($\varphi_V \circ F \circ (\varphi_U)^{-1}$ is a mapping from an open set of $\mathbb{R}^{n_M}$ into an open set of $\mathbb{R}^{n_N}$, hence its class of differentiability is defined.) If M is the open interval $]a, b[\, \in \mathbb{R}$ (with its standard manifold structure as an open submanifold of $\mathbb{R} = \mathbb{R}^1$; see the first example above), then $F : M \to N$ is called a smooth **parametrized curve** or simply a parametrized curve which is always assumed smooth if not otherwise explicitly mentioned.

If F is bijective and $F : M \to N$ and $F^{-1} : N \to M$ are both smooth mappings, then F is called a **diffeomorphism** of manifolds. The complete atlases $\mathcal{A}_M$ and $\mathcal{A}_N$ are called **isomorphic**, $\mathcal{A}_M \approx \mathcal{A}_N$, if a diffeomorphism $F : M \to N$ exists. (Just to mention, diffeomorphism is more than homeomorphism; there are homeomorphic C^∞-manifolds which are not diffeomorphic.)

3.3 Tangent Spaces

Before the general case is treated, a simple example is discussed which every physicist should be familiar with.

A simple (one-dimensional) manifold is a smooth curve $\boldsymbol{x}(t)$ in $\mathbb{R}^n$ given by n equations $x^i = x^i(t),\ i = 1, \ldots, n,\ a<t<b$ with respect to some base $\{\boldsymbol{e}_i\}$ of the vector space $\mathbb{R}^n$. It is the special case of a parametrized curve defined at the end of the last section, where the manifold N of that definition is $\mathbb{R}^n$. Consider the point $\boldsymbol{x}_0$ at $t = t_0$ on this curve. As is well known, the tangent vector in $\mathbb{R}^n$ on the curve $\boldsymbol{x}(t)$ at the point $\boldsymbol{x}_0$ is the vector $\hat{\boldsymbol{X}}_{\boldsymbol{x}_0} = (dx^i/dt|_{t_0})$. *Any* vector of the $\mathbb{R}^n$ is tangent vector at any point of the $\mathbb{R}^n$ on some smooth curve passing though that point, in other words, $\mathbb{R}^n$ is the tangent space to itself at any of its points. In this connection, any given vector $\hat{\boldsymbol{X}} \in \mathbb{R}^n$ is tangent vector at $\boldsymbol{x}_0 \in \mathbb{R}^n$ to a whole bunch of curves, for instance thought of as all paths of motion through $\boldsymbol{x}_0$ with velocity vector $\hat{\boldsymbol{X}}$ at that point. Above, a coordinate system in $\mathbb{R}^n$ was used from the outset by choosing a particular base $\{\boldsymbol{e}_i\}$. In vector analysis, analytic relations are defined and considered independent of the choice of a coordinate system, for instance by defining $\hat{\boldsymbol{X}}_{\boldsymbol{x}_0} = d\boldsymbol{x}/dt|_{t_0}$ in an invariant way. Consider next any real-valued smooth function $F : \mathbb{R}^n \to \mathbb{R}$. (Class C^1 would suffice here, but for later considerations C^∞ is assumed from the outset.) By composing it with $\boldsymbol{x}(t)$ it defines a function $\hat{F}(t) = F \circ \boldsymbol{x}(t)$ with derivative

$$\left.\frac{d\hat{F}}{dt}\right|_{t_0} = \sum_i \left.\frac{dx^i}{dt}\right|_{t_0} \left.\frac{\partial}{\partial x^i}\right|_{\boldsymbol{x}_0} F = \left.\frac{d\boldsymbol{x}}{dt}\right|_{t_0} \cdot \left.\frac{\partial}{\partial \boldsymbol{x}}\right|_{\boldsymbol{x}_0} F = \boldsymbol{X}_{\boldsymbol{x}_0} F. \tag{3.2}$$

It is just the directional derivative (Sect. 2.3) of F with respect to the vector $\hat{\boldsymbol{X}}_{\boldsymbol{x}_0}$ for which the operator of differentiation $\boldsymbol{X}_{\boldsymbol{x}_0}$ acting on F has been introduced in (3.2). As is seen from this chain of equations, $\boldsymbol{X}_{\boldsymbol{x}_0}$ may be thought of as a vector in a vector space with base $\{\partial/\partial x^i\}$, the components of which with respect to that base are dx^i/dt. Indeed, any vector operator $\boldsymbol{X}_{\boldsymbol{x}_0} = \sum_i \xi^i \partial/\partial x^i$ defines a directional derivative at $\boldsymbol{x}_0$ corresponding for instance to the smooth curve (straight line) $\xi(t) = \boldsymbol{x}_0 + t\sum_i \xi^i \boldsymbol{e}_i$.

A change of the base $\boldsymbol{e}_i$ in the $\mathbb{R}^n$ on which F was defined causes a change of the base $\partial/\partial x^i$ so that (3.2) remains invariant. Here, $d\hat{F}/dt$ is the scalar product of the tangent vector $d\boldsymbol{x}/dt$ with the gradient vector $\partial F/\partial \boldsymbol{x}$. In this chapter, differentials are more important than derivatives. By writing $dF_{x_0} = \sum_i (\partial F/\partial x^i) dx^i$, and understanding $\{dx^i\}$ as a base in the dual space to the tangent space, later introduced as the cotangent space, with the relation $\langle dx^i, \partial/\partial x^k\rangle = \delta^i_k$ one has $d\hat{F}_{x_0} = \boldsymbol{X}_{\boldsymbol{x}_0} F dt = \boldsymbol{X}_{\boldsymbol{x}_0} F = \langle dF_{x_0}, \boldsymbol{X}_{\boldsymbol{x}_0}\rangle$ where dt has been put equal to unity by definition. These are many details for the simple relation (3.2), but hopefully they help in understanding the precise meaning of the following. Note in particular that all considerations above need the functions involved only locally in any (arbitrarily small) neighborhood of the point $\boldsymbol{x}_0$.

If M is an arbitrary smooth manifold of n dimensions, the coordinates of its points $x \in U_\alpha \subset M$ are locally defined by using a chart $(U_\alpha, \boldsymbol{\varphi}_\alpha)$ out of the complete atlas $\mathcal{A}$ of $M : x^i_\alpha = \varphi^i_\alpha(x) = \pi^i \circ \boldsymbol{\varphi}_\alpha(x)$. The only demand on $\boldsymbol{\varphi}_\alpha$ is that it is a homeomorphism from U_α to $\boldsymbol{U}_\alpha \subset \mathbb{R}^n$ and that the transition functions $\boldsymbol{\psi}_{\beta\alpha}$ between charts are smooth. Linear coordinates in $\boldsymbol{U}_\alpha$ for instance do not have any preference any more since in general M is not a vector space and hence linear relations between its points are not defined any more. Within the complete atlas (differentiable structure) of M there is a huge arbitrariness not only of choosing the coordinates of a point $x \in M$ but also of choosing the neighborhood U_α of x in which those coordinates are defined. Since an arbitrarily small neighborhood suffices for considerations of the tangent space, the local behavior of a function is introduced by the concept of a **germ** of function. Consider a point $x_0 \in M$ and the family $C^\alpha_{x_0}$ of smooth real-valued functions F_α defined in some neighborhood of $\boldsymbol{\varphi}_\alpha(x_0) \in \boldsymbol{U}_\alpha = \boldsymbol{\varphi}_\alpha(U_\alpha)$ for some chart for M containing the point x_0 (coordinate neighborhood of x_0). Since the composition of smooth functions $F_\alpha \circ \boldsymbol{\psi}_{\alpha\beta} = F_\beta$ is smooth, F_α defines a smooth function F_β in some neighborhood of $\boldsymbol{\varphi}_\beta(x_0) \in \boldsymbol{U}_\beta$ for every local coordinate system $(U_\beta, \boldsymbol{\varphi}_\beta)$ centered at x_0. In other words, $C^\alpha_{x_0}$ may be considered as the family of all smooth real-valued functions on any local coordinate system of M centered at x_0, and apart from their smoothness which is only defined in connection with a local coordinate system, each of the functions F_α of $C^\alpha_{x_0}$ together with a local coordinate system defines a function $F = F_\alpha \circ \boldsymbol{\varphi}_\alpha$ on a neighborhood of $x_0 \in M$. This allows for the introduction of the family $\mathcal{C}_{x_0}$ of all real-valued functions F defined in some neighborhood of $x_0 \in M$ and smoothly depending on the coordinates of any local coordinate system of M centered at x_0.

Two functions $F, G \in \mathcal{C}_{x_0}$ are considered equivalent, $F \simeq G$, if there exists a neighborhood U of x_0 so that $F|_U = G|_U$. (Note that two non-identical smooth real functions still may coincide on some domain; smoothness is less than analyticity, where functions are uniquely continued from any open domain.) Given any local coordinate system of M centered at x_0, if $F \simeq G$, then obviously $\partial F_\alpha/\partial x^i_\alpha|_{\boldsymbol{x}=0} = \partial G_\alpha/\partial x^i_\alpha|_{\boldsymbol{x}=0}$ where without loss of generality the coordinates $\boldsymbol{x}$ of x_0 are put to zero. This is always done in what follows. An equivalence class $[F]$ of a function $F \in \mathcal{C}_{x_0}$ is called a **germ** at x_0 on M. The set of germs at x_0 on M is denoted by

$$\mathcal{F}_{x_0} = \mathcal{C}_{x_0}/\simeq = \{[F] | F \in \mathcal{C}_{x_0}\} \tag{3.3}$$

(quotient set with respect to the equivalence relation $\simeq$ in $\mathcal{C}_{x_0}$). Why is the concept of germs needed instead of simply considering the family of functions defined on some (fixed) neighborhood of x_0? The point is that in order to decide which functions are admissible in $\mathcal{C}_{x_0}$, local coordinate systems have to be used and their domain of definition cannot be fixed, it depends on the used charts and can in particular become arbitrarily small. Note also that the *same* function $F \in \mathcal{C}_{x_0}$ corresponds to infinitely many *different* functions $F_\alpha \in C^\alpha_{x_0}, F_\alpha = F_\alpha(x^1_\alpha, \dots, x^n_\alpha)$ for different local coordinate systems.

Next, the family of (smooth) parametrized curves $x(t)$, $t \in]a, b[$ in M passing through x_0 is considered (Fig. 3.4). Again without loss of generality it is assumed that $t = 0$ is an inner point of the interval $]a, b[$ and $x(0) = x_0$. This time smoothness is to be considered with respect to local coordinate systems of the

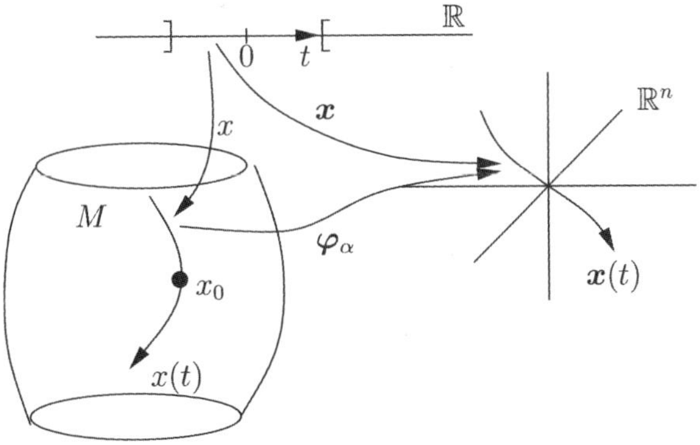

Fig. 3.4 A path $x(t)$ through x_0 in M and its image $\boldsymbol{x}(t)$ through 0 in a coordinate chart

target space M of the mapping $t \mapsto x(t)$: $\boldsymbol{x}_\alpha(t) = \boldsymbol{\varphi}_\alpha \circ x(t)$ must be smooth for some $\boldsymbol{\varphi}_\alpha$ for which $x_0 \in U_\alpha$; it then is smooth for any such chart, $\boldsymbol{x}_\beta(t) = \boldsymbol{\varphi}_\beta \circ x(t) = \boldsymbol{\psi}_{\beta\alpha} \circ \boldsymbol{\varphi}_\alpha \circ x(t)$. (More precisely, an appropriate restriction of the curve $x(t)$ which fits into U_α and U_β is meant with $x(t)$ in the above composite mapping.) Consider now any function $F \in \mathcal{C}_{x_0}$, any parametrized curve $x(t)$ passing through x_0 and any local coordinate system of M centered at x_0. The latter defines a function $F_\alpha(x^1_\alpha, \dots, x^n_\alpha) = (F \circ \boldsymbol{\varphi}_\alpha^{-1})(\boldsymbol{x}_\alpha)$ and a curve $\boldsymbol{x}_\alpha(t) = (x^1_\alpha(t), \dots, x^n_\alpha(t))$ corresponding to (some restriction of) $x(t)$. Furthermore, $\hat{F}_\alpha(t) = F_\alpha \circ \boldsymbol{x}_\alpha(t) = (F \circ \boldsymbol{\varphi}_\alpha^{-1} \circ \boldsymbol{\varphi}_\alpha \circ x)(t) = F(x(t))$ is the function F_α on the curve $\boldsymbol{x}_\alpha(t)$ which by construction is the same function of t as the original function F on the curve $x(t)$, and (3.2) with F replaced by F_α is valid for the directional derivative of F_α with respect to the tangent vector $\hat{\boldsymbol{X}}^\alpha_0 = d\boldsymbol{x}_\alpha/dt|_0$ on the curve $\boldsymbol{x}_\alpha(t)$:

$$\left.\frac{dF(x(t))}{dt}\right|_0 = \left.\frac{dF_\alpha(\boldsymbol{x}_\alpha(t))}{dt}\right|_0 = \sum_i \left.\frac{dx^i_\alpha}{dt}\right|_0 \left.\frac{\partial}{\partial x^i_\alpha}\right|_0 F_\alpha = \left.\frac{d\boldsymbol{x}_\alpha}{dt}\right|_0 \cdot \left.\frac{\partial}{\partial \boldsymbol{x}_\alpha}\right|_0 F_\alpha = X^\alpha_0 F_\alpha. \tag{3.4}$$

(It will be seen that the vector operators X^α_0 on a manifold form a vector space but do not in general any more form a Euclidean space, therefore it is not any more denoted in bold face.) Since the value of the third expression in this chain of equations depends on the partial derivatives of F_α at $\boldsymbol{x} = 0$ only, it is the same within a class $[F] \in \mathcal{F}_{x_0}$ independent of its representative F. Moreover, a change of the local coordinate system changes X^α_0 and F_α in such a way that (3.4) remains unchanged.

Consider such a change of the local coordinate system in more detail. The corresponding coordinates are

$$\begin{aligned} x^i_\alpha &= \pi^i \circ \boldsymbol{\varphi}_\alpha(x), \\ x^i_\beta &= \pi^i \circ \boldsymbol{\varphi}_\beta(x) = \pi^i \circ \boldsymbol{\psi}_{\beta\alpha} \circ \sum_j \boldsymbol{\pi}_j \circ \pi^j \circ \boldsymbol{\varphi}_\alpha(x) = \psi^i_{\beta\alpha}(x^1_\alpha, \dots, x^n_\alpha), \end{aligned} \tag{3.5}$$

where $\boldsymbol{\pi}_j$ maps the number x^j to a vector in $\mathbb{R}^n$ with the jth component as the only non-zero component equal to x^j; $\boldsymbol{\pi}_j(x^j) = (0, \dots, 0, x^j, 0, \dots, 0)$, $\sum_j \boldsymbol{\pi}_j \circ \pi^j = \mathrm{Id}_{\mathbb{R}^n}$.

Hence,

$$\frac{\partial x^i_\beta}{\partial x^j_\alpha} = (\psi_{\beta\alpha})^i_j, \quad (\psi_{\alpha\beta})^j_i = (\psi^{-1}_{\beta\alpha})^j_i. \tag{3.6}$$

The (ij)-matrix $((\psi_{\beta\alpha})^i_j)$ is the **Jacobian matrix** of the coordinate transformation $\psi_{\beta\alpha}$. The last relation of (3.6) considers the property 5 of the pseudo-group S of transformations. Now,

$$\begin{aligned}\frac{dx^i_\beta}{dt} &= \sum_j \frac{\partial x^i_\beta}{\partial x^j_\alpha}\frac{dx^j_\alpha}{dt} = \sum_j (\psi_{\beta\alpha})^i_j \frac{dx^j_\alpha}{dt},\\ \frac{\partial}{\partial x^i_\beta} &= \sum_j \frac{\partial x^j_\alpha}{\partial x^i_\beta}\frac{\partial}{\partial x^j_\alpha} = \sum_j (\psi_{\alpha\beta})^j_i \frac{\partial}{\partial x^j_\alpha} = \sum_j (\psi^{-1}_{\beta\alpha})^j_i \frac{\partial}{\partial x^j_\alpha},\end{aligned} \tag{3.7}$$

which again demonstrates the invariance of (3.4). A vector transforming according to the first transformation rule of (3.7) is called a **contravariant vector** and one transforming according to the second transformation rule of (3.7) is called a **covariant vector**.

An abstract vector X_{x_0} may be introduced, translated by a local coordinate system into the differential vector operator $X^\alpha_0 = \sum_i \xi^i_\alpha(\partial/\partial x^i_\alpha)$ where the ξ^i_α form the components of a contravariant vector and the operators $\partial/\partial x^i_\alpha$ form a base in the space of vectors X_{x_0}. The base vectors transform like the components of a covariant vector. According to (3.4), X_{x_0} provides a mapping

$$X_{x_0} : \mathcal{C}_{x_0} \to \mathbb{R} : F \mapsto X_{x_0}F = \sum_i \xi^i_\alpha(\partial F_\alpha/\partial x^i_\alpha). \tag{3.8}$$

This mapping has the obvious properties

1. $X_{x_0}(\lambda F + \mu G) = \lambda X_{x_0}F + \mu X_{x_0}G$, that is, it is linear,
2. $X_{x_0}(FG) = (X_{x_0}F)G(x_0) + F(x_0)(X_{x_0}G)$, **Leibniz rule**.

Any vector X_{x_0} is called a **tangent vector** on M at the point x_0. The vector space of all tangent vectors X_{x_0} is the **tangent space** $T_{x_0}(M)$ on M at the point x_0. It is also denoted by T_{x_0} if there is no doubt about the manifold M.

Given any local coordinate system centered at point x_0, consider the relation $\sum_i \xi^i_\alpha(\partial/\partial x^i_\alpha) = 0$, that is, $\sum_i \xi^i_\alpha(\partial F_\alpha/\partial x^i_\alpha) = 0$ for all $F \in \mathcal{C}_{x_0}$. Since $F^i \in \mathcal{C}_{x_0}$ for $F^i_\alpha(x) = x^i_\alpha$, it follows that $\xi^i_\alpha = 0$ for all $i = 1, \ldots, n$. This proves that the base vectors $\partial/\partial x^i_\alpha$ are linearly independent in T_{x_0}, and the dimension of T_{x_0} is equal to n, that is, equal to the dimension of M. Note that although this result seems to be obvious it is due to the differentiability of the pseudo-group of transition functions only; differentiability directly on M cannot be defined. It is natural *to provide T_{x_0} with a topology to be homeomorphic with $\mathbb{R}^n$*.

Coming back to the set (3.3) of germs, the definition of linear operations and of point wise multiplication of functions in $\mathcal{F}_{x_0}$,

$$\lambda[F] = [\lambda F], \quad [F] + [G] = [F + G], \quad [F][G] = [FG] \tag{3.9}$$

makes $\mathcal{F}_{x_0}$ into a commutative algebra over $\mathbb{R}$ (which means that it is also a real vector space, and as such is in fact a functional space and hence infinite dimensional: for instance all distinct polynomials in the coordinates of a fixed local coordinate system are linearly independent). On the right hand sides of (3.9) inside the square brackets the functions $F + G$ and FG are understood on the intersection of their domains of definition, this is why $\mathcal{C}_{x_0}$ is not an algebra: there is no common domain of definition of *all* functions $F \in \mathcal{C}_{x_0}$. The mapping (3.8) induces a corresponding mapping $X_{x_0} : \mathcal{F}_{x_0} \to \mathbb{R} : [F] \mapsto X_{x_0}[F] = X_{x_0}F$ which inherits the same mapping properties

$$\begin{aligned} &1. \quad X_{x_0}(\lambda[F] + \mu[G]) = \lambda X_{x_0}[F] + \mu X_{x_0}[G], \\ &2. \quad X_{x_0}([F][G]) = (X_{x_0}[F])G(x_0) + F(x_0)(X_{x_0}[G]), \end{aligned} \tag{3.10}$$

expressed by saying that X_{x_0} is a **linear derivation of the algebra** $\mathcal{F}_{x_0}$. The subset $\mathcal{F}^0_{x_0}$ of all germs $[F_0]$ vanishing at x_0 forms an ideal of the multiplicative ring of vectors of the algebra $\mathcal{F}_{x_0} : \mathcal{F}^0_{x_0}\mathcal{F}_{x_0} = \mathcal{F}_{x_0}\mathcal{F}^0_{x_0} = \mathcal{F}^0_{x_0}$. (The point wise product of any function F with a function F_0 yields another function $G_0 \in \mathcal{F}^0_{x_0}$.) Given a fixed coordinate neighborhood α of x_0, $\mathcal{F}^0_{x_0}$ contains in turn the germ of the function which is identical to zero, germs corresponding to all linear functions F_α with respect to the coordinates of a local coordinate system, germs of all quadratic (more precisely bilinear) such functions, and so on. Since the product of two linear functions is a bilinear function, $(\mathcal{F}^0_{x_0})^2$ contains in turn the germ of the function which is identical to zero, the germs due to quadratic functions F_α, the germs due to cubic functions, and so on. This holds true for any coordinate neighborhood α, hence, $\mathcal{F}_{x_0} \supset \mathcal{F}^0_{x_0} \supset (\mathcal{F}^0_{x_0})^2 \supset \cdots$.

From the properties (3.10) it is readily seen that every tangent vector X_{x_0} maps every germ from $(\mathcal{F}^0_{x_0})^2$ to zero:

$$\begin{aligned} X_{x_0}([F_0][G_0]) &= (X_{x_0}[F_0])G_0(x_0) + F_0(x_0)(X_{x_0}[G_0]) \\ &= (X_{x_0}[F_0]) \cdot 0 + 0 \cdot (X_{x_0}[G_0]) = 0. \end{aligned}$$

Hence, the action of X_{x_0} on $\mathcal{F}_{x_0}$ is completely determined by its action on the quotient vector space $\mathcal{F}^0_{x_0}/(\mathcal{F}^0_{x_0})^2$ represented by linear functions with respect to the coordinates of any local coordinate system. The members $F^0 \in \mathcal{C}_{x_0}$ of an equivalence class which constitutes an element of $\mathcal{F}^0_{x_0}/(\mathcal{F}^0_{x_0})^2$ differ between each other by functions having zero partial derivatives at x_0 in all local coordinate systems. These equivalence classes are denoted by dF_{x_0}, $\mathcal{F}^0_{x_0}/(\mathcal{F}^0_{x_0})^2 = \{dF_{x_0}\}$, and are called **differentials** of the functions $F = F_0 + \text{const.}$, since they are precisely what for functions in $\mathbb{R}^n$ are ordinary differentials: the linear part of a function (tangent hyperplane to the graph of the function). Recall again that linearity is not directly defined for functions on M since M is in general not a vector space. Moreover, the linear part of a function $F \in \mathcal{C}_{x_0}$ with respect to local coordinates is in general different for different local coordinate systems. However, given a local coordinate system, by construction all functions within an equivalence class dF_{x_0} differ from each other by additive terms which are higher than first order in the coordinates. Hence, from (3.8) it is also clear that for every given tangent vector X_{x_0} the value $X_{x_0}F$ is uniquely determined by its action on the differential dF_{x_0}. Moreover, it is easily seen now, that conversely any linear derivation of $\mathcal{F}_{x_0}$ defines a tangent vector on M at x_0, *there is a one–one correspondence between linear derivations of* $\mathcal{F}_{x_0}$ *defined by* (3.10) *and tangent vectors on* M *at* x_0 *defined by* (3.8), $T_{x_0}(M) = (\mathcal{F}^0_{x_0}/(\mathcal{F}^0_{x_0})^2)^*$ *is the dual space to* $\mathcal{F}^0_{x_0}/(\mathcal{F}^0_{x_0})^2$.

Also from (3.8), $(\lambda X_{x_0} + \mu Y_{x_0})F = \lambda X_{x_0}F + \mu Y_{x_0}F$, and therefore dF_{x_0} is a linear functional on the tangent vector space: $dF_{x_0} : T_{x_0}(M) \to \mathbb{R} : X_{x_0} \mapsto \langle dF_{x_0}, X_{x_0}\rangle \in \mathbb{R}$ or $dF_{x_0} \in T^*_{x_0}(M)$ where the **cotangent space** $T^*_{x_0}(M)$ on M in the point x_0 is the dual to the tangent space $T_{x_0}(M)$. The differentials dF_{x_0} form the **cotangent vectors** on M at the point x_0. As the dual of the real n-dimensional tangent vector space, the cotangent vector space $T^*_{x_0}(M)$ has the same dimension $n = \dim M$. Both vector spaces are isomorphic to $\mathbb{R}^n$ *as a vector space, not in general as a Euclidean space; tangent and cotangent vectors are carefully to be distinguished.* While tangent and cotangent vectors have a well defined meaning independent of a given local coordinate system, angles between two tangent vectors or between two cotangent vectors are not defined independent from local coordinates.

Given a local coordinate system centered at x_0 and the corresponding functions $F^i_\alpha(x) = x^i_\alpha$, the respective differentials denoted by dx^i_α form the base of the cotangent vector space dual to the base $\{\partial/\partial x^i_\alpha\}$:

$$\langle dx^i_\alpha, \partial/\partial x^j_\alpha\rangle = \delta^i_j, \quad dx^i_\beta = \sum_j dx^j_\alpha (\psi_{\beta\alpha})^i_j, \quad \frac{\partial}{\partial x^i_\beta} = \sum_j (\psi^{-1}_{\beta\alpha})^j_i \frac{\partial}{\partial x^j_\alpha}. \tag{3.11}$$

With respect to that local coordinate system,

$$dF_{x_0} = \sum_i \frac{\partial F_\alpha}{\partial x^i_\alpha}\bigg|_0 dx^i_\alpha, \quad \langle dF_{x_0}, X_{x_0}\rangle = X_{x_0}F, \quad X_{x_0} = \sum_i \xi^i_\alpha \frac{\partial}{\partial x^i_\alpha}. \tag{3.12}$$

Hence, the components ω^α_i of a general cotangent vector with respect to the base $\{dx^i_\alpha\}$,

$$\omega_{x_0} = \sum_i \omega^\alpha_i dx^i_\alpha, \tag{3.13}$$

transform between local coordinate systems as a covariant vector and the base vectors themselves transform like a contravariant vector. Equations (3.11, 3.12) together with the transformation rules for the components,

$$\omega^\beta_i = \sum_j \omega^\alpha_j (\psi^{-1}_{\beta\alpha})^j_i, \quad \xi^i_\beta = \sum_j (\psi_{\beta\alpha})^i_j \xi^j_\alpha, \tag{3.14}$$

completely determine the calculus with tangent and cotangent vectors.

3.4 Vector Fields

In the previous section, local entities on (smooth) manifolds M were considered which depend on the local structure of the manifold only. To this end, germs $[F]$ of functions F were introduced and their directional derivatives as the application of tangent vectors as well as their differentials as cotangent vectors containing the information on all directional derivatives of F at x_0 (see 3.12).

Now, global entities are introduced which have a meaning on the whole manifold M. The relation between the local entities and those global ones can be highly non-trivial and depends on the properties of the manifold itself. The study of those interrelations is one of the central tasks of the theory of manifolds.

A **smooth real function** on the manifold M is a smooth mapping $F: M \to \mathbb{R}$, considered as a smooth mapping between the manifolds M and $\mathbb{R}$ (see the end of Sect. 3.2). Since the real variable $t \in \mathbb{R}$ forms a local (and global, atlas of a single chart) coordinate on the real line $\mathbb{R}$ as a manifold, F is smooth, iff $F_\alpha(\varphi_\alpha(x)) = F_\alpha(\boldsymbol{x}_\alpha)$ is a smooth function of the local coordinates $\boldsymbol{x}_\alpha = (x^1_\alpha, \dots, x^n_\alpha)$ for every chart $(U_\alpha, \varphi_\alpha)$ of the complete atlas of M. The class of smooth real functions on M is denoted by $\mathcal{C}(M)$. Since, contrary to $\mathcal{C}_x$, all functions of $\mathcal{C}(M)$ have the same domain of definition M, linear combinations with real coefficients and point wise products of smooth real functions are again in $\mathcal{C}(M)$. In other words, $\mathcal{C}(M)$ is a real algebra (of infinite dimension; see below and the remark on $\mathcal{F}_{x_0}$, p. 65). Clearly, if $F \in \mathcal{C}(M)$, then $F \in \mathcal{C}_x$ at every point $x \in M$. The first question that arises is whether $\mathcal{C}(M)$ is non-empty at all. The answer is positive:

Every $[F] \in \mathcal{F}_x$ at any $x \in M$ can be continued into a smooth real function $F \in \mathcal{C}(M)$, that is, there is a locally defined function $F_x \in \mathcal{C}_x$ so that $[F] = [F_x]$ and F_x can be smoothly continued onto M.

Proof Consider a coordinate neighborhood U_α of x on which some F_α is defined and smooth for which $[F] = [F_\alpha]$. Consider the open set $\boldsymbol{U}_\alpha \in \mathbb{R}^n$. Since open cubes form a base of topology for the $\mathbb{R}^n$, there is an open cube $\boldsymbol{V}_\alpha$ the closure of which is contained in $\boldsymbol{U}_\alpha$ and another open cube $\boldsymbol{W}_\alpha$ the closure of which is in $\boldsymbol{V}_\alpha$ ($\mathbb{R}^n$ is a regular topological space). Let $W_\alpha = \varphi_\alpha^{-1}(\boldsymbol{W}_\alpha)$. Then, $[F] = [F_x]$ for $F_x = F_\alpha|_{W_\alpha}$. Let G^α be a smooth function, defined on $\boldsymbol{U}_\alpha$, which is equal to unity on $\boldsymbol{W}_\alpha$ and zero outside $\boldsymbol{V}_\alpha$ (see p. 34). Denote the corresponding function on $U_\alpha \subset M$ by G. Let F be equal to $F_\alpha G$ (point wise multiplication) on U_α and equal to zero on $M \backslash U_\alpha$. Obviously $F \in \mathcal{C}(M)$ and F smoothly continues $[F]$: F is smooth on U_α and every point $x \notin U_\alpha$ has a coordinate neighborhood disjoint with V_α (since the closure of V_α is in U_α) on which F is zero. □

This situation is in stark contrast to the situation for analytic functions for which the possibility of a continuation onto the whole manifold strongly limits the class of admissible analytic manifolds.

A **tangent vector field** on a manifold M is a specification of a tangent vector $X_x \in T_x(M)$ at every point x of M. For every smooth real function F on M, the tangent vector field defines another real function XF on $M : (XF)(x) = X_xF$. (X defines a real function even for all functions F for which F_α is C^1 for every local coordinate system centered at any point x in M; in this treatise only smooth functions are, however, considered.) A tangent vector field is called a **smooth tangent vector field**, if XF is smooth for every smooth function F, that is, $X : \mathcal{C}(M) \to \mathcal{C}(M)$. Since smoothness is a local property, for tangent vector fields it can again be expressed with the help of local coordinate systems: X is smooth, iff for every local coordinate system the components $\xi^i_\alpha(\boldsymbol{x}_\alpha) = \xi^i_\alpha(\varphi_\alpha(x))$ of $X = \sum_i \xi_\alpha(\boldsymbol{x}_\alpha)(\partial/\partial x^i_\alpha)$ are smooth functions of the local coordinates $\boldsymbol{x}_\alpha = (x^1_\alpha, \dots, x^n_\alpha)$. It is clear that this is necessary and sufficient for $XF = \sum_i \xi^i_\alpha(\boldsymbol{x}_\alpha)(\partial F/\partial x^i_\alpha)$ to be smooth for every smooth F. Moreover,

$$\begin{aligned} &1. \quad X(\lambda F + \mu G) = \lambda XF + \mu XG, \quad \lambda, \mu \in \mathbb{R} \\ &2. \quad X(FG) = (XF)G + F(XG), \end{aligned} \tag{3.15}$$

that is, X is a linear derivation of the algebra $\mathcal{C}(M)$.

Consider the set $\mathcal{X}(M)$ of all smooth tangent vector fields on M. The question whether it is non-empty is answered in the same way as for $\mathcal{C}(M)$, this time for each component of X with respect to a local coordinate system. $\mathcal{X}(M)$ is obviously a real vector space with respect to point wise addition of tangent vector fields and multiplication of tangent vector fields by real numbers. Point wise multiplication of tangent vector fields in the sense of multiplication of differential operators, however, does in general not lead again to a tangent vector field. (Check it.) Nevertheless, if X and Y are two smooth tangent vector fields, then the **commutator**

$$[X, Y] = XY - YX \in \mathcal{X}(M) \quad \text{for } X, Y \in \mathcal{X}(M) \tag{3.16}$$

is always again a tangent vector field: $\mathcal{X}(M)$ is a **Lie algebra**. The commutator or **Lie product** of vector fields has the following properties characterizing a Lie algebra:

$$\begin{aligned} &1. \quad [X, Y] = -[Y, X], \\ &2. \quad [X + Y, Z] = [X, Z] + [Y, Z], \\ &3. \quad [X, [Y, Z]] + [Y, [Z, X]] + [Z, [X, Y]] = 0. \end{aligned} \tag{3.17}$$

The last of these relations is called **Jacobi's identity**. All relations (3.16, 3.17) are easily proved by means of a local coordinate system. For instance, if on some chart (for the sake of simplicity of writing the chart index α is sometimes omitted, if no misunderstanding can arise) $X = \sum_i \xi^i(\partial/\partial x^i)$, $Y = \sum_i \eta^i(\partial/\partial x^i)$, then

$$[X, Y]F = X(YF) - Y(XF) = \sum_{ij} \left(\xi^j \frac{\partial \eta^i}{\partial x^j} - \eta^j \frac{\partial \xi^i}{\partial x^j} \right) \frac{\partial}{\partial x^i} F. \tag{3.18}$$

(The terms with second derivatives of F cancel in the commutator, they prevent a simple product from being a vector field. Exercise: Show that if X and Y obey

(3.15), then $[X, Y]$ also obeys these relations while XY does in general not.) Hence, in this coordinate neighborhood,

$$[X,Y]=\sum_i \zeta^i\frac{\partial}{\partial x^i},\quad \zeta^i=\sum_j\left(\xi^j\frac{\partial\eta^i}{\partial x^j}-\eta^j\frac{\partial\xi^i}{\partial x^j}\right). \tag{3.19}$$

The components ζ^i of $[X, Y]$ are smooth, if the ξ^i and η^i are smooth. (For $\mathcal{X}(M)$ to be an algebra, smoothness is essential; class C^m would not suffice, since then ζ^i would be only of class C^{m-1}.)

Let X be *any* linear derivation of $\mathcal{C}(M)$, that is, let X be a mapping $X:\mathcal{C}(M)\to\mathcal{C}(M)$ obeying (3.15). Consider the constant function $F\equiv 1$ on M. Then, the second relation (3.15) reads $XG=(XF)G+XG$, and it must hold for any $G\in\mathcal{C}(M)$, hence $XF=0$, and, by linearity (first relation 3.15), $XF=0$ for every $F\equiv$ const. on M. Now, let $U\subset M$ be any open set, let $\operatorname{supp}F=\overline{U}$ and let $\operatorname{supp}G=M\backslash U$. Then, $FG\equiv 0$ on M and $0=X(FG)=(XF)G+F(XG)$. Since $F=0$ on $M\setminus\overline{U}$ and $G\neq 0$ there, it follows that $\operatorname{supp}XF\subset\overline{U}=\operatorname{supp}F$ for any F. From that it follows easily that the value of XF at $x\in M$ is completely determined by the germ $[F]\in\mathcal{F}_x$ of F at x on M. Together with the equivalence of linear derivations X_x of $\mathcal{F}_x$ and tangent vectors $X_x\in T_x(M)$ this shows that *any linear derivation X of the algebra* $\mathcal{C}(M)$ *defines a tangent vector field* $X\in\mathcal{X}(M)$.

$\mathcal{X}(M)$ may also be considered as a module over the algebra (ring) $\mathcal{C}(M)$: For $X,Y\in\mathcal{X}(M)$ and $F,G\in\mathcal{C}(M)$, the linear combination $FX+GY$ is again a smooth vector field $\in\mathcal{X}(M)$ which is locally defined as $(FX+GY)(x)=F(x)X_x+G(x)Y_x$, that is, the components are $\zeta^i_\alpha(\boldsymbol{x}_\alpha)=F_\alpha(\boldsymbol{x}_\alpha)\xi^i_\alpha(\boldsymbol{x}_\alpha)+G_\alpha(\boldsymbol{x}_\alpha)\eta^i_\alpha(\boldsymbol{x}_\alpha)$. Now, one finds

$$[FX,GY]=F(XG)Y-G(YF)X+(FG)[X,Y],\quad F,G\in\mathcal{C}(M),\ X,Y\in\mathcal{X}(M) \tag{3.20}$$

by straightforward calculation of the action of $[FX, GY]$ on another smooth function H in a local coordinate system, using the second rule (3.15) (Leibniz rule).

Later on, a geometric interpretation will be given of the Lie product of tangent vector fields (Sect. 3.6).

Analogous to a tangent vector field, a cotangent vector field ω on a manifold M is a specification of a cotangent vector $\omega_x\in T^*_x(M)$ at every point $x\in M$, that is, at every point x a real linear function on the tangent space $T_x(M)$ is specified: $(\omega(X))_x=\langle\omega_x,X_x\rangle$. A cotangent vector field is smooth, if it defines a smooth real function on M for every smooth tangent vector field X. By repeating previous reasoning, ω is smooth, if for every local coordinate system centered at every point $x\in M$ the components ω^α_i of

$$\omega=\sum_i\omega^\alpha_i(\boldsymbol{x}_\alpha)dx^i_\alpha,\quad \omega(X)_x=\langle\omega,X\rangle_x=\sum_i\omega^\alpha_i(\boldsymbol{x}_\alpha)\xi^i_\alpha(\boldsymbol{x}_\alpha) \tag{3.21}$$

are smooth functions of the local coordinates $\boldsymbol{x}_\alpha = (x_\alpha^1, \ldots, x_\alpha^n)$. A smooth cotangent vector field is called a **differential 1-form** or in short a 1-form. It may also be considered as a $\mathcal{C}(M)$-linear mapping from the $\mathcal{C}(M)$-module $\mathcal{X}(M)$ into $\mathcal{C}(M)$:

$$\omega(FX + GY) = F\omega(X) + G\omega(Y) \in \mathcal{C}(M), \quad F, G \in \mathcal{C}(M), \quad X, Y \in \mathcal{X}(M) \tag{3.22}$$

which is directly seen from the second relation (3.21).

Based on this consideration, an **exterior product** (wedge product) $\omega \wedge \sigma$ of two 1-forms ω and σ may be introduced with the properties (so far $r = s = 1$)

$$\begin{aligned} &1. \quad \omega \wedge \sigma = (-1)^{rs} \sigma \wedge \omega, \\ &2. \quad \omega \wedge (F\sigma + G\tau) = F\omega \wedge \sigma + G\omega \wedge \tau, \\ &3. \quad (\omega \wedge \sigma) \wedge \tau = \omega \wedge (\sigma \wedge \tau), \end{aligned} \tag{3.23}$$

which (except for 3) defines an alternating (skew-symmetric) $\mathcal{C}(M)$-bilinear mapping from the direct product $\mathcal{X}(M) \times \mathcal{X}(M)$ into $\mathcal{C}(M)$: $(\omega \wedge \sigma)(X, Y) = (1/2)(\omega(X) \cdot \sigma(Y) - \omega(Y) \cdot \sigma(X))$. It is called an exterior differential 2-form. More generally, an **exterior differential *r*-form**, or in short an r-form, is an alternating $\mathcal{C}(M)$-r-linear mapping from the direct product $\mathcal{X}(M) \times \cdots \times \mathcal{X}(M)$ (r factors) into $\mathcal{C}(M) : (\omega_1 \wedge \cdots \wedge \omega_r)(X_1, \ldots, X_r) = (1/r!) \det(\omega_i(X_k))$ in the special case where the ω_i are 1-forms. In a coordinate neighborhood (index α dropped) the general expression of an r-form is

$$\omega = \sum_{i_1 < \cdots < i_r} \omega_{i_1 \ldots i_r}(\boldsymbol{x}) dx^{i_1} \wedge \cdots \wedge dx^{i_r}, \quad \omega = 0 \quad \text{if} \quad r > n, \tag{3.24}$$

where the $\omega_{i_1 \ldots i_r} \in \mathcal{C}(M)$. Since dx^i is a 1-form, the above determinant rule can now be applied to each item of (3.24).

With the exterior product defined by its properties (3.23), an $(r + s)$-form is obtained by wedge-multiplying an r-form with an s-form. From (3.24) it can be inferred that if ω is an r-form and $F \in \mathcal{C}(M)$, then $F\omega$ is again an r-form. On this basis, $F \in \mathcal{C}(M)$ is called a 0-form, and the real vector space $\mathcal{D}^0(M) = \mathcal{C}(M)$ is introduced together with the real vector spaces $\mathcal{D}^r(M)$ of r-forms. (For $r > n, \mathcal{D}^r(M)$ consists of the null-vector only, see Sect. 4.2.) Within this concept, $F\omega$ may be written as $F \wedge \omega$. The direct sum $\mathcal{D}(M) = \sum_{r=0}^{\infty} \mathcal{D}^r(M) = \sum_{r=0}^{n} \mathcal{D}^r(M)$ forms an **exterior algebra** which is a graded algebra, graded by the degree r of r-forms.

Recall that 0-forms are functions and 1-forms are (total) differentials of functions on M. A general **exterior differentiation** d is introduced which maps an r-form into an $(r + 1)$-form with the defining rules (using the known rule of forming dF_x at point $x \in M$)

1. dF for $F \in \mathcal{D}^0(M)$ is the total differential on M,
2. d is real-linear and $d(\mathcal{D}^r(M)) \subset \mathcal{D}^{r+1}(M)$,
3. $d(\omega \wedge \sigma) = (d\omega) \wedge \sigma + (-1)^r \omega \wedge (d\sigma), \quad \omega \in \mathcal{D}^r(M), \quad \sigma \in \mathcal{D}^s(M)$,
4. $d^2 = 0$.

$$(3.25)$$

The last rule means that a double application of d to any exterior differential form yields the null-vector, that is, the form that is identical zero on all M.

Within a coordinate neighborhood, if ω is given by (3.24), then

$$d\omega = \sum_{i_1<\cdots<i_r} d\omega_{i_1\ldots i_r} \wedge dx^{i_1} \wedge \cdots \wedge dx^{i_r}. \tag{3.26}$$

As is discussed later on (Sect. 5.1), the exterior differentiation generalizes the grad, rot (curl) and div operations of vector analysis. Note also that further on every $\mathcal{D}^r(M)$ may be understood as a $\mathcal{C}(M)$-r-linear mapping from $\mathcal{X}(M) \times \cdots \times \mathcal{X}(M)$ (r factors) into $\mathcal{C}(M)$. This is related to the scalar (contracting) product of tensors and will be generalized in the next chapter.

3.5 Mappings of Manifolds, Submanifolds

At the end of Sect. 3.2 the concept of smooth mappings of manifolds into each other was introduced. A smooth mapping $F: M \to N$ of a manifold M into a manifold N induces at every point $x \in M$ a linear mapping $F_*^x : T_x(M) \to T_{F(x)}(N)$ of the tangent space on M at point x into the tangent space on N at point $F(x)$. F_*^x is called the **push forward** or the **tangent map** of the mapping F at point x.

For any tangent vector $X_x \in T_x(M)$ its image $F_*^x(X_x) \in T_{F(x)}(N)$ is formed in the following natural way: Let G be a smooth real function on N in a neighborhood of $F(x)$. Then, $G \circ F$ is a smooth real function on M in a neighborhood of x. For every G, by definition,

$$(F_*^x(X_x))G = X_x(G \circ F). \tag{3.27}$$

This definition ensures the following: Given any parametrized curve through x in M, it is mapped by F into a parametrized curve through $F(x)$ in N (which could degenerate in the point $F(x)$ only, if F is constant along the curve in M). The directional derivative at point $F(x)$ along the curve in N of any real function G on N is obtained as the directional derivative at point x along the corresponding curve in M of the real function $G \circ F$. (If F is constant along the considered curve in M, this directional derivative is zero no matter what G in (3.27) is. Hence, (3.27) means in that case that the projection of the tangent vector $F_*^x(X_x)$ onto the direction of the considered curve in N is zero.) Because of this interpretation the mapping F_*^x is also called the **differential** at x of the mapping F of the manifold M into the manifold N.

Now, the natural question arises, given tangent vector fields X on the manifold M, under which conditions do the tangent mappings F_*^x for all $x \in M$ result in a mapping F_* of tangent vector fields X on M to tangent vector fields Y on the manifold N. This is obviously not the case, if F is not onto, because then the mapping would not define a tangent vector field Y on all N. Even if F is surjective but not injective, if for instance $F(x) = F(x') = y$ for $x \neq x'$, then any tangent vector field X with different vectors at x and x' would not give a uniquely defined result at $y \in N$ and hence not define a tangent vector field Y on N. Obviously, F must be onto and one–one, that is, it must be a bijection of manifolds in order that F_* may be defined as a push forward of F to a mapping of tangent vector fields to tangent vector fields. But even then, the image by F_* of a smooth tangent vector field need not be smooth again. Consider for example $M = N = \mathbb{R}$ and $(F : \mathbb{R} \to \mathbb{R} : x \mapsto y = x^3) \in C^\infty(\mathbb{R}, \mathbb{R})$. Take the smooth (constant) tangent vector field $X_x = \partial/\partial x$ and a smooth real function $G : y \mapsto G(y)$. One has $Y_y G = (F_*^x(X_x))G = X_x(G \circ F) = (\partial/\partial x)G(x^3) = 3x^2 \partial G/\partial y = 3y^{2/3} \partial G/\partial y$. Now, $Y_y = F_*^x(X_x) = 3y^{2/3}\partial/\partial y$ is not smooth at $y = 0$.

By duality, another linear mapping $F^*_{F(x)} : T^*_{F(x)}(N) \to T^*_x(M)$ of the cotangent space on N at point $F(x)$ to the cotangent space on M at point x is obtained, defined so that for every $X_x \in T_x(M)$ the relation

$$(F^*_{F(x)}(\omega_{F(x)}))(X_x) = \langle F^*_{F(x)}(\omega_{F(x)}), X_x \rangle = \langle \omega_{F(x)}, F_*^x(X_x) \rangle = \omega_{F(x)}(F_*^x(X_x)), \tag{3.28}$$

holds where $\omega_{F(x)} \in T^*_{F(x)}(N)$ is a cotangent vector (1-form) on N at point $F(x)$ and $F^*_{F(x)}(\omega_{F(x)}) \in T^*_x(M)$ is the corresponding cotangent vector on M at point x. $F^*_{F(x)}$ is called the **pull back** of F at x. As is easily seen (next page), this time *for every smooth mapping $F : M \to N$ there is a mapping F^* which maps 1-forms on N to 1-forms on M so that smooth 1-forms are mapped to smooth 1-forms*. In Chap. 7 all (co)tangent spaces of a smooth manifold M will be glued together to form another smooth manifold which is called the (co)tangent bundle $(T^*(M))$ $T(M)$ on M. The mapping F^* of the cotangent bundle $T^*(N)$ to the cotangent bundle $T^*(M)$ is called the ***pull back*** by the smooth mapping F of M to N.

Now, let $F : M \to N$ be a diffeomorphism of manifolds, that is, $F^{-1} : N \to M$ is also smooth. Then, one can pull back 1-forms from M to N by $(F^{-1})^*$ which by duality between tangent vector fields and 1-forms means also to push forward smooth tangent vector fields on M to smooth tangent vector fields on N. Then, *for a diffeomorphism $F : M \to N$ of manifolds F^* is a mapping from the tangent bundle $T(M)$ to the tangent bundle $T(N)$ which is called the **push forward** by the diffeomorphism F of M onto N.*

Consider the mappings $F_*^{x_0}$ and $F^*_{F(x_0)}$ in terms of local coordinates. Choose local coordinate systems of charts $(U_\alpha, \varphi_\alpha) \in \mathcal{A}_M$ and $(U_\beta, \varphi_\beta) \in \mathcal{A}_N$ with local coordinates $x^i_\alpha = \pi^i_\alpha \circ \varphi_\alpha(x), x \in M$ and $y^j_\beta = \pi^j_\beta \circ \varphi_\beta(y), y \in N$, where U_α is a coordinate neighborhood of $x_0 \in M$ and U_β is a coordinate neighborhood of

$F(x_0) \in N$, both neighborhoods chosen such that $F(U_\alpha) \subset U_\beta$. The mapping F induces a real vector function $\boldsymbol{F}_{\beta\alpha} = \boldsymbol{\varphi}_\beta \circ F|_{U_\alpha} \circ \boldsymbol{\varphi}_\alpha^{-1}$ of n_M real variables by the following commutative diagram:

$$\begin{array}{ccc} M \supset U_\alpha & \xrightarrow{F|_{U_\alpha}} & U_\beta \subset N \\ \varphi_\alpha \big\downarrow & & \big\downarrow \varphi_\beta \\ \boldsymbol{U}_\alpha & \xrightarrow[\boldsymbol{F}_{\beta\alpha}]{} & \boldsymbol{U}_\beta \end{array}$$

It consists of n_N real functions $F^j_{\beta\alpha} = \pi^j_\beta \circ \boldsymbol{F}_{\beta\alpha}$,

$$y^j_\beta = F^j_{\beta\alpha}(x^1_\alpha, \ldots, x^{n_M}_\alpha), \quad j = 1, \ldots, n_N, \tag{3.29}$$

of n_M real variables. Any real function G on N generates a real function $G_\beta = G|_{U_\beta} \circ \boldsymbol{\varphi}_\beta^{-1} = G_\beta(y^1, \ldots, y^{n_N})$ of n_N real variables y^j_β and a real function $(G \circ F)_\alpha = G|_{U_\beta} \circ \boldsymbol{\varphi}_\beta^{-1} \circ \boldsymbol{\varphi}_\beta \circ F|_{U_\alpha} \circ \boldsymbol{\varphi}_\alpha^{-1} = G_\beta(y^1_\beta(\boldsymbol{x}_\alpha), \ldots, y^{n_N}_\beta(\boldsymbol{x}_\alpha))$ of n_M real variables x^i_α. Now take the base vectors $X_{\alpha i} = \partial/\partial x^i_\alpha$ of the vector space $T_{x_0}(M)$ and find

$$(F^{x_0}_*(X_{\alpha i}))G = \frac{\partial}{\partial x^i_\alpha}(G \circ F)_\alpha = \frac{\partial}{\partial x^i_\alpha} G_\beta(F^1_{\beta\alpha}(\boldsymbol{x}_\alpha), \ldots, F^{n_N}_{\beta\alpha}(\boldsymbol{x}_\alpha)) = \sum_j \frac{\partial G_\beta}{\partial y^j_\beta} \frac{\partial F^j_{\beta\alpha}}{\partial x^i_\alpha},$$

which means

$$F^{x_0}_*\left(\frac{\partial}{\partial x^i_\alpha}\right) = \sum_{j=1}^{n_N} \frac{\partial F^j_{\beta\alpha}}{\partial x^i_\alpha} \frac{\partial}{\partial y^j_\beta}, \tag{3.30}$$

that is, the images of the base vectors $\partial/\partial x^i_\alpha$ have components $\partial F^j_{\beta\alpha}/\partial x^i_\alpha$ with respect to the base vectors $\partial/\partial y^j_\beta$ of the tangent space $T_{F(x_0)}$, or in other words, the matrix of the linear mapping $F^{x_0}_*$ (as matrix transformation of the vector components) is the transposed of $(\partial F^j_{\beta\alpha}/\partial x^i_\alpha)$, the Jacobian matrix of the transformation $\boldsymbol{y}_\beta(\boldsymbol{x}_\alpha)$. For a diffeomorphism F, the derivatives on the right hand side can smoothly be expressed by derivatives with respect to $\boldsymbol{y}$ to yield a smooth vector field on N.

Taking a base covector $\omega^j_{F(x_0)} = dy^j_\beta \in T^*_{F(x_0)}(N)$, and a base vector $\partial/\partial x^i_\alpha \subset T_{x_0}(M)$, (3.28, 3.30) and (3.11) yield

$$\begin{aligned}
(F^*_{F(x_0)}(\omega^j_{F(x_0)}))\left(\frac{\partial}{\partial x^i_\alpha}\right) &= \omega^j_{F(x_0)}\left(F^{x_0}_*\left(\frac{\partial}{\partial x^i_\alpha}\right)\right) \\
&= \omega^j_{F(x_0)}\left(\sum_k \frac{\partial F^k_{\beta\alpha}}{\partial x^i_\alpha}\frac{\partial}{\partial y^k_\beta}\right) = \left\langle dy^j_\beta, \sum_k \frac{\partial F^k_{\beta\alpha}}{\partial x^i_\alpha}\frac{\partial}{\partial y^k_\beta}\right\rangle = \frac{\partial F^j_{\beta\alpha}}{\partial x^i_\alpha},
\end{aligned}$$

that is,

$$F^*_{F(x_0)}(dy^j_\beta) = \sum_{i=1}^{n_M} \frac{\partial F^j_{\beta\alpha}}{\partial x^i_\alpha} dx^i_\alpha. \tag{3.31}$$

The mapping $F^*_{F(x_0)}$ is dual to the mapping $F^{x_0}_*$ between tangent spaces: it is in the opposite direction and between the duals of the tangent spaces and its matrix is the transposed to the matrix of the mapping $F^{x_0}_*$. For every smooth $\omega = \sum \omega_j(\mathbf{y})dy^j$ (3.31) together with the smooth function $\mathbf{y} = \mathbf{F}(\mathbf{x})$ yields a smooth 1-form on M.

If $F : M \to N$ and $G : N \to P$, then for the composite mapping $G \circ F : M \to P$ the mappings of tangent and cotangent spaces are $(G \circ F)^x_* = G^{F(x)}_* \circ F^x_*$ and $(G \circ F)^*_{G(F(x))} = F^*_{F(x)} \circ G^*_{G(F(x))}$, that is, F_* composes covariantly with F, and F^* contravariantly. (This is expressed by push forward and pull back.)

The mapping (3.28) may be generalized to r-forms at point $F(x)$:

$$(F^*_{F(x)}(\omega^r_{F(x)}))(X_{1x}, \ldots, X_{rx}) = \omega^r_{F(x)}(F^x_*(X_{1x}), \ldots, F^x_*(X_{rx})). \tag{3.32}$$

The expressions in local coordinates are directly obtained from (3.24) and (3.31). Hence, F^* is also a linear mapping from $\mathcal{D}(N)$ into $\mathcal{D}(M)$ (pull back).

A simple result is the following [4]:

Let M be a connected manifold and let $F : M \to N$ be such that $F^x_* = 0$ at every point $x \in M$. Then F is a constant map.

Proof Since M is connected, it is the only non-empty subset of M which is open and closed. Fix some point $y \in F(M) \subset N$. $F^{-1}(y)$ is closed as the preimage of a closed set in a continuous mapping. Choose coordinate neighborhoods of some $x \in F^{-1}(y)$ and of y. Since $\partial F^j_{\beta\alpha}/\partial x^i_\alpha = 0$ at every $x \in U_\alpha$, F is constant in U_α which is open. Since $x \in F^{-1}(y)$ was chosen arbitrarily, $F^{-1}(y)$ is open and closed, hence it is M. □

In a certain sense the opposite case is governed by the following **inverse function theorem**:

Let $F : M \to N$ and let $x_0 \in M$ be some point in the manifold M.

1. *If $F^{x_0}_*$ is injective (one–one), then there exists a local coordinate system $x^1_\alpha, \ldots, x^{n_M}_\alpha$ in a coordinate neighborhood U_α of $x_0 \in M$ and a local coordinate system $y^1_\beta, \ldots, y^{n_N}_\beta$ in a coordinate neighborhood of $F(x_0) \in N$ so that $y^i_\beta(F(x)) = x^i_\alpha(x)$ for all $x \in U_\alpha$ and $i = 1, \ldots, n_M$ and $F|_{U_\alpha} : U_\alpha \to F(U_\alpha)$ is a diffeomorphism of manifolds (one–one and onto).*
2. *If $F^{x_0}_*$ is surjective (onto), then there exists a local coordinate system $x^1_\alpha, \ldots, x^{n_M}_\alpha$ in a coordinate neighborhood U_α of $x_0 \in M$ and a local coordinate system $y^1_\beta, \ldots, y^{n_N}_\beta$ in a coordinate neighborhood of $F(x_0) \in N$ so that $y^i_\beta(F(x)) = x^i_\alpha(x)$ for all $x \in U_\alpha$ and $i = 1, \ldots, n_N$ and $F|_{U_\alpha} : U_\alpha \to N$ is an open mapping. (It maps open sets to open sets.)*

3. *If $F_*^{x_0}$ is a linear isomorphism from $T_{x_0}(M)$ to $T_{F(x_0)}(N)$, then F defines a diffeomorphism of some coordinate neighborhood of $x_0 \in M$ to some coordinate neighborhood of $F(x_0) \in N$.*

The last statement means that $F|_{U_\alpha}$ has a smooth inverse function $(F|_{U\alpha})^{-1}$: $U_\beta \to U_\alpha$. Since for $n_M = n_N = n$, local coordinates translate F into a mapping $\boldsymbol{F}_{\beta\alpha} = \boldsymbol{\varphi}_\beta \circ F|_{U_\alpha} \circ \boldsymbol{\varphi}_\alpha^{-1} : \boldsymbol{U}_\alpha \to \boldsymbol{U}_\beta$ from an open set of $\mathbb{R}^n$ into an open set of $\mathbb{R}^n$, the push forward $F_*^{x_0}$ to be a linear isomorphism means a non-zero Jacobi determinant of the mapping $\mathbf{F}_{\beta\alpha}$ at $\boldsymbol{\varphi}_\alpha(x_0)$. Case 3 immediately follows from the well known inverse function theorem of calculus (see any textbook of Analysis, e.g. [5]). The cases 1 and 2 then follow easily also from the corresponding variants of calculus.

If F is a smooth mapping of a manifold M into a manifold N (recall that all manifolds in this volume are supposed smooth), for which F_*^x is injective at every point $x \in M$, then F is called an **immersion**. One also says that M is immersed into N by F. $F(M)$ is locally diffeomorphic to M ($F(U_\alpha)$ is diffeomorphic to $U_\alpha \in M$ for sufficiently small U_α), but F is not necessarily globally injective: there may by self-intersections of $F(M)$ so that $F(M)$ is not necessarily a manifold. (See examples below.)

If $F : M \to N$ itself is additionally injective, then F is called an **embedding** and (M, F) is called an **embedded submanifold** of N. M is embedded into N by F.

Great care is needed to distinguish the topology of the embedding (M, F) from $F(M)$ as a subsct of N with its relative topology. Except for open submanifolds defined earlier and closed submanifolds, both considered below in more detail, the topology of an embedded submanifold is in general different from the relative topology of $F(M)$ as a subset of the topological space N: it is in general finer. The point is that embedded submanifolds are understood to inherit their complete atlases from M: they are generated by charts $(F(U_\alpha), \varphi_\alpha \circ F^{-1}|_{F(U_\alpha)})$ for $U_\alpha \in M$ small enough so that U_α and $F(U_\alpha)$ are diffeomorphic. (Some authors, e.g. Warner, use a slightly more special terminology of embedding.)

Examples

Open submanifolds of N: $M \subset N$ is open in N and $F = \mathrm{Id}_M$. Its manifold structure (atlas) was considered previously on p. 60. The topology of M as a topological space is the relative topology as a subspace of N. Note that although M is open in the topology of N, it is open and closed in the relative topology (as every topological space as a whole is open and closed by definition of topology.) Since $F_*^x = \mathrm{Id}_{T_x}$ for every $x \in M$, the dimension of M is always that of N.

Closed submanifolds of N: Let $G^i : N \to \mathbb{R}$, $i = 1, \ldots, k$ and $M = \cap_i (G^i)^{-1}(0)$, that is $M \subset N$ is the set of all points $x \in N$ for which $G^i(x) = 0$, $i = 1, \ldots, k$. Suppose $dG_x^1, \ldots, dG_x^k$ linearly independent in a neighborhood of M. Then M is a closed subset of N and (M, Id_M) is a closed submanifold of N of dimension $\dim N - k$. Again the topology of M is the relative topology as a subspace of N. For $k = 1$, M is called a **hypersurface**.

Fig. 3.5 The immersed submanifold of $N = \mathbb{R}^2$ of Example 3

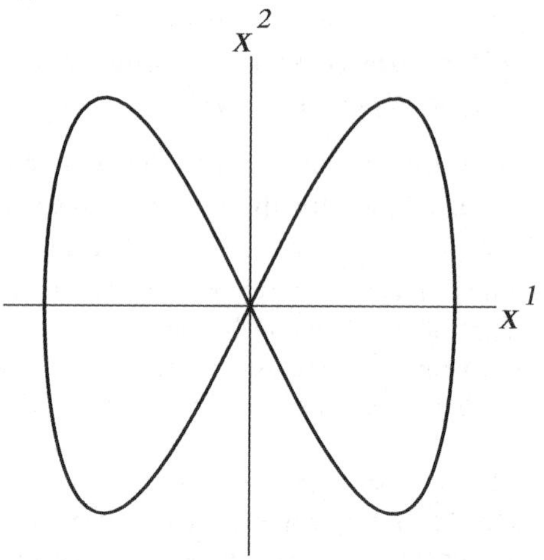

Example 3 Let $M = \{t \mid 0 \leq t \leq 2\pi \equiv 0\}$ (closed loop of length 2π) $N = \mathbb{R}^2$, and $F : M \rightarrow N : t \mapsto F(t) = (\cos t, \sin 2t)$ (see Fig. 3.5). It is an immersion since it is self-intersecting at (0, 0) $\in N$. Note that (M, F) is not a manifold since it inherits charts for each of the two branches through (0, 0) implying different tangent spaces at the same point (0, 0). It is also not a submanifold of $\mathbb{R}^2$ in the relative topology, since a neighborhood of (0, 0) is not homeomorphic to an open set of $\mathbb{R}$. In (M, F), pieces of the two branches through the origin (0, 0) are open sets (since charts are open sets) while in the relative topology induced from $N = \mathbb{R}^2$ only pieces of both branches together are open sets (intersections of $F(M)$ with open sets of the plane). Hence, the topology of $F(M)$ as an immersion is finer (has more open sets) than the relative topology in N.

Example 4 M and N as in Example 3, and $F : t \mapsto F(t) = (\cos t, \sin t)$ (see Fig. 3.6). M is just the unit circle in the plane N. It is an embedded submanifold since this time $F : M \rightarrow N$ is an injection. It is also a closed submanifold (one-dimensional 'hypersurface') given by $G(x^1, x^2) = (x^1)^2 + (x^2)^2 - 1 = 0$. Note that as a topological space itself and also in the relative topology induced from N, $F(M)$ is closed and also open. (It is the intersection of $F(M)$ with an open set of N.)

Example 5 $M = \{t \mid 0 < t < 2\pi\}, N = \mathbb{R}^2$, and $F : t \mapsto (\sin t, \sin 2t)$. It looks like in Fig. 3.5, but this time it is an embedded submanifold since the origin of N is only the image of $t = \pi$. There is no continuous branch from left to right upwards through the origin of N. Hence, there is only one tangent space on (M, F) at (0, 0) from right to left upwards. Pieces of this branch containing (0, 0) are open sets of (M, F) but not of $F(M)$ which is the same as in Example 3. Again the topology of (M, F) is finer than the relative topology of $F(M) \subset N$.

The discussion of the various topologies leads to a natural definition: If (M, F) is an embedded submanifold of N and $F(M) \subset N$ with the relative topology is homeomorphic to M, then (M, F) is called a **regular embedding** of M into N.

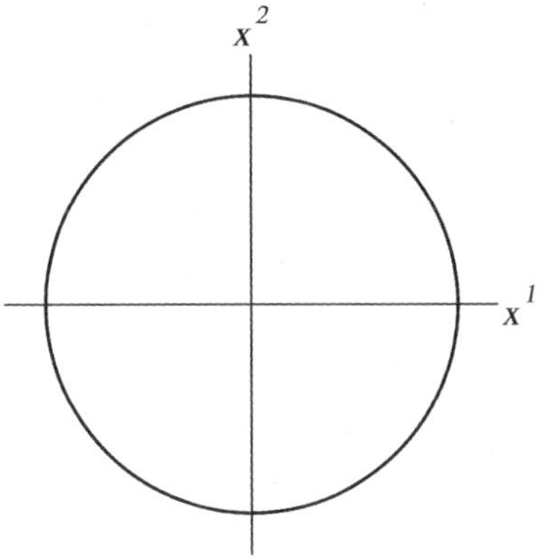

Fig. 3.6 The embedded submanifold of $N = \mathbb{R}^2$ of Example 4

(M, F) is a regular embedding, iff it is a closed submanifold of an open submanifold of N.

Also:

If (M, F) is an embedded submanifold of N and M is compact, then (M, F) is a regular embedding.

See e.g. [3] for proofs.

It can be shown that, if only the structure of a smooth manifold is observed, then any n-dimensional manifold can be embedded as a submanifold into the $\mathbb{R}^{2n+1}$.

Here, a general comment is in due place: A circle in $\mathbb{R}^3$ and a loop with a knot in it are homeomorphic and homotopy equivalent. However, they cannot continuously be deformed into each other *by only homeomorphic maps*: In order to open the knot either the loop must be cut or at a stage of deformation it must be self-intersecting. The same holds true for two linked circles (into a piece of chain) and two unlinked circles. Knots and links are properties of embeddings of loops into higher dimensional spaces, not of loops as such.

3.6 Frobenius' Theorem

A very important issue is the interrelation of smooth tangent and cotangent vector fields and smooth mappings of manifolds. Again, only smooth entities are considered in the sequel and the adjective smooth is dropped throughout. As considered in the last section, given a tangent vector field X on a manifold M and a bijective mapping F of M into N, $F_*(X)$ that defines a tangent vector $Y_{F(x)} = F_*^x(X_x)$ at every point $F(x) \in F(M) \subset N$, need not be a tangent vector field: $Y = F_*(X)$ need not be smooth in a neighborhood of points $F(x)$ for which $Y_{F(x)} = 0$. For a tangent vector field X on M, a point $x \in M$ for which $X_x = 0$ is called a **singular point** of X.

Let X be a tangent vector field on M and let $X_{x_0} \neq 0$, that is, $x_0 \in M$ is a non-singular point of X. Then there exists a local coordinate system $(U_\alpha, \varphi_\alpha)$ centered at x_0 in which $X|_{U_\alpha} = \partial/\partial x^1$.

(Since in this section the coordinate neighborhood is always denoted U_α, the index α at local coordinates is dropped as in $x^1 = x^1_\alpha$.) The technical proof by standard analysis of this natural proposition is skipped, see for instance [3].

Let F be a mapping from M into N. The tangent vector fields X on M and Y on N are called **F-related** tangent vector fields, if $F^x_*(X_x) = Y_{F(x)}$ for every $x \in M$.

Let $F : M \to N$ and let X_1, X_2 be tangent vector fields on M and Y_1, Y_2 tangent vector fields on N. If X_i and Y_i, $i = 1, 2$, are F-related, then $[X_1, X_2]$ and $[Y_1, Y_2]$ are F-related.

Apply straightforwardly (3.18) and (3.27) (exercise).

More interesting is the following problem: Given a set of tangent vector fields on N, is there a submanifold of N for which these vectors span the tangent space at every point? Let N be an n-dimensional manifold and m, $1 \le m \le n$, an integer. A selection of an m-dimensional subspace D_x of the tangent space $T_x(N)$ at every point $x \in N$ is called a (smooth) **distribution** D on N, if every point $x_0 \in N$ has a neighborhood U and m tangent vector fields $X_1, \ldots, X_m$ of which the tangent vectors $X_{1x}, \ldots, X_{mx}$ span D_x for every $x \in U$. The tangent vector fields $X_1, \ldots, X_m$ are said to form a **local base of the distribution** D. A tangent vector field X on N is said to **belong to a distribution** D, if $X_x \in D_x$ at every $x \in N$. A distribution D is called **involutive**, if whenever the tangent vector fields X and Y belong to D then also $[X, Y]$ belongs to D. Finally, a connected submanifold (M, F) of N is called an **integral manifold** of a distribution D on N, if $F^x_*(T_x(M)) = D_{F(x)}$ for every $x \in M$, that is, at every point $F(x)$ the vector space $D_{F(x)}$ is the tangent space on $F(M)$.

The solution to the problem posed above is now given by the generalization to manifolds of the **Frobenius theorem** of classical analysis:

Let D be an m-dimensional distribution on the n-dimensional manifold N, $1 \le m \le n$. There is a uniquely defined maximal connected (even pathwise connected) integral manifold (M_x, F_x) through every point $x \in N$, iff D is involutive: Every connected integral manifold of D through x is an open submanifold of (M_x, F_x).

Of course, the case $m = 1$ is special. In this case, D is just given by a tangent vector field which is nowhere singular (since D is one-dimensional at every point $x \in N$). Moreover, since trivially $[X, X] = 0$, a non-singular tangent vector field yields always an involutive one-dimensional distribution. A one-dimensional submanifold is a parametrized curve, it is called an **integral curve** of X, if it is an integral manifold of $D = \{\lambda X | \lambda \in \mathbb{R}\}$. Consider an integral curve through $x \in N$. There may be chosen an open interval $M = \{t \,|\, a<t<b\} \subset \mathbb{R}$ of the real line (a may be $-\infty$ and b may be ∞) containing $t = 0$ and a mapping F: $M \to N$ so that (M, F) is the integral curve of X in N through $x_0 = F(0)$. It was stated above that for every X there is a coordinate neighborhood $(U_\alpha, \varphi_\alpha)$ of x_0 so that $X|_{U_\alpha} = \partial/\partial x^1$. It is easily seen that

$$F_\alpha(M) \cap \boldsymbol{U}_\alpha = \{(x^1, 0, \ldots, 0)\} \cap \boldsymbol{U}_\alpha \tag{3.33}$$

represents the integral curve of X in that coordinate neighborhood and that it is unique in U_α. To prove the Frobenius theorem for $m = 1$, it remains to prove that the integral curve (3.33) is contained in a maximal integral curve. Order all possible domains $M \subset \mathbb{R}$ of integral curves through x partially by inclusion. The existence of a maximal element follows from Zorn's lemma.

Proof of the Frobenius theorem by induction Consider a local base $X_1, \ldots, X_m$ of D. The base vectors as tangent vectors on M_x and on N, respectively, are F_x-related. Hence, the vectors $[X_i, X_j]$ are also F_x-related, which proves necessity in the theorem. Sufficiency was proved for $m = 1$ above. Assume it holds for $m - 1$, and assume that for every involutive D' of dimension $m - 1$ and for every $x_0 \in N$ there is a local coordinate system $U_\alpha \subset N$ so that D' is spanned by $\partial/\partial x^1, \ldots, \partial/\partial x^{m-1}$ and hence $F'_{\alpha x_0}(M'_{x_0}) \cap \boldsymbol{U}_\alpha = \{(x^1, \ldots, x^{m-1}, 0, \ldots, 0)\} \cap \boldsymbol{U}_\alpha$.

Given $x_0 \in N$, there exists a local coordinate system U_α centered at x_0 and such that $X_m = \partial/\partial x^m$. The vectors $X_1, \ldots, X_{m-1}$ span an $(m-1)$-dimensional distribution D'. Let i, j, k run from 1 to $m - 1$. Let $X'_i = X_i - (X_i \varphi^m_\alpha) X_m$, then $X_m \varphi^m_\alpha = 1$, $X'_i \varphi^m_\alpha = 0$. In view of the involutivity of D, $[X'_i, X'_j] = \sum_k c^k_{ij} X'_k + d_{ij} X_m$, and from $[X'_i, X'_j]\varphi^m_\alpha = 0$ it follows $d_{ij} = 0$, that is, D' is involutive. Therefore, assuming X_m linearly independent of D' (otherwise nothing is to be proved), there exist local coordinates $y^1, \ldots, y^{m-1}$ in $M'_{x_0} \cap U_\alpha$ so that $X'_i = \partial/\partial y^i$. Again by the involutivity of D, $[\partial/\partial y^i, X_m] = \sum_k c^k_i \partial/\partial y^k$. (A term with X_m does not appear on the right hand side since further on $\partial x^m/\partial y^i = X'_i \varphi^m_\alpha = 0$.)

Now complete the coordinates y^i to a local coordinate system $y^1, \ldots, y^n$ in N. Then, $X_m = \sum_{l=1}^n \xi^l (\partial/\partial y^l)$ with certain functions ξ^l defined in U_α. This implies $[(\partial/\partial y^i), X_m] = \sum_{l=1}^n (\partial \xi^l/\partial y^i)(\partial/\partial y^l)$, and comparison with the c^k_i above yields $\partial \xi^l/\partial y^i = 0$ for $m \le l \le n$. Put $X'_m = \sum_{l=m}^n \xi^l (\partial/\partial y^l)$ for which still D is spanned by $\partial/\partial y^i, X'_m$. In the submanifold of U_α of points with coordinates $y^i = 0$, $i = 1, \ldots, m-1$, there are new coordinates $y'^l, m \le l \le n$, so that $X'_m = \partial/\partial y'^m$. Hence, for every $x_0 \in N$ there is a coordinate neighborhood U_α so that D is spanned by $\partial/\partial z^1, \ldots, \partial/\partial z^m, z^i = y^i, z^m = y'^m$, so that $F_{\alpha x_0}(M_{x_0}) \cap \boldsymbol{U}_\alpha = \{(z^1, \ldots, z^m, 0, \ldots, 0)\} \cap \boldsymbol{U}_\alpha$.

The existence of a maximal integral manifold (M_x, F_x) is proved by introduction of a partial order in the set of integral manifolds similarly as in the one-dimensional case. □

There is a dual variant of Frobenius' theorem which is equally important. Given an m-dimensional distribution D on N which in a neighborhood U_α of the point $x_0 \in N$ is spanned by the m tangent vector fields $X_1, \ldots, X_m$ and which defines an m-dimensional subspace D_x of each tangent space $T_x(N)$ in that neighborhood, for any $x \in U_\alpha$ there is an $(n-m)$-dimensional **annihilator subspace** of $T^*_x(N)$,

$$D^\perp_x = \{\omega_x \in T^*_x(N) | \langle \omega_x, X_x \rangle = 0 \text{ for any } X_x \in D_x\}, \qquad (3.34)$$

which in a neighborhood $V_\alpha \subset U_\alpha$ of x_0 is spanned by $n - m$ linearly independent differential 1-forms $\omega^{m+1}, \ldots, \omega^n$. Complete these sets of tangent vector fields and 1-forms to linearly independent sets $X_1, \ldots, X_n$ and $\omega^1, \ldots, \omega^n$ forming bases of

$T_x(N)$ and $T^*_x(N)$, respectively, at points x in the neighborhood V_α of x_0. Then, for $x \in V_\alpha$, $D^\perp_x$ is characterized by the set of $n - m$ total differential equations

$$\omega^i_x = 0 \quad \text{or} \quad \sum_{l=1}^{n} \omega^i_l(\boldsymbol{x}) dx^l = 0, \quad m < i \le n, \quad \boldsymbol{x} = (x^1, \ldots, x^n) \in \boldsymbol{V}_\alpha, \tag{3.35}$$

which is called a **Pfaffian equation system**.

Consider $d\omega^i = \sum_{l=1}^n d\omega^i_l \wedge dx^l$ and $d\omega^i(X_j, X_k), 1 \le j, k \le m$. From the definitions given after (3.23), $d\omega^i(X_j, X_k) = (1/2)\sum_{l=1}^n (d\omega^i_l(X_j) dx^l(X_k) - d\omega^i_l(X_k) dx^l(X_j)) = (1/2)\sum_{l=1}^n (d\omega^i_l(X_j)\xi^l_k - d\omega^i_l(X_k)\xi^l_j)$. Now, $d\omega^i_l(X_j) = \sum_k (\partial\omega^i_l/\partial x^k) dx^k(\sum_{k'} \xi^{k'}_j (\partial/\partial x^{k'})) = \sum_k \xi^k_j (\partial\omega^i_l/\partial x^k) = X_j \omega^i_l$ and, since $\omega^i = \sum_l \omega^i_l dx^l$, $\sum_l \omega^i_l \xi^l_j = \omega^i(X_j)$. All that together yields $d\omega^i(X_j, X_k) = (1/2)(X_j\omega^i(X_k) - X_k\omega^i(X_j) - \omega^i([X_j, X_k]))$. The last term appears since in the preceding terms the first X differentiates also the components of the second X which has to be subtracted since it does not appear in the previous expressions.

Since D_x is spanned by the X_{xj}, $j \le m$ and $D^\perp_x$ is spanned by the ω^i_x, $i > m$, for $i > m$ and j, $k \le m$ it holds that $\omega^i(X_j) = \omega^i(X_k) = 0$, and hence $d\omega^i(X_j, X_k) = -(1/2)\omega^i([X_j, X_k])$. The equations $\omega^i = 0$ imply $d\omega^i = 0$. Hence, if the system (3.35) has a solution, then $[X_j, X_k] \in D$, that is, D is involutive. If D is involutive, then $d\omega^i = 0$, $i > m$, on D, that is $d\omega^i \equiv 0 \bmod (\omega^{m+1}, \ldots, \omega^n)$, $i > m$, which means $d\omega^i = \sum_{j=m+1}^n \sigma^{ij} \wedge \omega^j$, where the σ^{ij} are arbitrary 1-forms. Since generally $\omega \wedge \omega = 0$, this condition may also by expressed as $d\omega^i \wedge \omega^{m+1} \wedge \cdots \wedge \omega^n = 0$, $i > m$. In this case there is an integral manifold of D. In summary, the **dual Frobenius theorem** reads:

The Pfaffian equation system (3.35) *describes a submanifold* (M, F) *of* N, *iff for* $i > m$ $d\omega^i \equiv 0 \bmod (\omega^{m+1}, \ldots, \omega^n)$ *or equivalently* $d\omega^i \wedge \omega^{m+1} \wedge \cdots \wedge \omega^n = 0$.

In that case, the Pfaffian system is called **completely integrable**. (See examples in the next section.)

The section is closed with a continuation of the discussion of integral curves of tangent vector fields X.

Consider an open set $U_\alpha \in N$ so that the construction leading to (3.33) exists for every point $x_0 \in U_\alpha$ with a function $F_{\alpha x_0}$ defined on a *fixed* interval $M = I_\varepsilon =]-\varepsilon, \varepsilon[\in \mathbb{R}$. Define a mapping $\phi : I_\varepsilon \times U_\alpha \to N : (t, x) \mapsto \phi(t, x) = \phi_t(x) = F_{\alpha x}(t)$ so that obviously

1. $\phi : I_\varepsilon \times U_\alpha \to N : (t, x) \mapsto \phi_t(x)$,
2. For each fixed t, ϕ_t is a diffeomorphism of U_α onto $\phi_t(U_\alpha)$ with the inverse $(\phi_t)^{-1} = \phi_{-t}$, (3.36)
3. $\phi_{s+t}(x) = \phi_t \circ \phi_s(x) = \phi_t(\phi_s(x))$.

Since for $t \in I_\varepsilon$ also $-t \in I_\varepsilon$, the expression for $(\phi_t)^{-1}$ follows directly from 3. A mapping with these three properties is called a **local 1-parameter group** of X. (Due to the restriction to I_ε it is not really a group.)

A tangent vector field X on N is called **complete** if $\phi_t(x)$ defines an integral curve of X for every $x \in N$ and for $-\infty < t < \infty$. In this case (3.36) holds with U_α replaced by N and I_ε replaced by $\mathbb{R}$. The transformations $\phi_t(x)$ of N now form indeed a group which is called the **1-parameter group** of X.

On a compact manifold N every tangent vector field is complete.

Proof The family of all sets U_α centered at all points $x_0 \in N$ of local 1-parameter groups of X form an open cover of N, of which a finite subcover may be selected. Let $\varepsilon > 0$ be the minimal ε-value on that finite subcover. Then, $\phi_t(x)$ is defined on $I_\varepsilon \times N$ and hence on $\mathbb{R} \times N$. □

Let ϕ be any **transformation** of N, that is, a diffeomorphism of N to itself.

If X creates the local 1*-parameter group* $\phi_t(x)$, *then* $\phi_*(X)$ *creates the local 1-parameter group* $\phi \circ \phi_t \circ \phi^{-1}$. *X is invariant under the transformation* ϕ, $\phi_*(X) = X$, *iff* $\phi \circ \phi_t = \phi_t \circ \phi$.

This is rather obvious.

For real-valued functions $f(t, x)$ and $g_t(x)$ so that $f = tg_t$ on $I_\varepsilon \times N$ consider the identities

$$f(t,x) = tg_t(x) = \int_0^1 \frac{\partial f(ts,x)}{\partial s} ds, \quad \left.\frac{\partial f(t,x)}{\partial t}\right|_{t=0} = g_0(x).$$

For an arbitrary real-valued function $F(x)$ on N and a local 1-parameter group $\phi_t(x)$, put $f(t, x) = F(\phi_t(x)) - F(x)$ and find

$$\lim_{t\to 0} \frac{F(\phi_t(x)) - F(x)}{t} = \lim_{t\to 0} \frac{1}{t} f(t,x) = \lim_{t\to 0} g_t(x) = g_0(x).$$

Now, take two tangent vector fields X and Y on N and the local 1-parameter group $\phi_t(x)$ created by X and find from the above

$$g_0(x) = X_x F, \quad ((\phi_t)_*(Y))_x F = Y_{\phi_{-t}(x)}(F \circ \phi_t) = Y_{\phi_{-t}(x)} F + tY_{\phi_{-t}(x)} g_t$$

and

$$\begin{aligned} \lim_{t\to 0} \frac{(Y - (\phi_t)_*(Y))_x}{t} F &= \lim_{t\to 0} \frac{Y_x F - Y_{\phi_{-t}(x)} F}{t} - \lim_{t\to 0} Y_{\phi_{-t}(x)} g_t \\ &= X_x(YF) - Y_x g_0 = [X, Y]_x F. \end{aligned}$$

Since F was arbitrary, the following proposition was demonstrated:

Let X and Y be two tangent vector fields on N and let the local 1*-parameter group* $\phi_t(x)$ *be created by X. Then*

$$[X, Y]_x = \lim_{t\to 0} \frac{Y_x - ((\phi_t)_*(Y))_x}{t} = \lim_{t\to 0} \frac{((\phi_{-t})_*(Y))_x - Y_x}{t} \tag{3.37}$$

for all $x \in N$.

The tangent vector field $[X, Y]$ describes the derivative of Y along the integral curve (flow) of X. A natural consequence is that the elements of the two local 1-parameter groups created by X and Y commute, iff $[X, Y] = 0$.

3.7 Examples from Physics

3.7.1 Classical Point Mechanics

(See e.g. any textbook on Mechanics, or, quite advanced, [6].) An assembly of mass points (particles) is described by their positions as functions of time. At any time, the positions are described by a collection of coordinates q^i on an m-dimensional manifold M, the **configuration space**. If n particles can occupy positions independently from each other, then m is three times their number n and M is the topological vector space $\mathbb{R}^m$. If there are constraints, the dimension may be reduced. If for instance two particles form a molecule with a fixed bond length, the configuration space has five dimensions instead of six. It is the product $\mathbb{R}^3 \times S^2$ of an Euclidean space with a sphere. If n particles form a molecule with $n(n-1)/2$ fixed bond lengths, M is more involved. (For many problems it suffices to consider molecules as assemblies of point masses, atomic nuclei, in a rigid mutual geometry.)

At any time, each particle has a velocity $v^i = dq^i/dt$. The collection of all velocity components v^i for some configuration $q \in M$ forms an m-dimensional vector $V_q \in T_q(M)$, $V_q = \sum_i v^i(\partial/\partial q^i)$, in the tangent space on M at point q. The motion is governed by a **Lagrange function**, which for a conservative system is a real function of q and for each q of the tangent vector V_q, that is, it is a real function on the tangent bundle $T(M)$ on the configuration space, $L : T(M) \to \mathbb{R} : (q^1, \dots, q^m; v^1, \dots, v^m) \mapsto L(q^1, \dots, q^m; v^1, \dots, v^m)$. From the extremal principle of action $S = \int L dt$ along trajectories with $dq^i/dt = v^i$ with positions at the end points fixed it follows that $(d/dt)(\partial L/\partial v^i) = \partial L/\partial q^i$. These are Lagrange's equations of motion.

In the Hamilton formalism, momenta $P = \sum_i p_i dq^i$ as cotangent vectors on M are introduced instead of velocities V so that $\langle P, V\rangle = \sum_1^m p_i v^i \in \mathbb{R}$. As a cotangent vector on M, P has a meaning as a 1-form on M, independent of the chosen local coordinates of M. Likewise, for a cotangent field $P = P(q)$, $\omega = -dP$ has such an independent meaning, which in every local coordinate system of M expresses as the **canonical 2-form** $\omega = \sum_{i=1}^m dq^i \wedge dp_i$. Coordinate transformations in the configuration space M of mechanics are called point transformations.

The **Hamilton function** H is a real function on the cotangent bundle $T^*(M)$ which is defined by the **Legendre transformation**

$$H(q^1, \dots, q^m; p_1, \dots, p_m) = \sup_{\{v^k\}}\{\langle P, V\rangle - L(q^1, \dots, q^m; v^1, \dots, v^m)\}, \qquad (3.38)$$

where it is assumed that L is a strictly convex C^1-function of the v^j. Then,

$$p_j = \frac{\partial L}{\partial v^j} \tag{3.39}$$

and

$$dH = \sum_i \left(\frac{\partial H}{\partial q^i} dq^i + \frac{\partial H}{\partial p_i} dp_i \right) = \sum_i \left(v^i dp_i - \frac{\partial L}{\partial q^i} dq^i \right).$$

From Lagrange's equations now Hamilton's equations of motion

$$\frac{dq^i}{dt} = \frac{\partial H}{\partial p_i}, \quad \frac{dp_i}{dt} = -\frac{\partial H}{\partial q^i} \tag{3.40}$$

follow. Note that the first set of equations forms a tangent vector equation in $T(M)$ while the second set forms a cotangent vector equation in $T^*(M)$.

The cotangent bundle $T^*(M)$ on M is a special $2m$-dimensional manifold Ω, the local coordinates of which may be chosen as the collection of local coordinates q^i of the configuration space M and for each set (q^i) of the components p_j of the momentum cotangent vector $P_q \in T_q^*(M)$. In a chart $(U_\alpha, \varphi_\alpha)$ of $\Omega = T^*(M)$ the points $x \in \Omega$ are send by $\boldsymbol{\varphi}_\alpha$ to $\boldsymbol{x} = (q_\alpha^1, \ldots, q_\alpha^n; p_\alpha^1, \ldots, p_\alpha^n) \in \boldsymbol{U}_\alpha$. The manifold Ω itself is called the **phase space** of the mechanical system. While up to (3.40) the p_j were understood as components of a cotangent vector on M and hence as depending on the chosen local coordinates q^i, they are now understood as independent local coordinates of Ω; for that reason p^j was now written instead of p_j.

Of course, it cannot be expected that the form of the equations of motion (3.40) would be the same in arbitrarily chosen local coordinates of Ω with q^i and p^i independently chosen. They will have this form for all point transformations in M with the components p_j of (3.39) and H of (3.38). The natural question arises, what are the most general coordinate transformations (diffeomorphisms) $\bar{q}^i(q^1, \ldots, q^m; p^1, \ldots, p^m)$, $\bar{p}^i(q^1, \ldots, q^m; p^1, \ldots, p^m)$ that leave the form (3.40) unchanged. These are the **canonical transformations** which leave the canonical (symplectic) 2-form $\bar{\omega} = \sum_{i=1}^m d\bar{q}^i \wedge d\bar{p}^i$ invariant. Obviously these transformations form a subgroup of the automorphism group of Ω. At the end of the next chapter Hamilton's equations of motion will be cast into a form from which it is readily seen that canonical transformations leave them invariant.

Introduce on Ω a tangent vector field W which in local coordinates has the general form $W = \sum_{i=1}^m (v^i(\partial/\partial q^i) + a^i(\partial/\partial p^i))$, and put in given local coordinates $v^i = \partial H/\partial p^i$, $a^i = -\partial H/\partial q^i$. For this special vector field $W = W_H$,

$$W_H = \sum_{i=1}^m \left\{ \frac{\partial H}{\partial p^i} \frac{\partial}{\partial q^i} - \frac{\partial H}{\partial q^i} \frac{\partial}{\partial p^i} \right\}, \tag{3.41}$$

consider the local 1-parameter group $\phi_t(x)$ created by W_H. It is obtained by integration of the Hamilton equations (3.40). Since $\langle dH, W_H \rangle = W_H H = 0$ (cf. 3.12),

$H(x)$ is constant along every integral curve $x(t) = \phi_t(x)$ of W_H: $H \circ \phi_t = H$. It is the energy of the conservative system. More generally, any real function F on the phase space Ω for which $W_H F = 0$ is constant on integral curves: $F \circ \phi_t = F$. F is a conserved quantity. $W_H F = \{H, F\}$ is called the **Poisson bracket**[1] of H and F. The vector field W_H on Ω is called the Hamiltonian vector field, in statistical physics it is called the Liouvillian. The corresponding local 1-parameter group $\phi_t(x)$ is called the Hamiltonian flow, in statistical physics the Liouvillian flow.

Like in (3.41) for H, a tangent vector field W_F may be defined for any C^1-function F on the phase space Ω, and for functions F, $G \in C^1(\Omega)$ a Poisson bracket $\{F, G\}$ is defined. Poisson brackets have the following algebraic properties (with real numbers λ_i):

1. $\{F, G\} = -\{G, F\}$,
2. $\{F, \lambda_1 G_1 + \lambda_2 G_2\} = \lambda_1 \{F, G_1\} + \lambda_2 \{F, G_2\}$, together with 1 bilinearity,
3. $\{F, \{G, K\}\} + \{G, \{K, F\}\} + \{K, \{F, G\}\} = 0$, Jacobi identity,
4. $\{F, G_1 G_2\} = G_2 \{F, G_1\} + G_1 \{F, G_2\}$, Leibniz rule.

(3.42)

Comparison of 1–3 with (3.17) shows that the Poisson brackets form a Lie algebra; 4. holds since $\{F, \cdot\} = W_F$ is a derivation, cf. (3.41) for $F = H$.

If $2m - l$ conserved quantities $F_k, k = l + 1, \ldots, 2m$ are given, then the equations $dF_k = 0$ form a Pfaffian system for the 1-forms dF_k which is completely integrable since $d(dF_k) = d^2 F_k = 0$. Hence, in this case the motion takes place on a submanifold of Ω of lower dimension l. For $m > 1$, only in very special cases enough conserved quantities can be found so that $l = 1$ and the motion takes place on curves which are regular embeddings in Ω, and little is known on general conditions under which this takes place. In most cases the motion in some submanifold Φ of Ω is chaotic, the closure of the orbit of any $x \in \Phi$ in the relative topology of $\Phi \subset \Omega$, $\overline{\{\phi_t(x) | -\infty < t < \infty\}}$, is all Φ.

3.7.2 Classical and Quantum Mechanics

Useful as the introduction of the phase space is, it looses track of important features of the inner structure of this manifold as the cotangent bundle on the configuration space M. Canonical transformations may interchange position coordinates and momentum components, while in a curved manifold M the position coordinates do not form a vector at all. This becomes a real problem of still ongoing research if one wants to quantize a general mechanical theory on a curved configuration manifold M (see [7] and citations therein).

[1] We use the traditional definition of Poisson brackets in standard Physics textbooks; in Mathematics it is more standard to call $W_H F$ a Poisson bracket $\{F, H\}$.

Consider canonical local coordinates (q^i, p_j) in any coordinate neighborhood of the cotangent bundle $T^*(M)$ over the configuration space M, $\dim M = m$, of a mechanical system. Let $F, G, \ldots$ be smooth functions on M with compact support, that is, the $F, G, \ldots \in C_0^\infty(M)$ depend on the q^i only. Then one finds for the Poisson brackets

$$\{F, G\} = 0, \quad \{p_j, F\} = \frac{\partial F}{\partial q^j}, \quad \{p_j, p_k\} = 0.$$

The second relation says that p_j acts on F via the Poisson bracket like the tangent vector $\partial/\partial q^j$. Recall from (3.20) that the tangent bundle, instead being considered locally spanned by the m base vectors $\partial/\partial q^j$, may be considered as a module over the algebra of smooth functions on M. Its subalgebra $C_0^\infty(M)$ then refers to the submodul $T_0(M)$ of (smooth) tangent vector fields with compact support. Let $X, Y, \ldots \in T_0(M)$, and let the just mentioned module structure be denoted by the mapping (called Rinehart product) $(C_0^\infty(M), T_0(M)) \to T_0(M) : (F, X) \mapsto F \odot X$, where $F \odot X$ means FX of (3.20), and

$$F \odot (G \odot X) = (FG) \odot X. \tag{3.43}$$

With the notation XF and $XY - YX$ from Sect. 3.4, we have

$$\{F, G\} = 0, \quad \{X, F\} = XF = -\{F, X\}, \quad \{X, Y\} = XY - YX \tag{3.44}$$

and

$$\{X, F \odot Y\} = \{X, F\} \odot Y + F \odot \{X, Y\}, \quad \text{Leibniz rule.} \tag{3.45}$$

If M itself is not compact, add a unity function 1 to $C_0^\infty(M)$ so that

$$1 \odot F = F, \quad 1 \odot X = X, \quad \{X, 1\} = 0, \tag{3.46}$$

it is the constant function on M equal to unity (and hence not $C_0^\infty(M)$ if M itself is not compact). The algebra $(\{1\}+)C_0^\infty(M) + T_0(M) \ni A, B, \ldots$ with the products $\{\cdot, \cdot\}$ and $\odot$ obeying (3.43–3.46) is called the Lie–Rinehart algebra $\mathcal{L}_R(M)$ of the manifold M. Any element $A \in \mathcal{L}_R(M)$ is a linear combination of 1 and elements from $C_0^\infty(M)$ and $T_0(M)$. With $F \odot X = FX$ it obeys (3.42) and hence is a Poisson algebra.

The elements of the Lie–Rinehart algebra $\mathcal{L}_R(M)$ are at most linear in the tangent vector fields. For quantum mechanics one wants the momenta also to form an associative polynomial algebra for the operator product, in particular to treat spectra, and with an involution $(^*)$ leaving the variables invariant to guarantee real spectra (Hermitian operators). Therefore such an algebraic structure with an (in general not commutative) dot-product is introduced as a second algebraic structure replacing the $\odot$-product in (3.43–3.46), with

$$
\begin{aligned}
&1.\quad A\cdot(B\cdot C)=(A\cdot B)\cdot C=A\cdot B\cdot C,\quad 1\cdot A=A,\\
&2.\quad [A,B]=A\cdot B-B\cdot A,\\
&3.\quad (A\cdot B)^*=B^*\cdot A^*,\qquad \text{but } \{A,B\}^*=\{A^*,B^*\},\\
&4.\quad \{A,B\cdot C\}=\{A,B\}\cdot C+B\cdot\{A,C\},\quad \text{Leibniz rule.}
\end{aligned}
\tag{3.47}
$$

Such an algebra with the Poisson bracket $\{\cdot,\cdot\}$ with properties (3.42) extended to all its elements is called a Poisson *-algebra. Per se the commutator $[\cdot,\cdot]$ and the Poisson bracket are independent, however, they are intertwined by the Leibniz rule, property 4. above, which ensures that $\{A,\cdot\}$ is further on a derivation. Repeated application of this rule and the property 1 in (3.42) to the identity $\{A\cdot C,B\cdot D\}+\{B\cdot D,A\cdot C\}=0$ yields straightforwardly (exercise)

$$[A,B]\cdot\{C,D\}=\{A,B\}\cdot[C,D]. \tag{3.48}$$

For any commutative Poisson *-algebra, $[\cdot,\cdot]\equiv 0$, this is trivially true.

The Poisson–Rinehart algebra $\Lambda_R(M)$ of a manifold M is the unique enveloping Poisson *-algebra of M in which the Lie–Rinehart algebra $\mathcal{L}_R(M)$ is injected, $J:\mathcal{L}_R(M)\to\Lambda_R(M)$, so that $(A,B,\dots,F,G,\dots,X,Y,\dots\in\mathcal{L}_R(M))$

$$
\begin{aligned}
&1.\quad J(\{A,B\})=\{J(A),J(B)\},\\
&2.\quad J(1)=1,\\
&3.\quad J(FG)=J(F)\cdot J(G),\\
&4.\quad J(F\odot X)=\frac{1}{2}(J(F)\cdot J(X)+J(X)\cdot J(F))
\end{aligned}
\tag{3.49}
$$

and so that $\Lambda_R(M)$ is universal, that is, if $J':\mathcal{L}_R(M)\to\Lambda'$ satisfies 1–4, then there is a unique homomorphism $\rho:\ \Lambda_R(M)\to\Lambda'$ so that $J'=\rho\circ J$.

It has been shown [7] that there exists (or may be added with a simple limiting process if M is not compact) an element Z in the center of both algebraic structures of $\Lambda_R(M)$, unique up to a real constant factor, so that

$$[A,B]=Z\cdot\{A,B\},\quad \{Z,A\}=0=[Z,A],\quad Z=-Z^*. \tag{3.50}$$

Classical physics is obtained with $Z=0$ (resulting in the quotient algebra $\Lambda_R(M)/I$ where I is the ideal generated by the elements $[A,\ B]$). In standard phase space quantization on a flat M, as is well known, $Z=i\hbar\cdot 1$. The value of $\hbar$ is of course phenomenology. It is interesting that, the above structure accepted, the existence of Z follows from this structure alone, and it is up to a constant factor (multiple of 1) unique for each configuration manifold M.

3.7.3 Classical Point Mechanics Under Momentum Constraints

In what follows, the constraints which are called primary constraints below are linear in the canonical momenta. They are called momentum constraints here in order to distinguish this case from constraints in a totally different context considered afterwards.

The so far outlined theory presupposes that the Lagrange function L is strictly convex in V. If as usual L has second derivatives with respect to the v^i, this means that for all $q \in M$ and for any base in $T_q(M)$ the symmetric $m \times m$-matrix of those second derivatives (the Hessian) has maximal rank:

$$\operatorname{rank}\left(\frac{\partial^2 L}{\partial v^i \partial v^j}\right) = m, \quad \det\left(\frac{\partial^2 L}{\partial v^i \partial v^j}\right) > 0. \tag{3.51}$$

This is not always the case. The theory considered now was pioneered by Dirac in the 1950 and 1960 [8], essentially to the goal of canonical quantization of gauge field theories (which itself is beyond the scope of the present text). With the rise of importance of Yang-Mills theories it was a very active subject of research in the 1980 and 1990 (see, e.g. [9]), and it holds unsolved problems till now.

For the sake of simplicity of notation the theory is usually presented for a finite number of degrees of freedom (m in our text), although corresponding systems of that type with a finite number of degrees of freedom look rather academic. However, the results readily transfer to fields (with a continuum of degrees of freedom), and all gauge fields are standard cases of this transfer.

Consider, in appropriate coordinates, a **Lagrange function**

$$\begin{aligned} &L(q,V) = \langle f(q,\overline{V}), \underline{V}\rangle + \underset{\sim}{L}(q,\overline{V}),\\ &f = (f_{\overline{m}+1},\ldots,f_m), \quad V = (\overline{V},\underline{V}) = (\overline{v}^1,\ldots,\overline{v}^{\overline{m}},\underline{v}^{\overline{m}+1},\ldots,\underline{v}_m),\\ &\operatorname{rank}\left(\frac{\partial^2 L}{\partial v^i \partial v^j}\right) = \overline{m} < m, \quad \det\left(\frac{\partial^2 L}{\partial \overline{v}^i \partial \overline{v}^j}\right) > 0. \end{aligned} \tag{3.52}$$

Define the action integral as

$$S[q(t),V(t),P(t)] = \int_{t_1}^{t_2} (L(q,V) + \langle P, dq/dt - V\rangle)dt \tag{3.53}$$

where the $P_i, i = 1,\ldots,m$, are Lagrange multipliers for the conditions $dq/dt = V$ along the trajectories $q(t)$ through M. The variation with fixed end points q yields the usual Lagrange equations of motion

$$\frac{dq}{dt} - V = 0, \quad \left(\frac{\partial L}{\partial V} - P\right)_{V=dq/dt} = 0, \quad \left(\frac{\partial L}{\partial q} - \frac{d}{dt}\frac{\partial L}{\partial V}\right)_{V=dq/dt} = 0 \tag{3.54}$$

with a self-explanatory vector notation of the derivatives of L (remark after 3.40).

Now, for instance,

$$\sup_{\underline{v}^m}\left\{\underline{p}_m \underline{v}^m - f_m(q,\overline{V})\underline{v}^m\right\} = \begin{cases} 0 & \text{for } \underline{p}_m = f_m(q,\overline{V}) \\ +\infty & \text{else,} \end{cases} \tag{3.55}$$

(exercise), and

$$\sup_{\overline{V}}\{\langle\overline{P},\overline{V}\rangle - L(q,V)\} \quad\Rightarrow\quad \overline{P}-\frac{\partial L}{\partial\overline{V}}=0 \quad\Rightarrow\quad \overline{V}=\mathcal{V}(q,\overline{P},\underline{V}). \tag{3.56}$$

From the last line of (3.52) it follows that the middle relations of (3.56) can be resolved for $\overline{V}$, and for $P=(\overline{P},\underline{P})$ the same notation as for V was used. Inserting the last result into f yields

$$\underline{p}_i = f_i(q,\mathcal{V}(q,\overline{P},\underline{V})) = \underset{\sim}{f_i}(q,\overline{P}), \quad i=\overline{m}+1,\ldots,m. \tag{3.57}$$

If the $\underline{V}$ would not drop out from the functions $\underset{\sim}{f_i}$, then obviously the rank in (3.52) would be larger than $\overline{m}$. Altogether (3.38) results in

$$H(q,P)=\begin{cases}\underset{\sim}{H}(q,\overline{P}) & \text{for } \underline{P}=\underset{\sim}{f}(q,\overline{P}),\\ +\infty & \text{else}\end{cases} \tag{3.58}$$

with

$$\underset{\sim}{H}(q,\overline{P}) = \langle\overline{P},\mathcal{V}(q,\overline{P})\rangle - \underset{\sim}{L}(q,\mathcal{V}(q,\overline{P})). \tag{3.59}$$

There are $\underline{m}=m-\overline{m}$ **primary constraints**

$$\Phi_i^{(1)}(q,P)=\underline{p}_i-\underset{\sim}{f_i}(q,\overline{P})=0, \quad i=\overline{m}+1,\ldots,m, \tag{3.60}$$

for the momenta P on the trajectories.

The trajectories are now obtained from the extremum of the action integral

$$S'[q(t),P(t),\lambda(t)]=\int_{t_1}^{t_2}(\langle P,dq/dt\rangle - \underset{\sim}{H}(q,\overline{P}) - \langle\Phi^{(1)}(q,P),\lambda\rangle)dt, \tag{3.61}$$

where λ^i, $i=\overline{m}+1,\ldots,m$, are the Lagrange multipliers for the constraints. This may be abbreviated as

$$\begin{aligned} S' &= \int\Big(\langle P,dq/dt\rangle - H^{(1)}(q,P,\lambda)\Big)dt,\\ H^{(1)}(q,P,\lambda) &= \underset{\sim}{H}(q,\overline{P}) + \langle\Phi^{(1)}(q,P),\lambda\rangle. \end{aligned} \tag{3.62}$$

Again with a notation $q=(\overline{q},\underline{q})$, the variation yields

$$\frac{dq}{dt}=\frac{\partial H^{(1)}}{\partial P}\left(\text{or } \frac{d\overline{q}}{dt}=\frac{\partial H^{(1)}}{\partial\overline{P}},\ \frac{d\underline{q}}{dt}=\lambda\right),\quad \frac{dP}{dt}=-\frac{\partial H^{(1)}}{\partial q},\quad \underline{P}=\underset{\sim}{f}\,. \tag{3.63}$$

The Lagrange multipliers turn out to be the velocities on the trajectories which remained unresolved in (3.56). It is seen that the time derivative along a trajectory of any function $A(q, P)$ is obtained from the Poisson bracket,

$$A(q,P):\frac{dA}{dt}=\{H^{(1)},A\}_{\Phi^{(1)}=0}=\left(\left\langle\frac{\partial A}{\partial q},\frac{\partial H^{(1)}}{\partial P}\right\rangle-\left\langle\frac{\partial H^{(1)}}{\partial q},\frac{\partial A}{\partial P}\right\rangle\right)_{\Phi^{(1)}=0} \quad (3.64)$$

where the constraints must be taken after the calculation of the brackets.

Note that the constraints (3.60) form a Pfaffian equation system defining an $\overline{m}$-dimensional distribution in $T(M)$. Any trajectory being solution of (3.54) must be kept in this distribution at all times, whence the time derivatives of the constraints should vanish,

$$\frac{d\Phi^{(1)}}{dt}=\{H^{(1)},\Phi^{(1)}\}_{\Phi^{(1)}=0}=\{\underset{\sim}{H},\Phi^{(1)}\}_{\Phi^{(1)}=0}+\left\langle\{\Phi^{(1)},\Phi^{(1)}\}_{\Phi^{(1)}=0},\lambda\right\rangle=0. \quad (3.65)$$

Here, $\Phi^{(1)}$ and $\{\underset{\sim}{H},\Phi^{(1)}\}$ are understood as cotangent vector fields on M while $\{\Phi^{(1)},\Phi^{(1)}\}$ is a 2-form on M (a q-dependent antisymmetric $\underline{m}\times\underline{m}$-matrix in local coordinates q^i).

Let, as a matrix in local coordinates q^i,

$$\operatorname{rank}\left(\{\Phi^{(1)},\Phi^{(1)}\}_{\Phi^{(1)}=0}\right)=\rho_1,\quad 0\le\rho_1\le\underline{m}. \quad (3.66)$$

As the rank of an antisymmetric matrix, ρ_1 is even. Then, there exist $\underline{m}-\rho_1$ non-zero vector fields $U_{(k)}$ so that

$$\left\langle\{\Phi^{(1)},\Phi^{(1)}\}_{\Phi^{(1)}=0},U_{(k)}\right\rangle=0,\ U_{(k)}=(U^1_{(k)},\ldots,U^{\underline{m}}_{(k)}),\quad k=\rho_1+1,\ldots,\underline{m}. \quad (3.67)$$

Hence, Eq. 3.65 determines the $\underline{m}$ linearly independent vector functions $\lambda(q, P, t)$ modulo $U_{(k)}$; a number of $\underline{m}-\rho_1$ linear combinations remain undetermined. Instead, (3.65) comprise the conditions

$$\left\langle\{\underset{\sim}{H},\Phi^{(1)}\}_{\Phi^{(1)}=0},U_{(k)})\right\rangle=0, \quad (3.68)$$

some of which may be identities, some may not be independent of the constraints $\Phi^{(1)}$, but a number $\underline{m}_1$ of them may form new constraints $\Phi^{(2,1)}$. Inserting the ρ_1 determined expressions λ in terms of q, P (as well as of the suppressed variable t) and the remaining undetermined combinations λ' into (3.65) yields a combination of ρ_1 independent expressions denoted as $\Psi^{(1)}(q,P,\lambda')$, of $\underline{m}$ primary constraints $\Phi^{(1)}(q, P)$ and of $\underline{m}_1$ new constraints $\Phi^{(2,1)}(q, P)$. Since $\{H^{(1)}, \Phi^{(1)}\}$ is linear in the constraints, (3.65) has a structure, *with* $\Phi = (\Phi_l)$ *further on written as a column*,

$$\{H^{(1)}, \Phi^{(1)}\}_{\Phi^{(1)}=0} = 0 \quad \Leftrightarrow \quad \Lambda^{(1)}(q,P) \begin{pmatrix} \Psi^{(1)}(q,P,\lambda') \\ \Phi^{(1)}(q,P) \\ \Phi^{(2,1)}(q,P) \end{pmatrix} = 0, \tag{3.69}$$

where $\Lambda^{(1)}$ is an $(\underline{m} \times (\rho_1 + \underline{m} + \underline{m}_1))$-matrix function divided into three blocks with rank $\Lambda^{(1)}_{\Phi^{(1)},\Psi^{(1)}} = \rho_1$, $\Lambda^{(1)}_{\Phi^{(1)},\Phi^{(1)}} = 1_{\underline{m}}$, (unit matrix) and rank $\Lambda^{(1)}_{\Phi^{(1)},\Phi^{(2,1)}} = \underline{m}_1$ in an obvious block matrix notation. At the same step, introducing the determined expressions for the λ into the Hamiltonian (3.62) results in a new Hamiltonian $H_1^{(1)}(q,P,\lambda')$ for which (3.65) transforms into $\{H_1^{(1)}, \Phi^{(1)}\}_{\Phi^{(1)}=0} = 0$. Repeating this process with $H_1^{(1)}(q,P,\lambda')$ and with the new constraints results in a second step

$$\{H_1^{(1)}, \Phi^{(2,1)}\}_{\substack{\Phi^{(1)}=0 \\ \Phi^{(2,1)}=0}} = 0 \quad \Leftrightarrow \quad \Lambda^{(2)}(q,P) \begin{pmatrix} \Psi^{(2,1)}(q,P,\lambda'') \\ \Phi^{(1)}(q,P) \\ \Phi^{(2,1)}(q,P) \\ \Phi^{(2,2)}(q,P) \end{pmatrix} = 0. \tag{3.70}$$

After a finite number of l steps there appear no new independent constraints in the $l+1$st step although there may still remain unresolved multipliers λ. (The number of independent constraints cannot exceed the dimension of $T_p^*(M)$.) All constraints $\Phi^{(2,j)}$ are called **secondary constraints**.

This process of 'breeding constraints' results in a number $\mu = \underline{m} + \underline{m}_1 + \cdots + \underline{m}_k$, unique for a given Lagrange function, of constraints Φ, independent in the sense that the rank of the Jacobi matrix

$$\text{rank}\left(\frac{D(\Phi)}{D(q,P)}\bigg|_{\Phi=0} \right) = \mu. \tag{3.71}$$

Let (superscript t meaning the transposed, Φ^t being a row)

$$\nu = \mu - \text{rank}\big(\{\Phi, \Phi^t\}_{\Phi=0}\big). \tag{3.72}$$

These ranks are independent of a linear functional transformation with a $(\mu \times \mu)$-matrix V,

$$\Phi'(q,P) = V(q,P)\Phi(q,P), \quad \det V|_{\Phi=0} \neq 0. \tag{3.73}$$

There is such a transformation that

$$V\Phi = \begin{pmatrix} \chi \\ \varphi, \end{pmatrix}, \quad \chi = (\chi_1, \ldots, \chi_\nu)^t, \quad \varphi = (\varphi_{\nu+1}, \ldots, \varphi_\mu)^t, \tag{3.74}$$

with

$$\{\chi, \Phi\} = O(\Phi), \quad \det\bigl(\{\varphi, \varphi\}_{\Phi=0}\bigr) \neq 0, \tag{3.75}$$

where $O(\Phi)$ means order of Φ; $O(\Phi)|_{\Phi=0} = 0$. The constraints $(\chi, \varphi) = 0$ are equivalent to $\Phi = 0$. The χ are called **first-class constraints**, and the φ are called **second-class constraints**.

Consider first the simpler case if there are only second-class constraints. Compared to (3.60) we subtract $\overline{m}$ from all subscripts i of the constraints and of all superscripts of the λ in what follows. In the actually considered case, in any local coordinate system, the matrix $\{\Phi, \Phi^t\}$ has maximal rank in the neighborhood of $\Phi = 0$, so that its inverse $\{\Phi, \Phi^t\}^{-1}$ exists,

$$\{\Phi, \Phi^t\}^{-1}\{\Phi, \Phi^t\} = 1_\mu = \{\Phi, \Phi^t\}\{\Phi, \Phi^t\}^{-1}, \tag{3.76}$$

where 1_μ means the $(\mu \times \mu)$ unit matrix. This means that all λ may be determined from (3.65) to be

$$\lambda^k = -\sum_{l=1}^{\mu}(\{\Phi, \Phi^t\}^{-1})_{kl}\{\Phi_l, \underset{\sim}{H}\} + O(\Phi), \quad k = 1, \ldots, \underline{m}, \tag{3.77}$$

while

$$\sum_{l=1}^{\mu}(\{\Phi, \Phi^t\}^{-1})_{kl}\{\Phi_l, \underset{\sim}{H}\} = O(\Phi), \quad k = \underline{m} + 1, \ldots, \mu. \tag{3.78}$$

In the equations of motion (3.64), terms of higher than first order in the Φ may be added to the Hamiltonian before the Poisson brackets are calculated, since after their calculation now all Φ are put to zero. One hence may define a Hamiltonian

$$H^\Phi = \underset{\sim}{H} - \Phi^t\{\Phi, \Phi^t\}^{-1}\{\Phi_l, \underset{\sim}{H}\}, \tag{3.79}$$

and obtain canonical equations of motion

$$A(q, P): \quad \frac{dA}{dt} = \{H^\Phi, A\}_{\Phi=0}. \tag{3.80}$$

Alternatively, one may define **Dirac's brackets**[2] instead of the Poisson brackets by

$$\{A, B\}^\Phi = \{A, B\} - \{A, \Phi^t\}\{\Phi, \Phi^t\}^{-1}\{\Phi, B\} \tag{3.81}$$

and have

[2] Do not confuse them with Dirac's notation of Hilbert vectors by bras and kets.

$$\frac{dA}{dt} = \{\underset{\sim}{H}, A\}^{\Phi}_{\Phi=0}. \tag{3.82}$$

The brackets $\{\cdot,\cdot\}^{\Phi}$ have algebraic properties (3.42) like the Poisson brackets.

As the rank of the antisymmetric matrix $\{\varphi,\varphi\}$ of (3.75) must be even, the number of second-class constraint is always even. It can be shown [9] that a canonical transformation from (q, P) to new canonical variables (η, ϕ) exists so that the η and the ϕ separately consist of pairs of canonically conjugate variables and the ϕ form the constraints in the new coordinates. Then, the equations of motion take the form

$$\frac{d\eta}{dt} = \{H^{\phi}, \eta\}, \quad \phi = 0, \quad H^{\phi} = H|_{\phi=0} = H(\eta). \tag{3.83}$$

That means, the presence of only second-class constraints simply reduces the system to an ordinary Hamilton system on a submanifold of M defined by the distribution of the 1-forms ϕ.

First-class constraints cannot form in Dirac's 'breeding' process, if there were no primary first-class constraints present: among first-class constraints there are necessarily primary first-class constraints. In this case the existence of a canonical transformation has been shown [9] to new canonical variables (η, Q, Π, ϕ) in which the dynamics is described by the equations of motion

$$\frac{d\eta}{dt} = \{H^{\phi}, \eta\}, \ \Pi = 0 = \phi, \ \frac{dQ^{(1)}}{dt} = \lambda_{\Pi^{(1)}}, \ \frac{dQ^{(2)}}{dt} = A(\eta), \ H^{\phi} = H|_{(\Pi,\phi)=0}, \tag{3.84}$$

where $\Pi = (\Pi^{(1)}, \Pi^{(2)}), \Pi^{(1)}$ are the primary and $\Pi^{(2)}$ the secondary first-class constraints, and Q are the conjugate variables to first-class constraints. $A(\eta)$ are fixed functions appearing in Dirac's procedure, and $\lambda_{\Pi^{(1)}} = \lambda_{\Pi^{(1)}}(\eta, Q, t)$ are the remaining undetermined Lagrange multipliers which do not enter the other equations of motion. Hence, for any initial condition the time evolution contains $\underline{m}$ arbitrary functions of the canonical variables and of time t, that is, an $\underline{m}$-dimensional class of trajectories.

The only way to save causality, which is supported by experience, is to say that each trajectory out of this class describes the same physical process. The $\lambda_{\Pi^{(1)}}$ merely constitute redundancy in the description, for whatsoever good reasons. This redundancy is conventionally called **gauge freedom**. Since transitions between the trajectories of a given class can be performed one after another and the $\lambda_{\Pi^{(1)}}$ may continuously be varied, those transitions form a continuous transformation group the elements of which are continuously connected to the identity transformation. This is the **group of gauge transformations** (see Sect. 8.3).

There is a systematic actual way to arrive at the indicated special canonical variables which is in general, however, quite technical (see [10]).

3.7.4 Classical Mechanics Under Velocity Constraints

We now consider sketchily a case which was already initiated towards the end of nineteenth century, essentially by H.R. Hertz and S.A. Chaplygin, which developed into a huge branch of Mathematical Physics and of various disciplines of engineering over the last century, and which is a very active field of research even today including also aspects of quantization [11].

We first switch to a much broader understanding of the configuration space M. It may be any smooth manifold the points of which describe some mechanical setting, for instance describing the group elements of translation and rotation groups of rigid bodies. It may even describe the instantaneous state of some sensor-actuator system which need not be mechanical.

Let $L(q, V)$ be a regular Lagrange function on M,

$$L(q,V) : \det\left(\frac{\partial^2 L}{\partial v^i \partial v^j}\right) = m = \dim M. \tag{3.85}$$

Let be given $\mu < m$ linearly independent smooth 1-forms $f^{(k)}(q)$ on M so that the motion of the system is constrained by the conditions

$$\langle f^{(k)}, V(t)\rangle = 0 \quad \text{for all } t, \quad k = 1, \ldots, \mu. \tag{3.86}$$

According to (3.34) the $f^{(k)}$ form the annihilator $D^{\perp}$ of an $(m - \mu)$-dimensional distribution D of velocity vectors. Constraints nonlinear in V are also considered in literature, but here we limit ourselves to linear constraints. The constraints cause **constraint forces** during the motion which are not (and in general cannot be) described by the Lagrange function (3.85) and which have to be added to the equations of motion (understood as a cotangent vector equation as previously),

$$\frac{d}{dt}\frac{\partial L}{\partial V} - \frac{\partial L}{\partial q} = \sum_{k=1}^{\mu} f^{(k)}(q)\lambda_{(k)}(q,t) = f\lambda, \tag{3.87}$$

where the coefficient functions $\lambda(q, t)$ are uniquely determined by the conditions (3.86). (One has $\langle f, V\rangle\lambda = 0$, the constraint forces do not perform work along the trajectories of motion; d'Alembert's principle.) On transition to the Hamilton function (3.38) corresponding to the Lagrange function (3.85) and adding the constraint forces and the constraints to the canonical equations of motion one gets

$$\frac{dq}{dt} = \frac{\partial H}{\partial P}, \quad \frac{dP}{dt} = -\frac{\partial H}{\partial q} + f\lambda, \quad \left\langle f, \frac{\partial H}{\partial P}\right\rangle = 0. \tag{3.88}$$

The time derivative of the constraints must be zero. The time derivative of the last equation of (3.88) along trajectories is

$$\left\langle\left(\frac{\partial}{\partial q}\left\langle f, \frac{\partial H}{\partial P}\right\rangle\right), \frac{\partial H}{\partial P}\right\rangle + \left\langle f\left|\frac{\partial^2 H}{\partial P \partial P}\right| - \frac{\partial H}{\partial q} + f\lambda\right\rangle = 0,$$

where in the second term a second rank tensor $\partial^2 H/(\partial P \partial P)$ in $T(M)$ is projected on the 1-forms on both sides. This expression can uniquely be resolved for the λ, if

$$\left\langle f \left| \frac{\partial^2 H}{\partial P \partial P} \right| f \right\rangle \neq 0, \tag{3.89}$$

which condition indeed follows from (3.85) and from the linear independence of the $f^{(k)}$. Hence, the constraint forces are correctly determined.

If the distribution D defined by (3.86) is involutive, that is, if $df^{(k)} \wedge f^{(1)} \wedge \cdots \wedge f^{(\mu)} = 0, \; k = 1, \ldots, \mu$, then the constraints are called **holonomic**. In this case the constraints are completely integrable and the distribution of velocities D defines a submanifold N of M as its integral manifold to which the whole motion of the system is confined.

If D is not involutive, the constraints are called nonholonomic. The classical example is a rolling disc without slipping on an inclined plane (Chaplygin). Though nonholonomic motion is rather the standard case in everydays life (see the just mentioned example) it comprises a huge in large parts unexplored field, and even the motion of simple (but tricky) toys like the 'rattleback' is poorly understood. There is in general a 'bracket formulation' of the equations of motion [12], but the brackets do not obey property 3 of Poisson brackets, the Jacobi identity (3.42).

3.7.5 Thermodynamics

The thermodynamic equilibrium state of a gas of N particles is described by its volume v and temperature t. The thermodynamic phase space is $M = \mathbb{R}^2$ in this case. The amount of heat put into an ideal gas is

$$\delta Q = c(t)dt + \frac{Rt}{v}dv, \tag{3.90}$$

where $c(t)$ is the heat capacity at constant volume which is a function of temperature only and R is the gas constant. A change of the thermodynamic state is called **adiabatic**, if no heat is exchanged, that is, if $\delta Q = 0$.

The question is, whether $\delta Q = 0$ defines uniquely paths through the phase space. Since (3.90) is a 1-form in $\mathbb{R}^2$, $\delta Q = 0$ is a Pfaffian equation, for which $d\delta Q = (\partial(Rt/v)/\partial t)dt \wedge dv \neq 0$, but since $d\delta Q \wedge \delta Q = 0$, the equation is completely integrable (as any 1-form in two dimensions, since $d\omega \wedge \omega$ is a 3-form for every 1-form ω). Hence, the answer to the question is positive, and there is an adiabatic flow $\phi_\tau^{\text{ad}}(t, v)$ ($\tau \in \mathbb{R}$ is some curve parameter) through the phase space: through every point (t, v) there is exactly one adiabatic trajectory. Consequently, there are functions, constant on trajectories and hence invariant under the adiabatic flow.

Nature or microscopic reasoning in Statistical Physics tells us, that as part of Second Law of thermodynamics it always holds that **entropy** s, given by $ds = \delta Q/t$ is such an adiabatic invariant, that is, ds is always a total differential, and $s \circ \phi^{\mathrm{ad}} = s$.

References

1. Hartshorne, R.: Algebraic Geometry. Springer, New York (1977)
2. Kobayashi, S., Nomizu, K.: Foundations of Differential Geometry, Interscience, vols. I and II. New York (1963, 1969)
3. Chern, S.S., Chen, W.H., Lam, K.S.: Lectures on Differential Geometry. World Scientific, Singapore (2000)
4. Warner, F.W.: Foundations of Differentiable Manifolds and Lie Groups. Springer, New York (1983)
5. Schwartz, L.: Analyse Mathématique. Hermann, Paris (1967)
6. Arnold, V.I., Kozlov, V.V., Neishtadt, A.I.: Dynamical Systems III. Encyclopedia of Mathematical Sciences, vol. 3. Springer, Berlin (1988)
7. Morchio, G., Strocchi, F.: Classical and quantum mechanics from the universal Poisson-Rinehart algebra of a manifold. Rep. Math. Phys. **64**, 33–48 (2009)
8. Dirac, P.A.M.: Lectures on Quantum Mechanics. Academic Press, New York (1964)
9. Gitman, D.M., Tyutin, I.V.: Quantization of Fields with Constraints. Springer, Berlin (1990)
10. Chitaia, N.P., Gogilidze, S.A., Surovtsev, Y.S.: Dynamical systems with first- and second-order constraints. I and II. Phys. Rev. D **56**, 1135–1155 (1997)
11. Bloch, A.M., Fernandez, O.E., Mestdag, T.: Hamiltonization of nonholonomorphic systems and the inverse problem of the calculus of variations. Rep. Math. Phys. **63**, 225–249 (2009)
12. van der Schaft, A.J., Maschke, B.M.: On the Hamiltonian Formulation of Nonholonomorphic Mechanical Systems. Rep. Math. Phys. **34**, 225–233 (1994).

Chapter 4
Tensor Fields

4.1 Tensor Algebras

Let V and V' be two arbitrary real vector spaces of any finite dimension each. Consider the product vector space $V \times V'$ consisting of all ordered pairs $(v, v'), v \in V, v' \in V'$. For instance, given $v \in V, v' \in V'$ and $\lambda, \mu \in \mathbb{R}$, $(\lambda v, \mu v')$ and $(\mu v, \lambda v')$ are two different vectors of $V \times V'$. Let W be the free real vector space generated by $V \times V'$, that is, the vector space consisting of all real linear combinations of vectors out of $V \times V'$. For instance, $2\lambda(v, v'), (\lambda v, v') + (v, \lambda v')$ are two more of different vectors of W. Let I be the subspace of W generated by all vectors of the forms

$$
\begin{aligned}
&(v_1 + v_2, v') - (v_1, v') - (v_2, v') \quad (v, v_1' + v_2') - (v, v_1') - (v, v_2') \\
&(\lambda v, v') - \lambda(v, v') \quad (v, \lambda v') - \lambda(v, v').
\end{aligned}
$$

The **tensor product** $V \otimes V'$ of the two vector spaces V and V' is the quotient space W/I. Its elements are linear combinations of the tensor products $v \otimes v'$ of vectors $v \in V, v' \in V'$ (image of the canonical mapping of $V \times V'$ to $V \otimes V'$) with the properties

$$
\begin{aligned}
(v_1 + v_2) \otimes v' &= v_1 \otimes v' + v_2 \otimes v', \\
v \otimes (v_1' + v_2') &= v \otimes v_1' + v \otimes v_2', \\
\lambda(v \otimes v') &= (\lambda v) \otimes v' = v \otimes (\lambda v').
\end{aligned}
\tag{4.1}
$$

Besides the elements of the form $v \otimes v'$, the space $V \otimes V'$ contains all their linear combinations. However, the canonical mapping $\phi : V \times V' \to V \otimes V' : (v, v') \mapsto v \otimes v'$ of ordered pairs (v, v') to their equivalence classes $v \otimes v'$ of the quotient space formation is a universal bilinear mapping in the following sense: If W is *any* vector space and $\psi : V \times V' \to W$ is *any* bilinear mapping, then there is a unique linear mapping $\psi' : V \otimes V' \to W$ so that $\psi = \psi' \circ \phi$. Up to isomorphisms, ϕ is uniquely determined by this universality property.

There are unique (obvious, canonical) isomorphisms between $V \otimes V'$ and $V' \otimes V$ and between $V \otimes (V' \otimes V'')$ and $(V \otimes V') \otimes V''$. Therefore it makes sense to write $V \otimes V' \otimes V''$ and accordingly for more factors, and the order of factors can be fixed by convention in these product constructions. (This does of course not mean that $v_1 \otimes v' + v_2 \otimes v'$ and $v_1 \otimes v' + v' \otimes v_2$ are equal; the second expression as a linear combination of elements of two different spaces is even not defined, if V and V' are different, and is different from the first expression, if $V = V'$. However, the first expression and $v' \otimes (v_1 + v_2)$ are conjugate by the canonical isomorphism.)

Let $\{e_1, \ldots, e_{n_V}\}$ and $\{e'_1, \ldots, e'_{n_{V'}}\}$ be bases of the vector spaces V and V'. Then $\{e_i \otimes e'_j\}$, $i = 1, \ldots, n_V$, $j = 1, \ldots, n_{V'}$ is a base of $V \otimes V'$.

All these are simple statements which can be proved as an exercise (cf. for instance [1]).

Let now V be any finite-dimensional real vector space and let V^* be its dual. Then the **tensor space** $V_{r,s}$ of type (r, s) is

$$V_{r,s} = \underbrace{V \otimes \cdots \otimes V}_{r \text{ copies}} \otimes \underbrace{V^* \otimes \cdots \otimes V^*}_{s \text{ copies}}. \tag{4.2}$$

A base in $V_{r,s}$ is

$$\{e_{i_1} \otimes \cdots \otimes e_{i_r} \otimes f^{j_1} \otimes \cdots \otimes f^{j_s}\}, \tag{4.3}$$

where the e_i form a base of V and the f^j form a base in V^* conveniently chosen dual to that of $V : \langle f^j, e_i \rangle = \delta^j_i$. Additionally one defines $V_{0,0} = \mathbb{R}$. If V is n-dimensional, then the tensor space $V_{r,s}$ is an n^{r+s}-dimensional vector space. A general element of $V_{r,s}$ is the tensor

$$t = \sum t^{i_1 \ldots i_r}_{j_1 \ldots j_s} e_{i_1} \otimes \cdots \otimes e_{i_r} \otimes f^{j_1} \otimes \cdots \otimes f^{j_s}, \tag{4.4}$$

where the summation runs from 1 to n over all indices which appear twice on the right hand expression, once as subscript and once as superscript. Further on, the summation sign will be omitted in tensor calculus, but not in exterior calculus for reasons becoming evident below, and **Einstein's summation convention** will be used which means the just described summation over pairs of indices always understood.

Now, the direct sum

$$\mathbb{T}(V) = \sum_{r,s \geq 0} V_{r,s}, \quad V_{0,0} = \mathbb{R}, \tag{4.5}$$

is called the **tensor algebra** of V. It is an associative but non-commutative (see remark in parentheses above on this page) graded (by r, s) algebra with unit. If $v_1 \otimes \cdots \otimes v_{r_1} \otimes w^{*1} \otimes \cdots \otimes w^{*s_1} \in V_{r_1,s_1}$ and $v'_1 \otimes \cdots \otimes v'_{r_2} \otimes w'^{*1} \otimes \cdots \otimes w'^{*s_2} \in V_{r_2,s_2}$, then their tensor product is defined as $v_1 \otimes \cdots \otimes v_{r_1} \otimes v'_1 \otimes \cdots \otimes v'_{r_2} \otimes w^{*1} \otimes \cdots \otimes w^{*s_1} \otimes w'^{*1} \otimes \cdots \otimes w'^{*s_2} \in V_{r_1+r_2,s_1+s_2}$ where the order of factors belongs to the definition of the product in the algebra (see above). The product of the two considered tensors as factors in inverse order is $v'_1 \otimes \cdots \otimes v'_{r_2} \otimes v_1 \otimes \cdots \otimes v_{r_1} \otimes w'^{*1} \otimes \cdots \otimes w'^{*s_2} \otimes w^{*1} \otimes \cdots \otimes w^{*s_1} \in V_{r_1+r_2,s_1+s_2}$. Tensors in some $V_{r,s}$ are called **homogeneous of degree** (r,s). If they are single tensor products of vectors and covectors as in the examples just considered, then they are called **decomposable**.

A change of the base and dual base,

$$e_i = \psi_i^k \tilde{e}_k, \quad f^j = (\psi^{-1})^j_l \tilde{f}^l, \quad \psi_i^k (\psi^{-1})^j_k = \delta_i^j, \tag{4.6}$$

with a regular $n \times n$ transformation matrix ψ, which should leave the tensor (4.4) unaffected, results in a transformation of the tensor components according to

$$\tilde{t}^{i_1 \dots i_r}_{j_1 \dots j_s} = \psi^{i_1}_{k_1} \cdots \psi^{i_r}_{k_r} t^{k_1 \dots k_r}_{l_1 \dots l_s} (\psi^{-1})^{l_1}_{j_1} \cdots (\psi^{-1})^{l_s}_{j_s}. \tag{4.7}$$

Hence, a tensor transforms like a **contravariant vector** with respect to its upper indices and like a **covariant vector** with respect to its lower indices. The **tensor product** of the tensors $t \in V_{r,s}$ and $t' \in V_{r',s'}$ has components

$$(t \otimes t')^{i_1 \dots i_{r+r'}}_{j_1 \dots j_{s+s'}} = t^{i_1 \dots i_r}_{j_1 \dots j_s} t'^{i_{r+1} \dots i_{r+r'}}_{j_{s+1} \dots j_{s+s'}}, \tag{4.8}$$

while the reversed order of the two factors leads to the reversed arrangement of the index groups in the tensor product and hence in general to a different result. Two tensors are equal if they have the same components with the order of indices observed.

Consider a decomposable tensor t of degree (r,s) and two integers $p, 1 \le p \le r$, and $q, 1 \le q \le s$. The **tensor contraction** $C_{p,q} : V_{r,s} \to V_{r-1,s-1}$ is defined as

$$\begin{aligned} C_{p,q}(t) &= C_{p,q}(v_1 \otimes \cdots \otimes v_r \otimes w^{*1} \otimes \cdots \otimes w^{*s}) \\ &= \langle v_p, w^{*q} \rangle v_1 \otimes \cdots \otimes v_{p-1} \otimes v_{p+1} \otimes \cdots \otimes v_r \otimes w^{*1} \\ &\quad \otimes \cdots \otimes w^{*q-1} \otimes w^{*q+1} \otimes \cdots \otimes w^{*s}, \end{aligned} \tag{4.9}$$

and for an arbitrary homogeneous tensor of degree (r,s) it is defined by linear continuation. For an arbitrary homogeneous tensor of degree (r,s) one has in components (summation over k)

$$C_{p,q}(t)^{i_1 \dots i_{r-1}}_{j_1 \dots j_{s-1}} = t^{i_1 \dots i_{p-1} k i_p \dots i_{r-1}}_{j_1 \dots j_{q-1} k j_q \dots j_{s-1}}. \tag{4.10}$$

There are various interrelations of tensors with mappings. First, every homogeneous tensor of degree (r,s) may be considered as a **multilinear mapping**

$$\begin{aligned} t : \underbrace{V^* \times \cdots \times V^*}_{r \text{ copies}} \times \underbrace{V \times \cdots \times V}_{s \text{ copies}} \to \mathbb{R} : \\ (w^{*1}, \dots, w^{*r}, v_1, \dots, v_s) \mapsto w^{*1}_{i_1} \cdots w^{*r}_{i_s} t^{i_1 \dots i_r}_{j_1 \dots j_s} v_1^{j_1} \cdots v_s^{j_s} \end{aligned} \tag{4.11}$$

of the indicated product vector space into the scalar field $\mathbb{R}$. Because of the universality of the canonical mapping ϕ of this product vector space onto $V_{r,s}$ as considered for two factors after (4.1), one has the following isomorphism:

$V_{r,s}$ is canonically isomorphic to the vector space of all $(r+s)$-linear mappings of $V^*\times\cdots\times V^*\times V\times\cdots\times V$ $((r+s)$factors) into $\mathbb{R}$.

There are simple variants of that proposition. For instance, there is a canonical isomorphism between $V_{1,s}$ and the s-linear mappings of $V\times\cdots\times V$ (s factors) into V. A symmetric (see below) tensor $g\in V_{0,2}$ with the property $g(v,v)\geq 0$ and $g(v,v)=0$ iff $v=0$ defines a scalar product in V and hence converts a general vector space V into a Euclidean space. Of course, all these mappings are mappings of vector spaces and do not depend on the actually chosen base in V. In this sense, scalars, vectors and tensors are called invariant and contra- and covariant, respectively, entities.

For $s=r$, one may also consider (4.11) as a bilinear mapping $V_{0,r}\times V_{r,0}\to\mathbb{R}$. It is easily seen that every bilinear mapping of these spaces into $\mathbb{R}$ has the form (4.11), hence the two spaces are dual to each other:

$$V_{0,r}=(V_{r,0})^*. \tag{4.12}$$

For $r=1$, this is just the duality of V and V^*, and, if V is a Euclidean space so that V^* is identified with V, then the mapping is the scalar product.

Next, consider mappings of V into another vector space V'. By duality, a homomorphism H from V to V' induces a homomorphism H^* from V'^* to V^* : $\langle w'^*,Hv\rangle=\langle H^*w'^*,v\rangle$, $v\in V$, $w'^*\in V'^*$. (Recall that a finite-dimensional vector space is reflexive. Given bases in V and V', H is represented by a matrix, and H^* by its transposed.) If V and V' are isomorphic, so are V^* and V'^*, and there is also an isomorphic mapping from V^* onto V'^* which is H^{*-1}. Let $\tilde{H}:\mathbb{T}(V)\to\mathbb{T}(V')$ be an isomorphism which in case of decomposable tensors acts like H on each vector factor and like H^{*-1} on each covector factor. It is easily seen that $\tilde{H}$ commutes with contractions, if it acts on $V_{0,0}=\mathbb{R}$ as the identity mapping (exercise). The following statement is now rather obvious:

There is a canonical one–one mapping between isomorphisms from V to V' and isomorphisms from $\mathbb{T}(V)$ to $\mathbb{T}(V')$ which preserve the degree and commute with tensor contraction. In particular, the automorphism group of V is isomorphic to the automorphism group of $\mathbb{T}(V)$.

The automorphisms of a vector space are also called (regular) **transformations**. By the canonical isomorphism between $V_{1,1}$ and the space of linear mappings (endomorphisms) of V into V (see top of this page) there is a one–one correspondence of tensors a of degree (1,1) with components given by *regular* matrices and automorphisms A of V. These tensors are called transformation tensors. Sometimes these transformations, which transform a given vector $v\in V$ in general in a different one $v'=Av$, in components related to a *fixed* base $v'^i=a^i_jv^j$, are called 'active coordinate transformations' while ordinary coordinate transformations (4.6), which leave all vectors on place and only switch to another base are called 'passive coordinate transformations'.

Accordingly, if $B : V \to V$ is any endomorphism of V, it corresponds to a tensor $b \in V_{1,1}$ whose components do not necessarily form a regular matrix. By an automorphism A, B is transformed into ABA^{-1} and the corresponding transformation of b in $\mathbb{T}(V)$ is in components $b^i_j \to a^i_k b^k_l (a^{-1})^l_j$. Endomorphisms of vector spaces form a ring: if B, B' are two endomorphisms, then $B + B'$ and BB' are again endomorphisms. With the Lie product $[B, B'] = BB' - B'B$ they form also a Lie algebra. An endomorphism D of $\mathbb{T}(V)$ is called a **derivation**, if

$$\begin{aligned} &1. \quad D \text{ preserves the degree: } DV_{r,s} \subset V_{r,s}, \\ &2. \quad D(t \otimes t') = (Dt) \otimes t' + t \otimes (Dt'), \\ &3. \quad DC_{p,q} = C_{p,q}D \text{ for every tensor contraction.} \end{aligned} \tag{4.13}$$

It is directly seen that, if D, D' are two derivations of $\mathbb{T}(V)$, then $[D, D']$ is again a derivation. The derivations of $\mathbb{T}(V)$ form another Lie algebra.

The Lie algebra of derivations of $\mathbb{T}(V)$ is isomorphic to the Lie algebra of endomorphisms of V; the isomorphism is provided by the restriction B of derivations D to $V \subset \mathbb{T}(V)$.

Proof As an endomorphism, $D\lambda t = \lambda Dt$, $\lambda \in \mathbb{R}$. However, $\lambda t = \lambda \otimes t$ in $\mathbb{T}(V)$. From property 2 of (4.13) it follows that $D\lambda = 0$ for every $\lambda \in \mathbb{R}$. Hence, with property 3, $0 = D\langle w^*, v\rangle = \langle Dw^*, v\rangle + \langle w^*, Dv\rangle$ for all $v \in V$ and $w^* \in V^*$ (exercise). By putting $Dv = Bv$, it follows $Dw^* = -B^* w^*$ where B^* is the endomorphism transposed to B, $\langle w^*, Bv\rangle = \langle B^* w^*, v\rangle$. Given B, from these relations D is determined for decomposable tensors and is uniquely extended by linearity to all $\mathbb{T}(V)$. It is easily seen that $B \mapsto D$ is a bijection. □

If B' is another endomorphism of V, it is transformed by the derivation D with $Dv = Bv$ into $DB' = BB' - B'B = [B, B']$.

Let $\mathcal{P}_r$ be the permutation group of the set $\{1, \ldots, r\}$ of r numbers and let $P \in \mathcal{P}_r$. Denote by the same letter a mapping $P : V_{r,0} \to V_{r,0} : t^{i_1 \ldots i_r} \mapsto (Pt)^{i_1 \ldots i_r} = t^{i_{P1} \ldots i_{Pr}}$ and an analogous mapping $P : V_{0,r} \to V_{0,r}$. This definition is obviously independent of the choice of a base in V. The symmetrization of a tensor of degree either $(r, 0)$ or $(0, r)$ is

$$\mathcal{S}t = \frac{1}{r!} \sum_{P \in \mathcal{P}_r} Pt \tag{4.14}$$

and the alternation is

$$\mathcal{A}t = \frac{1}{r!} \sum_{P \in \mathcal{P}_r} \operatorname{sign}(P) Pt, \tag{4.15}$$

where $\operatorname{sign}(P) = +1$ for an even permutation and $\operatorname{sign}(P) = -1$ for an odd permutation. A tensor $\mathcal{S}t$ is called a **symmetric tensor** and a tensor $\mathcal{A}t$ is called an

alternating tensor. It is directly seen that these tensors provide symmetric and alternating multilinear mappings of the product vector space of r factors V^* or r factors V into $\mathbb{R}$.

4.2 Exterior Algebras

Let $\underline{\mathbb{T}}(V) = \sum_{r=0}^{\infty} V_{r,0}$ be the subalgebra of contravariant tensors of $\mathbb{T}(V)$ and let $I(V)$ be the two-sided ideal of $\underline{\mathbb{T}}(V)$ generated by all elements of the form $v \otimes v$, $v \in V$, that is, $I(V)$ is the linear span of the sets $\underline{\mathbb{T}}(V) \otimes v \otimes v \otimes \underline{\mathbb{T}}(V)$ for all $v \in V$. The exterior algebra or **Grassmann algebra** of V is the graded algebra $\Lambda(V) = \underline{\mathbb{T}}(V)/I(V)$. Then, $\underline{\mathbb{T}}(V^*)$ is the subalgebra of covariant tensors of $\mathbb{T}(V)$ and $\Lambda(V^*) = \underline{\mathbb{T}}(V^*)/I(V^*)$. Grassmann was the first to introduce the exterior algebra for the study of subspaces of vector spaces.

$\Lambda(V)$ is graded in the following way: $\Lambda_0(V) = \mathbb{R}$, $\Lambda_1(V) = V$, $\Lambda_r(V) = V_{r,0}/I_r(V)$, $I_r(V) = I(V) \cap V_{r,0}$ for $r > 1$ and $\Lambda(V) = \sum_{r=0}^{\infty} \Lambda_r(V)$. Since for every $v_1, v_2 \in V$ the products $v_1 \otimes v_1$, $v_2 \otimes v_2$ and $(v_1 + v_2) \otimes (v_1 + v_2)$ are elements of $I_2(V)$, it follows that also $v_1 \otimes v_2 + v_2 \otimes v_1 \in I_2(V)$. Hence, if the product in $\Lambda(V)$ is denoted by $\wedge$, then $v_1 \wedge v_2 = -v_2 \wedge v_1$. It is easily seen that, if $\omega, \sigma, \tau \in \Lambda(V)$ and $F, G \in \mathbb{R} = \Lambda_0(V)$, then the **exterior product** $\wedge$ in $\Lambda(V)$ has all the properties (3.23) of a wedge-product.

Since every decomposable tensor $v_1 \otimes \cdots \otimes v_r$, $r > 1$, containing two consecutive equal factors is in $I(V)$ and reordering of the factors in $\mathbb{T}(V)$ leads to representatives of the same or of the reversed (reversed sign) equivalence class of $\mathbb{T}(V)/I(V) = \Lambda(V)$, the elements of the exterior algebra $\Lambda(V)$ may be represented as linear combinations of

$$1;\ e_{i_1} \wedge \cdots \wedge e_{i_r} = e_{i_1} \otimes \cdots \otimes e_{i_r}, \quad i_1 < \cdots < i_r, \quad r > 0, \tag{4.16}$$

for any given base $\{e_i\}$ of V. (The 'base vector' 1 spans $\Lambda_0(V) = \mathbb{R}$, and $1 \wedge e_i = 1e_i - e_i 1 = 0$.) Hence, (4.16) forms a base of $\Lambda(V)$. It immediately follows that, if $\dim V = n$, then

$$\dim \Lambda_r(V) = \binom{n}{r} = \frac{n!}{r!(n-r)!}, \quad \Lambda_n(V) \approx \mathbb{R}, \quad \Lambda_r(V) = \{0\} \quad \text{for } r > n \tag{4.17}$$

and hence $\dim \Lambda(V) = 2^n$. As opposed to the tensor algebra $\mathbb{T}(V)$, the exterior algebra $\Lambda(V)$ is finite-dimensional.

Again the canonical mapping $\phi' : V \times \cdots \times V \to \Lambda_r(V) : (v_1, \ldots, v_r) \mapsto v_1 \wedge \cdots \wedge v_r$ has the universality property that, if W is *any* vector space and $\psi : V \times \cdots \times V \to W$ is any alternating r-linear mapping, then there is a unique linear mapping $\psi' : \Lambda_r(V) \to W$ so that $\psi = \psi' \circ \phi'$. Up to isomorphisms, ϕ' is uniquely determined by this universality property.

From the base (4.16) and the properties (3.23) it is seen that the elements of the exterior algebra $\Lambda(V)$ may be represented by the linear combinations of alternating tensors $\sum_r \mathcal{A}t_r,\ t_r \in V_{r,0}$. Likewise, the elements of $\Lambda(V^*)$ may be represented by the linear combinations $\sum_r \mathcal{A}t^r,\ t^r \in V_{0,r}$. If $t \in \Lambda_r(V),\ r > 0$, then it may be represented as

$$t = \sum_{i_1,\dots,i_r} t^{i_1\dots i_r} e_{i_1} \otimes \cdots \otimes e_{i_r} = r! \sum_{i_1<\dots<i_r} t^{i_1\dots i_r} e_{i_1} \wedge \cdots \wedge e_{i_r}. \tag{4.18}$$

In the middle expressions, the i-sums run independently from 1 to n, but since $t^{i_1\dots i_r}$ is alternating, only items with distinct i are non-zero, and their $r!$ permutations appearing in the sums may be summed up into the right expressions.

Consider now $e_{i_1} \wedge \cdots \wedge e_{i_r}$, which is a special homogeneous tensor. According to (4.18), its components consist only of alternating sign factors:

$$e_{i_1} \wedge \cdots \wedge e_{i_r} = \frac{1}{r!} \sum_{P\in\mathcal{P}_r} \operatorname{sign}(P) e_{i_{P1}} \otimes \cdots \otimes e_{i_{Pr}}.$$

As an alternating r-linear mapping from $V^* \times \cdots \times V^*$ (r factors) into $\mathbb{R}$ like in (4.11) it yields (cf. the text after (3.23))

$$e_{i_1} \wedge \cdots \wedge e_{i_r}(w^{*1},\dots,w^{*r}) = \frac{1}{r!}\det(\langle w^{*i}, e_j\rangle). \tag{4.19}$$

By r-linearity and by the expansion rule for determinants, the base vectors in this relation may be replaced by any set of r linearly independent vectors $v_1,\dots,v_r$ of the vector space $V = \Lambda_1$. Using tensor contractions, one may also consider the bilinear mapping $\Lambda_r(V^*) \times \Lambda_r(V) \to \mathbb{R}$, for which again the right hand side of (4.19) would follow. However, in order to avoid nasty factorial prefactors in the exterior calculus, one *redefines* the prefactor of the latter mapping as

$$\langle w^{*1} \wedge \cdots \wedge w^{*r}, v_1 \wedge \cdots \wedge v_r\rangle = \det(\langle w^{*i}, v_j\rangle). \tag{4.20}$$

By linearity this mapping (4.20) can be extended to all $\Lambda_r(V^*) \times \Lambda_r(V)$, and this form comprises all linear functions from $\Lambda_r(V)$ into $\mathbb{R}$. Hence, $\Lambda_r(V^*)$ is isomorphic to the dual vector space to $\Lambda_r(V)$:

$$(\Lambda_r(V))^* \approx \Lambda_r(V^*) \text{ and hence } (\Lambda(V))^* \approx \Lambda(V^*). \tag{4.21}$$

Since the dual to a *finite* direct sum of vector spaces is isomorphic to the direct sum of their duals, the second relation follows. Note that this duality relation $\langle\cdot,\cdot\rangle$ differs from that transferred from (4.12) via the quotient algebra formation by an additional prefactor $r!$. For $r = 1$ they are equal. In practice, in the exterior calculus only the form (4.20) is used and no confusion can arise.

Consider again dual bases $\{e_j\}$ in V and $\{f^i\}$ in V^*, $\langle f^i, e_j\rangle = \delta^i_j$. Equation 4.20 yields

$$\langle f^{i_1} \wedge \cdots \wedge f^{i_r}, e_{j_1} \wedge \cdots \wedge e_{j_r}\rangle = \sum_{P\in\mathcal{P}_r} \operatorname{sign}(P)\delta^{i_1}_{j_{P1}} \cdots \delta^{i_r}_{j_{Pr}} = \delta^{i_1\dots i_r}_{j_1\dots j_r}. \tag{4.22}$$

This is called a **generalized Kronecker symbol**. It is zero, if the sets of integers $\{i_k\}$ and $\{j_k\}$ are not the same; it is equal to 1, if they are the same and the ordered set $(j_1,\ldots,j_r)$ is an even permutation of the ordered set $(i_1,\ldots,i_r)$, and equal to -1, if it is an odd permutation. Let now $t \in \Lambda_r(V)$ be any alternating tensor (4.18) or $t' \in \Lambda_r(V^*)$, then

$$t^{i_1\ldots i_r} = \frac{1}{r!}\langle f^{i_1} \wedge \cdots \wedge f^{i_r}, t\rangle, \quad t'_{j_1\ldots j_r} = \frac{1}{r!}\langle t', e_{j_1} \wedge \cdots \wedge e_{j_r}\rangle. \tag{4.23}$$

These are the general rules of calculating vector components by projection on the dual basis, applied to the cases of $\Lambda_r(V)$ and $\Lambda_r(V^*)$. For later use, an important consequence of (4.22) for the relation between tensor algebra and exterior algebra is

$$\frac{1}{r!}\langle \omega, X_1 \wedge \cdots \wedge X_r\rangle = C_{1,1}\cdots C_{r,r}(\omega \otimes X_1 \otimes \cdots \otimes X_r), \tag{4.24}$$

where ω is an arbitrary alternating tensor of type $(0,r)$ and X_i are vectors.

Next, important endomorphisms of $\Lambda(V)$ are considered. In (4.17) it was already noted that $\Lambda_r(V) = \{0\}$, if $r > \dim V$. In addition, by definition,

$$\Lambda_r(V) = \{0\} \quad \text{for } r<0, \quad \Lambda(V) = \sum_{-\infty<r<\infty} \Lambda_r(V). \tag{4.25}$$

Obviously, the last relation is the same as that given at the beginning of this section. An endomorphism L of a graded algebra Λ is called an **endomorphism of degree** s, if

$$L : \Lambda_r \to \Lambda_{r+s}, \quad -\infty<r,s<\infty. \tag{4.26}$$

For instance, for any $u \in \Lambda_s(V)$, the wedge-multiplication $L_u : \Lambda(V) \to \Lambda(V) : t \mapsto u \wedge t$ is an endomorphism of degree s. An endomorphism is called a **derivation**, if

$$D(t \wedge t') = (Dt) \wedge t' + t \wedge (Dt') \quad \text{for all } t, t' \in \Lambda, \tag{4.27}$$

it is called an **anti-derivation**, if

$$D(t \wedge t') = (Dt) \wedge t' + (-1)^r t \wedge (Dt') \quad \text{for all } t \in \Lambda_r,\ t' \in \Lambda. \tag{4.28}$$

By repeated application of (4.28) and realization of the associativity of the algebra $\Lambda(V)$ one gets for an antiderivation of a decomposable element

$$D(v_1 \wedge \cdots \wedge v_r) = \sum_{i=1}^{r}(-1)^{i+1} v_1 \wedge \cdots \wedge (Dv_i) \wedge \cdots \wedge v_r. \tag{4.29}$$

As opposed to the derivatives in analysis considered in Sect. 2.3, the derivations of an algebra (or a ring) are always endomorphisms. Examples of derivations of degree 0 are the linear derivation of the algebra $\mathcal{F}_{x_0}$ of germs of real functions, (3.10), (formally, any algebra can be considered as a graded algebra, where all subspaces of non-zero degree are $\{0\}$), the derivation of the algebra $\mathcal{C}(M)$ of smooth real functions, (3.15), by any tangent vector field and the derivation (4.13) of a tensor algebra. An example of an anti-derivation of degree 1 is the exterior differentiation d of r-forms, (3.25).

With regard to mutually dual algebras $\Lambda(V)$ and $(\Lambda(V))^* = \Lambda(V^*)$, the transposed of the above mentioned endomorphism L_u, $u \in \Lambda(V)$ may be considered. It is denoted by $\imath_u$:

$$\imath_u : \Lambda(V^*) \to \Lambda(V^*) : \langle \imath_u t^*, t' \rangle = \langle t^*, L_u t' \rangle = \langle t^*, u \wedge t' \rangle \quad \text{for every } t' \in \Lambda(V). \tag{4.30}$$

It is called the **interior multiplication** by u in $\Lambda(V^*)$. (For $t^*, t' \in \Lambda_0(V) = \mathbb{R}$ it follows from the remark in parentheses after (4.16) that $\imath_u t^* = 0$.) An example is the Hodge operator considered in Sect. 5.1.

For $v \in V$, the endomorphism $\imath_v$ is an anti-derivation of degree -1 on $\Lambda(V^*)$.

Proof Since L_v is of degree 1 according to its description after (4.26), it follows from (4.30) that $\imath_v$ is of degree -1. Consider decomposable elements. From the definition of $\imath_v$, $\langle \imath_{v_1}(w^{*1} \wedge \cdots \wedge w^{*r}), v_2 \wedge \cdots \wedge v_r \rangle = \langle w^{*1} \wedge \cdots \wedge w^{*r},\ v_1 \wedge v_2 \wedge \cdots \wedge v_r \rangle = \det(\langle w^{*i}, v_j \rangle)$. If one replaces in (4.29) D with $\imath_{v_1}$ and v_i with w^{*i} and inserts the right hand side of the obtained relation into $\langle \imath_{v_1}(w^{*1} \wedge \cdots \wedge w^{*r}),$ $v_2 \wedge \cdots \wedge v_r \rangle$, one obtains the expansion of the same determinant with respect to its first line. □

If $\psi : V \to V'$ is a homomorphism of vector spaces, it extends to a homomorphism (push forward) $\psi_* : \Lambda(V) \to \Lambda(V')$ of algebras as $\psi_*(v_1 \wedge \cdots \wedge v_r) = \psi(v_1) \wedge \cdots \wedge \psi(v_r)$ and further by linear extension. It also yields by duality a homomorphism (pull back) $\psi^* : \Lambda(V'^*) \to \Lambda(V^*)$ via $\langle \psi^*(u'^*), t \rangle = \langle u'^*, \psi_*(t) \rangle$ for all $u'^* \in \Lambda(V'^*)$ and all $t \in \Lambda(V)$.

The section is closed with three simple useful theorems which are easily proved by completion of the considered sets of vectors to a base of the vector space and by observing that for any base $\{v_i\}$ of an n-dimensional vector space V the wedge-product $v_1 \wedge \cdots \wedge v_n$ is non-zero, and that an expansion of a vector into a base has unique components.

A set of vectors $v_1, \ldots, v_r \in V$ is linearly dependent, iff $v_1 \wedge \cdots \wedge v_r = 0$.

Let $\{v_i\}$ and $\{v'_i\}$ be two sets each of r vectors of V so that $\sum_{i=1}^r v_i \wedge v'_i = 0$. If the v_i are linearly independent, then the v'_j may be linearly expanded into the v_i, $v'_j = \sum_{i=1}^r \psi_{ji} v_i$, with a symmetric coefficient matrix, $\psi_{ji} = \psi_{ij}$.

Let $\{v_i\}$ be a set of r linearly independent vectors of V and let $t \in \Lambda_s(V)$. Then, $t \equiv 0 \bmod (v_1, \ldots, v_r)$, that is $t = v_1 \wedge u_1 + \cdots + v_r \wedge u_r$ with certain $u_i \in \Lambda_{s-1}(V)$, iff $v_1 \wedge \cdots \wedge v_r \wedge t = 0$.

Prove the theorems as an exercise. For $s = 2$ the last one was considered in the Frobenius theorem for Pfaffian systems, Sect. 3.6.

4.3 Tensor Fields and Exterior Forms

In Sect. 3.4 tangent vector fields on a manifold M were introduced by assigning a tangent vector to every point $x \in M$. While smoothness of real functions on M could naturally be defined with the help of coordinate neighborhoods on the basis of a well-defined atlas structure for M (pseudo-group of transition functions), this is not so simple for a general vector field. Although an n dimensional vector field may be given by n real component functions, these components depend on the choice of a base in the vector space at each point of M, and for the concept of smoothness some rules are needed how the bases of vector spaces on neighboring points x of M should be related. This will be finally worked out in Chap. 7 with the concept of fiber bundles. For the special case of tangent space the problem was solved by relating the bases of the tangent spaces to the coordinates on coordinate neighborhoods of M via considering the action of a tangent vector on real functions on M. Then, the bases of cotangent spaces were related to those of tangent spaces via considering the action of cotangent vectors (1-forms) on tangent vectors. Since the base of a tensor algebra or an exterior algebra on a vector space is determined by the base of the vector space, tensor fields and exterior fields on tangent and cotangent spaces, which are sections of corresponding fiber bundles considered in Chap. 7, can be treated here without the concept of fiber bundles.

Let M be a manifold, T_x the tangent space and T_x^* the cotangent space at point $x \in M$. Consider the sets

$$T_{r,s}(M) = \bigcup_{x \in M} (T_x)_{r,s} : \text{tensor bundle of type } (r,s) \text{ over } M, \tag{4.31}$$

$$\Lambda_r^*(M) = \bigcup_{x \in M} \Lambda_r(T_x^*) : \text{exterior } r \text{ bundle over } M. \tag{4.32}$$

In Chap. 7 a topology will be introduced into these sets to provide them with the special manifold structure of bundles.

Consider sets of r 1-forms ω^i, $i = 1, \dots, r$ and s tangent vector fields X_j, $j = 1, \dots, s$ on M, and multilinear mappings $t : T^*(M) \times \cdots \times T^*(M) \times T(M) \times \cdots \times T(M) \to \mathcal{C}(M)$ with r factors $T^*(M)$ and s factors $T(M)$. According to (4.11), at each point $x \in M$ the mapping is given by $(\omega_x^1, \dots, \omega_x^r, \xi_{1x}, \dots, \xi_{sx}) \mapsto t^{i_1 \dots i_r}_{j_1 \dots j_s}(x) \omega^1_{xi_1} \cdots \omega^r_{xi_r} \xi^{j_1}_{1x} \cdots \xi^{j_s}_{sx} \in \mathbb{R}$. Recall that in coordinate neighborhoods in M the base forms dx^i in $T^*(M)$ and the base vectors $\partial/\partial x^j$ in $T(M)$ are smooth. Hence, the mapping is into $\mathcal{C}(M)$, if the component functions $t^{i_1 \dots i_r}_{j_1 \dots j_s}(x)$ are smooth functions of x. In this case,

$$t(x) = t^{i_1 \dots i_r}_{j_1 \dots j_s}(\mathbf{x}) \frac{\partial}{\partial x^{i_1}} \otimes \cdots \otimes \frac{\partial}{\partial x^{i_r}} \otimes dx^{j_1} \otimes \cdots \otimes dx^{j_s} \tag{4.33}$$

is called a **tensor field** of type (r, s) on M.

The collection of all tensor fields of type (r, s) on M forms again a real vector space $\mathcal{T}_{r,s}(M)$ with respect to point wise linear combinations as is easily seen from (4.33). It is an infinite-dimensional functional vector space. It can also be considered a $\mathcal{C}(M)$-module with respect to point wise multiplications with $\mathcal{C}(M)$-functions. The graded **algebra of tensor fields** on M is

$$\mathcal{T}(M) = \sum_{r,s=0}^{\infty} \mathcal{T}_{r,s}(M). \tag{4.34}$$

It is a real associative but non-commutative algebra with point wise tensor multiplication as the multiplication in $\mathcal{T}(M)$. Note that according to the definition (4.34) a tensor field of a certain type at some point $x \in M$ has the same type all over M. For connected components of M this is a consequence of the demand of smoothness, for distinct components of a multicomponent manifold it is just by definition. Also, tensor contractions of tensor fields on M are defined as the same contraction performed at every point $x \in M$.

An important example of a symmetric tensor field of type $(0, 2)$ is the Riemannian **metric tensor**, in a coordinate neighborhood given by $g(x) = g_{ij}(x) dx^i \otimes dx^j$, or, as a bilinear mapping, $g(X, Y) = g_{ij} \xi^i \eta^j$ with the properties $g(X, X) \geq 0, g(X, X) = 0$, iff $X = 0$, and $g(X, Y) = g(Y, X)$. It defines at every point $x \in M$ a scalar product and hence converts the tangent space $T_x(M)$ into an inner product space (cf. p. 19) and M into a Riemannian manifold, a concept which is considered in more detail in Chap. 9.

An endomorphism of $\mathcal{T}(M)$ is a real linear mapping from tensor fields to tensor fields. At every point $x \in M$, it induces an endomorphism of the tensor algebra $\mathbb{T}(T_x)$ of the tangent space T_x on M at that point x, which in a sense analyzed in Chap. 7 depends smoothly on x. The endomorphism of $\mathcal{T}(M)$ is again called a **derivation**, if at every x it has the properties (4.13). As an endomorphism, a derivation of $\mathcal{T}(M)$ again vanishes applied to a *constant* $\lambda \in \mathcal{C}(M) = \mathcal{T}_{0,0}(M)$ but not in general for a function $F \in \mathcal{C}(M)$. (An endomorphism of $\mathcal{T}(M)$ is an $\mathbb{R}$-linear mapping but not a $\mathcal{C}(M)$-linear mapping.)

The most important derivation is the **Lie derivative** L_X with respect to the tangent vector field X. From (3.15) on p. 68 it follows that for every tangent vector field $X \in \mathcal{X}(M) = \mathcal{T}_{1,0}(M)$ the mapping $X : \mathcal{C}(M) \to \mathcal{C}(M) : F \mapsto XF$ is a derivation of $\mathcal{T}_{0,0}(M) = \mathcal{C}(M)$. It maps F to the directional derivative of F in the directions of the integral curves of X on M. In (3.37) on p. 82 it was shown that the mapping $X : \mathcal{T}_{1,0}(M) \to \mathcal{T}_{1,0}(M) : Y \mapsto [X, Y]$ maps similarly a tangent vector field $Y \in \mathcal{T}_{1,0}(M)$ to its derivatives along the integral curves of X and hence is a derivation of $\mathcal{T}_{1,0}(M) = \mathcal{X}(M)$. By definition, in these two cases

$$L_X F = XF, \quad L_X Y = [X, Y]. \tag{4.35}$$

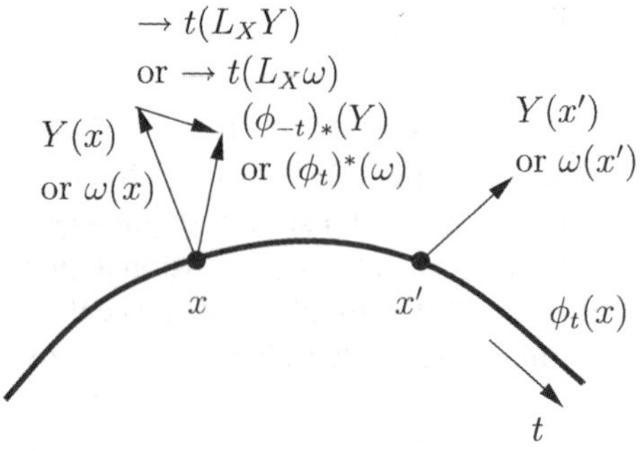

Fig. 4.1 The Lie derivative along $\phi_t(x)$ corresponding to the tangent vector field X

Both cases may be understood in the following way (Fig. 4.1): Fix a tangent vector field X on M. It defines a local 1-parameter group $\phi_t(x)$ in a neighborhood of every point $x \in M$. Fix x and consider the (unique maximal) integral curve $\phi_t(x)$ of X through x. Push the considered entity (F or Y) forward by $(\phi_{-t})_*$ (that is backward on the curve $\phi_t(x)$). This way its value originally at $x' = \phi_t(x)$ is brought to x, and there it is compared to the original value at x. A similar procedure can be applied to a cotangent vector field ω. This time it is pulled back from x' to x by $(\phi_t)^*$ (cf. (3.28) on p. 72). Hence, a derivation may be defined for any tensor field $u \in \mathcal{T}(M)$ as

$$L_X u = \lim_{t\to 0}\frac{\Phi_t u - u}{t}, \quad \Phi_t = \begin{cases} (\phi_{-t})_* & \text{for tangent vector fields} \\ (\phi_t)^* & \text{for cotangent vector fields} \end{cases} \tag{4.36}$$

which declares Φ_t for decomposable tensor fields, and then extended by linearity. The proof that (4.36) is indeed a derivation is the same as for the product rule of any derivative. Obviously, $L_X : \mathcal{T}_{r,s}(M) \to \mathcal{T}_{r,s}(M)$, and L_X commutes with tensor contractions as expressions of the type (4.36) do commute with linear combination with constant coefficients.

It is again obvious that real linear combinations and Lie products $[D, D'] = DD' - D'D$ of derivations form again derivations; the set of all derivations of $\mathcal{T}(M)$ is a Lie algebra. Let D be a derivation of $\mathcal{T}(M)$. As it was shown in Sect. 3.4, its action on $\mathcal{T}_{0,0}(M) = \mathcal{C}(M)$ comes from a tangent vector field Y and can be localized at every point $x \in M$. According to the first relation (4.35) it may be denoted L_Y and acts as a Lie derivative. Let $U \subset M$ be an open set and consider all functions $F \in \mathcal{C}(M)$ with $\operatorname{supp} F \subset U$. A tangent vector field $X \in \mathcal{X}(M)$ may be said to be zero on U, if $XF = 0$ for all those F, and $\operatorname{supp} X$ may be defined as the smallest closed subset of M outside of which X is zero. Now, consider the action of any derivation D of $\mathcal{T}(M)$ on $FX = F \otimes X \in \mathcal{T}_{1,0}(M) = \mathcal{X}(M)$. From the second property (4.13), by arguments analogous to those in Sect. 3.4, it is seen that the action of D on $\mathcal{X}(M)$ can be localized: $\operatorname{supp} DX \subset \operatorname{supp} X$. Hence, at every point $x \in M$, any derivation D of $\mathcal{T}(M)$ which vanishes on $\mathcal{C}(M)$ induces an endomorphism of $T_x(M)$ which according to the analysis after (4.13) uniquely defines a derivation of the tensor algebra $\mathbb{T}(T_x)$. Let D and D' coincide on $\mathcal{C}(M)$ and on

$\mathcal{X}(M)$. Then, $D - D'$ vanishes there and hence vanishes on the whole $\mathcal{T}(M)$. The consequence is the first part of the following theorem:

Any derivation D of $\mathcal{T}(M)$ is uniquely determined by its restrictions to $\mathcal{T}_{0,0}(M) = \mathcal{C}(M)$ and to $\mathcal{T}_{1,0}(M) = \mathcal{X}(M)$. Moreover, it has the form $D = L_X + S$ with a uniquely determined tangent vector field X and a uniquely determined endomorphism S given by a tensor field s of type $(1,1)$.

Proof of the second statement of the theorem Fix a derivation D of $\mathcal{T}(M)$. The analysis of (3.15) on p. 68 shows that every derivation of $\mathcal{C}(M)$ is provided by a tangent vector field X. Take this X and consider the derivation $D - L_X$ of $\mathcal{T}(M)$. It is zero on $\mathcal{C}(M)$. Let D' be any derivation vanishing on $\mathcal{C}(M)$, let Y be any tangent vector field and $F \in \mathcal{C}(M)$. Then, $D'(FY) = FD'Y$ which is a linear mapping of $\mathcal{X}(M)$ into $\mathcal{X}(M)$ and hence, according to the theorem after (4.13), it defines uniquely a tensor field $s \in \mathcal{T}_{1,1}(M)$. Now, the second statement of the theorem follows from the first one. □

If again D is any derivation of $\mathcal{T}(M)$ and S is any tensor field of type $(1,1)$, then $[D,S]F = (DS)F + S(DF) - SDF = FDS$ which is another tensor field of type $(1,1)$: $[D,S] \in \mathcal{T}_{1,1}(M)$ for $S \in \mathcal{T}_{1,1}(M)$. In other words, $\mathcal{T}_{1,1}(M)$ is an ideal of the Lie algebra of all derivations of $\mathcal{T}(M)$. Now, $[L_X, L_Y]F = [X,Y]F$, and, for $Z \in \mathcal{X}(M)$, $[L_X, L_Y]Z = [X,[Y,Z]] - [Y,[X,Z]] = [[X,Y],Z]$ due to Jacobi's identity (3.17-3). Hence,

$$[L_X, L_Y] = L_{[X,Y]}. \tag{4.37}$$

The algebra of all Lie derivatives is itself a Lie subalgebra of the Lie algebra of all derivations of $\mathcal{T}(M)$.

In order to find the coordinate expressions of Lie derivatives, consider first $L_X(C_{1,1}(Y \otimes \omega)) = C_{1,1}((L_X Y) \otimes \omega + Y \otimes (L_X \omega))$ which equation can be rewritten as $C_{1,1}(Y \otimes L_X \omega) = XC_{1,1}(Y \otimes \omega) - C_{1,1}([X,Y] \otimes \omega)$. In a coordinate neighborhood in which $X = \sum \xi^i(\partial/\partial x^i)$ and $Y = \sum \eta^j(\partial/\partial x^j)$ with $\eta^j = \delta^j_k$, with (4.10) and (3.19) straightforwardly $(L_X\omega)_k = \omega_j(\partial\xi^j/\partial x^k)$ is obtained. In the same case, $(L_X Y)^i = [X,Y]^i = -\partial\xi^i/\partial x^k$ results. Now, for a general tensor field u of type (r,s),

$$\begin{aligned}
&L_X C_{1,1} \cdots C_{r+s,r+s}(u \otimes \omega^1 \otimes \cdots \otimes \omega^r \otimes X_1 \otimes \cdots \otimes X_s) \\
&= C_{1,1} \cdots C_{r+s,r+s}\Bigg((L_X u) \otimes \omega^1 \otimes \cdots \otimes \omega^r \otimes X_1 \otimes \cdots \otimes X_s \\
&\quad + \sum_{p=1}^{r} u \otimes \omega^1 \otimes \cdots \otimes (L_X \omega^p) \otimes \cdots \otimes \omega^r \otimes X_1 \otimes \cdots \otimes X_s \\
&\quad + \sum_{q=1}^{s} u \otimes \omega^1 \otimes \cdots \otimes \omega^r \otimes X_1 \otimes \cdots \otimes (L_X X_q) \otimes \cdots \otimes X_s \Bigg).
\end{aligned} \tag{4.38}$$

For $\omega^i = dx^i$ and $X_j = \partial/\partial x^j$, the left hand side of the equation (4.38) is $L_X(u^{i_1\dots i_r}_{j_1\dots j_s}) = \xi^i \partial u^{i_1\dots i_r}_{j_1\dots j_s}/\partial x^i$ and the first term on the right hand side is $(L_X u)^{i_1\dots i_r}_{j_1\dots j_s}$, altogether,

$$(L_X u)^{i_1\dots i_r}_{j_1\dots j_s} = \xi^i \frac{\partial u^{i_1\dots i_r}_{j_1\dots j_s}}{\partial x^i} - \sum_{p=1}^{r} \frac{\partial \xi^{i_p}}{\partial x^i} u^{i_1\dots i_{p-1} i i_{p+1}\dots i_r}_{j_1\dots j_s} + \sum_{q=1}^{s} \frac{\partial \xi^j}{\partial x^{j_q}} u^{i_1\dots i_r}_{j_1\dots j_{q-1} j j_{q+1}\dots j_s}. \quad (4.39)$$

It was shown in the last section that the elements of $\Lambda_r(T^*_x)$ are just the alternating tensors of type $(0, r)$. Hence, the sections out of $\Lambda^*_r(M)$ are just the alternating tensor fields of type $(0, r)$. They can be identified with the **exterior differential r-forms** already introduced in Sect. 3.4 on p. 70. They form the real vector space $\mathcal{D}^r(M)$ which is also a $\mathcal{C}(M)$-module with respect to point wise linear combinations. Their point wise exterior multiplication yields the graded **exterior algebra**

$$\mathcal{D}(M) = \sum_{r=-\infty}^{\infty} \mathcal{D}^r(M), \quad \mathcal{D}^r(M) = \{0\} \quad \text{for } r<0 \text{ and } r>n \quad (4.40)$$

studied in Sect. 3.4.

Consider a mapping $F : M \to N$ from the manifold M into the manifold N and at every $x \in M$ its push forward (differential) as a mapping $F^x_* : T_x(M) \to T_{F(x)}(N)$ from tangent vectors on M to tangent vectors on N, given by (3.27) on p. 71. It induces the mapping $F^* : \mathcal{D}(N) \to \mathcal{D}(M)$ between the exterior algebras, given point wise by (3.32) on p. 74, which pulls back any r-form on N to an r-form on M. As it was explained in Sect. 3.5, there is no such induced mapping between the tensor algebras $\mathcal{T}(M)$ and $\mathcal{T}(N)$. This is why the exterior algebra of r-forms plays such a central role in the theory of manifolds as was first realized by E. Cartan.

Notation: The same quantity as an alternating tensor field t or as an exterior form ω is conventionally written as

$$t = t_{i_1\dots i_r} dx^{i_1} \otimes \cdots \otimes dx^{i_r} = \sum_{i_1<\cdots<i_r} \omega_{i_1\dots i_r} dx^{i_1} \wedge \cdots \wedge dx^{i_r}, \quad \omega_{i_1\dots i_r} = r! t_{i_1\dots i_r}. \quad (4.41)$$

Not all authors use the factor $r!$ in this connection. Check up in each case.

4.4 Exterior Differential Calculus

This is the differential calculus for the algebra $\mathcal{D}(M)$ of exterior forms on M. By comparison of (3.25) on p. 71 with (4.28) it is seen that the **exterior differentiation** d is an anti-derivation of degree 1 (cf. (4.26)). Its action in a coordinate neighborhood on a general r-form (3.24) is repeated here:

$$
\begin{aligned}
d\omega &= \sum_{i_1<\cdots<i_r} d\omega_{i_1\ldots i_r} \wedge dx^{i_1} \wedge \cdots \wedge dx^{i_r} \\
&= \sum_{i,i_1<\cdots<i_r} \frac{\partial \omega_{i_1\ldots i_r}}{\partial x^i} dx^i \wedge dx^{i_1} \wedge \cdots \wedge dx^{i_r} \\
&= \sum_{s=1}^{r+1} (-1)^{s+1} \sum_{i_1<\cdots<i_{r+1}} \frac{\partial \omega_{i_1\ldots i_{s-1} i_{s+1}\ldots i_{r+1}}}{\partial x^{i_s}} dx^{i_1} \wedge \cdots \wedge dx^{i_{r+1}}.
\end{aligned} \tag{4.42}
$$

Any derivation or anti-derivation of $\mathcal{D}(M)$ is defined by its action on the space $\mathcal{D}^0(M) = \mathcal{C}(M)$ of functions F and on the space $\mathcal{D}^1(M)$ of 1-forms or differentials dF (exercise; consider first decomposable forms and then the linearity of the derivation). An anti-derivation of degree 1 on $\mathcal{D}(M)$ is uniquely defined by (3.25) (see for instance [2]).

Let $F : M \to N$ be a smooth mapping from the manifold M into the manifold N. Then, from (3.31) it follows that for every smooth 1-form ω on N the pulled back 1-form $F^*(\omega)$ on M is smooth. Equation 3.32 shows that a general r-form on N can be pulled back in a coordinate neighborhood according to (4.23) to a smooth r-form on M so that the pull back $F^* : \mathcal{D}(N) \to \mathcal{D}(M)$ is a homomorphism of algebras. Moreover, F^* *commutes with* d, *that is,*

$$
d(F^*(\omega)) = F^*(d\omega), \quad \omega \in \mathcal{D}(N). \tag{4.43}
$$

This holds due to (3.31) for 1-forms as the differentials of 0-forms (functions). For the general case it is straightforwardly demonstrated using a coordinate neighborhood.

From the statement proved after (4.30) on p. 105 it follows that for any vector field $X \in \mathcal{X}(M)$ by point wise application the **interior multiplication** $\iota_X(\omega)$ is an anti-derivation of degree -1 on $\mathcal{D}(M)$. Since it is of degree -1 and $\mathcal{D}^{-1} = \{0\}$, ι_X yields 0 if applied to any $F \in \mathcal{D}^0(M)$, see remark in parentheses after (4.30). On $\mathcal{D}^1(M)$, its action is given by (4.30) for the case $v = X$ and $t' = F$, which with (3.21) and (4.10) immediately yields $\iota_X(\omega) = \omega(X) = C_{1,1}(X \otimes \omega)$ for any $\omega \in \mathcal{D}^1(M)$. The general expression is

$$
\iota_X(\omega) = C_{1,1}(X \otimes \omega) = r \sum_{i,i_1<\cdots<i_r} \xi^i \omega_{ii_1\ldots i_{r-1}} dx^{i_1} \wedge \cdots \wedge dx^{i_{r-1}}, \quad \omega \in \mathcal{D}^r(M). \tag{4.44}
$$

For $D = \iota_X$ and decomposable forms, (4.28) is proved with (4.20) by Laplace's expansion formula for a determinant. Hence, (4.44) is the correct extension from $\mathcal{D}^0(M)$ and $\mathcal{D}^1(M)$. Since $\omega \in \mathcal{D}^r(M)$ is alternating,

$$
(\iota_X)^2 = 0 \text{ on } \mathcal{D}(M) \tag{4.45}
$$

follows immediately from (4.44).

As an example, let $M = \mathbb{R}^3, r = d = 3$ and $\omega = \varepsilon = dx^1 \wedge dx^2 \wedge dx^3$. Then,

$$\langle \iota_X(\varepsilon), Y \wedge Z \rangle = \varepsilon_{ijk}\xi^i \eta^j \zeta^k = (X, Y, Z), \quad \varepsilon_{ijk} = \delta^{123}_{ijk} \text{ (cf. (4.22))}, \tag{4.46}$$

is the triple scalar product of the vectors X, Y and Z with the components ξ^i, η^j and ζ^k, respectively.

Finally, since for every tangent vector field $X \in \mathcal{X}(M) = \mathcal{T}_{1,0}(M)$ the mapping $L_X : C(M) \to C(M) : F \mapsto XF$ is a derivation (of degree 0) of $C(M)$ and the mapping $L_X : \mathcal{T}_{0,1}(M) = \mathcal{D}^1(M) \to \mathcal{D}^1(M) : \omega \mapsto L_X\omega$ with $(L_X\omega)(Y) = X(\omega(Y)) - \omega([X, Y])$ for every $Y \in \mathcal{X}(M)$ is a derivation of degree 0 of $\mathcal{D}^1(M)$, the **Lie derivative** (4.36) for alternating tensors u of type $(0, r)$ is a derivation of degree 0 of $\mathcal{D}(M)$.

On $\mathcal{D}(M)$, the connection between d, ι_X and L_X is

$$L_X = d \circ \iota_X + \iota_X \circ d, \quad [d, L_X] = 0, \quad [\iota_Y, L_X] = \iota_{[Y,X]}, \quad d^2 = 0, \quad (\iota_X)^2 = 0. \tag{4.47}$$

Proof From (3.32) on p. 74 it is easily seen by operation with d on both sides that d commutes with F^* for every mapping F of manifolds. Hence, d commutes with $(\phi_t)^*$ of (4.36) and therefore also with L_X. Since $\iota_X\mathcal{D}^0(M) = 0$, for $F \in \mathcal{D}^0(M)$ the first equation reduces to $L_XF = \iota_X(dF) = dF(X) = XF$, which is true due to the definition of L_X. Now, let D and D' be two derivations of degree 0 of $\mathcal{D}(M)$ which coincide on $\mathcal{D}^0(M)$ and commute with d. From $(D - D')F = 0$ and $(D - D')dG = d(D - D')G = 0$ one has for a general 1-form $\omega = FdG$ that also $(D - D')\omega = ((D - D')F)dG + F(D - D')dG = 0$ and hence D and D' coincide on $\mathcal{D}^1(M)$. Consequently, both sides of the first relation (4.47) coincide on $\mathcal{D}^0(M)$ and on $\mathcal{D}^1(M)$ and thus on $\mathcal{D}(M)$ (cf. remark after (4.42)). The second relation is a direct consequence of the first and $d^2 = 0$.

Again because of $\iota_X\mathcal{D}^0(M) = 0$, both sides of the third equation are zero on $\mathcal{D}^0(M)$. Now, recall that for any 1-form ω and any tangent vector X, $\iota_X\omega = \omega(X) = C_{1,1}(X \otimes \omega) \in \mathcal{D}^0(M)$ and $L_XF = XF, L_XY = [X, Y]$. Then, $[\iota_Y, L_X]\omega = \iota_Y L_X\omega - L_X\iota_Y\omega = (L_X\omega)(Y) - L_X(C_{1,1}(Y \otimes \omega)) = \lim_{t\to 0}(1/t)(((\phi_t)^*\omega - \omega)(Y) - \lim_{t\to 0}(1/t)C_{1,1}\left(((\phi_{-t})_*Y - Y) \otimes \omega - Y \otimes ((\phi_t)^*\omega - \omega)\right) = -C_{1,1}(\omega \otimes \lim_{t\to 0}(1/t)$ $((\phi_{-t})_*Y - Y) = -C_{1,1}(\omega \otimes [X, Y]) = -\omega([X, Y]) = \omega([Y, X]) = \iota_{[Y,X]}\omega$ which proves the third equation (4.47) on $\mathcal{D}^1(M)$ and hence on all $\mathcal{D}(M)$.

The remaining two equations were considered previously and are only repeated here for completeness. □

For an r-form ω in place of the tensor u in (4.38), (4.38) and (4.24) yield

$$L_X\langle \omega, X_1 \wedge \cdots \wedge X_r \rangle = \langle L_X\omega, X_1 \wedge \cdots \wedge X_r \rangle + \sum_{p=1}^{r} \langle \omega, X_1 \wedge \cdots \wedge [X, X_p] \wedge \cdots \wedge X_r \rangle. \tag{4.48}$$

With this relation and the first equation (4.47), induction with respect to r yields

$$\langle d\omega, X_1 \wedge \cdots \wedge X_{r+1}\rangle$$
$$= \sum_{p=1}^{r+1}(-1)^{p+1}L_{X_p}\langle \omega, X_1 \wedge \cdots \wedge X_{p-1} \wedge X_{p+1} \wedge \cdots \wedge X_{r+1}\rangle$$
$$+ \sum_{p<q}^{r+1}(-1)^{p+q}\langle \omega, [X_p, X_q] \wedge X_1 \wedge \cdots \wedge X_{p-1} \wedge X_{p+1}$$
$$\wedge \cdots \wedge X_{q-1} \wedge X_{q+1} \wedge \cdots \wedge X_{r+1}\rangle. \tag{4.49}$$

From (4.48) the coordinate expression of $L_X\omega$ is obtained in the following manner. Put $X_q = \partial/\partial x^{i_q}$, then the left hand side is $L_X(\omega_{i_1\dots i_r}) = \sum_i \xi^i \partial\omega_{i_1\dots i_r}/\partial x^i$, and the first term on the right hand side is $(L_X\omega)_{i_1\dots i_r}$. In the rest use $[X, (\partial/\partial x^{i_q})] = -\sum_i(\partial\xi^i/\partial x^{i_q})(\partial/\partial x^i)$. The result is

$$(L_X\omega)_{i_1\dots i_r} = \sum_{i=1}^{n} \xi^i \frac{\partial\omega_{i_1\dots i_r}}{\partial x^i} + \sum_{p=1}^{r}\sum_{i=1}^{n} \frac{\partial\xi^i}{\partial x^{i_p}}\omega_{i_1\dots i_{p-1} i i_{p+1}\dots i_r} \tag{4.50}$$

which of course coincides with (4.39), if ω is treated as an alternating tensor of type $(0, r)$. Observe that in the last sum the subscripts of ω are not in ascending order; ordering them introduces additional sign factors. A similar treatment of (4.49) would make the last sum of this relation vanish since $[X_p, X_q]$ would be zero, and the rest would just recover (4.42).

As an example the **phase space** Ω of **classical point mechanics** is again considered. This $2m$-dimensional manifold has a **symplectic structure**. A symplectic structure on an even-dimensional manifold is defined by a symplectic 2-form, that is a 2-form (alternating tensor field) ω which has the properties

$$d\omega = 0, \quad (\omega_x(V, W) = 0 \text{ for all } V \in T_x(\Omega)) \to W = 0. \tag{4.51}$$

Instead of diffeomorphisms of a differentiable structure, symplectomorphisms which leave ω invariant are now admitted to form the pseudo-group S.

The symplectic 2-form of the phase space Ω is

$$\omega = \sum_{i=1}^{m} dq^i \wedge dp^i. \tag{4.52}$$

Clearly $d\omega = 0$, and $\omega(V, W) = \sum_{i=1}^{m}(v^i w^{m+i} - v^{m+i}w^i) = 0$ for all V implies $W = 0$. In local coordinates ω is given by the skew-symmetric $2m \times 2m$ matrix

$$\boldsymbol{\omega} = \begin{pmatrix} 0 & 1 \\ -1 & 0 \end{pmatrix}, \quad \omega(V, W) = \boldsymbol{V} \cdot \boldsymbol{\omega} \cdot \boldsymbol{W}, \tag{4.53}$$

where the entries 0 and 1 are $m \times m$ zero and unit matrices. Hamilton's equations of motion (3.40) and the Poisson brackets spell now

$$\imath_W \omega = dH \to W = W_H, \quad \{F, G\} = \imath_{W_F} \imath_{W_G} \omega = \langle \omega, W_G \wedge W_F \rangle, \tag{4.54}$$

where in a local coordinate system W_H is given by (3.41) and W_F correspondingly. With the first relation (4.47) it follows immediately from (4.54) and (4.51) that the Lie derivative of ω with respect to W_H vanishes: $L_{W_H}\omega = d(\imath_{W_H}\omega) + \imath_{W_H}(d\omega) = d^2H = 0$. This implies that the Hamiltonian flow ϕ_t of the vector field W_H leaves the symplectic form invariant:

$$\phi_t^* \omega = \omega. \tag{4.55}$$

In this context ω is called the **Poincaré invariant**. L_{W_H} is called the **Liouvillian**. (Applied to real C^1-functions on Ω it is just W_H.)

Before continuing and giving further examples of application in physics, integration over manifolds as an important application of the exterior calculus is treated in the next chapter.

References

1. Kobayashi, S., Nomizu, K.: Foundations of Differential Geometry. Interscience, vols. I and II. New York (1963 and 1969)
2. Warner, F.W.: Foundations of Differentiable Manifolds and Lie Groups. Springer, New York (1983)

Chapter 5
Integration, Homology and Cohomology

5.1 Prelude in Euclidean Space

To start from commonly familiar ground, the Euclidean space $\mathbb{R}^n$ is considered. Let $\boldsymbol{x} = (x^1, \ldots, x^n)$ be Cartesian coordinates in $\mathbb{R}^n$ so that the volume element (measure) is $\tau = dx^1 \cdots dx^n$, a real number equal to the volume of an n-dimensional brick with edge lengths dx^i. In (4.17) on p. 102 it was stated that $\Lambda_n(\mathbb{R}^n) \approx \mathbb{R}$ and hence $dx^1 \wedge \cdots \wedge dx^n$ is equivalent to a real number. Put

$$\tau = dx^1 \cdots dx^n = dx^1 \wedge \cdots \wedge dx^n. \tag{5.1}$$

Let $\boldsymbol{y} = \boldsymbol{\psi}(\boldsymbol{x}) = (\psi^1(\boldsymbol{x}), \ldots, \psi^n(\boldsymbol{x}))$ be arbitrary smooth coordinate functions, and let $\omega : \boldsymbol{\psi}(\mathbb{R}^n) \to \mathbb{R}$ be a real (piece wise continuous) function. It is well known from integral calculus that

$$\int\limits_{\boldsymbol{\psi}(U)} \omega(\boldsymbol{y}) dy^1 \cdots dy^n = \int\limits_{U} \omega(\boldsymbol{\psi}(\boldsymbol{x})) \left| \frac{D(\psi^1, \ldots, \psi^n)}{D(x^1, \ldots, x^n)} \right| dx^1 \cdots dx^n \tag{5.2}$$

with the Jacobian defined in (2.16). Here, $dy^1 \cdots dy^n$ is the volume element in the Euclidean target space $\boldsymbol{\psi}(\mathbb{R}^n)$ of the mapping $\boldsymbol{\psi}$ where the y^i form Cartesian coordinates. On the other hand, considering ψ^i as a 0-form and $d\psi^i$ as a 1-form on the original $\mathbb{R}^n$, one has according to (4.42) and with (4.22)

$$\begin{aligned} d\psi^1 \wedge \cdots \wedge d\psi^n &= \sum_{j_1, \ldots, j_n} \frac{\partial \psi^1}{\partial x^{j_1}} \cdots \frac{\partial \psi^n}{\partial x^{j_n}} dx^{j_1} \wedge \cdots \wedge dx^{j_n} \\ &= \sum_{j_1, \ldots, j_n} \frac{\partial \psi^1}{\partial x^{j_1}} \cdots \frac{\partial \psi^n}{\partial x^{j_n}} \delta^{j_1 \cdots j_n}_{1 \cdots n} dx^1 \wedge \cdots \wedge dx^n \\ &= \frac{D(\psi^1, \ldots, \psi^n)}{D(x^1, \ldots, x^n)} dx^1 \wedge \cdots \wedge dx^n. \end{aligned} \tag{5.3}$$

together with the definition (5.1), this justifies to write (5.2) as

$$\int_{\psi(U)} \omega\, dy^1 \wedge \cdots \wedge dy^n = \int_U \psi^*\left(\omega\, dy^1 \wedge \cdots \wedge dy^n\right), \tag{5.4}$$

where, besides ψ^i and $d\psi^i$ being treated as forms, ψ is also treated as a transformation which pulls back ω on $\psi(\mathbb{R}^n)$ to $\psi^*(\omega) = \omega \circ \psi$ on $\mathbb{R}^n$ and, according to (3.31) on p. 74, pulls back $d\psi^i$ to $\psi^*(d\psi^i) = \sum_j (\partial\psi^i/\partial x^j)dx^j$ on the corresponding cotangent spaces which in the considered Euclidean case are again $\psi(\mathbb{R}^n)$ and $\mathbb{R}^n$, respectively. U and $\psi(U)$ are supposed to have finite volume. *It is also assumed that the ψ^i are indexed in such an order that the Jacobian is positive.*

Since $\dim \mathcal{D}^n(\mathbb{R}^n) = 1$, the expression $\omega(\psi)d\psi^1 \wedge \cdots \wedge d\psi^n$ is the general expression of an n-form in $\psi(\mathbb{R}^n)$ expressed in coordinates ψ^i, if ω is smooth. One may consider it as a generalized volume element by giving the measure in $\psi(\mathbb{R}^n)$ a more flexible meaning. Any n-form

$$\omega = \omega(y)dy^1 \wedge \cdots \wedge dy^n, \quad \omega(y) > 0 \text{ everywhere} \tag{5.5}$$

is called a **volume form**. Since it transforms from one coordinate system to another (x being not necessarily the original Cartesian coordinates) by (5.3), a positive n-form (5.5) remains a positive n-form under all regular coordinate transformations with positive Jacobian. By writing the integral (5.2) as $\int_{\psi(U)} \omega$, according to (5.4) any smooth transformation ψ yields

$$\int_{\psi(U)} \omega = \int_U \psi^*(\omega) \tag{5.6}$$

with the meaning that in Cartesian coordinates (5.1) *holds.*

The $\mathbb{R}^n$ is **orientable** (p. 60). A mapping ψ of a part of $\mathbb{R}^n$ into a part of $\psi(\mathbb{R}^n)$ is said to preserve orientation, if the Jacobian of the mapping is positive, it reverses orientation, if the Jacobian is negative. For two domains U and U' which contain the same points but have reversed orientation, $U' = -U$ is written. According to the last expression (5.1), $\tau' = -\tau$ in this case. Therefore,

$$\int_{-U} \omega = -\int_U \omega. \tag{5.7}$$

If the *disjoint* sum of two domains U_1 and U_2 is denoted by $U_1 + U_2$, then moreover

$$\int_{U_1+U_2} \omega = \int_{U_1} \omega + \int_{U_2} \omega \tag{5.8}$$

holds, and accordingly for more items. Equations 5.7 and 5.8 provide together a homomorphism from the Abelian group of domains of oriented manifolds into the Abelian group of integrals over domains of a fixed form ω. One even may extend those groups and the homomorphism to real vector spaces by setting

$$\int_{\lambda_1 U_1 + \lambda_2 U_2} \omega = \lambda_1 \int_{U_1} \omega + \lambda_2 \int_{U_2} \omega, \quad \lambda_1, \lambda_2 \in \mathbb{R}. \tag{5.9}$$

If for instance domains and integrals have a physical meaning one may think of probability distributions over the integrals corresponding to probability distributions over domains.

Now, let $U = \{x \mid 0 \leq x^i \leq 1,\ i = 1, \ldots, n\} \in \mathbb{R}^n$, and let $\psi : \mathbb{R}^n \to \mathbb{R}^m$, $m \geq n$, be a regular embedding of a neighborhood of $U \in \mathbb{R}^n$ into the Euclidean space $\mathbb{R}^m$ of possibly higher dimension. Let ω be any n-form (not necessarily positive) on a neighborhood of $\psi(U)$ in $\mathbb{R}^m$. Recall, that as an embedding ψ is smooth and injective and ψ_x^* is injective at every point $x \in U$, that is, the Jacobi matrix has rank n (cf. p. 75). One defines now the left hand side of (5.6) as an integral over an embedded n-dimensional manifold in $\mathbb{R}^m$, $m \geq n$, by the right hand side which is an integral of an n-form over the coordinate cube U in $\mathbb{R}^n$.

Let $n = 1$ and $\omega = \sum_{i=1}^m \omega_i(y)dy^i$ a 1-form on an open subset of $\mathbb{R}^m$ containing $\psi(U)$ where U is the unit interval of $\mathbb{R}$ and $\psi(U)$ is a parametrized curve $y = \psi(x)$ in $\mathbb{R}^m$ of finite length with the curve parameter $x \in U$. Then, (5.6) reads

$$\int_{\psi(U)} \omega = \int_{\psi(U)} \sum_{i=1}^m \omega_i(y)dy^i = \int_0^1 \sum_{i=1}^m \omega_i(y(x)) \frac{d\psi^i}{dx} dx.$$

Replacing dx by $-dx$ and integrating from $x = 1$ to $x = 0$ reverses also the sign of $d\psi^i/dx$ and hence the sign of the value of the integral. This is the integral over $-\psi(U)$. Next, fix the point $y_0 = \psi(0)$ and let s be the arc length from y_0 along the curve $y = \psi(x)$. This yields a one–one function $x = x(s)$ with $x(S) = 1$ where S is the total length of the curve from $y_0 = \psi(0)$ to $y_1 = \psi(1)$. Then,

$$\int_{\psi(U)} \omega = \int_0^S \sum_{i=1}^m \omega_i(y(x(s))) \frac{d\psi^i}{dx}\frac{dx}{ds} ds = \int_0^S \sum_{i=1}^m \omega_i(y(s)) \frac{d\psi^i}{ds} ds,$$

and in this sense the integral is independent of the parametrization of the curve. Now, assume that there is a real function $u(y)$ on $\mathbb{R}^m$, which can be taken as a 0-form, and that $\omega = du = \sum_i (\partial u/\partial y^i)dy^i$. Then,

$$\int_{\psi(U)} \omega = u(y_1) - u(y_0), \quad \omega = du,$$

which is the **Fundamental Theorem of Calculus** for the case of a parametrized curve. Moreover, since ω may be considered as a linear mapping of tangent vector fields to scalar functions, the motion of a point in time t along the curve from $\boldsymbol{y}_0$ at $t=0$ to $\boldsymbol{y}_1$ at $t=T$ may be considered,

$$\int_{\psi(U)} \omega = \int_0^T \sum_{i=1}^m \omega_i(y(t)) \frac{d\psi^i}{dt} dt,$$

where now ω may be a force field and $d\psi/dt$ a velocity field, and the integral is the work performed on the point. Again, if the force field ω has a potential u, then the work depends only on the potential values at the boundary points $\boldsymbol{y}_0$ and $\boldsymbol{y}_1$ of the path.

Next, let $n=2$ and $\omega = \sum_{1\le i_1<i_2\le m} \omega_{i_1 i_2}(y) dy^{i_1} \wedge dy^{i_2}$ a 2-form on an open subset of $\mathbb{R}^m$ containing $\psi(\boldsymbol{U})$ where $\boldsymbol{U}$ is the unit square of $\mathbb{R}^2$ and $\psi(\boldsymbol{U})$ is a parametrized surface $\boldsymbol{y} = \psi(x^1,x^2)$ in $\mathbb{R}^m$ of finite area with parameters x^1,x^2, $(x^1,x^2)\in \boldsymbol{U}$. Then, (5.6) reads

$$\begin{aligned}
\int_{\psi(U)} \omega &= \int_{\psi(U)} \sum_{1\le i_1<i_2\le m} \omega_{i_1 i_2}(y) dy^{i_1} \wedge dy^{i_2} \\
&= \int_0^1\int_0^1 \sum_{1\le i_1<i_2\le m} \omega_{i_1 i_2}(y(x)) \sum_{j_1,j_2=1}^{2} \frac{\partial \psi^{i_1}}{\partial x^{j_1}} dx^{j_1} \wedge \frac{\partial \psi^{i_2}}{\partial x^{j_2}} dx^{j_2} \\
&= \int_0^1\int_0^1 \sum_{1\le i_1<i_2\le m} \omega_{i_1 i_2}(y(x)) \left(\frac{\partial \psi^{i_1}}{\partial x^1}\frac{\partial \psi^{i_2}}{\partial x^2} - \frac{\partial \psi^{i_1}}{\partial x^2}\frac{\partial \psi^{i_2}}{\partial x^1} \right) dx^1 \wedge dx^2 \\
&= \int_0^1\int_0^1 \sum_{i_1,i_2=1}^{m} \omega_{i_1 i_2}(y(x)) \frac{\partial \psi^{i_1}}{\partial x^1}\frac{\partial \psi^{i_2}}{\partial x^2} dx^1 dx^2.
\end{aligned}$$

In the last equality it was used that $\omega_{i_1 i_2}$ is an alternating tensor.

Let $u = \sum_{i=1}^m u_i(y) dy^i$ be a 1-form and $\omega = du$, that is (cf. (4.42))

$$\begin{aligned}
\omega = \sum_{i=1}^m du_i \wedge dy^i &= \sum_{i_1,i_2=1}^{m} \frac{\partial u_{i_2}}{\partial y^{i_1}} dy^{i_1} \wedge dy^{i_2} \\
&= \sum_{1\le i_1<i_2\le m} \left(\frac{\partial u_{i_2}}{\partial y^{i_1}} - \frac{\partial u_{i_1}}{\partial y^{i_2}} \right) dy^{i_1} \wedge dy^{i_2}.
\end{aligned}$$

Inserting this expression in parentheses for $\omega_{i_1 i_2}$ into the last line of the integral of the previous paragraph and using the chain rule $\partial u_i/\partial x^j = \sum_k (\partial u_i/\partial y^k)(\partial \psi^k/\partial x^j)$ one finds

$$\int_{\psi(U)} \omega = \int_0^1 \int_0^1 \sum_{i=1}^m \left(\frac{\partial u_i}{\partial x^1} \frac{\partial \psi^i}{\partial x^2} - \frac{\partial \psi^i}{\partial x^1} \frac{\partial u_i}{\partial x^2} \right) dx^1 dx^2$$

$$= \int_0^1 \sum_{i=1}^m \Big(\big(u_i(1,x^2) - u_i(0,x^2)\big) \frac{\partial \psi^i}{\partial x^2} dx^2$$

$$- \big(u_i(x^1,1) - u_i(x^1,0)\big) \frac{\partial \psi^i}{\partial x^1} dx^1 \Big)$$

$$= \int_{y_{10} \to y_{11}} u - \int_{y_{00} \to y_{01}} u - \int_{y_{01} \to y_{11}} u + \int_{y_{00} \to y_{10}} u,$$

where in the second equation integrations per part over x^1 in the first item and over x^2 in the second item were done. The terms with the second derivative of ψ^i cancel. For the sake of simpler writing $u(\boldsymbol{x})$ stands for $u(\boldsymbol{y}(\boldsymbol{x}))$. If then $\boldsymbol{y}_{ij}$ denotes $\psi(i, j)$, $i, j = 0, 1$ (see Fig. 5.1), then the four terms of the integral are in fact curvilinear integrations along the boundaries of $\psi(\boldsymbol{U})$ as depicted in Fig. 5.1. By observing (5.7) they constitute an integral around $\psi(\boldsymbol{U})$ with consecutive orientation in the mathematical positive sense with regard to the x^1, x^2-plane. Interchanging x^1 with x^2 would reverse this positive sense and also the sign of the integral.

In general, the integral over an n-dimensional regularly embedded manifold in the $\mathbb{R}^m$, $m \geq n$, as described above of an n-form ω is obtained as

$$\int_{\psi(U)} \omega = \int_0^1 \cdots \int_0^1 \sum_{1 \leq i_1 < \cdots < i_n \leq m} \omega_{i_1 \cdots i_n}(\boldsymbol{y}(\boldsymbol{x})) \det \left(\frac{\partial \psi^{i_j}}{\partial x^k} \right) dx^1 \wedge \cdots \wedge dx^n$$

$$= \int_0^1 \cdots \int_0^1 \sum_{i_1, \ldots, i_n = 1}^m \omega_{i_1 \cdots i_n}(\boldsymbol{y}(\boldsymbol{x})) \frac{\partial \psi^{i_1}}{\partial x^1} \cdots \frac{\partial \psi^{i_n}}{\partial x^n} dx^1 \cdots dx^n. \quad (5.10)$$

What was above considered for $n = 1, 2$, if $\omega = du$, is **Stokes' theorem** for the case of a coordinate n-cube of the $\mathbb{R}^n$,

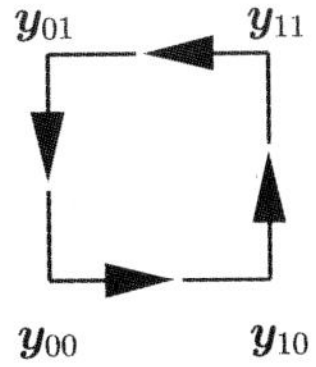

Fig. 5.1 The image ψ(U) of the unit square $\boldsymbol{U}$. For the notation of the corners see text

$$\int_{\psi(U)} du = \int_{\partial\psi(U)} u, \tag{5.11}$$

where $\partial\psi(U)$ means the boundary of $\psi(U)$, oriented in a way in general specified later. The boundary of a curve consists of its end points oriented outward, if the curve is not closed (otherwise it has no boundary). To treat the Fundamental Theorem of Calculus as the special case of Stokes' theorem for $n = 1$, the integral over a point of a 0-form is defined as the function value of the 0-form at that point, provided with an appropriate sign for the orientation of that end point. For $n > 1$ the proof goes along the same line as for $n = 1$. The orientations of the faces are obtained from the sign factors of (4.42) in combination with the integrations per part. Since any n-dimensional domain can be approximated by small n-cubes, and since cubes which touch each other by a face have this face with opposite orientation, (5.8) shows that all the integrals over inner faces of the covering of the domain by cubes cancel and only those of surface faces survive. Instead of cubes less regular polyhedra can be used. This makes it evident that Stokes' theorem does not only hold for cubes but for any shape of domains. This will be worked out in detail in the following sections.

As an example consider $m = 3$ and the n-forms ω^n, $n = 0, 1, 2, 3$ on $\mathbb{R}^3$. Let $\omega^n = d\omega^{n-1}$, $n = 1, 2, 3$, in particular (cf. (4.42)),

$$\omega^1 = \sum_{i=1}^{3} \frac{\partial\omega^0}{\partial y^i} dy^i = (\text{grad}\,\omega^0)\cdot \boldsymbol{dy},$$

$$\omega^2 = \sum_{1\le i_1,i_2\le 3} \left(\frac{\partial\omega^1_{i_2}}{\partial y^{i_1}} - \frac{\partial\omega^1_{i_1}}{\partial y^{i_2}}\right) dy^{i_1}\wedge dy^{i_2} = (\text{rot}\,\boldsymbol{\omega}^1)\cdot \boldsymbol{dS},$$

$$dS_i = \frac{1}{2!}\sum_{i,j,k=1}^{3} \delta^{123}_{ijk}\, dy^j \wedge dy^k,$$

$$\omega^3 = \left(\frac{\partial\omega^2_{23}}{\partial y^1} - \frac{\partial\omega^2_{13}}{\partial y^2} + \frac{\partial\omega^2_{12}}{\partial y^3}\right) dy^1\wedge dy^2\wedge dy^3 = (\text{div}\,\boldsymbol{\Omega})\cdot\tau,$$

$$\Omega^i = \frac{1}{2!}\sum_{i,j,k=1}^{3} \delta^{ijk}_{123}\,\omega^2_{jk}.$$

Stokes' theorem reads in these cases

$$\int_U (\text{grad}\,\omega^0)\cdot \boldsymbol{dy} = \int_{\partial U} \omega^0 = \omega^0(\boldsymbol{y}_1) - \omega^0(\boldsymbol{y}_0),$$

$$\int_U (\text{rot}\,\boldsymbol{\omega}^1)\cdot \boldsymbol{dS} = \int_{\partial U} \boldsymbol{\omega}^1\cdot \boldsymbol{dy},$$

$$\int_U (\text{div}\,\boldsymbol{\Omega})\cdot\tau = \int_{\partial U} \boldsymbol{\omega}\cdot \boldsymbol{dS}.$$

The first case is the classical Fundamental Theorem of Analysis, the second case is the classical Stokes theorem, and the third case is Gauss' theorem.

Differential forms are the equivalents to alternating covariant tensors. Hence they have a geometrical meaning independent of coordinate systems. The orientation of an oriented manifold may be changed by a mapping, in local coordinates expressed as $(y^1, y^2, \ldots, y^n) \to (-y^1, y^2, \ldots, y^n)$. Likewise, the orientation of $\mathbb{R}^n$ for odd n changes by inflection $\boldsymbol{y} \to -\boldsymbol{y}$ of the space coordinates. Hence, in an odd-dimensional space, tensors of odd degree change sign of their tensor components in an inflection of spatial coordinates and tensors of even degree do not. Additionally, **pseudo-tensors** are introduced with reversed sign-change behavior compared to tensors with respect to a change of orientation of space. If the above considered n-forms ω^n are tensor equivalents, then obviously $\operatorname{rot}\boldsymbol{\omega}^1$, $d\boldsymbol{S}$ and $\boldsymbol{\omega}$ are pseudo-vectors and $\operatorname{div}\boldsymbol{\Omega}$ and τ are pseudo-scalars. The other quantities are tensors (including vectors and scalars). Alternatively, the ω^n may be understood as pseudo-forms (pseudo-tensor equivalents), and then the roles of tensors and pseudo-tensors in these relations are reversed. One easily checks that all the above relations remain valid in this case. (Orientation and pseudo-character have only a relative meaning.)

In the above examples in $\mathbb{R}^3$, a 2-form was related to a pseudo-vector $\operatorname{rot}\boldsymbol{\omega}^1$ and a 3-form to a pseudo-scalar $\operatorname{div}\boldsymbol{\Omega}$. This has a generalization to any dimension. The Euclidean space $\mathbb{R}^m$ is an inner product space with the standard inner product $(\boldsymbol{a}\,|\,\boldsymbol{b}) = \sum_{i=1}^m a_i b_i$, if a_i and b_i are the components of the vectors $\boldsymbol{a}$ and $\boldsymbol{b}$ with respect to an *orthonormal basis* $\{\boldsymbol{f}^1, \ldots, \boldsymbol{f}^m\}$, $(\boldsymbol{f}^i\,|\boldsymbol{f}^j) = \delta_{ij}$. This inner product may be extended to an inner product of the exterior algebra $\Lambda(\mathbb{R}^m)$ by putting

$$(\boldsymbol{a}^1 \wedge \cdots \wedge \boldsymbol{a}^n | \boldsymbol{b}^1 \wedge \cdots \wedge \boldsymbol{b}^n) = \det((\boldsymbol{a}^i\,|\,\boldsymbol{b}^j)), \quad n \le m, \tag{5.12}$$

putting $\Lambda_n(\mathbb{R}^m)$ and $\Lambda_{n'}(\mathbb{R}^m)$ orthogonal to each other for $n \neq n'$, taking the ordinary product of numbers in $\Lambda_0(\mathbb{R}^m)$, and finally extending by bilinearity to all $\Lambda(\mathbb{R}^m)$. Note that in case of an inner product a bilinear form on the direct product of the space with itself is meant, not with its dual which may be a different space. The latter case is more general and was considered in (4.20). Nevertheless, as in (4.22),

$$(\boldsymbol{f}^{i_1} \wedge \cdots \wedge \boldsymbol{f}^{i_n}\,|\boldsymbol{f}^{j_1} \wedge \cdots \wedge \boldsymbol{f}^{j_n}) = \delta^{i_1 \cdots i_n}_{j_1 \cdots j_n}, \tag{5.13}$$

where the right hand side is -1, 0, or 1.

W. V. D. Hodge introduced as the anticipated generalization of the above situation the **star operator** or **Hodge operator** $* : \Lambda_n(\mathbb{R}^m) \to \Lambda_{m-n}(\mathbb{R}^m)$ defined as a linear operator by

$$\begin{aligned} &*(1) = \boldsymbol{f}^1 \wedge \cdots \wedge \boldsymbol{f}^m, \quad *(\boldsymbol{f}^1 \wedge \cdots \wedge \boldsymbol{f}^m) = 1, \\ &*(\boldsymbol{f}^{i_1} \wedge \cdots \wedge \boldsymbol{f}^{i_n}) = \delta^{i_1 \cdots i_m}_{1 \cdots m} (\boldsymbol{f}^{i_{n+1}} \wedge \cdots \wedge \boldsymbol{f}^{i_m}), \quad \text{all } 1 \le i_k \le m \text{ distinct}, \end{aligned} \tag{5.14}$$

and extended by linearity to arbitrary exterior forms. Note that an order of the orthonormal basis vectors f^i is to be fixed to define the positive orientation of $\mathbb{R}^m$. Changing the orientation of $\mathbb{R}^m$, for instance by replacing $f^1 \to -f^1$, introduces a minus sign in the right hand side of all three defining relations (5.14). In either case it immediately follows that

$$** = \delta^{i_1\cdots i_m}_{1\cdots m}\delta^{i_{n+1}\cdots i_m i_1\cdots i_n}_{1\cdots m} = (-1)^{n(m-n)}$$

on base forms, if all numbers i_k are distinct. (The sign is just the sign of the permutation from the superscripts of the first δ-factor into those of the second because for an equal order $\delta^2 = 1$ would result.) Since the right hand side is a constant sign for each n, the result is valid in general for the application of $**$ to an n-form:

$$** = (-1)^{n(m-n)}, \tag{5.15}$$

that is, up to a possible sign the Hodge operator is its own inverse. Also, since $\Lambda_n(\mathbb{R}^m)$ and $\Lambda_{m-n}(\mathbb{R}^m)$ have the same dimension, from the definition (5.14) it follows that $*$ is an isomorphism between these vector spaces.

On the basis of (5.13) one easily checks for two n-forms ω and σ

$$(\omega|\sigma) = *(\omega \wedge *\sigma) = *(\sigma \wedge *\omega). \tag{5.16}$$

Since σ is an n-form, $*\sigma$ is an $(m-n)$-form and $\omega \wedge *\sigma$ is an m-form which is equivalent to the number $(\omega\,|\,\sigma)$. Let $u \in \Lambda_l(\mathbb{R}^m)$, $\sigma \in \Lambda_{n+l}(\mathbb{R}^m)$ and $\omega \in \Lambda_n(\mathbb{R}^m)$. Then (cf. (4.30) on p. 105), $(\iota_u\sigma\,|\,\omega) = (\sigma\,|\,L_u\omega) = (\sigma\,|\,u \wedge \omega)$. Application of the second variant (5.16) yields $*(\omega \wedge *(\iota_u\sigma)) = *(u \wedge \omega \wedge *\sigma) = (-1)^{ln} * (\omega \wedge u \wedge *\sigma)$ for any $\omega \in \Lambda_n(\mathbb{R}^m)$. Hence, since $*$ is an isomorphism, $*(\iota_u\sigma) = (-1)^{ln}u \wedge *\sigma$. With (5.15), one more application of $*$ to this result yields

$$\iota_u\sigma = (-1)^{n(m-n-l)} * (u \wedge (*\sigma)), \quad \sigma \in \Lambda_{n+l}(\mathbb{R}^m), \quad u \in \Lambda_l(\mathbb{R}^m), \quad n+l \leq m.$$

The Hodge operator is mainly used to extend the Laplacian to manifolds (Sect. 5.9, see also Sect. 9.6).

5.2 Chains of Simplices

In order to analyze the boundary operation in a more general context, instead of coordinate cubes simplices as the simplest polyhedra in $\mathbb{R}^n$ are considered and their images in continuous or smooth mappings into manifolds, which are called singular simplices.

Let $(v_0, \ldots, v_r)$ be $r+1$ linear independent vectors of the $\mathbb{R}^n$, $n > r$. The set

$$S^r(v_0, \ldots, v_r) = \{x = \lambda^0 v_0 + \cdots + \lambda^r v_r \,|\, \lambda^i \geq 0, \lambda^0 + \cdots + \lambda^r = 1\} \tag{5.17}$$

of points of the $\mathbb{R}^n$ is an r-dimensional simplex, in short an **r-simplex** with vertices v_i. A 0-simplex is a point, a 1-simplex is a line segment, a 2-simplex is a triangle, a 3-simplex is a tetrahedron, higher dimensions are considered by analogy. For later use, an r-simplex for any $r<0$ is the empty set. It is clear that

$$\Delta^r = \{x = (\lambda^1, \dots, \lambda^r) \mid \lambda^i \geq 0, \lambda^1 + \dots + \lambda^r \leq 1\} \tag{5.18}$$

is an r-simplex (with its vertices the origin and the first r orthonormal base vectors of the $\mathbb{R}^n$) and that it is homeomorphic to any r-simplex $S^r(v_0, \dots, v_r)$. Δ^r is called the **standard r-simplex**. It is understood as oriented by the r-form $d\lambda^1 \wedge \dots \wedge d\lambda^r$.

An r-simplex has $r+1$ faces, each of which is an $(r-1)$-simplex. The faces of $S^r(v_0, \dots, v_r)$ are $S^{r-1}(v_0, \dots, v_{i-1}, v_{i+1}, \dots, v_r)$, $i = 0, \dots, r$. The faces of an r-simplex are empty for $r \leq 0$, a 1-simplex has two 0-simplices (points) as faces, a 2-simplex (triangle) has three 1-simplices (legs) as faces, a 3-simplex (tetrahedron) has four 2-simplices (triangles) as faces, and so on. The **oriented boundary of a simplex** is defined as

$$\partial S^r(v_0, \dots, v_r) = \sum_{i=0}^{r} (-1)^i S^{r-1}(v_0, \dots, v_{i-1}, v_{i+1}, \dots, v_r). \tag{5.19}$$

See Fig. 5.2 where $\partial S^1(v_0, v_1) = +S^0(v_1) - S^0(v_0)$, $\partial S^2(v_0, v_1, v_2) = +S^1(v_1, v_2) - S^1(v_0, v_2) + S^1(v_0, v_1)$, and the corresponding expression (5.19) for $r = 3$ is visualized.

Suppose an orientation of $S^r(v_0, \dots, v_r)$ has been fixed. As previously in (5.7), the same simplex with the opposite orientation is denoted by $-S^r(v_0, \dots, v_r)$. Obviously,

$$\partial(-S^r(v_0, \dots, v_r)) = -\partial S^r(v_0, \dots, v_r). \tag{5.20}$$

Moreover, like domains in (5.8), simplices may be added in which case the sums can be understood to be unions of disjoint sets. In this sense, faces of dimension less than $r-1$ are counted several times in (5.19). This will not be a problem in the following. (In $\mathbb{R}^{r-1}$ a set of dimension less than $r-1$ has zero measure and

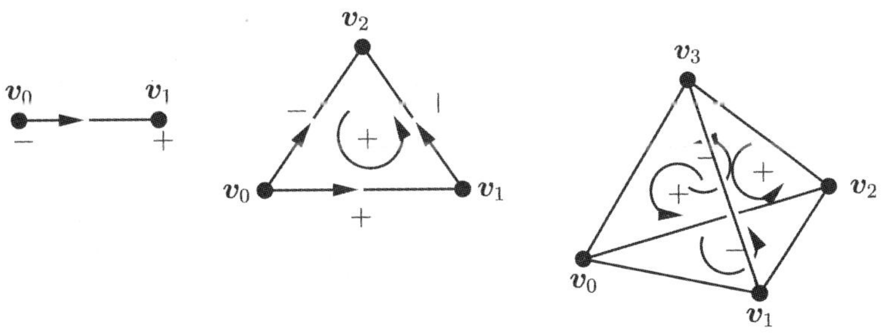

Fig. 5.2 Oriented boundaries of simplices. The $+$ and $-$ signs are those of (5.19), the *arrows* indicate the orientation of the faces as simplices entering (5.19)

does not change integrals.) Now, let $S_i^{r-1} = S^{r-1}(v_0, \ldots, v_{i-1}, v_{i+1}, \ldots, v_r)$ and $S_{ij}^{r-2} = S^{r-2}(v_0, \ldots, v_{i-1}, v_{i+1}, \ldots, v_{j-1}, v_{j+1}, \ldots, v_r)$, $j > i$. Then,

$$\partial\partial S^r(v_0, \ldots, v_r) = \sum_{j<i} (-1)^{i+j} S_{ji}^{r-2} + \sum_{i<j} (-1)^{j+i-1} S_{ij}^{r-2} = \varnothing. \tag{5.21}$$

For $r > 1$, the boundary of a simplex is closed and hence its boundary is empty. For $r \leq 1, S^{r-2}$ is by definition empty anyhow.

Let M be any manifold. A **continuous (smooth) singular r-simplex** σ in M is a continuous (smooth) mapping (of an open neighborhood in $\mathbb{R}^r$) of the standard r-simplex Δ^r into M (Fig. 5.3). M is supposed to be smooth in case of a smooth singular r-simplex. For $r<0, \sigma$ is the empty mapping from the empty set into the empty set. In order to define the boundary operation for σ, first a mapping $\lambda_i^{r-1} : \Delta^{r-1} \to \Delta^r$, $i = 0, \ldots, r$, of Δ^{r-1} onto the faces of Δ^r must be fixed. (Since Δ^r has no faces for $r \leq 0$, no λ_i^{r-1} for $r \leq 0$ is needed.) For $r = 1$, that is $\Delta^0 = \{0\}$ and $\Delta^1 = [0, 1], \lambda_0^0(0) = 1$ and $\lambda_1^0(0) = 0$. For $r > 1$ one finds

$$\begin{aligned} \lambda_0^{r-1}(\lambda^1, \ldots, \lambda^{r-1}) &= \left(\left(1 - \sum_{j=1}^{r-1} \lambda^j\right), \lambda^1, \ldots, \lambda^{r-1}\right), \\ \lambda_i^{r-1}(\lambda^1, \ldots, \lambda^{r-1}) &= (\lambda^1, \ldots, \lambda^{i-1}, 0, \lambda^i, \ldots, \lambda^{r-1}), \quad i = 1, \ldots, r. \end{aligned} \tag{5.22}$$

$\lambda_0^{r-1}(\Delta^{r-1})$ is the face of Δ^r opposite to the origin, and $\lambda_i^{r-1}(\Delta^{r-1})$ are the faces with $\lambda^i = 0$ in Δ^r (and λ^j of Δ^{r-1} is becoming λ^{j+1} of Δ^r for $j \geq i$).

Like it was done in the last section for domains of integration, formally linear combinations of singular simplices with integer coefficients may be introduced and usefully exploited. An **r-chain** of singular r-simplices σ_l in M is a finite linear

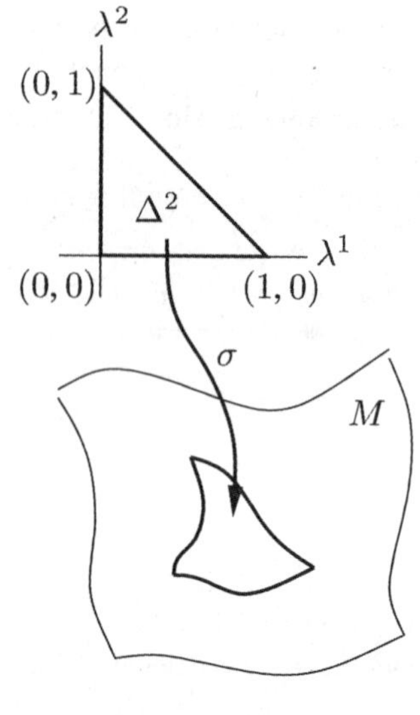

Fig. 5.3 A singular 2-simplex

combination $c = \sum_l k_l \sigma_l, k_l \in \mathbb{Z}$. The set of r-chains forms the free Abelian group generated by the set of all singular r-simplices. It can also be considered as a $\mathbb{Z}$-module. However, it turns out to be useful to allow for a field K instead of the ring $K = \mathbb{Z}$ of integers. Most important besides $K = \mathbb{Z}$ are $K = \mathbb{R}$ and $K = \mathbb{F}_2$, the field of integers modulo 2: $\mathbb{F}_2 = \{0, 1\}, 0 \pm 0 = 0, 0 \pm 1 = 1, 1 \pm 1 = 0, 0 \cdot 0 = 0 \cdot 1 = 0, 1 \cdot 1 = 1$. With a field K ($\mathbb{R}$ or $\mathbb{F}_2$), the r-chains form a vector space. The module or vector space over K of all r-chains in M is denoted as

$$
\begin{aligned}
{}_0C_r(M,K) &= \left\{ c = \sum_l k_l \sigma_l \,|\, k_l \in K, \quad \sigma_l \text{ continuous singular } r\text{-simplices in } M \right\},\\
{}_\infty C_r(M,K) &= \left\{ c = \sum_l k_l \sigma_l \,|\, k_l \in K, \quad \sigma_l \text{ smooth singular } r\text{-simplices in } M \right\}.
\end{aligned}
\tag{5.23}
$$

Since for every non-trivial manifold M there is an infinite set of distinct singular simplices σ, both spaces ${}_{(0,\infty)}C_r(M,K)$ are in general infinite-dimensional for $r \geq 0$ with the distinct singular simplices forming a base. For $r<0$, there is no base element, and hence

$$
{}_{(0,\infty)}C_r(M,K) = \{0\} \quad \text{for } r<0. \tag{5.24}
$$

Later, the direct sum of these spaces will be treated as a grated (by r) module (vector space).

Now, the boundary operation may be defined as

$$
\begin{aligned}
&\partial \text{ or } \partial_r : \ {}_{(0,\infty)}C_r(M,K) \to {}_{(0,\infty)}C_{r-1}(M,K),\\
&\partial c = \partial\left(\sum_l k_l \sigma_l\right) = \sum_l k_l \partial \sigma_l, \quad \partial \sigma = \sum_{i=0}^{r} (-1)^i \sigma \circ \lambda_i^{r-1}.
\end{aligned}
\tag{5.25}
$$

If it is necessary to indicate the dimension r of the chain to which the boundary operator is applied, the notation ∂_r will be used. The boundary of a chain is defined as the corresponding chain of boundaries of the singular simplices, and the boundary of a singular simplex is a chain (with integer coefficients ± 1) formed by first mapping by λ_i^{r-1} the standard $(r-1)$-simplex Δ^{r-1} onto the ith face of the standard r-simplex Δ^r, then mapping by σ this face of Δ^r into M, and finally linearly combining those mappings.

The standard r-simplex Δ^r is a special case of an r-simplex S^r and hence (5.21) holds for it. It is then obvious that

$$
\partial \circ \partial = 0 \tag{5.26}
$$

holds on every r-chain.

Let M be a smooth n-dimensional manifold and let the image of the r-chain c of *smooth* singular simplices be part of an open set in the topology of M on which an

r-form ω is defined, $r \leq n$. By the singular r-simplex σ, ω may be pulled back to Λ^r. In a coordinate neighborhood of $x = \sigma(\boldsymbol{x}) \in M$, $\boldsymbol{x} \in \Delta^r$, that is, $\sigma = (\sigma^1(\boldsymbol{x}), \ldots, \sigma^n(\boldsymbol{x}))$, ω may be given as $\omega = \sum_{1<i_1<\cdots<i_r<n} \omega_{i_1 \cdots i_r}(x) dx^{i_1} \wedge \cdots \wedge dx^{i_r}$. Then, with the orthonormal base in $\mathbb{R}^r \ni \Delta^r$ and the corresponding coordinates λ^i, $\sigma^*(\omega)$ may be given as $\sigma^*(\omega) = \sum_{1<i_1<\cdots<i_r<n} \omega_{i_1 \cdots i_r}(\sigma(\boldsymbol{x})) \sum_{j_1, \ldots, j_r} (\partial \sigma^{i_1}/\partial \lambda^{j_1}) \cdots (\partial \sigma^{i_r}/\partial \lambda^{j_r}) d\lambda^{j_1} \wedge \cdots \wedge d\lambda^{j_r} = \sum_{1<i_1<\cdots<i_r<n} \omega_{i_1 \cdots i_r}(\sigma(\boldsymbol{x})) D(\sigma^{i_1}, \ldots, \sigma^{i_r}) / D(\lambda^1, \ldots, \lambda^r) d\lambda^1 \wedge \cdots \wedge d\lambda^r$ (cf. (5.2)). The integral of the r-form ω over the image of the singular r-simplex $\sigma(\Delta^r)$ in M may now be defined as the ordinary $\mathbb{R}^r$-integral of the pull-back $\sigma^*(\omega)$ over Δ^r:

$$\int_\sigma \omega = \int_{\Delta^r} \sigma^*(\omega) \quad \text{for } r \geq 1, \quad \int_\sigma \omega = \omega(\sigma(0)) \quad \text{for } r = 0. \tag{5.27}$$

The integral over an r-chain $c = \sum_l k_l \sigma_l$ is defined as

$$\int_c \omega = \sum_l k_l \int_{\sigma_l} \omega. \tag{5.28}$$

Now, **Stokes' theorem** for r-chains,

$$\int_{\partial c} \omega = \int_c d\omega, \tag{5.29}$$

can be proved in the general r-dimensional case, where it obviously suffices to prove it for a smooth singular r-simplex. The proof is technical but straightforward with the above developed tools. It can be left as an exercise.

Stokes' theorem for r-chains is the key to the deepest interrelations between topology, algebra and analysis, the investigation of which in the middle of 20th century, but proposed mainly by Poincaré at its beginning, was initiated by de Rham's theorem (Sect. 5.4).

In the above considerations, σ must at least be of class $C^1(\mathbb{R}^r)$ in order that ω can be pulled back. For ω itself it would suffice for the integral to exist that it is a continuous r-form. However, in Stokes' theorem ω must also be C^1. In most applications both σ and ω may be assumed smooth. Note that in this section, σ was not assumed to be bijective; for that reason the simplices σ were called singular. For instance, σ might be constant: $\sigma(\Delta^r) = \{x\}$, $x \in M$. In this case the pull back of ω is the constant r-form equal to its value at x and the integral is the volume $|\Delta^r| = 1/r!$ of Δ^r times this constant ω. In that sense, integrals over r-chains are still integrals in $\mathbb{R}^r$. Nevertheless, these constructs are very useful. Before exploiting them, in the next section more natural integrals which may be understood more directly over domains of manifolds are considered.

5.3 Integration of Differential Forms

First, a **regular domain** Ω in a paracompact smooth orientable n-dimensional manifold M is defined: every point $x \in M$ is either an inner point of Ω or an inner point of $M \setminus \Omega$ or there is a coordinate neighborhood (U, φ) of x such that $\varphi(U \cap \Omega) = \varphi(U) \cap \mathbb{R}^n_-$ where $\mathbb{R}^n_-$ is the half-space of points $\boldsymbol{x} = (\lambda^1, \ldots, \lambda^n) \in \mathbb{R}^n$ with $\sum_i \lambda^i \leq 1$. In other words, the boundary of Ω is locally diffeomorphic to an $(n-1)$-dimensional hyperplane (the hyperplane $\sum_i \lambda^i = 1$ of $\mathbb{R}^n$). In this precise sense *a regular domain* Ω *is a domain with smooth boundary* $\partial\Omega$. Note, however, that a regular domain Ω need not have a boundary at all, it could for instance be all M.

Let Ω have a boundary. Consider a smooth real function F on a neighborhood of $\partial\Omega$, which is constant on $\partial\Omega$ and for which $F(x_i) < F(x_o)$ whenever x_i is an inner point of Ω and x_o is an inner point of $M \setminus \Omega$. Let $x \in \partial\Omega$. A vector X of the n-dimensional tangent space $T_x(M)$ is an **outer vector** to Ω, if $XF > 0$. Consider now the $(n-1)$-dimensional tangent space $T_x(\partial\Omega)$ at a boundary point $x \in \partial\Omega$. A base $X_1, \ldots, X_{n-1}$ in this tangent space is called **coherently oriented** with M, if with an outer vector X to Ω the base $X, X_1, \ldots, X_{n-1}$ of $T_x(M)$ defines the orientation of M, that is, the dual base $dx, dx^1, \ldots, dx^{n-1}$ in $T^*_x(M)$ defines the positive n-form $dx \wedge dx^1 \wedge \cdots \wedge dx^{n-1}$. It is clear that this definition of coherent orientation does not depend on the chosen outer vector X, and that there is a coordinate neighborhood U of x in $\partial\Omega$ and there are local coordinates $x'^1, \ldots, x'^{n-1}$ in U smoothly defining an orientation coherent with that of M. In other words, an orientation of $T_x(\partial\Omega)$ coherent with that of M is a smooth and hence all the more continuous function of x on $\partial\Omega$. Since an orientation can only have two discrete values, if the orientation on $\partial\Omega$ is coherent with that of M, it must be constant on each topological component of $\partial\Omega$.

Now, let Ω be a regular domain in M and let ω be an at least continuous n-form, $n = \dim M$, *with compact support*. In order to define the integral of ω over Ω, **regular n-simplices** are defined as *diffeomorphisms* σ from a neighborhood in $\mathbb{R}^n$ of the standard n-simplex Δ^n into M. If σ preserves orientation, it is called an oriented regular simplex.

A partition of unity on M (which exists since M is paracompact) is used to reduce the integral over Ω to a sum of integrals over oriented regular simplices covering $\operatorname{supp}\omega \cap \Omega$. Since $\operatorname{supp}\omega \cap \Omega$ is compact, it has a finite open cover $\{U_1, \ldots, U_m\}$. Let furthermore U be the open set $U = M \setminus (\operatorname{supp}\omega \cap \Omega)$, so that $\{U, U_1, \ldots, U_m\}$ is a finite open cover of M. Consider a partition of unity $\{\phi, \phi_1, \ldots, \phi_m\}$ subordinate to this open cover of M, that is, $\operatorname{supp}\phi \subset U$, $\operatorname{supp}\phi_i \subset U_i$, $i = 1, \ldots, m$ and $\phi(x) + \sum_i \phi_i(x) = 1$ on M. If $U_i \subset \Omega$, choose an oriented regular simplex σ_i the image of Δ^n of which contains U_i and is entirely in Ω (which is always possible since Ω is closed and U_i is open). If $U_j \cap \partial\Omega$ is non-empty, choose an oriented regular n-simplex σ_j with $\Omega \supset \sigma_j(\Delta^n) \supset U_j \cap \Omega$ and so that $\partial\Omega$ intersects only with the image of the face of Δ^n opposite to the origin (Fig. 5.4).

Fig. 5.4 Regular 2-simplices for a parition of unity of M

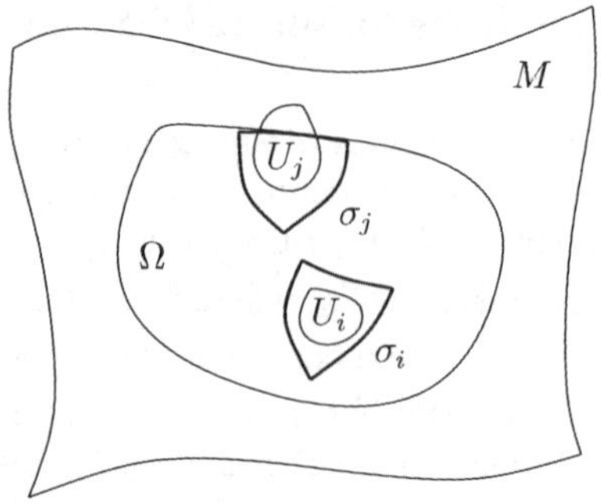

Although U may intersect Ω, no simplex need be chosen for U since $\operatorname{supp}\phi \cap \operatorname{supp}\omega \cap \Omega = \emptyset$ and hence $\phi(x)\omega(x) = 0$ on Ω. The images of the simplices σ_i, $i = 1\ldots, m$ form a closed (overlapping) cover of $\operatorname{supp}\omega \cap \Omega$, and $\sigma_i(\Delta^n) \cap (M \setminus \Omega) = \emptyset$ for all i. On the other hand, $\phi_i(x)\omega(x)$ are smooth n-forms with support in U_i, if ω is smooth (since the ϕ_i are smooth by definition of a partition of unity), and $\omega = \phi\omega + \sum_i \phi_i\omega$ on M. Since, however, $\phi\omega = 0$ on Ω, $\omega = \sum_i \phi_i\omega$ on Ω. Therefore, one may define

$$\int_\Omega \omega = \sum_{i=1}^m \int_{\sigma_i} \phi_i\omega = \sum_{i=1}^m \int_{\Delta^n} \sigma_i^*(\phi_i\omega). \tag{5.30}$$

The last sum is over well defined ordinary integrals in $\mathbb{R}^n$. It remains to show that this expression is unique in the sense that it does not depend on the used partition of unity.

Indeed, consider another partition $\{\psi, \psi_1, \ldots, \psi_l\}$ subordinate to the open cover $\{U, V_1, \ldots, V_l\}$ and correspondingly chosen simplices σ'_i, $i = 1, \ldots, l$. (U was defined by Ω and ω only, hence it is not changed.) On $\operatorname{supp}\omega \cap \Omega$, there holds $\phi = \psi = 0$. Hence,

$$\sum_{i=1}^m \int_{\sigma_i} \phi_i\omega = \sum_{ij} \int_{\sigma_i} \psi_j\phi_i\omega, \quad \sum_{j=1}^l \int_{\sigma'_j} \psi_j\omega = \sum_{ij} \int_{\sigma'_j} \phi_i\psi_j\omega.$$

By the above construction, for each pair (i,j) one has that $\operatorname{supp}(\psi_j\phi_i\omega) \cap \sigma_i(\Delta^n) = \operatorname{supp}(\phi_i\psi_j\omega) \cap \sigma'_j(\Delta^n)$. It may be empty. If it is non-empty, $\sigma_i^{-1} \circ \sigma'_j$ is an orientation preserving diffeomorphism on its open domain of definition in $\mathbb{R}^n$ (open neighborhood of part of Δ^n) which maps part of Δ^n into Δ^n. Therefore,

$$\int_{\sigma_i} \psi_j\phi_i\omega = \int_{\Delta^n} \sigma_i^*(\psi_j\phi_i\omega) = \int_{\Delta^n} (\sigma_i^{-1} \circ \sigma'_j)^*(\sigma_i^*(\psi_j\phi_i\omega))$$

$$= \int_{\Delta^n} \sigma'^*_j(\phi_i\psi_j\omega) = \int_{\sigma'_j} \phi_i\psi_j\omega,$$

and both double sums in the previous expressions are equal. ($((\sigma_i^{-1} \circ \sigma'_j)^* \circ \sigma_i^* = \sigma'^*_j \circ \sigma_i^{-1*} \circ \sigma_i^* = \sigma'^*_j$ was used, see p. 74).

The definition (5.30) may now justly be considered to be the integral of the n-form ω over the regular domain Ω in M. If, on the other hand, ω is a *smooth* $(n-1)$-form on M, $\dim M = n$, and ω has compact support, then $d\omega$ has also compact support, and **Stokes' theorem** holds:

$$\int_{\Omega} d\omega = \int_{\partial\Omega} \omega. \tag{5.31}$$

Proof Use the partition of unity as in the definition of $\int_\Omega$. Since $\sum_i \phi_i = 1$ on a neighborhood of $\operatorname{supp}\omega \cap \Omega$, $\sum_i d\phi_i = d\sum_i \phi_i = 0$ there and hence $\sum_i d(\phi_i\omega) = \sum_i \phi_i\, d\omega = d\omega$ on Ω. For $U_i \subset \Omega$, $\int_{\partial\Omega} \phi_i\omega = 0$ because $\phi_i = 0$ on $\partial\Omega$. $\phi_i = 0$ on the image of $d\sigma_i$ too, and since a regular simplex is all the more a smooth singular simplex, (5.29) applies, and $\int_\Omega \phi_i\omega = \int_{\sigma_i} \phi_i\omega = 0$. Let $U_i \cap \partial\Omega \neq \emptyset$. Then, $\phi_i \neq 0$ in the interior of $\sigma_i(\Delta_0^{n-1})$ only where Δ_0^{n-1} is the face of Δ^n opposite to the origin. Since Δ_0^{n-1} is coherently oriented with Δ^n and σ_i is orientation preserving, $\sigma_i(\Delta_0^{n-1})$ is coherently oriented with $\sigma_i(\Delta^n)$. Again (5.29) applies, and $\int_\Omega d(\phi_i\omega) = \int_{\sigma_i} d(\phi_i\omega) = \int_{\partial\sigma_i} \phi_i\omega = \int_{\partial\Omega} \phi_i\omega$. Hence in total, $\int_\Omega d\omega = \int_\Omega \sum_i d(\phi_i\omega) = \int_{\partial\Omega} \sum_i \phi_i\omega = \int_{\partial\Omega} \omega$. □

Observe for both the definition of the integral over Ω and the proof of Stokes' theorem: If Ω itself is compact, then ω need not have compact support in M.

5.4 De Rham Cohomology

Consider as an example $\dim M = 2$ and the equation $d\omega = \rho$ where a 2-form ρ is given and a 1-form ω, in local coordinates $\omega = \omega_1 dx^1 + \omega_2 dx^2$, is sought. One has $d\omega = (\partial\omega_2/\partial x^1 - \partial\omega_1/\partial x^2) dx^1 \wedge dx^2$. How must ρ behave in order that the equation has a solution ω? For any domain Ω of finite measure one has $\int_\Omega \rho = \int_\Omega d\omega = \int_{\partial\Omega} \omega$. Hence, if Ω has no boundary (for instance if $\Omega = S^2$ is the two-dimensional sphere), then $\int_\Omega \rho = 0$ must hold as a necessary condition for a solution ω to exist. If $M = \mathbb{R}^2$, then there are no such compact domains Ω without boundary, and no such condition need be posed on ρ.

In the latter case, $\mathbb{R}^2$ may be considered as the complex plane, $x^1 = \operatorname{Re} z$, $x^2 = \operatorname{Im} z$, and ω may be considered as a complex function, $\tilde{\omega} = i(\omega_1 + i\omega_2)$. For an analytic function $\tilde{\omega}$, in this notation $d\omega = 0$ by the Cauchy–Riemann equations, and hence $\int_{\partial\Omega} \omega = 0$, if Ω is an oriented domain of analyticity of $\tilde{\omega}$ the oriented boundary of which is $\partial\Omega$. This integral is in the adopted notation the imaginary part of the complex integral, and its vanishing is part of Cauchy's integral theorem. For the integral to vanish it is sufficient that $\partial\Omega$ is the complete oriented boundary of a domain of analyticity of $\tilde{\omega}$.

The two natural questions that arise in these considerations are: (i) which are the domains Ω of a manifold M that have no boundary, and (ii) which surfaces of M are complete oriented boundaries of oriented domains Ω. If for instance M is the two-dimensional torus of Fig. 1.3 on p. 3 (which is orientable), then non of the circles drawn in the figure is a complete oriented boundary, because as a boundary it would have to have both orientations simultaneously. Only pairs of oppositely oriented circles (winding around the torus in opposite directions are boundaries of domains of the torus.

A domain which has no boundaries is called a **cycle**. (For instance a circle is a one-dimensional cycle, an n-dimensional sphere S^n is an n-cycle.) Clearly, every boundary is a cycle, but, as the above example shows, the reverse need not be true. Not every cycle need be a boundary. The classification of cycles and boundaries of manifolds is the subject of homology theory. However, this theory turned out to be simpler in a more general setting.

In Sect. 5.2, as a certain generalization of domains of manifolds r-chains of singular r-simplices were introduced. Consider the real vector space ${}_{\infty}C_r(M,\mathbb{R})$ of r-chains $c = \sum_l k_l\sigma_l$ of linear combinations with real coefficients k_l of smooth singular r-simplices σ_l. Only the smooth case is treated in the sequel although most results hold also true in the continuous case. Therefore, the presubscript will be omitted, $C_r(M,\mathbb{R}) = {}_{\infty}C_r(M,\mathbb{R})$. Let $B_r(M,\mathbb{R})$ be the set of boundaries and $Z_r(M,\mathbb{R})$ the set of cycles of $C_r(M,\mathbb{R})$. Since linear combinations of boundaries are boundaries and linear combinations of cycles are cycles, both sets are linear subspaces of $C_r(M,\mathbb{R})$. The boundary operator (recall that its operation on $C_r(M,K)$ is sometimes denoted by ∂_r) maps $C_r(M,\mathbb{R})$ into $B_{r-1}(M,\mathbb{R}) \subset Z_{r-1}(M,\mathbb{R}) \subset C_{r-1}(M,\mathbb{R})$ (since $\partial\circ\partial = 0$), and by definition of cycles it maps $Z_r(M,\mathbb{R})$ to 0:

$$\operatorname{Im}\partial_r = B_{r-1}(M,\mathbb{R}), \quad \operatorname{Ker}\partial_{r-1} = Z_{r-1}(M,\mathbb{R}) \supset B_{r-1}(M,\mathbb{R}), \tag{5.32}$$

where $\operatorname{Im}\partial_r$ means the image of the boundary operator ∂_r defined on $C_r(M,\mathbb{R})$, and the **kernel** $\operatorname{Ker}\partial_{r-1}$ is defined as the preimage of the origin of $C_{r-2}(M,\mathbb{R})$ in $C_{r-1}(M,\mathbb{R})$. (See Compendium C.1 on homomorphisms.)

The direct sum of all $C_r(M,\mathbb{R})$, $r\in\mathbb{Z}$ may be considered as a graded (by r) vector space $C(M,\mathbb{R})$ with an endomorphism ∂ of degree -1:

$$C(M,\mathbb{R}) = \{\cdots \xrightarrow{\partial} C_{r+1}(M,\mathbb{R}) \xrightarrow{\partial} C_r(M,\mathbb{R}) \xrightarrow{\partial} C_{r-1}(M,\mathbb{R}) \xrightarrow{\partial} \cdots\}. \tag{5.33}$$

$C(M,\mathbb{R})$ is called a (real) **chain complex**. Recall that by definition $C_r(M,K) = \{0\}$ for $r<0$, hence $C(M,K)$ may be considered as an infinite sequence of mappings ∂_r of modules. Together with a collection of r-simplices (repetition allowed) the chain complex $C(M,\mathbb{Z})$ contains all their oriented faces as $(r-1)$-simplices, the oriented faces of the latter as $(r-2)$-simplices and so on down to the oriented edges of 2-simplices as line elements and their oriented endpoints as 0-simplices

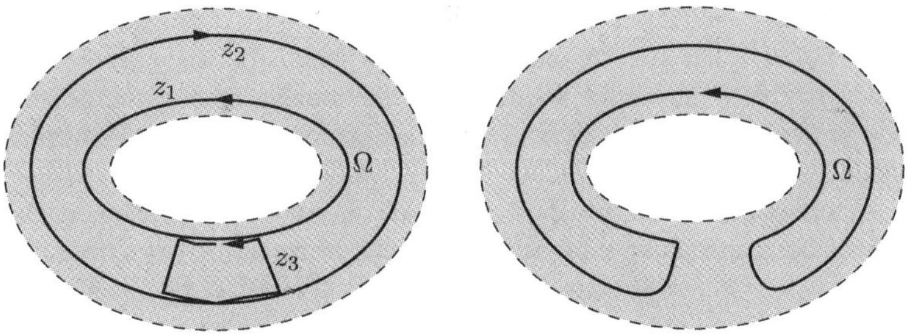

Fig. 5.5 1-cycles on a 2-torus. *Left*: z_1 and z_2 are not boundaries, z_3 is a boundary. *Right*: $b = z_1 + z_2 + z_3$ is a boundary

(set of all vertices of the original collection). This is how complexes originally were introduced in topology. In the case of $C(M, \mathbb{R})$ all those collections have in addition real coefficients.

As an example, consider the three 1-cycles on the 2-torus depicted on the left part of Fig. 5.5. As was already discussed, the cycles z_1 and z_2 are not boundaries. Depending on the orientation of the torus, $b = z_1 + z_2$ is the boundary of the visible domain Ω on the torus or of $-\Omega$. Let the first case be valid. Then, z_3 is the boundary of minus the visible domain enclosed by this cycle. The sum $b = z_1 + z_2 + z_3$ depicted on the right part of the figure is the boundary of the visible enclosed domain denoted again Ω on this figure. One realizes, if certain cycles are not boundaries, nevertheless their sums or differences may be boundaries. The alert reader also immediately realizes the relevance of considerations of that type for complex analysis, and indeed complex analysis of several variables was one of the early motivations to develop homology theory.

Two r-cycles z_1, z_2 are called **homologous**, $z_1 \sim z_2$, if their difference is an r-boundary:

$$Z_r(M, K) \ni z_1 \sim z_2 \quad \Longleftrightarrow \quad z_1 - z_2 \in B_r(M, K). \tag{5.34}$$

A boundary is called **homologically trivial**. Hence, two cycles are homologous, if their difference is homologically trivial. Clearly, the homology relation (5.34) is an equivalence relation. The Abelian group of equivalence classes in homology of r-cycles is called the rth **homology group** $H_r(M, K)$. It is the quotient group

$$H_r(M, K) = Z_r(M, K)/B_r(M, K) = \operatorname{Ker} \partial_r / \operatorname{Im} \partial_{r+1}. \tag{5.35}$$

In the case $K = \mathbb{Z}$, as every Abelian group it may also be considered a module; if K is a field (like $\mathbb{R}$), more specifically it is a vector space. Unlike $Z_r(M, K)$ and $B_r(M, K)$, $H_r(M, K)$ is finite-dimensional in most interesting cases.

The reader may convince himself from the above example that the main topological difference between the 2-torus $\mathbb{T}^2$ and the plane $\mathbb{R}^2$ is that in $\mathbb{R}^2$ every cycle is a boundary (of the encircled domain) while in $\mathbb{T}^2$ there are cycles which are not boundaries. If all cycles of a chain complex (5.33) are boundaries, this means that $\operatorname{Im}\partial_{r+1} = \operatorname{Ker}\partial_r$ for all r, or, equivalently, that all homology groups are trivial: $H_r(M,K) = \{0\}$. A sequence of morphisms between algebraic objects like (5.33) is called an **exact sequence**, if the image of one morphism in the sequence is the kernel of the next. Sequences of non-trivial chain complexes are not exact.

Yet, exact sequences form a powerful tool in algebra. For instance, the exact sequence of homomorphisms of Abelian groups or modules (in particular vector spaces)

$$0 \to G \xrightarrow{f} H$$

means that f is injective: Since the image of the first mapping can only consist of the zero element of G, the kernel of f must be $\{0\}$, and by linearity f must be injective. If, on the other hand, the sequence

$$G \xrightarrow{f} H \to 0$$

is exact, this means that f is surjective: since all of H is mapped to 0 by the right mapping, its kernel is all of H which also must be the image of f. Hence, the exact sequence

$$0 \to G \xrightarrow{f} H \to 0$$

means that the homomorphism f is indeed an isomorphism. Interpret as an exercise the so called **short exact sequence** for the case of Abelian groups or modules,

$$0 \to H \to G \to G/H \to 0,$$

where H is a subgroup or submodule of G and G/H is the quotient structure.

Coming back to the homology groups $H_r(M,K)$ of (5.35), it will be seen later that their dimensions are topological invariants.

$$\beta^r(M) = \dim H_r(M,\mathbb{R}) \tag{5.36}$$

is called the rth **Betti number** of M.

Recall that a 0-simplex is just a point of M, a 0-chain hence is a linear combination of points. Since $C_{-1}(M,\mathbb{R})$ is trivial, a 0-chain has zero boundary. Hence, every 0-chain is a 0-cycle. The standard 1-simplex is a line element, its image in M is a finite path between two points. Every pair of points (z_1,z_2) which can be connected by a path Ω in M yields a boundary as its difference: $\partial\Omega = z_2 - z_1$. Hence, all points which can be connected by a path in M are homologous: the pathwise connected components of M form a base of the vector space $H_0(M,\mathbb{R})$, and *the zeroth Betti number of any manifold M is equal to the number of pathwise connected components of M*.

If M is contractible (see Sect. 2.5), that is M may continuously be contracted into one point, then it is intuitively clear and will formally be proved in the next section that every r-cycle, $r > 0$ is a boundary, that is, $H_r(M, \mathbb{R}) = \{0\}$. Hence, $\beta^r(M) = 0$, $r > 0$, *if M is contractible*. In particular,

$$\beta^0(\mathbb{R}^n) = 1, \quad \beta^r(\mathbb{R}^n) = 0, \quad r > 0,\ n \geq 1. \tag{5.37}$$

Less trivial cases will be considered in the sequel.

Comparison of $\partial \circ \partial = 0$ with $d \circ d = 0$ for the exterior derivation d of degree +1 and consideration of Stokes' theorem suggest a duality between the chain complex $C(M, \mathbb{R})$ and the graded algebra $\mathcal{D}(M)$ of exterior forms:

$$\mathcal{D}(M) = \{\cdots \overset{d}{\longleftarrow} \mathcal{D}^{r+1}(M) \overset{d}{\longleftarrow} \mathcal{D}^r(M) \overset{d}{\longleftarrow} \mathcal{D}^{r-1}(M) \overset{d}{\longleftarrow} \cdots\}. \tag{5.38}$$

Again, as previously in Sect. 4.2, $\mathcal{D}^r = \{0\}$ if $r < 0$ or $r > \dim M$. An exterior r-form ω on M is called a **closed form**, if $d\omega = 0$, it is called an **exact form**, if there exists an $(r-1)$-form σ so that $\omega = d\sigma$. Two closed forms ω^1 and ω^2, $d\omega^i = 0$ are called **cohomologous** to each other, $\omega^1 \sim \omega^2$, if their difference is exact, that is, $\omega^1 - \omega^2 = d\sigma$ for some form σ. An exact form is called **cohomologically trivial**. Clearly, every exact form is closed, and clearly, closed forms as well as exact forms form linear subspaces of the vector spaces $\mathcal{D}^r(M)$. **De Rham's cohomology group** is the quotient group

$$H^r_{dR}(M) = \{\text{closed } r\text{-forms}\}/\{\text{exact } r\text{-forms}\} = \operatorname{Ker} d_r / \operatorname{Im} d_{r-1}. \tag{5.39}$$

Since $\dim H^r_{dR}(M) \leq \dim \mathcal{D}^r(M)$ because $H^r_{dR}(M)$ is a quotient space of a subspace of $\mathcal{D}^r(M)$, clearly $\dim H^r_{dR}(M) = 0$ for $r < 0$ or $r > \dim M$. Moreover, from $\dim \mathcal{D}^{-1} = 0$ it follows that $\operatorname{Im} d_{-1} = \{0\}$, and hence $H^0_{dR}(M) = \operatorname{Ker} d_0$. Now, a 0-form ω^0 is a real function on M, and hence $d\omega^0 = 0$ means that the function ω^0 is constant on each pathwise connected component of M (by integration of $d\omega^0$ along any path in M). If M has m components, then $\operatorname{Ker} d_0$ is the space of real m-tuples which means that $\dim H^0_{dR}(M) = \dim \operatorname{Ker} d_0$ *is equal to the number of pathwise connected components of M and hence equal to the Betti number $\beta^0(M)$*. It will be seen that this is not an accident.

Let ω be a closed r-from and let z be an r-cycle. Consider the real number

$$\langle \omega, z \rangle = \int_z \omega$$

given by the integral (5.27, 5.28). It is obviously bilinear in ω and z as suggested by the way of writing. Let $\omega' = d\sigma$ be any exact r-form and let $b = \partial z'$ be any r-boundary. Then, by virtue of Stokes' theorem (5.29) for singular chains,

$$\langle \omega + \omega', z \rangle = \int_z \omega + \int_z d\sigma = \int_z \omega + \int_{\partial z} \sigma = \langle \omega, z \rangle,$$

since $\partial z = 0$ for a cycle z. Likewise,

$$\langle \omega, z+b \rangle = \int_z \omega + \int_{\partial z'} \omega = \int_z \omega + \int_{z'} d\omega = \langle \omega, z \rangle,$$

since $d\omega = 0$ for a closed form ω. In effect, the considered integral depends on the homology class $[z]$ of the cycle z and on the cohomology class $[\omega]$ of the closed form ω only:

$$\langle [\omega], [z] \rangle = \int_z \omega \tag{5.40}$$

is a linear form on the space $H_r(M, \mathbb{R})$, that is, an element of the dual space $(H_r(M, \mathbb{R}))^*$, and every element $[\omega]$ of $H^r_{dR}(M)$ yields uniquely such a linear form. In other words, (5.40) yields a homomorphism of vector spaces

$$H^r_{dR}(M) \to (H_r(M, \mathbb{R}))^*. \tag{5.41}$$

This reflects the point of view of letting $[\omega]$ run through $H^r_{dR}(M)$ and considering (5.40) as linear functions on $H_r(M, \mathbb{R})$, that is, as elements of $(H_r(M, \mathbb{R}))^*$.

De Rham's theorem states that (5.41) *is in fact an isomorphism.*

In this connection the real number (5.40) is called the **period** of the r-form ω over the cycle z,

$$\mathrm{per}(z) = \langle [\omega], [z] \rangle = \int_z \omega. \tag{5.42}$$

De Rham's theorem implies that, if there exists a linear function per on $Z_r(M, \mathbb{R})$ with the property $\mathrm{per}(b) = 0$ for every boundary b, then there exists a closed r-form ω so that $\int_z \omega = \mathrm{per}(z)$. It also implies

$$\dim H^r_{dR}(M) = \dim(H_r(M, \mathbb{R}))^* = \dim H_r(M, \mathbb{R}) = \beta^r(M). \tag{5.43}$$

Two isomorphic vector spaces have the same dimension, and two spaces connected by a non-degenerate bilinear form have also the same dimension. Moreover, $H^r_{dR}(M) \approx H_r(M, \mathbb{R})$, if $\beta^r(M) < \infty$, since two real *finite-dimensional* vector spaces of the same dimension are isomorphic. Now, considering (5.40) as a bilinear form on $H^r_{dR}(M) \times H_r(M, \mathbb{R})$, for every $[\omega] \neq 0$ there is a $[z]$ so that (5.40) is non-zero. Otherwise (5.40) would yield the same result on all $[z]$ for $[\omega] \neq 0$ and for $[\omega'] = 0$ and (5.41) could not be an isomorphism. Likewise, for every $[z] \neq 0$ there is an $[\omega]$ so that (5.40) is non-zero. Otherwise for all $[\omega] \in \dim H^r_{dR}(M)$ (5.40) would yield the same value 0 on $[z] \neq 0$ and on $[z'] = 0$ and (5.41) would not be surjective.

An immediate consequence is

$$\beta^r(M) = 0 \quad \text{for } r > \dim M \tag{5.44}$$

and also that $\dim H^r_{dR}(M) = 0$ *for all* $r > 0$, *if* M *is contractible*. Hence in particular, on a contractible manifold M the necessary and sufficient condition for the equation $d\omega = \rho$ to have a solution ω is that ρ is closed, $d\rho = 0$, since $\dim H^r_{dR}(M) = 0$ means that every closed form is exact.

In these considerations both $H^r_{dR}(M)$ and $H_r(M, \mathbb{R})$ are treated as real vector spaces. (Recall that every vector space is an Abelian group with respect to vector addition. This justifies to retain the names homology group and cohomology group in the considered more special cases.) However, $\mathcal{D}(M)$ is also an algebra with respect to exterior multiplication. It is easily seen that the wedge product is compatible with the cohomology classes of $\mathcal{D}(M)$. Indeed, let v, v', ω, ω' be closed forms and let $v - v' = d\rho$, $\omega - \omega' = d\sigma$ for some forms ρ and σ, that is, $[v - v'] = 0$, $[\omega - \omega'] = 0$. Then, obviously $v \wedge \omega$ and $v' \wedge \omega'$ are also closed forms, and

$$\begin{aligned} v \wedge \omega - v' \wedge \omega' &= (v - v') \wedge \omega + v' \wedge (\omega - \omega') = d\rho \wedge \omega + v' \wedge d\sigma \\ &= d(\rho \wedge \omega + (-1)^{r_v} v' \wedge \sigma) + (-1)^{r_v} \rho \wedge d\omega - (-1)^{r_v} dv' \wedge \sigma. \end{aligned}$$

The last two terms vanish since ω and v' are closed forms. Hence, $v \wedge \omega - v' \wedge \omega'$ is an exact form. This implies that $v \wedge (\omega - \omega')$ and $(v - v') \wedge \omega$ as special cases of the just considered one are also exact forms. This altogether means that the cohomology class $[v \wedge \omega]$ does not depend on the representatives of the cohomology classes $[v]$ and $[\omega]$, and one may define a wedge product in $H^r_{dR}(M)$ by

$$[v] \wedge [\omega] = [v \wedge \omega]. \tag{5.45}$$

Therefore, the de Rham cohomology $H_{dR}(M)$, the direct sum of all $H^r_{dR}(M)$, is indeed again a graded algebra.

5.5 Homology and Homotopy

The alert reader may anticipate from the last section that there is a close connection between the homology of chain complexes and homotopy.

Let $F : M \to N$ be a (smooth) mapping from the (smooth) manifold M into the (smooth) manifold N. (Recall that generally smooth entities are considered in this volume.) Let $C(M, \mathbb{R})$ be the chain complex on M. A (smooth) singular r-simplex $\sigma \in C(M, \mathbb{R})$ is a mapping of a neighborhood of the standard r-simplex in $\mathbb{R}^n$ into M. Clearly, $F_*(\sigma) = F \circ \sigma$ is a singular r-simplex in N. Since the oriented boundary of σ was defined in (5.25) as the push forward by σ of the oriented boundary of the standard r-simplex, it is clear that F_* maps cycles on M into cycles on N and boundaries on M into boundaries on N. These mappings need of course not be one–one, also, M and N need not have the same dimension. Recall that a singular r-simplex in N even may consist of a single point. Nevertheless, and that is one of the main advantages of *singular* chains, it is clear that $F_* \circ \partial = \partial \circ F_*$ and that $F_* : H_r(M, \mathbb{R}) \to H_r(N, \mathbb{R})$ *is a homomorphism of vector spaces*. Indeed, if $z_1 \sim z_2$ are homologous chains of $C(M,\mathbb{R})$ then $F_*(z_1) \sim F_*(z_2)$ are homologous chains of $C(N,\mathbb{R})$.

Now, let F_1 and F_2 be two homotopic mappings from M into N, that is (Sect. 2.5), there is a continuous mapping $H : [0,1] \times M \to N$ with $H(0,\cdot) = F_1$ and $H(1,\cdot) = F_2$. H may be extended to $\tilde{I} \times M$ where $\tilde{I}$ is an open neighborhood in $\mathbb{R}$ of the closed interval $[0,1]$. Together with M, $\tilde{I} \times M$ is also a smooth manifold. Hence, H may be assumed to be smooth since F_1 and F_2 are smooth and a continuous function on a smooth manifold (which latter is locally diffeomorphic with $\mathbb{R}^n$) may be arbitrarily closely approximated by a smooth function.

Let $z \in C(M,\mathbb{R})$ be an r-cycle. Then, $(\mathrm{Id}_{\tilde{I}}, z) \in C(\tilde{I} \times M, \mathbb{R})$ is a singular $(r+1)$-chain, which is the image of an $(r+1)$-cylinder in $\mathbb{R}^{n+1}$ of height 1 whose basis and top is the same cycle of ordinary simplices. Clearly its boundary is $(1,z) - (0,z)$ (Fig. 5.6, z itself as a cycle has no boundary). Hence, $H_*(\mathrm{Id}_{\tilde{I}}, z) \in C(N,\mathbb{R})$ is also a chain whose boundary is $(F_2)_*(z) - (F_1)_*(z)$. Since the latter difference is a boundary, $(F_1)_*(z) \sim (F_2)_*(z)$ are homologous:

The homomorphisms in homology $(F_1)_*$ and $(F_2)_*$ of homotopic maps F_1 and F_2 from M into N are the same: $(F_1)_* = (F_2)_*$.

Finally, let M and N be homotopy equivalent, that is, there exist mappings $F : M \to N$ and $G : N \to M$ so that $G \circ F \cong \mathrm{Id}_M$ and $F \circ G \cong \mathrm{Id}_N$ (Sect. 2.5). Since $(\mathrm{Id}_M)_* : H_r(M,\mathbb{R}) \to H_r(M,\mathbb{R})$ is the identity homomorphism and $(G \circ F)_* = G_* \circ F_*$ (cf. p. 73), it follows that $G_* = (F_*)^{-1}$ and hence $H_r(M,\mathbb{R})$ and $H_r(N,\mathbb{R})$ are isomorphic:

Homotopy equivalent manifolds have isomorphic homology groups.

Consider now a contractible manifold, that is, a manifold that is homotopy equivalent to the one-point manifold $\{x\}$. In the latter manifold, every singular r-simplex is a constant mapping σ^r of the standard r-simplex to x. Hence, every r-chain is given as $k\sigma^r$, $k \in \mathbb{R}$, $r \geq 0$. From (5.25) it follows that for $r > 0$ the boundary of $k\sigma^r$ is $k\sigma^{r-1}$ if r is even and is the zero r-chain, if r is odd. That means that for $r > 0$, r odd, every r-chain is a cycle and at the same time is a boundary of an $(r+1)$-chain, while for $r > 0$, r even, there are no non-zero cycles. In summary, all homology groups $H_r(\{x\},\mathbb{R})$ for $r > 0$ are trivial (consist of the zero element only and hence are also zero-dimensional). In view of the last theorem the same is true for any contractible manifold, which also proves (5.37).

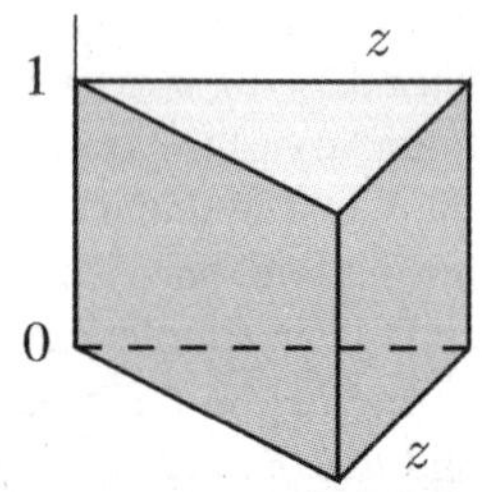

Fig. 5.6 A cylinder of height 1 with a cycle z (boundary of a triangle) as basis

Coming back to the homotopy H of the two homotopic mappings F_1 and F_2 from M into N, consider first two cohomologous r-forms $\omega^1 \sim \omega^2$ on N, $d(\omega^2 - \omega^1) = 0$. They are pulled back to M by $(F_i)^*$, and, since according to (4.43) any F^* commutes with d, cohomologous r-forms on N are pulled back to cohomologous r-forms on M. Moreover, since any $F^* : \mathcal{D}(N) \to \mathcal{D}(M)$ is a homomorphism of algebras (see p. 111), one finds:

The pull back F^* due to a smooth mapping F from M into N provides a homomorphism from the de Rham cohomology algebra $H_{dR}(N)$ into $H_{dR}(M)$.

With the definition (5.27, 5.28) of integrals of singular chains, the above considerations of the functors F_* and F^* immediately imply

$$\int_c F^*(\omega) = \int_{F_*(c)} \omega, \quad c \in C(M,\mathbb{R}),\ \omega \in \mathcal{D}(N). \tag{5.46}$$

With this relation, from the definition (5.40) it follows that

$$\langle [F^*(\omega)], [z] \rangle = \langle [\omega], [F_*(z)] \rangle, \quad z \in Z(M,\mathbb{R}),\ \omega \in \mathcal{D}(N),\ d\omega = 0, \tag{5.47}$$

for the homology and cohomology classes. With the non-degeneracy of the bilinear form $\langle \cdot, \cdot \rangle$ which was deduced from de Rham's theorem on p. 134, one arrives at the result that *homotopic mappings $F_1 \cong F_2$* (which yield the same homomorphisms $(F_1)_* = (F_2)_*$ in homology) *yield also the same homomorphisms $(F_1)^* = (F_2)^*$ in cohomology*. Historically, the latter result was in an earlier context proved independently from de Rham's theorem by Poincaré by a direct analysis using coordinate neighborhoods and was used by de Rham to prove his theorem.

Directly from de Rham's theorem and the situation with homology it follows:

Homotopy equivalent manifolds have isomorphic cohomology groups.

As an example consider again a one-point manifold $\{x\}$. It is zero-dimensional, and hence all $\mathcal{D}^r(\{x\})$ for $r > 0$ are zero-dimensional. The above theorem yields in an extremely simple way that all groups $H^r_{dR}(M) \approx H_r(M,\mathbb{R})$ for $r > 0$ are trivial for a contractible manifold M. Generally, cohomology is easier to handle than homology which circumstance substantiates the central role of de Rham's theorem in algebraic topology.

These interrelations between homology and homotopy have a very important consequence. At the beginning of this section the fact was used that every homotopy can arbitrarily closely be approximated by a smooth homotopy provided the manifold is smooth, that is, the manifold is locally diffeomorphic to $\mathbb{R}^n$. With the same homotopic approximations of continuous mappings by smooth mappings it can be proved that *all homology and cohomology results obtained for smooth mappings between manifolds hold true for only continuous mappings provided only that the considered manifolds themselves are smooth. In particular, the ${}_0H_r(M,K)$ are isomorphic to the ${}_\infty H_r(M,K)$* (therefore the presubscript ∞ was already omitted) *and for $K = \mathbb{R}$ both are isomorphic to $H^r_{dR}(M)$.*

In order to emphasize the duality between homology and cohomology, the algebra $\mathcal{D}(M)$ is also called a **cochain complex** C^*, the closed forms are then called **cocycles**, forming sets $Z^r(C^*) \subset \mathcal{D}^r(M)$, and exact forms are called **coboundaries**, forming sets $B^r(C^*) \subset Z^r(C^*) \subset \mathcal{D}^r(M)$. The derivation operator d is called a **coboundary operator** in this context.

It would be desirable to have also a pure cohomology notion of homotopy. Let again F_1 and F_2 be two homotopic mappings from a *pathwise connected manifold* M into N, let d^M and d^N be the exterior derivations in $\mathcal{D}(M)$ and $\mathcal{D}(N)$. Suppose there exist linear mappings $h_r : \mathcal{D}^r(N) \to \mathcal{D}^{r-1}(M)$ ($h_r = 0$ for $r \leq 0$) so that for every $\omega \in \mathcal{D}^r(N)$

$$h_{r+1}(d^N\omega) + d^M h_r(\omega) = (F_2)^*(\omega) - (F_1)^*(\omega).$$

If ω is closed, the first term is zero and the second is exact. Hence, the left hand side is exact for every closed ω, which is precisely the property of the right hand side, if F_1 and F_2 yield the same homomorphism in de Rham cohomology from $H_{dR}(N)$ to $H_{dR}(M)$. This is the case since $F_1 \cong F_2$. (The first term on the left hand side is needed since for a general ω not every form $(F_2)^*(\omega) - (F_1)^*(\omega)$ is closed even for homotopic F_i.) Specifically, for $r = 0$ a closed form is a constant on every connected component, hence the right hand side is zero for homotopic mappings from a pathwise connected manifold. In the above relation, h may be considered as an endomorphism from $\mathcal{D}(N)$ into $\mathcal{D}(M)$ of degree -1, and, in an operator notion, the relation may be written as

$$h \circ d + d \circ h = (F_2)^* - (F_1)^*. \tag{5.48}$$

If such an operator h exists it is called a **homotopy operator** for F_1 and F_2. This compares with the mappings $z \mapsto (\mathrm{Id}_{\bar{I}}, z)$ and $H_*(\mathrm{Id}_{\bar{I}}, z)$ and the boundary operators $\partial_{\bar{I}} \times M$ and ∂_N of homology which were combined to yield $(F_2)_*(z) - (F_1)_*(z)$ on p. 136.

As an example consider the mappings $F_1 : M \to \{x\} \subset M$ and $F_2 = \mathrm{Id}_M$ for a contractible manifold M. Then, $F_1 \cong F_2$ and $(F_1)^* = 0$ for $r > 0$ as previously for a pull back from the one-point manifold while $(F_2)^* = \mathrm{Id}_{\mathcal{D}(M)}$. The existence of a homotopy operator h in this case,

$$h \circ d + d \circ h = \mathrm{Id}_{\mathcal{D}(M)}$$

was proved by Poincaré and its explicit form in coordinate neighborhoods was given [1, paragraph 4.19]. This way he proved that on contractible manifolds the condition $d\rho = 0$ is not only necessary but also sufficient for the differential equation $d\omega = \rho$ to have a solution ω.

5.6 Homology and Cohomology of Complexes

The algebraic structure of a cochain complex has a variety of applications in algebra and topology. Let K be a field (for instance $K = \mathbb{R}$, or more generally let K

be a ring, for instance $K = \mathbb{Z}$) and let $C^* = (\ldots, C^{-1}, C^0, C^1, \ldots)$ be a sequence of vector spaces over K (more generally a sequence of K-modules; an example of a $\mathbb{Z}$-module is a crystal lattice; unlike the case of a vector space, the equation $\lambda x + \mu y = 0$, $\lambda, \mu \in K$, $x, y \in C^i$ need not have a solution x for all y in a module C^i).

As already mentioned, a **cochain complex** C^* is a sequence

$$\cdots \to C^{r-1} \xrightarrow{d^{r-1}} C^r \xrightarrow{d^r} C^{r+1} \to \cdots, \quad \operatorname{Im} d^{r-1} \subset \operatorname{Ker} d^r. \tag{5.49}$$

As previously, instead of d^r often d is written for all r. It is called the **coboundary operator** and has obviously the property $d \circ d = 0$ which is equivalent to the right relation of (5.49). $B^r(C^*) = \operatorname{Im} d^{r-1}$ is the set of degree r **coboundaries**, and $Z^r(C^*) = \operatorname{Ker} d^r$ is the set of degree r **cocycles**.

The quotient module (space)

$$H^r(C^*) = Z^r(C^*)/B^r(C^*) \tag{5.50}$$

is called the rth **cohomology module** (or cohomology group). One also introduces the direct sums

$$C^* = \oplus_r C^r, \quad H^*(C^*) = \oplus_r H^r(C^*) \tag{5.51}$$

as graded modules (vector spaces, sometimes even algebras as in the de Rham cohomology). A graded morphism f of degree s from a graded module C^* into a graded module D^* is a sequence of homomorphisms f^r from C^r into D^{r+s}. (d is a graded morphism of degree 1 from C^* to C^*.)

A **cochain mapping** $f : C^* \to D^*$ is a graded morphism of degree 0 for which each diagram

$$\begin{array}{ccc} C^{r+1} & \xrightarrow{f^{r+1}} & D^{r+1} \\ {\scriptstyle d^r}\uparrow & & \uparrow{\scriptstyle d^r} \\ C^r & \xrightarrow{f^r} & D^r \end{array} \tag{5.52}$$

commutes. Because of this commutativity, f sends cocycles into cocycles and coboundaries into coboundaries (exercise). Hence, it canonically induces a graded morphism (also denoted by f)

$$H^*(C^*) \xrightarrow{f} H^*(D^*). \tag{5.53}$$

One could denote the cohomology mappings by $H(d)$ instead of d and by $H(f)$ instead of f, and consider H a functor from the category of cochain complexes into the category of graded K-modules (see C.1).

With respect to their algebraic structure, homology and cohomology are totally symmetric. One may drop all prefixes 'co' in the above text and reverse all arrows (or equivalently reverse all degrees of grading) and obtain the completely analogous homology structure. Hence, all *algebraic* statements on cohomology transfer to homology, and in algebra both names cohomology and homology are used synonymously. The preference of 'co' comes from applications in topology.

Consider a short exact sequence (p. 132) of cochain mappings

$$0 \to C^* \xrightarrow{f} D^* \xrightarrow{g} E^* \to 0 \tag{5.54}$$

which expands in detail into the diagram

$$\begin{array}{ccccccccc}
 & & \vdots & & \vdots & & \vdots & & \\
 & & \uparrow & & \uparrow & & \uparrow & & \\
0 & \longrightarrow & C^{r+2} & \xrightarrow{f} & D^{r+2} & \xrightarrow{g} & E^{r+2} & \longrightarrow & 0 \\
 & & d\uparrow & & d\uparrow & & d\uparrow & & \\
0 & \longrightarrow & C^{r+1} & \xrightarrow{f} & D^{r+1} & \xrightarrow{g} & E^{r+1} & \longrightarrow & 0 \\
 & & d\uparrow & & d\uparrow & & d\uparrow & & \\
0 & \longrightarrow & C^{r} & \xrightarrow{f} & D^{r} & \xrightarrow{g} & E^{r} & \longrightarrow & 0 \\
 & & \uparrow & & \uparrow & & \uparrow & & \\
 & & \vdots & & \vdots & & \vdots & &
\end{array} \tag{5.55}$$

where every cell of arrows is a commutative diagram. The horizontal short exact sequences mean that $E^r = D^r/C^r$ are quotient modules and f^r is the canonical injection of C^r into D^r as a submodule, while g^r is the canonical surjection of D^r onto E^r by mapping the elements of D^r to their equivalence classes in E^r.

Pick any cocycle $z \in E^r$, that is, $0 = dz \in E^{r+1}$. Since g^r is surjective, one finds (not uniquely) an element $c \in D^r$ so that $gc = z$. Commutativity means $d^r_E g^r = g^{r+1} d^r_D$. (Superscripts and subscripts are used occasionally for the sake of clarity.) Therefore, $gdc = dgc = dz = 0$ must hold implying $dc \in \operatorname{Ker} g^{r+1} = \operatorname{Im} f^{r+1}$ or, in other words, there is an element $c' \in C^{r+1}$ for which $fc' = dc$ and hence $dfc' = ddc = 0$. Now, from the commutativity $d^{r+1}_D f^{r+1} = f^{r+2} d^{r+1}_C$ it follows that $0 = dfc' = fdc'$, and the injectivity of f implies $dc' = 0$. Hence, c' is a cocycle, $c' \in Z^{r+1}(C^*)$. In this sequence of mappings, $c' = f^{-1}dc \in f^{-1}dg^{-1}z$, the element $c \in g^{-1}z$ was not necessarily uniquely determined, because g need not be injective. However, since $E^r = D^r/C^r$, the element c is determined modulo an additive element $\tilde{c} \in D^r$ for which there is an element $\tilde{c}'' \in C^r$ with $f\tilde{c}'' = \tilde{c}$ and, because of the commutativity $d^r_D f^r = f^{r+1} d^r_C$, it holds that $fd\tilde{c}'' = df\tilde{c}'' = d\tilde{c}$. Surjectivity of f finally guarantees an element $\tilde{c}'$ for which $f\tilde{c}' = d\tilde{c} = fd\tilde{c}''$ and hence $\tilde{c}' = d\tilde{c}''$, that is, $\tilde{c}'$ is a coboundary, $\tilde{c}' \in B^{r+1}(C^*)$. To summarize, $c' \in Z^{r+1}(C^*)$ is determined by $z \in Z^r(E^*)$ up to a coboundary, or, the mapping $c' \in f^{-1}dg^{-1}z$ induces homomorphisms $\delta^r : Z^r(E^*) \to Z^{r+1}(C^*)/B^{r+1}(C^*)$. Now, specifically pick $z = b \in B^r(E^*)$ to be a coboundary. Then, there are elements $c_b \in C^{r-1}$ so that

$dgc_b = b = gdc_b = gc$ and hence $b = gc$ where now $c = dc_b$ itself is a coboundary. Hence, by the above reasoning, δ^r maps a coboundary $b \in B^r(E^*)$ into a coboundary $\tilde{c}' \in B^{r+1}(C^*)$. Thus, it induces canonically a graded morphism (also denoted δ) of degree 1

$$H^*(E^*) \xrightarrow{\delta} H^*(C^*). \tag{5.56}$$

As it was shown, δ is uniquely determined by the short exact sequence (5.54), that is, by the quotient structure of the cochain complex $E^* = D^*/C^*$.

By similar tedious but straightforward chasing around the diagram (5.55) it can be shown that the sequence

$$\begin{aligned} \cdots \xrightarrow{\delta} H^r(C^*) \xrightarrow{f} H^r(D^*) \xrightarrow{g} H^r(E^*) \\ \xrightarrow{\delta} H^{r+1}(C^*) \xrightarrow{f} H^{r+1}(D^*) \xrightarrow{g} H^{r+1}(E^*) \xrightarrow{\delta} \cdots \end{aligned} \tag{5.57}$$

is exact.

The link between this purely algebraic (co)homology theory and topology is provided by sheaf theory. A sheaf of modules is a topological space X each point of which is attached with a K-module (a stalk) and a quite fine topology is extended from X to the sheaf. (The germs $[F]$ of real functions F on open sets $U \subset M$ with $x \in U$ form the stalks $\mathcal{F}_x$ of a sheaf of $\mathbb{R}$-algebras on M, see Sect. 3.3. Sheaf theory is mainly a rather abstract application of diagrams of commuting and exact parts (sometimes in a positive sense called 'abstract nonsense'). It is used to prove the de Rham theorem and the equivalence of many homology theories. It is not considered here since it would digress from the main goal of this text. The interested reader is referred to the concise and clear introduction by Warner [1, Chap. 5].

The central role of (co)homology in topology derives from the fact that the homology groups are the best understood topological invariants. It was seen in Sect. 5.5 that even homotopy equivalent manifolds have up to isomorphy the same homology and cohomology groups. (Recall from Sect. 2.5 that topologically equivalent spaces, that is, homeomorphic spaces are homotopy equivalent; the inverse is not in general true, e.g. a single point and a contractible space are homotopy equivalent.) The (co)homology groups for the same topological space depend, however, in general on the ring K. In this respect, most important is the case $K = \mathbb{Z}$, because from the (co)homology groups of this case those for all other rings K may be straightforwardly calculated by applying results of algebra. On the other hand, the de Rham theory holds for the case $K = \mathbb{R}$. As another example of the above algebra, the classical theory of polyhedra in combinatorial topology is shortly considered.

A **polyhedron** $|c|$ in $\mathbb{R}^n$ is the union (of sets of points of the $\mathbb{R}^n$) of a collection of r-simplices S_i^r of (5.17) in regular mutual position. If $\{v_0, \ldots, v_r\}$ is the set of vertices of the simplex S_i^r, then any proper subset of $s+1, s<r$, vertices spans an s-face of the r-simplex S_i^r which itself is an s-simplex. (The vertices themselves

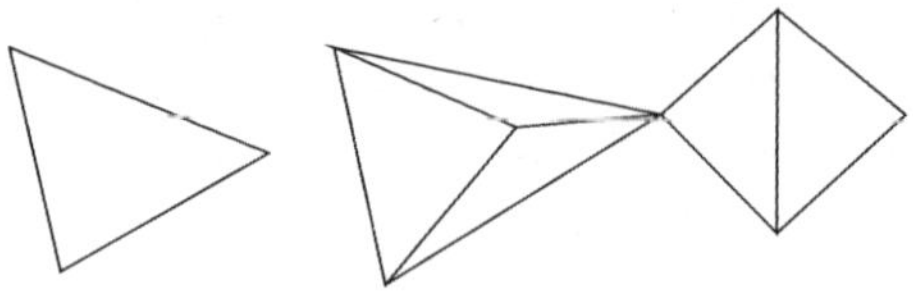

Fig. 5.7 A polyhedron consisting of one tetrahedron and three triangles

are 1-faces.) Regular mutual position of simplices of the polyhedron means that any two of the simplices of the polyhedron either are disjoint or intersect precisely by some faces of either simplex. The collection of all *distinct* vertices $v = \{v_j \mid j = 0,\ldots,l\}$ of the simplices of the polyhedron are put into a fixed order. Then, there is a one–one correspondence between simplices S_i^r of the polyhedron and subsets c_i of the set v consisting of r vertices in an order derived from v. A set c is formed the elements of which are all those subsets c_i corresponding to the simplices S_i^r of the polyhedron, and to all *distinct* faces of simplices contributing to c. For instance, in Fig. 5.7 a polyhedron consisting of one tetrahedron and three triangles is shown. Into its set c one four-point set, 7 three-point sets corresponding to the 7 triangles including the four faces of the tetrahedron, 14 two-point sets corresponding to all distinct legs of the triangles, and 10 one-point sets corresponding to the 10 vertices of the polyhedron enter. The set c is called the **abstract complex** corresponding to the polyhedron. It is easily seen that by the given convention there is a one–one correspondence between actual realizations of polyhedra by simplices and abstract complexes. However, for a given polyhedron there is an infinite many of possibilities of realizations by simplices. For instance, a triangle may be given by a set of smaller triangles in regular mutual position. The set of simplices corresponding to the c_i of the abstract complex is called the geometrical complex. The geometrical complex of the polyhedron of Fig. 5.7 consists of one tetrahedron, 7 triangles, 14 line segments and 10 points (vertices). An orientation is defined in both the abstract and the geometrical complexes by defining the simplices in the fixed order of their vertices as positively oriented. An odd permutation of the vertices reverses orientation. Linear combination of the elements c_i of an abstract complex with coefficients of some ring K and introduction of the boundary operator derived from (5.19) make it into a chain complex, which is isomorphic to a subset of the complex of continuous singular chains ${}_0C(|c|, K)$ considered in Sect. 5.4. Indeed it can be shown that the homology groups of this complex and those of chains of the abstract complex of the polyhedron $|c|$ are isomorphic.

Before considering the homology of chains of an abstract complex, a simple result on embedding of polyhedra is considered. The **dimension of a polyhedron** is the largest dimension $r = m$ of a simplex entering the polyhedron. $m + 1$ points $\boldsymbol{v}_0,\ldots,\boldsymbol{v}_m$ of the $\mathbb{R}^n (n \geq m)$ are linearly independent, if the vectors from $\boldsymbol{v}_0$ to the $\boldsymbol{v}_i$, $i = 1,\ldots,m$ are linearly independent. This does not depend on the order of the $\boldsymbol{v}_i$ and on which of them is taken to be $\boldsymbol{v}_0$. For an arbitrary number m, m points of the $\mathbb{R}^n$ are in general position, if any $n + 1$ of them are linearly independent.

An m-dimensional polyhedron with l vertices may be embedded into $\mathbb{R}^{2m+1}$ by choosing arbitrarily l vertices in general position.

Proof Obviously, all polyhedra with the same number of vertices grouped in the same way into simplices are homeomorphic. (Recall that geometrical simplices of polyhedra are in regular mutual position.) Distribute the vertices of the polyhedron in general position over the $\mathbb{R}^{2m+1}$ and consider the geometrical simplices with these vertices corresponding to the simplices of the polyhedron. Let S_1^r and S_2^s be two simplices of the polyhedron, hence $r, s \leq m$. Consider the $(r+s+1)$-dimensional simplex spanned by all $r+s+2$ vertices of the former two simplices of the polyhedron. It may not be part of the polyhedron, but some part of its boundary is. All simplices of the boundary of a given simplex are in regular mutual position. Hence, the obtained embedding is homeomorphic to the originally given polyhedron. □

This result may be used to prove that every compact smooth m-dimensional manifold may be embedded in the $\mathbb{R}^{2m+1}$ [2]. Much harder is it to prove the fact that the same also holds for non-compact manifolds. (Note also that much higher dimensions may be needed to embed a metric manifold *isometrically* into some $\mathbb{R}^n$.)

Now, let a polyhedron $|c|$ of dimension m in $\mathbb{R}^n$ be given and consider the corresponding abstract complex c. The collection of all abstract simplices c_i of c with dimension $\leq r$ is called the rth **skeleton** c^r of $c = c^m$. Let $C_r(c, \mathbb{R})$ be the chain module (in fact a vector space in this case) over $K = \mathbb{R}$ generated by all simplices $c_i \in c^r \setminus c^{r-1}$, that is by all r-dimensional simplices of c. This implies $C_r(c, \mathbb{R}) = \{0\}$ for $r<0$ and for $r > m$. Let $C(c, \mathbb{R}) = \oplus_r C_r(c, \mathbb{R})$ be the chain complex of the abstract complex c corresponding to the polyhedron $|c|$ of dimension m.

The boundary operator ∂ induced in c by (5.19) obviously has the properties $\partial c^r \subset c^{r-1}$ and $\partial \circ \partial = 0$. By linearity it generalizes to a boundary operator ∂ in $C(c, \mathbb{R})$ which is a graded morphism of degree -1 of the graded vector space $C(c, \mathbb{R})$. $B(c, \mathbb{R}) = \partial C(c, \mathbb{R})$ contains the boundaries of $C(c, \mathbb{R})$, and $B(c, \mathbb{R}) \subset Z(c, \mathbb{R}) = \operatorname{Ker} \partial$, the set of cycles of $C(c, \mathbb{R})$. The homology groups (vector spaces) of this chain complex are $H_r(c, \mathbb{R}) = Z_r(c, \mathbb{R})/B_r(c, \mathbb{R})$.

Among the polyhedra $|c|$ of dimension m there are in particular polyhedra which are also C^0-manifolds M of dimension m. In a quite similar manner as for $H_0(M, \mathbb{R})$ on p. 133 it is easily seen that $\dim H_0(c, \mathbb{R})$ is equal to the number of components of the polyhedron (two in the case of Fig. 5.7). Hence, for a polyhedron $|c| = M$ which is a manifold, both groups are isomorphic, $H_0(c, \mathbb{R}) \approx H_0(M, \mathbb{R})$. Assume further $|c| = M$, $\dim |c| = \dim M = m$, and consider an m-cycle of singular simplices, which is a chain of mappings of standard m-simplices into $|c|$. If its image would contain only part of a given m-simplex of c, it could not be a cycle, since it would have a boundary in the sense of singular simplices. Hence, its image can only consist of whole m-simplices of c, and to be a cycle in $C_m(M, \mathbb{R})$ these m-simplices must form also a cycle in $C(c, \mathbb{R})$. Since in both chain complexes there are no m-boundaries ($C_{m+1} = \{0\}$), one has $H_m(M, \mathbb{R}) = Z_m(M, \mathbb{R}) \approx Z_m(c, \mathbb{R}) = H_m(c, \mathbb{R})$.

Next consider a singular $(m-1)$-boundary the image of which lies entirely in the $(m-1)$-skeleton c^{m-1} of c. In order to do so it must be the boundary of a singular m-chain the image of which consists of entire m-simplices of $|c|$. Consequently it is also a boundary in $C_{m-1}(c,\mathbb{R})$. Of course, a general singular $(m-1)$-boundary need not have this property that its image lies entirely in c^{m-1}. This image may instead intersect the interior of an m-simplex of $|c|$ as a hypersurface of dimension smaller than m. But then obviously it can homotopically be moved into c^{m-1}. In summary, two homology equivalent cycles of $Z_{m-1}(c,\mathbb{R})$ correspond to homology equivalent cycles of $Z_{m-1}(M,\mathbb{R})$. An arbitrary cycle of $Z_{m-1}(M,\mathbb{R})$ may likewise homotopically be moved into the skeleton $|c^{m-1}|$. Then, by repeating the above consideration with $|c|$ replaced by the skeleton $|c^{m-1}|$, one finally has $H_{m-1}(M,\mathbb{R}) \approx Z_{m-1}(|c^{m-1}|,\mathbb{R})/B_{m-1}(|c^{m-1}|,\mathbb{R}) \approx Z_{m-1}(c,\mathbb{R})/B_{m-1}(c,\mathbb{R}) = H_{m-1}(c,\mathbb{R})$.

By repeating these considerations for skeletons $|c^r|$ of lower dimensions one finds that indeed *for* $|c| = M$ *the homologies* $H(M,\mathbb{R})$ *and* $H(c,\mathbb{R})$ *are isomorphic*.

Of course there exists an abstract formal proof replacing these plausibility considerations which however needs further technical tools.

$C_r(c,\mathbb{R})$ is a finite-dimensional real vector space of which the r-simplices of c form a basis. A linear functional f on this vector space maps every r-simplex c_i to a real number $\langle f, c_i\rangle$ and extends to all $C_r(c,\mathbb{R})$ by linearity. The set of all linear functionals forms the dual vector space $C^r(c,\mathbb{R}) = (C_r(c,\mathbb{R}))^*$ of the same dimension as $C_r(c,\mathbb{R})$. Let d be the operator in $C^r(c,\mathbb{R})$ adjoint to ∂, that is, $\langle df, c_i\rangle = \langle f, \partial c_i\rangle$, and extension by linearity. From $\langle ddf, c_i\rangle = \langle f, \partial\partial c_i\rangle = 0$ it follows immediately that $d \circ d = 0$. Clearly, $C^*(c,\mathbb{R}) = \oplus_r C^r(c,\mathbb{R})$ is a graded vector space and $d : C^r(c,\mathbb{R}) \to C^{r+1}(c,\mathbb{R})$ is a graded morphism of degree $+1$. Hence, $C^*(c,\mathbb{R})$ is a cochain complex. Consider two homologous cycles $z, z' = z + \partial u$, $\partial z = \partial z' = 0$ and two cohomologous cocycles $f, f' = f + dg$, $df = df' = 0$. It follows $\langle f, z'\rangle = \langle f, z\rangle$ and $\langle f', z\rangle = \langle f, z\rangle$, hence $\langle f, z\rangle = \langle [f], [z]\rangle$ where $[z]$ and $[f]$ are the (co)homology classes of z and f and the statement is that the linear functional is independent of the representatives within these classes. This, however, means $H^*(c,\mathbb{R}) = (H(c,\mathbb{R}))^*$, and, since dual finite-dimensional vector spaces are isomorphic, the cohomology $H^*(c,\mathbb{R})$ is isomorphic to the homology $H(c,\mathbb{R})$. Together with the de Rham theorem one has

$$H_{dR}(|c|) \approx H(|c|,\mathbb{R}) \approx H(c,\mathbb{R}) \approx H^*(c,\mathbb{R}). \tag{5.58}$$

However, the relation between polyhedra $|c|$ and abstract complexes c is as already mentioned not one–one. It immediately follows that abstract complexes c related to the same polyhedron $|c|$ have the same (co)homology groups. The reader should check this in a few simple cases of small complexes c by direct verification.

On p. 133 it was already stated (and proved in Sect. 5.5) for real singular chain complexes that all homology groups for $r > 0$ of a contractible space are trivial. Since the n-dimensional unit ball B^n is contractible, $H_r(B^n,\mathbb{R}) = \{0\}$, $r > 0$; and $H_0(B^n,\mathbb{R}) = \mathbb{R}$, since B^n is pathwise connected. Hence,

$$B^n = \{\boldsymbol{r} \in \mathbb{R}^n | |\boldsymbol{r}| \leq 1\}, \quad H_0(B^n, \mathbb{R}) = \mathbb{R}, \quad H_r(B^n, \mathbb{R}) = \{0\},\ r \neq 0. \tag{5.59}$$

The same is true for a single n-simplex since it is homeomorphic and hence all the more homotopy equivalent to B^n. Let c be the complex of a single n-simplex. Then, for $0<r<n$ the skeleton c^r consists of all faces of all simplices of c^{r+1}. Hence, $c^r \setminus c^{r-1}$ consists of the r-faces of a collection of $(r+1)$-simplices. The $r+2$ r-faces of an $(r+1)$-simplex form an r-cycle. It can now be inferred from (5.58, 5.59) that every r-cycle $(0<r<n)$ is the boundary of a collection of $(r+1)$-simplices. This fact is hard to prove directly with polyhedra.

Consider now the polyhedron $|c^{n-1}|$ where c is again the complex of a single n-simplex. This polyhedron is homeomorphic to the $(n-1)$-sphere S^{n-1}. (Find as an exercise a continuous one–one mapping.) For $n-1<r<0$, the same holds true as above. However, the single $(n-1)$-cycle $c^{n-1}\setminus c^{n-2}$ of this case is not any more a boundary because c does not any more belong to the polyhedron. Hence, $H_n(|c^{n-1}|, \mathbb{R}) = \mathbb{R}$ and in total

$$\begin{aligned} &S^{n-1} = \{\boldsymbol{r} \in \mathbb{R}^n | |\boldsymbol{r}| = 1\}, \\ &H_0(S^{n-1}, \mathbb{R}) = H_{n-1}(S^{n-1}, \mathbb{R}) = \mathbb{R}, \quad H_r(S^{n-1}, \mathbb{R}) = \{0\},\ r \neq 0, n-1. \end{aligned} \tag{5.60}$$

In both cases, the arguments and the results remain the same, if $K = \mathbb{R}$ is replaced by $K = \mathbb{Z}$.

The interior of an n-simplex is homeomorphic to an open n-ball and its boundary is homeomorphic to an $(n-1)$-sphere. The latter is homeomorphic to an open $(n-1)$-ball compactified by a point. A point is considered as an open 0-ball. Spaces, homeomorphic to an open ball are called **cells**. Instead of building a topological space which is homeomorphic to a polyhedron out of simplices, it can be build out of cells. Then, **cell complexes** and **(co)homologies of cell chains** are obtained which latter can again be shown to be isomorphic to (5.58). They often provide an even simpler approach to the (co)homology groups. For the calculation of (co)homology groups, all kinds of isomorphies and of homotopies are extensively exploited.

For instance, *for compact oriented n-dimensional manifolds M*, **Poincaré's duality** *is the isomorphism*

$$H_{n-r}(M, \mathbb{R}) \approx H^r(M, \mathbb{R}), \tag{5.61}$$

where H^r means the dual of H_r. In view of (5.58) this also means $\beta^r = \beta^{n-r}$ for the Betti numbers in this case. Poincaré studied this duality (and coined the name Betti numbers in honor of the Italian pioneer of topology, Enrico Betti), but it was proved in general only with the help of so-called cup and cap products which extend the (co)homologies of simplicial chain complexes into graded algebras like the de Rham algebra (see p. 135) and which are not considered here. It is only mentioned that in view of de Rham's theorem (5.61) implies $H^r_{dR}(M, \mathbb{R}) \approx H^{n-r}_{dR}(M, \mathbb{R})$ which implies that for every r-form ω on M there is an $(n-r)$-form τ so that $\langle\omega, \tau\rangle = \int_M \omega \wedge \tau \neq 0$.

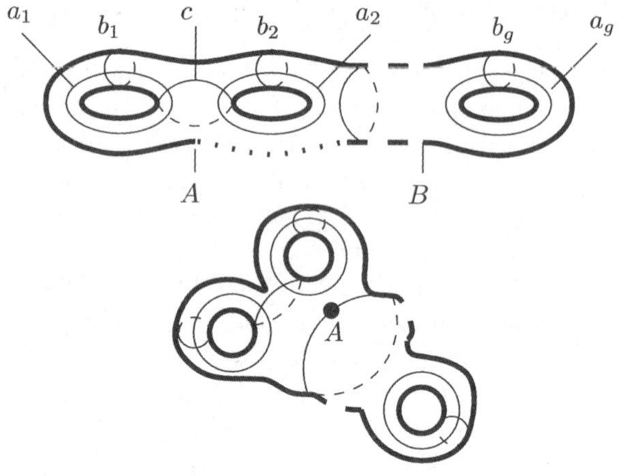

Fig. 5.8 A two-dimensional manifold of genus g. The *upper part* shows a torus $\mathbb{T}_g^2$ with g holes, the *lower part* shows a ball with g handles. See text for further explanations

As another example, a two-dimensional pretzel $\mathbb{T}_g^2$ with g holes is sketched in the upper part of Fig. 5.8. To each hole, there correspond two cycles a_i, b_i, $i = 1, \ldots, g$ which are not boundaries as was already discussed previously for the torus. Each of them represents a homology class of similar cycles, and any other cycle which is not a boundary as for instance c may be represented as a combination of the cycles a_i, b_i, for instance $c = b_1 - b_2$. By homotopically deforming this torus (and thereby contracting a path from point A to point B, dotted in the upper part of the figure, into a single point A of the lower part), the torus is deformed into a topologically equivalent 'sphere with g handles'. Both surfaces are homology equivalent and called surfaces of genus g. This can be summarized into

$$H_0(\mathbb{T}_g^2, \mathbb{R}) = H_2(\mathbb{T}_g^2, \mathbb{R}) = \mathbb{R}, \quad H_1(\mathbb{T}_g^2, \mathbb{R}) = \mathbb{R}^{2g}. \tag{5.62}$$

It can be shown that any *connected compact oriented* two-dimensional manifold is homology equivalent either to a sphere ($g = 0$) or to one of these spheres with handles and is homologically characterized by its genus.

At the end of Chap. 2 it was stated that our knowledge about the homotopy groups $\pi_m(S^n)$, $m > n$ is limited; however, unlike this largely unsolved problem on mappings between spheres of high dimensions, all (co)homology groups $H_m(S^n, K)$ are known and are trivial for $m > n$. Discovered half a century later than regular simplicial homology and homotopy, (co)homology nevertheless turned out to be a much simpler concept than homotopy to find topological invariants (but also providing less of them).

5.7 Euler's Characteristic

Consider a polyhedron $|c|$ or a manifold homeomorphic to a polyhedron, and consider an abstract complex c corresponding to that polyhedron. Denote the number of r-simplices in c by $\alpha_r(c)$. Then, the **Euler–Poincaré theorem** states

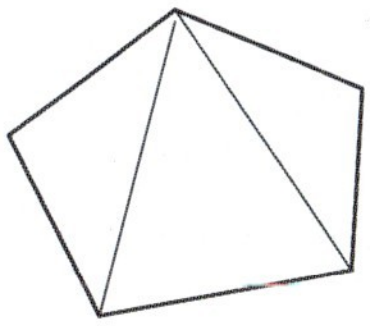

Fig. 5.9 A pentagon subdivided into three triangles

$$\chi(|c|) = \sum_r (-1)^r \alpha_r(c) = \sum_r (-1)^r \beta^r(|c|). \tag{5.63}$$

The middle expression of this relation is called **Euler's characteristic** χ of the polyhedron $|c|$. Leonard Euler observed that for any closed three-dimensional ordinary polyhedron $\alpha_0(c) - \alpha_1(c) + \alpha_2(c) = 2$ holds (number of corners − number of edges + number of faces), even if arbitrary polygons are considered as faces of the polyhedron (their number being $\alpha_2(c)$). If a polygon with n corners is divided into triangles, $n-3$ more edges (legs of the triangles) and $n-3$ more faces ($n-2$ triangles instead of the single polygon) are introduced cancelling in the first sum (5.63). Figure 5.9 shows as an example the subdivision of a pentagon into three triangles by introducing two more legs. The number of triangles equals the number of pentagons (one) plus $2 = 5 - 3$.

The surface of an ordinary polyhedron is homeomorphic to a 2-sphere. The latter has Betti numbers $\beta^0 = 1$, $\beta^1 = 0$, $\beta^2 = 1$ yielding 2 in the second sum of (5.63). Thus, Euler found a topological invariant in the 18th century.

Proof of the second equality of (5.63) Observe the simple facts that for a homomorphism of vector spaces $f : X \to Y$ the relation $\dim X = \dim \mathrm{Im} f + \dim \mathrm{Ker} f$ holds, and that for a chain complex

$$0 \to Z_r = \mathrm{Ker}\, \delta_r \to C_r \to B_{r-1} = \mathrm{Im}\, \delta_r \to 0$$

is an exact sequence. Hence, $\dim C_r = \dim Z_r + \dim B_{r-1}$ where for an abstract complex $\dim C_r(c, \mathbb{R}) = \alpha_r$ since $C_r(c, \mathbb{R})$ is spanned by the abstract r-simplices. On the other hand (again for vector spaces), $H_r = Z_r / B_r$ implies $\dim H_r = \dim Z_r - \dim B_r$, and Poincaré's original definition of Betti numbers was $\beta^r = \dim H_r(c, \mathbb{R})$. Inserting all that into (5.63) proves equality there. Agreement with (5.36) comes from the isomorphy (5.58) between singular simplicial chain homology and abstract simplicial chain homology. □

Since the Betti numbers are topological invariants, Euler's characteristic is also a topological invariant, and the first sum of (5.63) is independent of the 'triangulation' of the polyhedron, its subdivision into simplices, and also is the same for homeomorphic polyhedrons. The second sum is defined for any space homotopy equivalent to a polyhedron, and so is Euler's characteristic. Moreover, combining the last sum of (5.63) with Poincaré's duality (5.61) and using of de Rham's theorem leads immediately to the result that *Euler's characteristic of a compact orientable manifold of odd dimension is zero.*

Euler's characteristic has many applications in topology and geometry. The above proof fits into a much more general scheme. Consider the category of K-modules and a fixed Abelian group A. Consider further an Euler–Poincaré mapping ϕ of K-modules into A, that is a mapping with the following property: For every exact sequence $0 \to M' \to M \to M'' \to 0$ of K-modules, if $\phi(M')$ and $\phi(M'')$ exist, then $\phi(M)$ exists and $\phi(M) = \phi(M') + \phi(M'')$. (For $\mathbb{R}$-vector spaces M, $A = \mathbb{Z}$ and $\phi(M) = \dim M$ this is obviously the case of the proof above.) Now, let $0 \to C \xrightarrow{f} D \xrightarrow{g} E \to 0$ be an exact sequence of chain complexes of K-modules with morphisms f and g of degree 0 like in (5.54) (there it was written for the cochain case which just means a sign reversion of the degree of grading). Define the characteristic $\chi_\phi(C) = \sum_r (-1)^r \phi(C_r)$. Then, *if the characteristic* χ_ϕ *is defined for two of the complexes* C, D, E, *then it is defined for the third one and* $\chi_\phi(D) = \chi_\phi(C) + \chi_\phi(E)$. This results from the existence of the long exact sequence H of (5.57) which can be viewed as a chain complex of K-modules with trivial homology (all $H_r(H) = \{0\}$) in which each $H_r(C)$ or $H_r(D)$ or $H_r(E)$ is placed between modules H_r or $H_{r\pm 1}$ of the two other chains C, D, E. Since because of the triviality $0 = \chi_\phi(H) = \chi_\phi(C) - \chi_\phi(D) + \chi_\phi(E)$ (all $\phi(H_r(H)) = 0$), the above statement follows.

If one defines $B^+_r(c,\mathbb{R}) = B_{r-1}(c,\mathbb{R})$ and considers the exact sequence $0 \to Z(c,\mathbb{R}) \xrightarrow{\iota} C(c,\mathbb{R}) \xrightarrow{\partial} B^+(c,\mathbb{R}) \to 0$ of chain complexes of real vector spaces with the canonical injection ι and the boundary operator ∂ of the simplicial complex which due to the definition of the grading of B^+ is of degree 0, then without thinking one obtains (with $\phi(M) = \dim M$) $\chi_{\dim}(C) = \chi_{\dim}(Z) + \chi_{\dim}(B^+) = \chi_{\dim}(Z) - \chi_{\dim}(B) = \chi_{\dim}(Z/B)$ which is again (5.63).

Given a set $\mathcal{A}$ of modules over the same ring, defined up to isomorphism and such that for every exact sequence $0 \to M' \to M \to M'' \to 0$ the module M belongs to $\mathcal{A}$, if M' and M'' belong to $\mathcal{A}$, the set of Euler–Poincaré mappings (ϕ, A) has a universal element $(\gamma, K(\mathcal{A}))$, that is, an Abelian group $K(\mathcal{A})$ and a mapping γ so that every A is a subgroup of $K(\mathcal{A})$ with injection ι and $\phi = \iota \circ \gamma$. $K(\mathcal{A})$ is Grothendieck's K-group of $\mathcal{A}$. K-theory is another powerful tool to prove theorems in algebra and topology.

5.8 Critical Points

As an application of (co)homology theory of great relevance in physics, the **Morse theory** of critical points of smooth real functions on smooth manifolds is considered. Again the adjective smooth is dropped in the text.

Let M be an m-dimensional manifold and $F \in \mathcal{C}(M)$ be a real function on M. Let $x_0 \in M$ and let $U_\alpha \subset M$ be a coordinate neighborhood of x_0 with coordinates $(x^1, \ldots, x^m)$. The subscript α is dropped where there is no risk of clarity.

The restriction $F|_{U_\alpha}$ is given by $F_\alpha(x^1,\ldots,x^m)\circ\varphi_\alpha$. A **critical point** x_0 of F is a point where the differential dF vanishes, in local coordinates

$$\left.\frac{\partial F_\alpha}{\partial x^i}\right|_{x_0} = 0, \quad i = 1,\ldots,m. \tag{5.64}$$

This definition, like dF, is independent of the actual local coordinate system α, since the Jacobian matrix $(\psi_{\beta\alpha}^{-1})_i^j$ of the second line of (3.7) is regular between two local coordinate systems. The real value of F at the critical point is a **critical value** $F(x_0)$.

The critical point x_0 is **non-degenerate**, if the Hessian of F_α at x_0,

$$\left.\frac{\partial^2 F_\alpha}{\partial x^i \partial x^j}\right|_{x_0}, \tag{5.65}$$

is a non-degenerate (i,j)-matrix. Again, this condition is independent of the used local coordinate system. The **index of the non-degenerate critical point** is the number λ_{x_0} of negative eigenvalues of the Hessian of F_α at x_0. Since the Hessian is non-degenerate, it has all eigenvalues non-zero. A family of local coordinate transformations with regular Jacobian matrices, smoothly depending on parameters, cannot transform the determinant of the Hessian to zero, hence the eigenvalues cannot smoothly change sign depending on local coordinate systems. The index again is uniquely defined for a function F. Clearly, if the index is 0, F has a minimum, if the index is m, F has a maximum, and if the index has another value, F has a **saddle point**.

Consider a function F that has at most finitely many critical points *on a compact manifold M without boundary*. By the Weierstrass theorem, F takes on its minimal and maximal values on M. (Would that happen on a boundary, the corresponding points would not necessarily be critical.) Denote by M_c the subset $F^{-1}((-\infty,c))$ of M, that is, the set of points $x\in M$ with values $F(x)<c$, and denote by S_c its boundary given by $F(x)=c$. Clearly, $M_{c'}\subset M_c$ for $c'\leq c$. One may think of M as an m-dimensional generalization of a geographical surface of a porous ground and F as a gravitational potential. If this geography is gradually flooded up to sea level c (in terms of a gravitational potential level), then M_c is the part of the geography under water. (In fact just this problem was analyzed for $m=2$ in a paper by J. C. Maxwell in 1870 that can be regarded as the early root of Morse theory.)

In the following, local coordinates are used and the subscript α is dropped throughout. In Chap. 9 it will be shown, that in every smooth manifold a Riemannian metric g^{ij}, $g^{ij}g_{jk}=\delta^i_k$, (p. 107) can be introduced. Consider the tangent vector field, in local coordinates given by

$$\frac{dx^i}{dt} = -\varphi(x)g^{ij}\frac{\partial F}{\partial x^j}\left(g^{kl}\frac{\partial F}{\partial x^k}\frac{\partial F}{\partial x^l}\right)^{-1} \tag{5.66}$$

(Einstein summation understood). At non-critical points it is parallel to the tangent vector $g^{ij}\partial F/\partial x^j$ and hence in the Riemannian metric orthogonal to $S_{F(x)}$.

(The Riemannian metric is needed to ensure that this expression is regular at non-critical points.) At critical points the right hand side of (5.66) is not defined. Therefore, the smooth non-negative prefactor $\varphi(x)$ is introduced which is defined to be unity outside 2δ-balls centered at all critical points and zero inside the corresponding δ-balls. The right hand side of (5.66) is defined to vanish inside those δ-balls. This vector field can be integrated to a local 1-parameter group $\phi_t(x)$ with the obvious property

$$\frac{dF(\phi_t(x))}{dt} = \frac{\partial F}{\partial x^i}\frac{dx^i}{dt} = -\varphi(x) \leq 0 \tag{5.67}$$

for which purpose it was constructed. For $t \geq 0$, ϕ_t maps every set M_c into itself.

Let $c' > c$ such that for some small ϵ the interval $(c-\epsilon, c'+\epsilon)$ does not contain critical values of F. Then, δ can be chosen small enough so that $\varphi(x) = 1$ on $M_{c'} \setminus M_c$ since for any real interval of F-values there are at most finitely many critical points. Take $t \in [0, c'-c]$ and integrate (5.67) to $F(\phi_{c'-c}(x)) - F(x) = c - c'$ for $x \in S_{c'}$. Hence, $\phi_{c'-c}$ maps $S_{c'}$ into S_c. Likewise it is seen that it maps continuously (by the integral flow of a smooth tangent vector field) $M_{c'}$ into M_c. Generally, from $0 \leq \varphi \leq 1$ in (5.67) it follows that $|F(\phi_t(x)) - F(x)| \leq |t|$, hence $\phi_{c-c'} = (\phi_{c'-c})^{-1}$ maps continuously M_c into $M_{c'}$. It follows that $M_{c'}$ and M_c are homeomorphic.

A topological space M is called of category $k = \mathrm{cat}(M)$, if it can be covered with k contractible subsets of M but not with fewer number. A sphere S^n, $n > 0$ for instance is of category 2, $\mathrm{cat}(S^n) = 2$, since it is not itself contractible, but can be covered with two contractible half-spheres. Category is a topological property, homeomorphic spaces like for instance M_c and $M_{c'}$ above have the same category.

If c is a critical value corresponding to r critical points, then for small enough ϵ so that there are no more critical values in the interval $(c-2\epsilon, c+2\epsilon)$, by the same analysis a flow ϕ_t from $M_{c+\epsilon}$ into itself is constructed. Choose φ such that the 2δ-balls B_i, $i = 1,\ldots r$, around the r critical points do not overlap $M_{c-\epsilon}$ and each other and are inside $M_{c+\epsilon}$. Then, ϕ_t provides a flow of parts of $M_{c+\epsilon}$ into all of $M_{c-\epsilon} \subset M_{c+\epsilon}$ and of parts into the $B_i \subset M_{c+\epsilon}$. Take the contraction by this flow to see that $\mathrm{cat}(M_{c-\epsilon}) + r$ contractible sets cover $M_{c+\epsilon}$. Of course, several of them may be covered by one contractible set, hence $\mathrm{cat}(M_{c+\epsilon}) \leq \mathrm{cat}(M_{c-\epsilon}) + r$. Now, start with $c_0 < \min_x F(x)$. Then, $M_{c_0} = \emptyset$, $\mathrm{cat}(M_{c_0}) = 0$. By continuously increasing c to a value $c_1 > \max_x F(x)$, for which $M_{c_1} = M$, $\mathrm{cat}(M_c)$ may jump at most $C(F: M \to \mathbb{R})$ times by one, where $C(F: M \to \mathbb{R})$ is the number of critical points of F. (Up to here they need not be non-degenerate.) The result is

$$C(F: M \to \mathbb{R}) \geq \mathrm{cat}(M). \tag{5.68}$$

It is easily seen that for this result it would suffice that M is any manifold, with $\mathrm{cat}(M)$ either finite or $+\infty$, and that F would have a minimum and in every finite real interval at most finitely many critical values each corresponding to finitely many critical points.

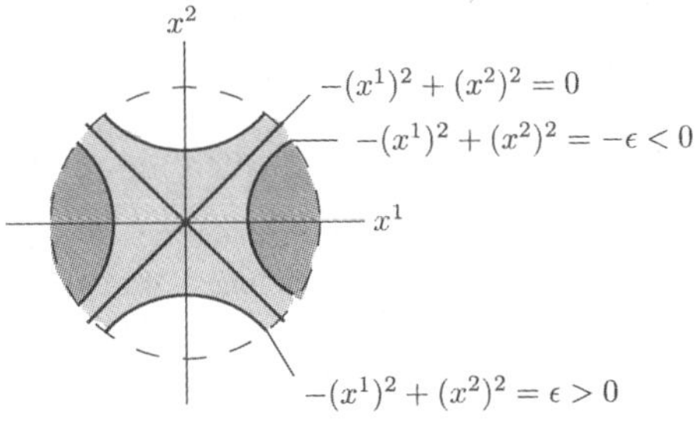

Fig. 5.10 $F_\alpha - c$ in the image of a coordinate neighborhood. The image of $U_\alpha \cap M_{c-\epsilon}$ is the *dark shadowed area*, and that of $U_\alpha \cap M_{c+\epsilon}$ consists of both the *dark* and the *light shadowed areas*

Let x_0 be a non-degenerate critical point of F of index λ with critical value c. A coordinate neighborhood $U_\alpha \subset M$ centered at x_0 exists the coordinate image of which may be chosen to be the open unit ball of $\mathbb{R}^m$ with

$$F_\alpha = c - (x^1)^2 - \cdots - (x^\lambda)^2 + (x^{\lambda+1})^2 + \cdots + (x^m)^2 + \text{higher terms.}$$

(Morse observed that always coordinates can be found that all higher terms vanish.) Figure 5.10 shows the case $m = 2$ and $\lambda = 1$.

$M_{c'}$ as an open subset of M is an m-dimensional manifold or empty. The change of its homology at a critical value c can be studied by the change of homology of the image of $U_\alpha \cap M_{c'}$ in $\mathbb{R}^m$. $U_\alpha \cap M_{c-\epsilon}$ is empty for $\lambda = 0$, or homotopy equivalent to a sphere $S^{\lambda-1}$ (S^0, that is two points, in the case of Fig. 5.10), while $U_\alpha \cap M_{c+\epsilon}$ is homotopy equivalent to a point for $\lambda = 0$, or to a ball B^λ (B^1, that is a horizontal line segment, in the case of Fig. 5.10; think of a third axis x^3 in that figure replacing x^2 while the new x^2-axis should point into the drawing plane, and put $\lambda = 2$, which makes the figure rotational symmetric around the x^3-axis and leads to a circle S^1 and a disk B^2 in the (x^1, x^2)-plane instead of S^0 and B^1 along the x^1-axis).

To proceed, the concept of **relative homology** is helpful. Let N be a submanifold of M and consider the singular chain complexes on both manifolds. For short they will be called M-chains and N-chains. Clearly, every N-chain is also an M-chain, hence $C_r(N, \mathbb{R})$ is a subspace of $C_r(M, \mathbb{R})$, and the quotient spaces $C_r(M, \mathbb{R})/C_r(N, \mathbb{R})$ may be considered which form a chain complex in which M-chains which differ only by an N-chain are identified. An M-chain whose boundary is an N-chain represents a cycle in the quotient complex, and an M-chain which combines with an N-chain to an M-boundary represents a boundary in the quotient complex. It is readily seen that the boundary operator ∂ for M-chains induces the boundary operator

$$\partial_r : C_r(M, \mathbb{R})/C_r(N, \mathbb{R}) \to C_{r-1}(M, \mathbb{R})/C_{r-1}(N, \mathbb{R}) \qquad (5.69)$$

and that

$$0 \to C(N, \mathbb{R}) \to C(M, \mathbb{R}) \to C(M, \mathbb{R})/C(N, \mathbb{R}) \to 0 \qquad (5.70)$$

is an exact sequence, which according to (5.57) induces the long exact sequence

$$\cdots \to H_r(C(N,\mathbb{R})) \to H_r(C(M,\mathbb{R})) \to H_r(C(M,\mathbb{R})/C(N,\mathbb{R})) \\ \to H_{r-1}(C(N,\mathbb{R})) \to \cdots \tag{5.71}$$

$H(C(M,\mathbb{R})/C(N,\mathbb{R}))$ is called the relative homology of $M \bmod N$.

The exactness of the sequence $A \xrightarrow{f} B \xrightarrow{g} C$ of vector spaces implies $\dim B = \dim \operatorname{Im} g + \dim \operatorname{Ker} g = \dim \operatorname{Im} g + \dim \operatorname{Im} f$ (recall $\operatorname{Im} f = \operatorname{Ker} g$). Since $\dim \operatorname{Im} g \leq \dim C$ and $\dim \operatorname{Im} f \leq \dim A$, it follows $\dim B \leq \dim A + \dim C$. Applied to (5.71) this yields for the Betti numbers

$$\beta^r(M) \leq \beta^r(N) + \beta^r(M \bmod N), \quad \beta^r(M \bmod N) = \dim H_r(C(M,\mathbb{R})/C(N,\mathbb{R})). \tag{5.72}$$

Coming back to the case of a single non-degenerate critical point with index λ on an m-dimensional manifold, it was seen that $(U_\alpha \cap M_{c+\epsilon}) \bmod (U_\alpha \cap M_{c-\epsilon})$ was homotopy equivalent to $B^\lambda \bmod S^{\lambda-1}$ which is homotopy equivalent to S^λ. (For instance a two-disc whose circumference is contracted into a point yields a two-sphere: a piece of textile is tightened into a bag by going the left two steps of Fig. 2.8 on p. 43 backwards.) The Betti numbers of an empty set are all zero, and the Betti numbers of S^λ are $\beta^0 = \beta^\lambda = 1$ and all others zero (see (5.60)). Starting with $M_{-\infty} = \varnothing$ and proceeding to $M_\infty = M$, it is readily seen that

$$C_\lambda(F : M \to \mathbb{R}) \geq \beta^\lambda(M), \tag{5.73}$$

where C_λ is the total number of critical points of index λ of a function F *on a compact manifold M provided F has only finitely many non-degenerate critical points*. This is the **weak Morse inequality**.

Next, following in (5.71) the equalities of dimensions in exact sequences given before (5.72) from $r+1$ down to $r = 0$, one finds (we drop the field $\mathbb{R}$ for the sake of shorter writing)

$$\begin{aligned}
&\dim \operatorname{Im}(H_{r+1}(C(M)/C(N)) \to H_r(C(N))) \\
&= \dim H_r(C(N)) - \dim \operatorname{Im}(H_r(C(N)) \to H_r(C(M))) \\
&= \beta^r(N) - (\dim H_r(C(M)) - \dim \operatorname{Im}(H_r(C(M)) \to H_r(C(M)/C(N))) \\
&= \beta^r(N) - \beta^r(M) + \dim H_r(C(M)/C(N)) \\
&\quad - \dim \operatorname{Im}(H_r(C(M)/C(N)) \to H_{r-1}(C(N))) \\
&= \beta^r(N) - \beta^r(M) + \beta^r(M \bmod N) \\
&\quad - (\beta^{r-1}(N) - \beta^{r-1}(M) + \beta^{r-1}(M \bmod N)) + - \cdots
\end{aligned}$$

Realizing that the leftmost expression of this chain of equations is non-negative and that all β^{-1} are zero one may cast this result into

$$\gamma^r(M) \leq \gamma^r(N) + \gamma^r(M \bmod N), \quad \gamma^r = \sum_{s=0}^{r} (-1)^{r-s} \beta^r, \tag{5.74}$$

which analogously to the above consideration yields the **strong Morse inequality**

$$\sum_{s=0}^{\lambda}(-1)^{\lambda-s}C_s(F:M\to\mathbb{R})\geq\sum_{s=0}^{\lambda}(-1)^{\lambda-s}\beta^s(M). \tag{5.75}$$

Finally, (5.44) says that the long exact sequence (5.71) becomes trivial at the left end for $r>m$, which yields instead of the inequality (5.74) now an equality for the Euler characteristics,

$$\chi(M)=\chi(N)+\chi(M \bmod N). \tag{5.76}$$

This makes (5.75) also into an equality for that case, the so called **algebraic number of critical points** of F,

$$\sum_{s=0}^{m}(-1)^{s}C_s(F:M\to\mathbb{R})=\chi(M). \tag{5.77}$$

Recall that in all these results it was assumed that M is compact and F has only finitely many non-degenerate singular points.

However, meanwhile Morse theory has been widely generalized to be exploited in the theory of non-linear equations.

5.9 Examples from Physics

As a first example, **classical point mechanics** is revisited again (cf. Sect. 3.7 and the end of Sect. 4.4): It is easily checked, that the **Liouville measure** τ_Ω on the phase space Ω can be expressed via the canonical (symplectic) 2-form ω as

$$\tau_\Omega=dq^1\wedge\cdots\wedge dq^m\wedge dp^1\wedge\cdots\wedge dp^m=\frac{(-1)^{(m-1)m/2}}{m!}\underbrace{\omega\wedge\cdots\wedge\omega}_{m\ \text{factors}}. \tag{5.78}$$

In the wedge product of m factors ω all $m!$ terms with all dq^i, dp^i distinct survive, all with the same sign due to the pairing. The (not very important) reordering according to Liouville's definition then yields the sign factor.

Since $\omega=-dP$ and $d\omega=0$, the canonical 2-form is exact (coboundary) and hence closed. The same is true for the $2m$-form τ_Ω of the Liouville measure:

$$\tau_\Omega=-\frac{(-1)^{(m-1)m/2}}{m!}d(P\wedge\underbrace{\omega\wedge\cdots\wedge\omega}_{m-1\ \text{factors}}),$$

and $d\tau_\Omega=0$ as a consequence, but also in general as a $(2m+1)$-form on the $2m$-dimensional phase space. Hence,

$$\int_U \tau_\Omega = -\frac{(-1)^{(m-1)m/2}}{m!} \int_{\partial U} P \wedge \underbrace{\omega \wedge \cdots \wedge \omega}_{m-1 \text{ factors}})$$

for every subset U of the phase space Ω.

Most importantly, from the invariance (4.55) of ω under the time-evolution ϕ_t (Hamilton flow), the same invariance of the Liouville measure follows immediately:

$$\phi_t^* \tau_\Omega = \tau_\Omega. \tag{5.79}$$

This is **Liouville's theorem** which has well known important applications in statistical physics, and which bears the well known danger of misinterpretation too.

For more details on classical point mechanics see [3, 4].

Next, **Maxwell's electrodynamics** on a four-dimensional manifold of space-time is considered. The electromagnetic field is a 2-form which in Minkowski coordinates $(y^1, y^2, y^3, y^4) = (t, x^1, x^2, x^3)$ is given as

$$\begin{aligned} F = \frac{1}{2} F_{\mu\nu} dy^\mu \wedge dy^\nu &= E_1 dt \wedge dx^1 + E_2 dt \wedge dx^2 + E_3 dt \wedge dx^3 \\ &- B_1 dx^2 \wedge dx^3 - B_2 dx^3 \wedge dx^1 - B_3 dx^1 \wedge dx^2, \end{aligned} \tag{5.80}$$

where units with $\epsilon_0 = \mu_0 = c = 1$ are used here and in the following (ϵ_0 is the vacuum permittivity, μ_0 the vacuum permeability and c the vacuum velocity of light, all in flat space-time).

Maxwell theory was brought into its most concise form half a century ago based on E. Cartan's exterior calculus and on Hodge's duality (p. 121). This form holds likewise in global Minkowski geometry of a flat space-time as well as in general.

A **pseudo-Riemannian geometry** is introduced by the non-degenerate symmetric covariant **metric tensor** or **fundamental tensor** g, in local Minkowski coordinates given as (in the following we use Einstein's summation convention)

$$g_{ij} dy^i dy^j = (dt)^2 - (dx^1) - (dx^2)^2 - (dx^3)^2 \quad \text{or} \quad (g_{ij}) = \begin{pmatrix} 1 & 0 \\ 0 & -\mathbf{1}_3 \end{pmatrix}. \tag{5.81}$$

It defines an indefinite scalar product, sign carrying lengths, and angles in the tangent space in the usual way,

$$(X|Y) = g_{ij} X^i X^j, \quad |X| = (X|X)^{1/2}, \quad \angle(X, Y) = \arccos \frac{(X|Y)}{|X||Y|} \quad \text{for } |X| \neq 0 \neq |Y|. \tag{5.82}$$

It further provides a bijection between tangent and cotangent spaces,

$$\omega_i = g_{ij} X^j, \quad g^{-1} = g^{ij} \frac{\partial}{\partial y^i} \frac{\partial}{\partial y^j} \quad X^i = g^{ij} \omega_j, \quad g^{ij} g_{jk} = \delta^i_k, \tag{5.83}$$

for which $\langle\omega, Y\rangle = (X|Y)$ holds (cf. (4.24)). A more detailed treatment is given in Chap. 9.

For the sake of generality, the next relations are given for m dimensions. The coordinate independence of the scalar product in (5.82) implies $g_{ij}X^iX^j = g'_{kl}X'^kX'^l = g'_{kl}\psi^k_i\psi^l_jX^iX^j$ and hence $g_{ij} = g'_{kl}\psi^k_i\psi^l_j$ where the determinant of the transformation (3.6) from the y^i to the y'^i is the Jacobian $J = \det\psi$. Taking the determinant of g yields $\det g = \det g' J^2$ and hence $J = |\det g/\det g'|^{1/2}$. Therefore, instead of (5.1),

$$\tau = |\det g|^{1/2} dy^1 \wedge \cdots \wedge dy^m \tag{5.84}$$

yields a coordinate independent volume form in the present case. The corresponding alternating covariant tensor is the **Levi-Civita pseudo-tensor** $E_{1\cdots m} = |\det g|^{1/2}$. Its general components are

$$E_{i_1\cdots i_m} = |\det g|^{1/2}\delta^{1\cdots m}_{i_1\cdots i_m}, \quad E^{i_1\cdots i_m} = \frac{s_g}{|\det g|^{1/2}}\delta^{i_1\cdots i_m}_{1\cdots m}, \quad s_g = \text{sign}\det g, \tag{5.85}$$

where the contravariant form according to the bijection (5.83) follows from $E^{1\cdots m} = g^{1i_1}\cdots g^{mi_m}E_{i_1\cdots i_m} = g^{1i_1}\cdots g^{mi_m}\delta^{1\cdots m}_{i_1\cdots i_m}|\det g|^{1/2} = \det g^{-1}|\det g|^{1/2}$.

In order to adjust **Hodge's star operator** (5.14) to the present more general case, the second line of (5.14) is written as

$$*\omega = \iota_\omega E, \tag{5.86}$$

which according to (4.30) for an n-form ω and any $(m-n)$-form σ means

$$\langle\iota_\omega E, \sigma\rangle = \langle E, \omega\wedge\sigma\rangle = E^{i_1\cdots i_m}\omega_{i_1\cdots i_n}\sigma_{i_{n+1}\cdots i_m}$$

and hence (cf. (4.23))

$$\begin{aligned}(*\omega)^{i_{n+1}\cdots i_m} &= \frac{1}{(m-n)!}E^{i_1\cdots i_m}\omega_{i_1\cdots i_n},\\ (*\omega)_{i_{n+1}\cdots i_m} &= \frac{1}{(m-n)!}E_{i_1\cdots i_m}\omega^{i_1\cdots i_n} = \frac{E_{i_1\cdots i_m}}{(m-n)!}g^{i_1j_1}\cdots g^{i_nj_n}\omega_{j_1\cdots j_n}.\end{aligned} \tag{5.87}$$

A second star operation results in

$$\begin{aligned}(**\omega)_{k_1\cdots k_n} &= \frac{1}{n!}E_{k_{n+1}\cdots k_m k_1\cdots k_n}(*\omega)^{k_{n+1}\cdots k_m}\\ &= \frac{1}{n!(m-n)!}E_{k_{n+1}\cdots k_m k_1\cdots k_n}E^{i_1\cdots i_n k_{n+1}\cdots k_m}\omega_{i_1\cdots i_n}\\ &= s_g\delta^{1\cdots m}_{k_{n+1}\cdots k_m k_1\cdots k_n}\delta^{i_1\cdots i_n k_{n+1}\cdots k_m}_{1\cdots m}\omega_{i_1\cdots i_n}\\ &= s_g(-1)^{n(m-n)}\omega_{k_1\cdots k_n}.\end{aligned}$$

First, all items of the $i_1 \cdots i_n$-sums are equal to $\delta^{k_1 \cdots k_m}_{1 \cdots m} \omega_{k_1 \cdots k_n}$ (no summation) which cancels the factor $1/n!$, and then the summation over the distinct indices $k_{n+1} \cdots k_m$ cancels the other prefactor $1/(m-n)!$. The final answer is

$$** = s_g(-1)^{n(m-n)}. \tag{5.88}$$

For consistency, the second relation (5.14) is also redefined while the first relation remains in effect:

$$*(1) = dy^1 \wedge \cdots \wedge dy^m, \quad *(dy^1 \wedge \cdots \wedge dy^m) = s_g. \tag{5.89}$$

(In (5.14) $g_{ij} = \delta_{ij}$ was assumed, then, for $m = 3, E_{ijk} = E^{ijk} = \delta^{123}_{ijk}$. For example, for the three-dimensional 2-form $-F_{ij}$ build by the spatial part of (5.80) one has from the first equation (5.87) $-(*F)^k = -E^{ijk}F_{ij} = B^k$.)

Besides the metric independent derivatives d and L_X of exterior forms, the **codifferential operator** $\delta : \mathcal{D}(M) \to \mathcal{D}(M)$ is introduced, which applied to an n-form ω is given by (sign factors are nasty but unavoidable in oriented manifolds)

$$\begin{aligned} \delta\omega &= (-1)^n *^{-1} d * \omega = s_g(-1)^{m(n+1)+1} * d * \omega, \quad \omega \in \mathcal{D}^n(M), \\ d\omega &= (-1)^{m-n} * \delta *^{-1} \omega. \end{aligned} \tag{5.90}$$

The second equation of the first line is obtained with (5.88) and $(-1)^{(n+1)n} = 1$ for every n. One has $*\omega \in \mathcal{D}^{m-n}(M), d * \omega \in \mathcal{D}^{m-n+1}(M)$ and hence $\delta\omega \sim * d * \omega \in \mathcal{D}^{n-1}(M)$. This implies that $\delta F = 0$ for a 0-form (function) F. For a 1-form $\omega \mathrel{\hat{=}} X$ (cf. (5.82)) and $g_{ij} = \delta_{ij}$ one has $\delta\omega = \operatorname{div} \boldsymbol{X}$, hence the codifferential operator generalizes the divergence of a vector field. The second line of (5.90) is obtained by substituting $*^{-1}\omega \in \mathcal{D}^{m-n}(M)$ for $\omega \in \mathcal{D}^n(M)$ and hence $m - n$ for n in the first equation of the first line and operating with $*$ from the left. It is readily seen that $\delta^2 \sim *^{-1} d * *^{-1} d * = *^{-1} d^2 * = 0$.

The **Laplace–Beltrami operator** is defined as

$$\Delta = \delta d + d\delta \tag{5.91}$$

which applied to a function in a flat metric reduces to the ordinary Laplace operator, $\Delta F = \operatorname{div} \operatorname{grad} F$. Simple rules are

$$d\Delta = \Delta d, \quad \delta\Delta = \Delta\delta, \quad * \Delta = \Delta * . \tag{5.92}$$

The first two follow readily from the definition (5.91) and $d^2 = 0 = \delta^2$. The last one follows from corresponding commutation rules $*\delta d = d\delta *$ and $*d\delta = \delta d *$ which demand just a bit more of straightforward calculations. One also finds for $\omega \in \mathcal{D}^{n-1}(M), \sigma \in \mathcal{D}^n(M)$

$$d(\omega \wedge * \sigma) = d\omega \wedge * \sigma + (-1)^{n-1} \omega \wedge d * \sigma = d\omega \wedge * \sigma - \omega \wedge * \delta\sigma \tag{5.93}$$

where in the last equation the first relation (5.90) was used.

For a compact manifold M or in general on the exterior algebra $\mathcal{D}_c(M)$ of forms with compact support, a scalar product

$$[\omega|\sigma] = \int_M \omega \wedge *\sigma \quad \text{for } \omega, \sigma \in \mathcal{D}^n_c(M), \quad [\omega|\sigma] = 0 \quad \text{for } \omega \in \mathcal{D}^n_c(M) \not\ni \sigma \tag{5.94}$$

is introduced. Since (5.93) for $\omega \in \mathcal{D}^{n-1}_c(M), \sigma \in \mathcal{D}^n_c(M)$ implies

$$0 = \int_M d(\omega \wedge *\sigma) = [d\omega|\sigma] - [\omega|\delta\sigma],$$

one has

$$[d\omega|\sigma] = [\omega|\delta\sigma]. \tag{5.95}$$

In this sense d and δ are mutually adjoint operators in $\mathcal{D}_c(M)$ considered as a functional space, normed by the scalar product (5.94).

Finally, *for a positive metric g it follows from* (5.16) *that the scalar product* (5.94) *is positive and symmetric*. Now, $d\omega = 0$ and $\delta\omega = 0$ obviously implies $\Delta\omega = 0$. Inversely, if $\Delta\omega = 0$, one has

$$0 = [\Delta\omega|\omega] = [(d\delta + \delta d)\omega|\omega] = [\delta\omega|\delta\omega] + [d\omega|d\omega]$$

and hence, in the case of a positive norm, $\delta\omega = 0$ and $d\omega = 0$.

Coming now back to Maxwell's electrodynamics, **Maxwell's equations** in modern form read

$$dF = 0 \quad \text{and} \quad \delta F = J \quad \text{or} \quad d * F = *J, \tag{5.96}$$

where

$$J = J_\mu dy^\mu = g_{\mu\nu} J^\mu dy^\nu \tag{5.97}$$

is related to the four-current density of electric charges as analyzed below, and F was given in (5.80). These equations are valid independently of the chosen local coordinate system and, more importantly, independently of a possible curvature of space-time due to the presence of a gravitational field. It is remarkable that for the formulation of Maxwell's equations no connections (Christoffel symbols, see Chap. 7) and no curvature tensor (Chap. 9) are needed explicitly.

The homogeneous Maxwell equations, $dF = 0$, contain the 3-form dF which in a four-dimensional manifold has four independent components. Hence they comprise four equations (of which due to their particular structure only three are independent, since they are connected by the one condition $d^2F = 0$ for a 4-form in four-dimensions). These four equations may be written as $*\, dF = 0$ which with (5.87) means $E^{\mu\nu\sigma\tau}(dF)_{\nu\sigma\tau} = 0$ and hence $\delta^{\mu\nu\sigma\tau}_{1234}\partial F_{\sigma\tau}/\partial y^\nu = 0$, the common tensorial writing of the homogeneous Maxwell equations. With (5.80), the time component ($\mu = 1$) of this latter four-vector equation is $\delta^{ijk}_{123}\partial F_{jk}/\partial x^i = \operatorname{div}\boldsymbol{B} = 0$, which *in three dimensions* also means $*dF = 0$. (Caution! Note that the definition of $*$ depends on g and on the dimensions.) In the remaining three equations for $i \hat{=} \mu \neq 1$ the ν-sum contains the $\nu = 1$ contribution which may be written as $\delta^{ijk}_{123}\partial F_{jk}/\partial t = -\partial \boldsymbol{B}/\partial t$ and the contribution with $\nu \neq 1$ which is $\delta^{ijk}_{123}\partial F_{0k}/\partial x^j = \operatorname{rot}\boldsymbol{E}$. Hence, these equations read in 3-vector notation $\operatorname{rot}\boldsymbol{E} = -\partial \boldsymbol{B}/\partial t$.

If N_3 is any three-dimensional hypersurface in M with boundary ∂N_3, then

$$\int_{\partial N_3} F = \int_{N_3} dF = 0. \tag{5.98}$$

These are a two-dimensional integral over a 2-form and a three-dimensional integral over a 3-form. Hence Stokes' theorem applies. If in particular N_3 is any space-like finite volume, $t = \text{const.}$, then ∂N_3 has no time component, and (5.98) reads $\int_{\partial N_3} \boldsymbol{B} \cdot d\boldsymbol{S} = 0$, where $d\boldsymbol{S}$ is the surface normal vector to ∂N_3. This is Gauss' law for magnetism, and it expresses the absence of magnetic charges (monopoles). The total magnetic flux through a closed surface is always zero. Next take N_3 to be a cylinder (Fig. 5.11) with base S in the (x^1, x^2)-plane, $x^3 = 0$, and extending in t-direction from $t = t_1$ to $t = t_2$. Now, the first equation (5.98) reads $\int_{t_1}^{t_2}\int_{\partial S} d\boldsymbol{s} \cdot \boldsymbol{E} - \int_{S_{t=t_1}} d\boldsymbol{S} \cdot \boldsymbol{B} + \int_{S_{t=t_2}} d\boldsymbol{S} \cdot \boldsymbol{B} = 0$ where $d\boldsymbol{s}$ is the line element of the bounding contour ∂S of the cylinder base. The first integral is over the cylinder mantle and the two others are over the base and the top plane. Differentiation with respect to time yields Faraday's law of induction $\oint_{\partial S} d\boldsymbol{s} \cdot \boldsymbol{E} = -(d/dt)\int_S d\boldsymbol{S} \cdot \boldsymbol{B}$.

Finally, *if* $H^2_{dR}(M) = 0$ *but not in general*, then the relation $dF = 0$, which means that F is closed, also implies that it is exact:

$$F = dA, \quad A \in \mathcal{D}^1(M). \tag{5.99}$$

This is in particular true in a contractible manifold M and *locally* in every manifold since every point of a manifold has a contractible neighborhood. A is the four-potential of the electromagnetic field F. It is never unique, since A and $A' = A + d\chi$ with any real smooth function χ on M lead obviously to the same electromagnetic field F. This is the **gauge freedom** of the electromagnetic four-potential. A is a cohomology class of 1-forms rather than a single 1-form.

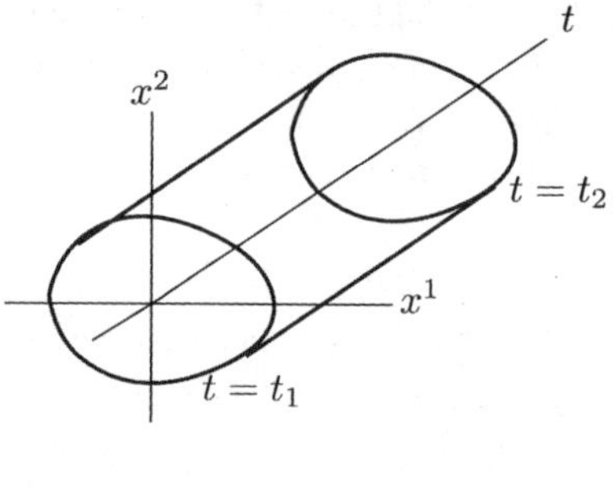

Fig. 5.11 Hypersurface N_3 as a cylinder in t-direction with base in the (x^1, x^2)-plane

However, as just discussed, the first equation (5.96) is more fundamental than (5.99). The equation (5.99) has a 2-form on both sides, hence Stokes' theorem applies for two-dimensional hypersurfaces in M and their boundary curves. Integrals over the fields on surfaces are related to contour integrals over potentials.

The inhomogeneous Maxwell equations in the form $\delta F = J$ have a 1-form on both sides since δ is a graded morphism of degree -1. Equating 1-forms, they comprise again four equations (of which because of $0 = \delta^2 F = \delta J$ again only three are independent, see below). However, since J has the physical meaning of a charge and current density, three-dimensional volume integrals are more relevant, which are obtained by integrating $d * F = * J$ which has 3-forms on both sides:

$$\int_{\partial N_3} * F = \int_{N_3} d * F = \int_{N_3} * J. \tag{5.100}$$

From the fact that three-volume integrals over J are demanded from physics and that they involve the star operator it follows that electrodynamics requires a metric of space–time. Among the consequences there is bending of light propagation by gravitational fields.

In local coordinates, $(* F)_{\sigma\tau} = (1/2)E_{\sigma\tau\alpha\beta}F^{\alpha\beta}$ and $(d * F)_{\nu\sigma\tau} = (1/2)(\partial E_{\sigma\tau\alpha\beta} \cdot F^{\alpha\beta}/\partial y^\nu)$. Therefore, $(\delta F)^\mu = (* d * F)^\mu = E^{\mu\nu\sigma\tau}(d * F)_{\nu\sigma\tau} = (1/2)E^{\mu\nu\sigma\tau}(\partial E_{\sigma\tau\alpha\beta} \cdot F^{\alpha\beta}/\partial y^\nu) = -1/(2|\det g|^{1/2})\delta^{\mu\nu\sigma\tau}_{1234}\delta^{1234}_{\alpha\beta\sigma\tau}(\partial|\det g|^{1/2}F^{\alpha\beta}/\partial y^\nu) = -|\det g|^{-1/2}(\delta^\mu_\alpha\delta^\nu_\beta - \delta^\nu_\alpha\delta^\mu_\beta)$ $(\partial|\det g|F^{\alpha\beta}/\partial y^\nu) = 2|\det g|^{-1/2}(\partial|\det g|^{1/2}F^{\nu\mu}/\partial y^\nu)$. Again it is seen that δ is related to the divergence of vector analysis as was mentioned after (5.90). Now it is readily seen from (5.80) and (5.97) that in local coordinates $\delta F = J$ reads $2|\det g|^{-1/2}(\partial|\det g|^{1/2}F^{\nu\mu}/\partial y^\nu) = J^\mu$ which again as for the homogeneous equations is the common tensorial writing of the inhomogeneous Maxwell equations. In tensor notation one usually omits the factor 2 here and the factor $1/2$ in (5.80).

In order to find the physical meaning of J^μ consider the 3-form $* J = (1/3!)|\det g|^{1/2}\delta^{1234}_{\mu\nu\sigma\tau}J^\tau dy^\mu \wedge dy^\nu \wedge dy^\sigma$. If N_3 is a spatial three-dimensional hypersurface of M, then the contained charge is $Q = \int_{N_3} * J = \int_{N_3} |\det g|^{1/2} J^0 dx^1 \wedge dx^2 \wedge dx^3 = \int_{N_3} \rho dx^1 dx^2 dx^3$. Hence $J^0 = \rho/|\det g|^{1/2}$ where ρ is the charge density in locally flat coordinates, and in these coordinates the inhomogeneous Maxwell equation for $\mu = 1$ reads $\operatorname{div}\boldsymbol{E} = \rho$ or, in the form (5.100), $\int_{\partial N_3} \boldsymbol{E} \cdot d\boldsymbol{S} = Q$ which is Gauss' law of electrostatics. If N_3 is the finite hypersurface of Fig. 5.11 with $x^3 = 0$, then $\int_{N_3} j^3 dt dx^1 dx^2 = \int_{N_3} * J = \int_{N_3} |\det g|^{1/2} J^3 dt \wedge dx^1 \wedge dx^2$ with the electric current density $\boldsymbol{j}$ in locally flat coordinates, hence $\boldsymbol{J} = \boldsymbol{j}/|\det g|^{1/2}$. Now, after differentiation with respect to t, with the same notation as on the previous page (5.100) reads $\oint_{\partial S} d\boldsymbol{s} \cdot \boldsymbol{B} = \int_S d\boldsymbol{S} \cdot \boldsymbol{J} + (d/dt) \int_S d\boldsymbol{S} \cdot \boldsymbol{E}$ which is Ampere's law with Maxwell's extension by the displacement current.

The inhomogeneous Maxwell equations (5.95) cannot hold for an arbitrary form J. Because of $\delta^2 = 0 = d^2$, it must hold that

$$\delta J = 0 = d * J, \qquad \int\limits_{\partial N_4} * J = \int\limits_{N_4} d * J = 0, \tag{5.101}$$

which expresses the charge conservation law. N_4 is any compact four-dimensional submanifold of M. Equation 5.101 means that $* J$ is closed, and *for a contractible* M, from that it already follows that it is also exact, that is, that there exists a form $* F$ so that the last equation (5.96) holds. In general this last equation has a deeper meaning. If for instance the universe for $t = 0$, N_3, is closed, $\partial N_3 = 0$, then from the last equation (5.96) it follows that $Q = \int_{N_3} * J = \int_{\partial N_3} * F = 0$: *a closed universe must be exactly electrically neutral.* In a closed universe, to every positive charge as a source of electric field lines there must correspond a negative charge as a sink of electric field lines.

If the punctured three-space $\mathbb{R}^3 \setminus \{0\}$ is considered or the Minkowski space with the time axis $x^1 = x^2 = x^3 = 0$ removed, then the field

$$F = \frac{m}{8\pi|\boldsymbol{r}|^3} \delta^{123}_{ijk} x^i dx^j \wedge dx^k = \frac{m}{|\boldsymbol{r}|^3} \boldsymbol{r} \cdot d\boldsymbol{S}$$

is closed but not exact. There is no vector potential $\boldsymbol{A}$ on $\mathbb{R}^3 \setminus \{0\}$ from which this field derives. However, for the part of $\mathbb{R}^3 \setminus \{0\}$ with the positive x^3-axis removed, it is easily checked (exercise) that

$$\boldsymbol{A}_+ = m \frac{x^1 dx^2 - x^2 dx^1}{4\pi|\boldsymbol{r}|(x^3 - |\boldsymbol{r}|)}$$

is a potential of the above field, and likewise

$$\boldsymbol{A}_- = m \frac{x^1 dx^2 - x^2 dx^1}{4\pi|\boldsymbol{r}|(x^3 + |\boldsymbol{r}|)}$$

for the part of $\mathbb{R}^3 \setminus \{0\}$ with the negative x^3-axis removed. In the overlap of the two domains of definition both potentials are cohomologous and hence gauge equivalent. Integrating F over a large sphere S^2 around the origin of the three-space, one obtains

$$\int\limits_{S^2} F = \int\limits_{S^2} \boldsymbol{B} \cdot d\boldsymbol{S} = m.$$

This is the magnetic charge of the magnetic Dirac monopole in the origin. In classical electrodynamics F must be a 2-form in the whole Minkowki space and no magnetic monopole is possible. However, in unified theories of particle physics the above case might be allowed [5, Chap. 12].

More details on Maxwell theory can for instance be found in [6].

Consider next the **dynamics in ideal crystalline solids**. An ideal crystal is assumed to be infinitely extended in the position space $\mathbb{R}^3$ with non-degenerate vectors of lattice periods $\boldsymbol{a}_1, \boldsymbol{a}_2, \boldsymbol{a}_3$. In classical approximation, in the ground state the atoms forming the crystal are at rest in positions $\boldsymbol{R}_n + \boldsymbol{S}_i$ where $\boldsymbol{R}_n = n_1\boldsymbol{a}_1 + n_2\boldsymbol{a}_2 + n_3\boldsymbol{a}_3$, n_l integer, $-\infty < n_l < \infty$, is the (arbitrarily chosen) reference point of the periodicity volume or **unit cell of the crystal lattice** and $\boldsymbol{S}_i$ are finitely many atom positions within a unit cell. In quantum theory the ground state is a many-particle wave function the absolute square of which has the periodicity of the lattice. The atoms have a probability distribution centered at $\boldsymbol{S}_i$ which is conceived as 'zero point vibrations' around $\boldsymbol{S}_i$. The set of all lattice vectors $\boldsymbol{R}_n$ forms an Abelian module over the ring (integral domain) $\mathbb{Z}$ with the three generators $\boldsymbol{a}_l$. This module will be denoted $\mathbb{L}_{\boldsymbol{r}}$ and is called the **Bravais lattice** of the crystal.

If a particle is localized at position $\boldsymbol{r}$, it makes no physical difference in which unit cell $\boldsymbol{r}$ is chosen; due to the infinite extension, the physics of the crystal looks identically the same when considered from position $\boldsymbol{r}$ or from position $\boldsymbol{r} + \boldsymbol{R}_n$ for any lattice vector $\boldsymbol{R}_n$. Hence, *for a single excited particle over the ground state* the physical configuration space is the 3-torus $\mathbb{T}^3_{\boldsymbol{r}} = \mathbb{R}^3/\mathbb{L}_{\boldsymbol{r}}$ in which positions $\boldsymbol{r}$ and $\boldsymbol{r} + \boldsymbol{a}_l$, $l = 1, 2, 3$ are identified. Figure 5.12 shows how a 2-torus is formed out of a two-dimensional unit cell; a 3-torus is formed analogously, it is only hard to draw. If a particle on motion leaves a unit cell it may be considered to enter it immediately at the equivalent position on the opposite face, keeping up its direction and velocity of motion. All physically measurable quantities must be real-valued periodic functions on $\mathbb{R}^3$ with the periods $\boldsymbol{a}_l$ or, which means the same, real single-valued functions on the torus $\mathbb{T}^3_{\boldsymbol{r}}$. This situation holds for both classical and quantum physics which is important since kinetic processes in solids are often treated sufficiently well quasi-classically.

A quantum wave function of a particle (for instance moving in the mean potential field of the crystal; quasi-particle theory yields a more general theoretical basis for this picture) is by itself not measurable, it may have a phase factor on it different in different unit cells. Group theory says that this phase factor can always be chosen so that the wave function obeys $\phi(\boldsymbol{r} + \boldsymbol{R}_n) = \exp(i\boldsymbol{p} \cdot \boldsymbol{R}_n/\hbar)\phi(\boldsymbol{r})$. This is the content of **Bloch's theorem**. The parameter $\boldsymbol{p}$, which has the physical meaning of the quasi-momentum of the quasi-particle wave, is determined by this relation only as

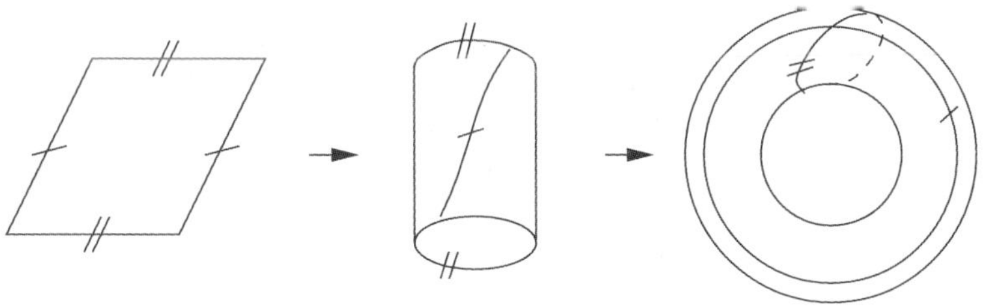

Fig. 5.12 The formation of a 2-torus out of a two-dimensional unit cell (parallelogram)

$$\boldsymbol{p} = \boldsymbol{p} \bmod \hbar \boldsymbol{G_m}, \quad \boldsymbol{G_m} \cdot \boldsymbol{R_n} = 2\pi \cdot \text{integer}. \tag{5.102}$$

This implies $\boldsymbol{G_m} = \sum_{k=1}^{3} m_k \boldsymbol{b}_k$, m_k integer, $\boldsymbol{b}_k \cdot \boldsymbol{a}_l = 2\pi\delta_{kl}$. The $\boldsymbol{G_m}$ form the reciprocal lattice $\mathbb{L}_{\boldsymbol{p}}$ to $\mathbb{L}_{\boldsymbol{r}}$. The space of quasi-momenta $\boldsymbol{p}$ is again a 3-torus, $\mathbb{T}^3_{\boldsymbol{p}}$, formed out of the unit cell of the reciprocal lattice $\mathbb{L}_{\boldsymbol{p}}$. This unit cell is called the first **Brillouin zone**. Alternatively, the torus may be again unfolded into the space $\mathbb{R}^3_{\boldsymbol{p}} = \mathbb{T}^3_{\boldsymbol{p}} \oplus \mathbb{L}_{\boldsymbol{p}}$, this time of momenta, in which all physical quantities must be periodic functions. This is called a repeated zone scheme.

Note that the choice of the unit cell as the periodicity volume in $\mathbb{R}^3$ is not unique. In Fig. 5.13 two different choices are sketched for a two-dimensional lattice. Often the choice of the right pattern of the figure is made which has the point symmetry of the Bravais lattice and is called the Wigner–Seitz cell. Nevertheless, the tori $\mathbb{T}^3_{\boldsymbol{r}}$ and $\mathbb{T}^3_{\boldsymbol{p}}$ are defined by the quotient spaces with respect to the lattices only and inherit the quotient topology from the $\mathbb{R}^3$. Figure 5.14 shows how the torus of Fig. 5.12 which corresponds to the left choice of Fig. 5.13 is also obtained from the right choice of Fig. 5.13.

In both cases, $\mathbb{T}^3_{\boldsymbol{r}}$ and $\mathbb{T}^3_{\boldsymbol{p}}$, the space $\mathbb{R}^3$ is the simply connected covering space (next chapter) of the tori: it winds an infinite number of times around the torus in $\boldsymbol{a}_1(\boldsymbol{b}_1)$-direction and an infinite number of times opposite to the $\boldsymbol{a}_1(\boldsymbol{b}_1)$-direction, an infinite number of times in (against) $\boldsymbol{a}_2(\boldsymbol{b}_2)$-direction and an infinite number of times in (against) $\boldsymbol{a}_3(\boldsymbol{b}_3)$-direction. A closed loop on the tori is characterized by its three winding numbers (n_1, n_2, n_3) until it closes. This triple of integers classifies

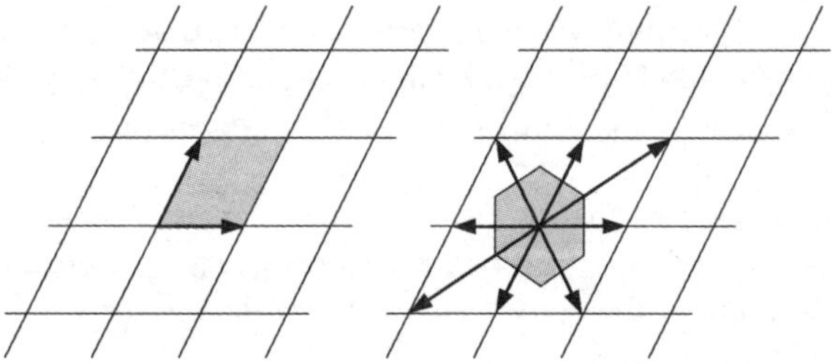

Fig. 5.13 Unit cell of a lattice. *Left*: spanned by some smallest linear independent lattice vectors, *right*: Wigner–Seitz cell enclosed by the bisectrices of all lattice vectors

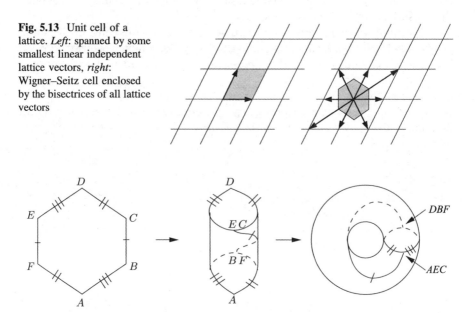

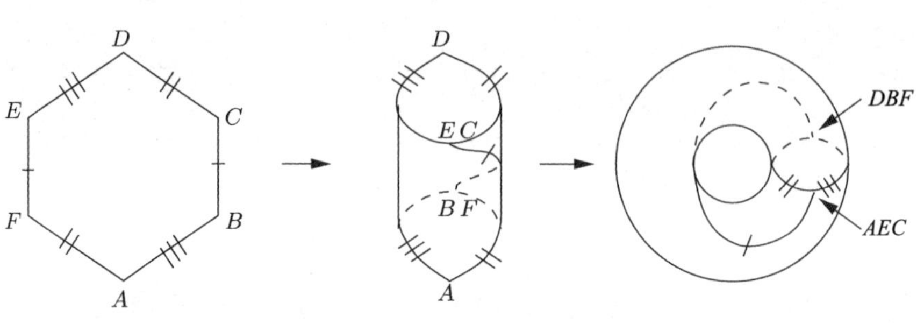

Fig. 5.14 The formation of a 2-torus out of a Wigner–Seitz cell in $\mathbb{R}^2$

the homotopy classes of loops in the fundamental group of the 3-torus which is $\pi_1(\mathbb{T}^3) = \mathbb{Z}^3$ (Sect. 2.5).

This approach treats a crystalline solid in the **thermodynamic limit** $V \to \infty$, $V/N_\alpha = \text{const.}$ which focusses on so-called bulk properties of the solid and neglects surface effects. Here, V is the crystal volume and N_α its particle number of sort α. Since the limit $V \to \infty$ is subtle in many respects, instead one often chooses a large but finite crystal volume consisting of a large number of unit cells in a large parallelepiped with edge lengths $L_l \boldsymbol{a}_l$, $l = 1, 2, 3$, L_l large integers, and puts periodic boundary conditions, also called Born–van Kármán boundary conditions, on that volume by closing it into a large 3-torus. Then, the above periodicity requirements in $\boldsymbol{r}$-space for singly excited quasi-particles now even are in effect for a finite total volume of the large 3-torus because no unit cell on that torus is distinguished. Since a quantum wavefunction is always required to be single-valued in the whole $\boldsymbol{r}$-space (this requirement yields for instance all quasi-classical quantization conditions), the allowed momenta (5.102) now take on discrete values $\boldsymbol{p}_\mu = \sum_{k=1}^{3} \mu_k \boldsymbol{b}_k$, $L_k \mu_k$ integer, only. These discrete momentum values are forming a lattice on the torus $\mathbb{T}^3_{\boldsymbol{p}}$ and also on the still infinite momentum space $\mathbb{R}^3_{\boldsymbol{p}}$ of the repeated zone scheme. The thermodynamic limit now means to let the lattice spacings of this discrete momentum lattice go to zero.

In particular the **dispersion relation**, that is the energy quasi-momentum relation of a single excited quasi-particle (lattice phonon, Bloch electron, ...), is a multi-valued periodic function of quasi-momentum in the repeated zone scheme or equivalently a smooth (with few exceptions) multi-valued function on the Brillouin zone $\mathbb{T}^3_{\boldsymbol{p}}$. Its derivative at $\boldsymbol{p}$ is the group velocity of a wave pocket concentrated around $\boldsymbol{p}$ in quasi-momentum space:

$$\boldsymbol{v}_\nu = \frac{\partial \varepsilon_\nu}{\partial \boldsymbol{p}}, \qquad \varepsilon = \varepsilon_\nu(\boldsymbol{p} \bmod \hbar \boldsymbol{G}_{\boldsymbol{m}}).$$

The subscript ν is the band index, including a polarization or spin index if necessary. The terminology 'quasi'-... refers to that meaning of $\boldsymbol{p}$ modulo $\hbar$ times a reciprocal lattice vector (Bragg reflection on the lattice $\mathbb{L}_{\boldsymbol{r}}$) and to the corresponding multi-valued energies of a lattice vibration spectrum, a band structure, ...

The smoothness of a multi-valued function needs an explanation [7, Paragraph II.5]. It always suffices to consider a *finite number N of bands*, if necessary by cutting off the band structure at the upper energy end. Then, the values $\{\varepsilon_\nu(\boldsymbol{p}),\ \nu = 1, \ldots, N\}$ are considered as a non-ordered set $E(\boldsymbol{p})$. If band energies are degenerate, they are counted according to their multiplicity. A metric is introduced into the space of sets E with the distance function (exercise)

$$d(E, E') = \min_{\mathcal{P}} \max_{\nu} |\varepsilon_\nu - \varepsilon'_{\mathcal{P}\nu}| \tag{5.103}$$

where $\mathcal{P}$ runs over all permutations of the N subscripts of the second set E'. Smoothness is now understood with respect to that metric. Alternatively, *in the* *case when all ε_ν are real* (by neglecting their imaginary parts which describe quasi-particle lifetime), sometimes it is appropriate to use an ordered set $\tilde{E} = \{\tilde{\varepsilon}_\nu(\boldsymbol{p}),\ \nu = 1,\ldots,N\}$ of single-valued functions $\tilde{\varepsilon}(\boldsymbol{p})$, for each value $\boldsymbol{p}$ ordered in ascending order of energies. These functions are in most cases continuous but not smooth.

As a consequence of the general dispersion law (5.103), the angle between the quasi-momentum vector $\boldsymbol{p}$ and the group velocity vector $\boldsymbol{v}$ can be quite arbitrary, they even can point in opposite directions (negative effective mass) or $\boldsymbol{v}$ can be zero for non-zero momentum (standing waves with non-zero momentum and hence non-zero phase velocity). Points $\boldsymbol{p}$ of zero group velocity lead to so-called **van Hove singularities** in the quasi-particle density of states

$$D(\varepsilon) = \sum_\nu \int d^3p\delta(\varepsilon - \varepsilon_\nu(\boldsymbol{p})) = \sum_\nu \int_{\varepsilon_\nu(\boldsymbol{p})=\varepsilon} \frac{d^2p}{|\partial\varepsilon_\nu(\boldsymbol{p})/\partial\boldsymbol{p}|}$$

where the last integral runs over an iso-energy surface in $\mathbb{T}^3_{\boldsymbol{p}}$. *Note, that for all lattices the tori $\mathbb{T}^3_{\boldsymbol{r}}$ and $\mathbb{T}^3_{\boldsymbol{p}}$ are compact, and hence so are all iso-energy surfaces on the latter torus.*

The van Hove singularities arise from the zero in the denominator of the integral over the iso-energy surface, hence each band $\tilde{\varepsilon}_\nu(\boldsymbol{p})$, if it has a critical point in the sense of Morse theory, contributes a singularity. In order to apply Morse theory, the Betti numbers of the 3-torus are needed. They are found in textbooks of topology and follow from the Künneth theorem: If $M = M_1 \times M_2$, then

$$H^r_{dR}(M,\mathbb{R}) = \oplus_{p+q=r} H^p_{dR}(M_1,\mathbb{R}) \otimes H^q_{dR}(M_2,\mathbb{R}),$$

which may be condensed into a product formula for the graded cohomology algebras as $H^*(M) = H^*(M_1) \otimes H^*(M_2)$ and which yields for the Euler characteristics $\chi(M) = \chi(M_1)\chi(M_2)$. For an n-torus $\mathbb{T}^n = S^1 \times \cdots \times S^1$ (n factors), by induction this results in (exercise)

$$\beta^r(\mathbb{T}^n) = \dim H^r(\mathbb{T}^n) = nr, \quad \chi(\mathbb{T}^n) = (1-1)^n = 0. \tag{5.104}$$

For the 3-torus the sequence of Betti numbers is $1,3,3,1,0,0,\ldots$.

For a single analytic band with only non-degenerate critical points there are minima of index $\lambda = 0$, two kinds of saddle points of signature $(++-)$ and $(+--)$ of indices $\lambda = 1$ and $\lambda = 2$, respectively, and maxima of index $\lambda = 3$. The weak Morse inequalities say $C_\lambda(\varepsilon_\nu : \mathbb{T}^3 \to \mathbb{R}) \geq \beta^\lambda(\mathbb{T}^3)$ which means in turn

$$C_0 \geq 1, \quad C_1 \geq 3, \quad C_2 \geq 3, \quad C_3 \geq 1.$$

Stronger estimates are provided by the strong Morse inequalities, resulting in

$$C_0 \geq 1, \quad C_1 - C_0 \geq 2, \quad C_2 - C_1 + C_0 \geq 1, \quad C_3 - C_2 + C_1 - C_0 \geq 0.$$

The left hand side of the last inequality is in fact the negative of the algebraic number of critical points, and hence even equality holds there:

$$C_0 - C_1 + C_2 - C_3 = \chi(\mathbb{T}^3) = 0.$$

There must be at least one minimum and one maximum and three saddle points of each type, but of course there can be many more relative minima and maxima and many more saddle points for a general dispersion low. Even then their numbers are not independent. They must fulfil the strong Morse inequalities, and their algebraic number must be zero.

These are the estimates for the corresponding numbers of van Hove singularities of a smooth single non-hybridized band in three dimensions. Analogous results are easily found for two- and one-dimensional cases. For acoustic branches, the minimum at $\boldsymbol{p} = 0$ is often not smooth. Nevertheless there is a singularity (non-analyticity) of the density of states there, possibly of a more soft type. In the case of hybridizing bands there may be zero-, one- and two-dimensional band crossings which may be minima or maxima or saddle points of $\tilde{\varepsilon}_\nu(\boldsymbol{p})$, but which do not lead to van Hove singularities since $\boldsymbol{v}$ is non-zero there. (Again they may lead to softer singularities.) In that case the number of van Hove singularities per band may be reduced. The above estimates then give the minimum numbers for a whole band group as a smooth multi-valued function.

In simple models of dispersion there may occur degenerate critical points. For instance in the nearest-neighbor tight-binding model for an s-band in the bcc lattice there appears a degenerate saddle point, and in the corresponding model for the fcc and hcp lattices there appears a degenerate maximum. Similar degenerate critical points appear in the d-band complexes of such models. They all lead to stronger van Hove singularities in lesser number compared to non-degenerate critical points.

Next, the quasi-classical dynamics of Bloch electrons of metals in an external homogeneous magnetic field is considered. This problem was essentially solved for all physics-relevant situations without use of topological methods in the late fifties of 20th century by I. M. Lifshits and coworkers [8]. The topological treatment is due to S. P. Novikov and coworkers [9, Chap. 2].

In these processes, only electrons in a vicinity of the Fermi level of negligible width on the scale of $\varepsilon_\nu(\boldsymbol{p})$ are involved. In three dimensions, the **Fermi surface**, $FS = \{\boldsymbol{p} | \varepsilon_\nu(\boldsymbol{p}) = \varepsilon_F \text{ for some } \nu\}$, is a compact two-dimensional surface (oriented submanifold) in $\mathbb{T}^3_{\boldsymbol{p}}$ under the assumption that it does not contain critical points of $\varepsilon_\nu(\boldsymbol{p})$ and that there are no band crossings (degeneracies) at the Fermi energy ε_F. If critical points on the Fermi surface or band crossings at the Fermi energy appear, they can be removed by a small perturbing potential, and afterwards the limes may be considered in which the amplitude of this perturbing potential approaches zero. Since the Fermi surface separates the domains in $\mathbb{T}^3_{\boldsymbol{p}}$ with $\varepsilon_\nu(\boldsymbol{p}) < \varepsilon_F$ from the rest for smooth functions $\varepsilon_\nu(\boldsymbol{p})$, it is a boundary with orientation defined by the velocity vector (5.103). It is also a closed submanifold of $\mathbb{T}_p^3$ (see second example on p. 75) and as a closed subset of a compact set it is compact.

The number of connected components (number of 'sheets') of the Fermi surface is $\beta^0(FS) = \dim H_0(FS, \mathbb{R})$, and the genus g of each connected component FS_μ is $g = \beta^1(FS_\mu)/2 = (\dim H_1(FS_\mu, \mathbb{R}))/2$ (cf. (5.62) and Fig. 5.8, sphere, 2-torus, pretzel with g holes).

Consider the homotopy of sheets of Fermi surfaces. The sheet index μ is suppressed in the following. If a sheet has genus $g = 0$, that is, it is homotopy equivalent to a sphere and hence contractible on the torus $\mathbb{T}_p^3$, then $\pi_1(FS) = 0$. If it has genus $g = 1$, that is, it is a 2-torus, then a loop may have two independent windings, $\pi_1(FS) = \mathbb{Z}^2$ (cf. the end of Sect. 2.5). If the genus of a sheet in general is g, then the same arguments as in connection with Fig. 5.8 on p. 146 yield $\pi_1(FS) = \mathbb{Z}^{2g}$. It is a peculiarity of a two-dimensional compact oriented manifold that $\pi_1(FS) = H_1(FS, \mathbb{Z})$.

Next, consider the embedding map F of a Fermi surface sheet into the Brillouin zone, $F : FS \to \mathbb{T}_p^3$, that is, a point on FS in an arbitrary surface parametrization is mapped by F onto the corresponding quasi-momentum $\boldsymbol{p}$. This mapping induces a mapping of any loop on FS onto a loop on $\mathbb{T}_p^3$ and also a mapping of homotopy classes of loops on FS into homotopy classes of loops in $\mathbb{T}_p^3$. If two loops are homotopic on FS, that is, they can continuously be deformed into each other on FS, then they can a fortiori be continuously deformed into each other in $\mathbb{T}_p^3$ where the deformation need not be kept on FS. Hence, the push forward $F_* : \pi_1(FS) \to \pi_1(\mathbb{T}_p^3)$ is a homomorphism of groups. Therefore, the image of the mapping F_* is a subgroup of $\pi_1(\mathbb{T}^3) = \mathbb{Z}^3$ which has 0, $\mathbb{Z}$, $\mathbb{Z}^2$ and $\mathbb{Z}^3$ as subgroups of rank $0, 1, 2$ and 3. Generator of the subgroup $\mathbb{Z}$ for instance can be any element (n_1, n_2, n_3) of the original group $\mathbb{Z}^3$, where $n(n_1, n_2, n_3)$, $n \in \mathbb{Z}$ are the elements of $\mathbb{Z}$; accordingly for the other subgroups. The rank r of $F_*(\pi_1(FS))$ is also called the rank of the Fermi surface sheet FS.

Now, the relation between the genus g and the rank r of a Fermi surface sheet is studied. The details are depicted in Fig. 5.15. From left to right in the first row the following cases are shown: First, an FS is shown which is homotopic to a sphere. This was discussed above to yield $\pi_1(FS) = 0$, hence, trivially $F_*(\pi_1(FS)) = 0$ and $r = 0$. Next, a torus is shown, $\pi_1(FS) = 2$, $g = 1$, of which however both winding loops, a and a loop around the hole of the torus, are contractible in $\mathbb{T}_p^3$. Hence, $F_*(\pi_1(FS)) = 0$ and $r = 0$. In the right picture another torus is shown as FS which, unfolded in the covering space $\mathbb{R}^3$, yields a corrugated cylinder. Here, a loop around the cylinder is still contractible in $\mathbb{T}_p^3$, but the loop a is not any more contractible, it winds around one closure of the torus $\mathbb{T}_p^3$. The loop b winds two times around that closure, there are loops winding n times around it or n times in the opposite winding direction (counted $-n$). Hence, $F_*(\pi_1(FS)) = \mathbb{Z}$ and $r = 1$. In the second row from left to right, first a pretzel with two holes and hence $g = 2$

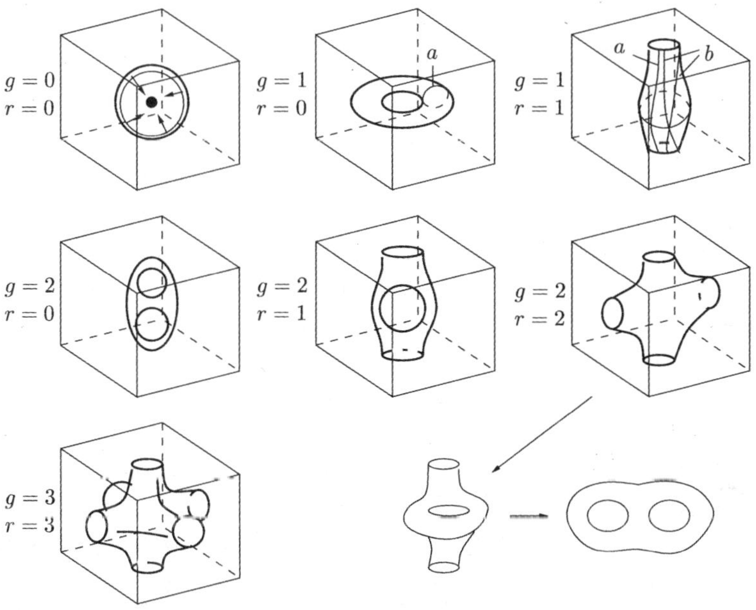

Fig. 5.15 Genus g and rank r of Fermi surface sheets. The Brillouin zone $\mathbb{T}_p^3$ is shown as a cube of which opposite faces have to be identified. Further explanations are given in the text

is shown where again all loops are contractible in $\mathbb{T}_p^3$ which means $r = 0$. Next, a pretzel is shown of which one hole is again inside $\mathbb{T}_p^3$ and the other one coincides with one hole of $\mathbb{T}_p^3$ (that one closing the top and bottom face). There is only one type of loops not contractible in $\mathbb{T}_p^3$ and hence $r = 1$. Why the last picture in this row shows a pretzel with two holes in $\mathbb{T}_p^3$ is indicated in the sketches below, where first the left and right faces are closed and then the top and bottom faces. Here there are two types of loops on FS, one from bottom to top and one from left to right, which are not contractible in $\mathbb{T}_p^3$. Hence, $F_*(\pi_1(FS)) = \mathbb{Z}^2$ and $r = 2$. Finally, in the bottom row only the case $r = 3$ for a FS which is a pretzel with tree holes and hence $g = 3$ is shown. As an exercise the reader may draw sketches for FS with any $g > 2$ for cases $r = 0, 1, 2, 3$.

The conjecture from these consideration is

$$r \leq 3, \quad r \leq g.$$

It was already shown by the homomorphism argument for F_* that $r \leq 3$ and that $g = 0$ implies $r = 0$, hence $r > 0$ implies $g > 0$. That means that only the second inequality for $r = 2, 3$ remains to be proved. To that goal, the homology groups H_2 of FS and of $\mathbb{T}_p^3$ are considered. Suppose that a single FS sheet is a boundary in the

Brillouin zone. The alternative is considered below. Here, the push forward of the embedding F is another homomorphism $F_* : H_2(FS, \mathbb{R}) \to H_2(\mathbb{T}^3_p, \mathbb{R})$, which is trivial, $F_*(H_2(FS, \mathbb{R})) = 0$, since FS is a boundary in $\mathbb{T}^3_p$. Therefore, for any closed 2-form ω, $d\omega = 0$, on $\mathbb{T}^3_p$ the bilinear form $\int_{FS} \omega = \langle [\omega], [FS] \rangle = \int_{\partial^{-1} FS} d\omega = 0$ (cf. (5.40), $\partial^{-1} FS$ is the domain in $\mathbb{T}^3_p$ to which FS is the boundary), which implies that the pull back $F^* : H^2_{dR}(\mathbb{T}^3_p) \to H^2_{dR}(FS)$ is also trivial. Moreover, since $\mathbb{Z}^r = F_*(\pi_1(FS)) = F_*(H_1(FS, \mathbb{Z}))$, there are r mutually non-homologous cycles $c(FS)$ (not combined into boundaries) on FS which remain non-homologous on $\mathbb{T}^3_p$. Again exploiting the non-degeneracy of (5.40), there must be r linearly independent cohomology classes $[\sigma]$ of closed 1-forms σ on $\mathbb{T}^3_p$ so that $\langle [\sigma], [c(FS)] \rangle \neq 0$. Hence, $F^*(H^1_{dR}(\mathbb{T}^3_p, \mathbb{R})) = \mathbb{R}^r \subset H^1_{dR}(FS) = \mathbb{R}^{2g}$. Now, in $\mathbb{R}^{2g}$ a symplectic structure may be introduced with the non-degenerate closed 2-form $\omega = \sum_{i=1}^{g} dq^i \wedge dp^i$ (cf. p. 113), where q^i and p^i may be, roughly speaking, local coordinates along the cycles a_i and b_i of Fig. 5.8. Assume for some i that dq^i and dp^i are both pull-backs of some closed 1-forms σ_i, τ_i on $\mathbb{T}^3_p$. Then, $F^*([\sigma_i \wedge \tau_i]) \subset F^*(H^2_{dR}(\mathbb{T}^3_p)) = 0$ which contradicts the non-degeneracy of ω. Hence, at most one of each pair of 1-forms in the symplectic form ω can be a pull-back of a closed 1-form on $\mathbb{T}^3_p$, and consequently $2r \leq 2g$. (Accordingly, in Fig. 5.15 at least one of the cycles a_i, b_i for each pretzel hole of the FS is contractible in the Brillouin zone $\mathbb{T}^3_p$.)

Note that in these considerations a central point was that the considered single FS sheet is a boundary. The only alternative is a pair of corrugated planes 'in average' parallel to each other which are not pathwise connected in $\mathbb{T}^3_p$ but which only together form a boundary. For that reason they must always appear in pairs, since the total FS is necessarily a boundary as shown earlier. According to their orientation, the two partners have homology classes opposite to each other. They are heuristically seen to form two 2-tori ($g = 2$) with $r = 2$ each, which also can be proved formally.

On Fig. 5.16, the development of a real Fermi surface of YCo_5 under increasing pressure is shown where sheets of all ranks except $r = 3$ appear. In the third upper panel there are small sheets with $g = r = 0$ centered at the top and bottom faces of the Brillouin zone while in the lower panels the emergence under pressure of a sheet with $g = r = 0$ around the center of the Brillouin zone is shown. In the second upper panel there are small sheets (tori) with $g = 2$, $r = 0$ centered on the edges of the hexagonal faces. The sheet of the left upper panel is a corrugated cylinder with $g = r = 1$. The large sheet of the second panel has $g = 3$, $r = 1$: the six holes around the vertical edges of the hexagonal Brillouin zone yield, after closing the sides of the Brillouin zone as shown in Fig. 5.14 on p. 162, two holes centered at the points AEC and DBF of Fig. 5.14 (each of the six holes belongs to three zones). Cycles around these holes are, however, obviously contractible (into the above mentioned points) in $\mathbb{T}^3_p$. Hence the only non-contractible class of trajectories appears due to closing the top and bottom face of the Brillouin zone

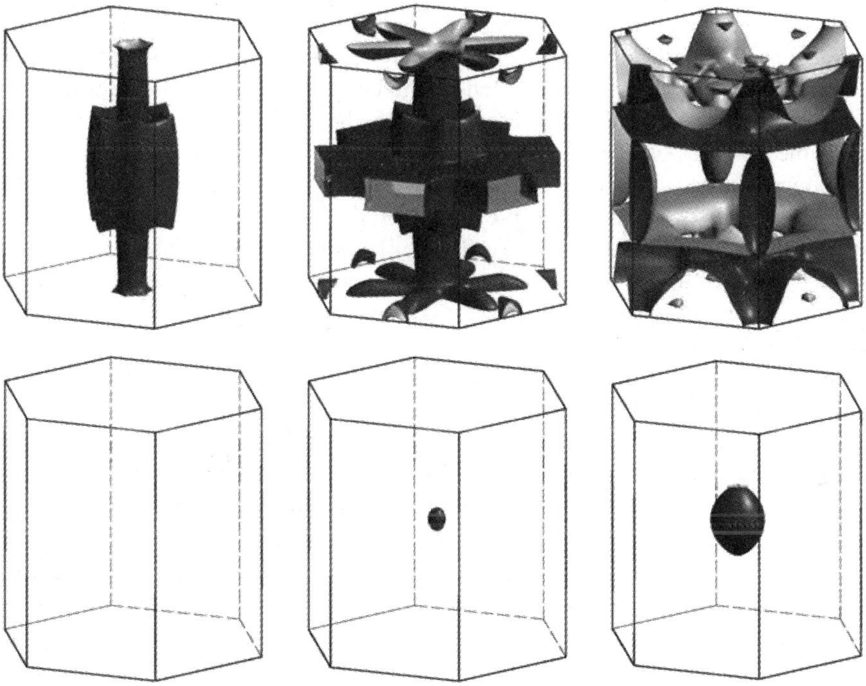

Fig. 5.16 Real Fermi surface of two majority spin conduction bands of hexagonal YCo_5 under increasing pressure from left, 0 GPa, to right, about 25 GPa (courtesy of H. Rosner)

into a torus, like in the first band arising from trajectories parallel to the hexagonal axis. The monster like Fermi surface sheet of the third panel has $g = 7$, $r = 2$: the two holes centered on the hexagonal axis and the tree holes centered on the edges of the hexagonal faces (one from two opposite edges) all yield contractible cycles in $\mathbb{T}^3_{\boldsymbol{p}}$. The only two relevant holes are those appearing by closing the two uninterrupted horizontal edge lines of the Fermi surface sheet around the mantle of the Brillouin zone as in Fig. 5.14. They yield two classes of cycles non-contractible in $\mathbb{T}^3_{\boldsymbol{p}}$, from trajectories in three directions in the hexagonal plane two of which are linearly independent. No trajectory perpendicular to the hexagonal plane remains non-contractible.

The rank r of a Fermi surface is defined to be equal to the maximal rank of its sheets. Why is the rank of a Fermi surface interesting? It for instance governs the magneto-resistivity of pure samples at low temperatures in strong magnetic fields. In this case the dynamics of the electrons can be treated quasi-classically. If no voltage is applied, the equations of motion are

$$\frac{dp^i_\nu}{dt} = \frac{e}{c}F^{ij}\frac{\partial}{\partial p^j}\varepsilon_\nu(\boldsymbol{p}) = \frac{e}{c}\left[\frac{\partial \varepsilon_\nu}{\partial \boldsymbol{p}} \times \boldsymbol{B}\right]^i = \frac{e}{c}[\boldsymbol{v}_\nu \times \boldsymbol{B}]^i, \quad \frac{d\boldsymbol{r}}{dt} = \boldsymbol{v}_\nu = \frac{\partial \varepsilon_\nu}{\partial \boldsymbol{p}}, \qquad (5.105)$$

where e is the electron charge and c is the velocity of light. $\boldsymbol{B}$ is a spatially homogeneous applied magnetic field and $\boldsymbol{v}_\nu$ is the Fermi velocity of a Bloch electron on the Fermi surface of band ν. The motion in quasi-momentum space is all the time perpendicular to the Fermi velocity, hence the Bloch electron stays all the time at constant energy on the Fermi surface. If Cartesian coordinates are introduced in both the quasi-momentum space and the position space with the z-axis in $\boldsymbol{B}$-direction, then the projection of the motion onto the x, y-plane in quasi-momentum space is geometrically similar with the motion in position space rotated by 90° in the mathematically positive direction (due to the negative sign of e) as compared to the motion in quasi-momentum space. While $p_\nu^z =$ const., $dz/dt = v_\nu^z(\boldsymbol{p}_\nu(t))$ is a periodic function of $\boldsymbol{p}_\nu$ which in the simplest model case of a Fermi sphere is a constant. Recall, however, that in general $\boldsymbol{v}$ is an arbitrary function of $\boldsymbol{p}$, and both are in general not collinear.

While in the considered case with a homogeneous magnetic field the motion in quasi-momentum space is always on closed orbits, the position space is a torus only in the idealized model of an infinite perfect crystal. In reality the distance of a unit cell from a boundary of a sample is measurable, and hence the physical motion is in the universal covering space $\mathbb{R}^3$. Then, *the trajectory of the Bloch electrons through the crystal is a closed orbit, if the corresponding trajectory in the Brillouin zone is contractible. It is running through the whole crystal as an open trajectory, if the corresponding trajectory in the Brillouin zone is not contractible, which can be the case in r linearly independent directions for each Fermi surface sheet the rank r of which is non-zero.* The directions are given by the generators (n_1^j, n_2^j, n_3^j), $j = 1, \ldots, r$ of the subgroup $F_*(\pi_1(FS)) \subset \pi_1(\mathbb{T}_{\boldsymbol{p}}^3)$, that is, by the reciprocal lattice vectors $\boldsymbol{G}^j = n_1^j \boldsymbol{b}_1 + n_2^j \boldsymbol{b}_2 + n_3^j \boldsymbol{b}_3$. Only in the case of pairs of corrugated planes there are open trajectories in all directions on those planes.

If the mean scattering time of Bloch electrons (defined by the purity of the crystal and by the temperature) is τ, then the length of the trajectory in quasi-momentum space between two scattering events is on average $\Delta p = |(ev_{\mathrm{F}}/c)B\tau|$ where v_{F} is the average Fermi velocity. Strong magnetic fields are those for which $\Delta p \gg \hbar|\boldsymbol{b}_i|$, but $\mu_{\mathrm{Bohr}} B \ll \varepsilon_{\mathrm{F}}$ in order that the quasi-classical treatment applies. If only closed trajectories in position space are present, then the conductivity tensor σ and the resistivity tensor ρ are [8]

$$\sigma^{ij} \sim \begin{pmatrix} B^{-2} & B^{-1} & B^{-1} \\ B^{-1} & B^{-2} & B^{-1} \\ B^{-1} & B^{-1} & a^{zz} \end{pmatrix}, \quad \rho^{ij} \sim \begin{pmatrix} b^{xx} & B & b^{xz} \\ B & b^{yy} & b^{yz} \\ b^{zx} & b^{zy} & b^{zz} \end{pmatrix}$$

while, if open trajectories are present, then

$$\sigma^{ij} \sim \begin{pmatrix} B^{-2} & B^{-1} & B^{-1} \\ B^{-1} & a^{yy} & a^{yz} \\ B^{-1} & a^{zy} & a^{zz} \end{pmatrix}, \quad \rho^{ij} \sim \begin{pmatrix} B^2 & B & B \\ B & b^{yy} & b^{yz} \\ B & b^{zy} & b^{zz} \end{pmatrix}$$

The field direction is assumed to be the z-direction and in the second case the direction of open trajectories *in the quasi-momentum space* is the x-direction. The entries a^{ij} and b^{ij} in the matrices mean that these components stay constant in the limit $B \to \infty$ of the quasi-classical theory. The quotient $\rho^{yx}/B = R$ is the Hall constant.

Further analysis now needs for a given Fermi surface to find the directions of magnetic field $\boldsymbol{B}$ with respect to the reciprocal lattice for which open trajectories may occur. This task can again be solved with topological methods [9, Chapter 2]. Even if $r = 1$ there need not be directions for the field so that open orbits appear: if the 'corrugated cylinder' is a spiral shaped tube, there may be no plane intersecting it in open trajectories.

References

1. Warner, F.W.: Foundations of Differentiable Manifolds and Lie Groups. Springer, New York (1983)
2. Pontryagin, L.S.: Foundations of Combinatorial Topology. Dover, London (1996)
3. Thirring, W.: A Course in Mathematical Physics, 2nd ed., vol. 1. Springer, Wien (1992)
4. Scheck, F.: *Mechanics*, 2nd corrected and enlarged ed. Springer, Berlin (1994)
5. t'Hooft, G. (ed.): 50 Years of Yang-Mills Theory. World Scientific, Singapore (2005)
6. Thirring, W.: A Course in Mathematical Physics, 2nd ed., vol. 2. Springer, New York (1986)
7. Kato, T.: Perturbation Theory for Linear Operators. Springer, Berlin (1966)
8. Lifshits, I.M., Azbel, M.Y., Kaganov, I.M.: Electron Theory of Metals. Nauka, Moscow (1971) (Translated: Consultants Bureau, New York, 1973)
9. Monastyrsky, M.I.: Topology in Condensed Matter. Springer, Berlin (2006)

Chapter 6
Lie Groups

6.1 Lie Groups and Lie Algebras

A Lie group is a smooth manifold that is also a group. Lie groups play a central role in the geometry of manifolds and in the theory of invariants of dynamical systems in physics. They are named in honor of S. Lie, their theory was much developed by E. Cartan.

A Lie group G is a smooth manifold with a group structure such that for all $g, h \in G$ the mapping $G \times G \to G : (g, h) \mapsto gh^{-1}$ is smooth. Then, the mapping $h \mapsto h^{-1}$ is also smooth, since it can be considered as a case of the previous mapping with $g = e$, the unit element of the group. The composition of these two mappings yields the mapping $(g, h) \mapsto gh$, which hence is also smooth. In summary, all group operations are smooth as a consequence of the smoothness of $(g, h) \mapsto gh^{-1}$.

Since a Lie group is a special case of a topological group, all the arguments used on p. 46 for topological groups are valid. In particular, if the group consists of more than one pathwise connected component as a topological space, then each pathwise connected component is diffeomorphic to the pathwise connected component G^e containing the unit element e of the group, and $G/G^e = \pi_0(G)$, the zeroth homotopy group of G whose elements are in one–one correspondence with the pathwise connected components of G.

A simple case of a Lie group is $\mathbb{R}^n$ with its usual topology and vector addition as group operation. It is Abelian and additively written. The product $G \times G'$ of two Lie groups G and G' with the product manifold structure (p. 60) and the direct product group structure (Compendium C.1) is a Lie group. For instance $\mathbb{R}^n = \mathbb{R} \times \cdots \times \mathbb{R}$ (n factors).

Let $\{\boldsymbol{a}_1, \ldots, \boldsymbol{a}_n\}$ be a base of the vector space $\mathbb{R}^n$, and consider the lattice $L = \{\sum_{i=1}^n n_i \boldsymbol{a}_i \,|\, n_i \in \mathbb{Z}\}$. L is a subgroup of the Lie group $\mathbb{R}^n$. The quotient group $\mathbb{T}^n = \mathbb{R}^n/L$ is also a topological space with the quotient topology. $\mathbb{T}^n$ is a Lie group and is called the n-**torus group**. The 1-torus group can be viewed as the multiplicative group $S^1 = \{e^{i2\pi t} \,|\, t \in \mathbb{R}\}$. Then, $\mathbb{T}^n \approx S^1 \times \cdots \times S^1$ (n factors).

Write the points of $\mathbb{R}^{n^2}$ as real $n \times n$-matrices A, and consider the subset of $\mathbb{R}^{n^2}$ of non-singular matrices, $\det A \neq 0$, with the relative topology from $\mathbb{R}^{n^2}$. Since $\det A$ is a polynomial and hence a smooth function on $\mathbb{R}^{n^2}$, $Gl(n,\mathbb{R}) = \{A \mid \det A \neq 0\}$ is an open subset of dimension n^2 of $\mathbb{R}^{n^2}$ and hence an n^2-dimensional smooth submanifold. It becomes a Lie group under matrix multiplication where AB^{-1} is a set of rational functions with non-zero denominators and hence smooth. It is called the **general linear group** of n dimensions (linear transformations of the vector space $\mathbb{R}^n$). In particular $Gl(1,\mathbb{R}) = \mathbb{R} \setminus \{0\}$ is the multiplicative group of non-zero real numbers. It consists of two pathwise connected components, the positive and the negative real numbers. Likewise, the points of $\mathbb{C}^{n^2} \approx \mathbb{R}^{2n^2}$ (p. 21) may be written as complex $n \times n$-matrices C, and the submanifold $Gl(n,\mathbb{C}) = \{C \mid \det C \neq 0\}$ forms the **complex general linear group** of n dimensions. In particular $Gl(1,\mathbb{C}) = \mathbb{C} \setminus \{0\}$ is the multiplicative group of non-zero complex numbers. It is pathwise connected but not simply connected.

Consider the product manifold $Gl(n,\mathbb{R}) \times \mathbb{R}^n$, but instead of the ordinary direct product group structure define the group operations by $(A,\boldsymbol{x})(A',\boldsymbol{x}') = (AA', A\boldsymbol{x}' + \boldsymbol{x})$. This is the Lie group of **affine motions** or **affine linear transformations** of $\mathbb{R}^n$. By defining the action of the group elements $(A,\boldsymbol{x})$ on any point $\boldsymbol{y} \in \mathbb{R}^n$ by $(A,\boldsymbol{x})\boldsymbol{y} = A\boldsymbol{y} + \boldsymbol{x}$, any group element performs an affine motion of $\mathbb{R}^n$ and group multiplications correspond to compositions of affine motions (exercise). Formally, $(A,\boldsymbol{x})$ may be represented by a special $(n+1) \times (n+1)$-matrix for which the group operation is now the matrix product:

$$(A,\boldsymbol{x}) \mapsto \begin{pmatrix} A & \boldsymbol{x} \\ 0 & 1 \end{pmatrix}, \quad \begin{pmatrix} A & \boldsymbol{x} \\ 0 & 1 \end{pmatrix}\begin{pmatrix} A' & \boldsymbol{x}' \\ 0 & 1 \end{pmatrix} = \begin{pmatrix} AA' & A\boldsymbol{x}' + \boldsymbol{x} \\ 0 & 1 \end{pmatrix}.$$

The action on $\boldsymbol{y} \in \mathbb{R}^n$ becomes a matrix multiplication by appending an $n+1$st unit element to $\boldsymbol{y}$.

Returning to the general theory, every element $g \in G$ defines mappings $l_g : h \mapsto gh$ and $r_g : h \mapsto hg$ of G into itself. They are called **left and right translations** by g. If H is a subset of G, then $l_g(H) = gH, r_g(H) = Hg$. These mappings are injective: $gh = gk$ yields $h = k$ after left translation by g^{-1}. They are also surjective: any element $k \in G$ is $g(g^{-1}k)$ and hence image of some element $g^{-1}k \in G$ with respect to the left translation by g. These simple considerations apply likewise to right translations. Being group operations the translations are smooth transformations of the manifold G.

Let $X \in \mathcal{X}(G)$ be a tangent vector field on the manifold G. At every point $h \in G$ it defines a tangent vector $X_h \in T_h(G)$. Let G be n-dimensional and let $\boldsymbol{x}(h') = (x^1(h'), \dots, x^n(h'))$ be a local coordinate system centered at h, $\boldsymbol{x}(h) = 0$. For every smooth real function $F : G \to \mathbb{R}$ it yields $X_h F = \sum_i \xi^i(h)(\partial F/\partial x^i)_{\boldsymbol{x}=0}$. For every $h \in G$, a left translation by g induces as a push forward a mapping $l^h_{g*} : T_h(G) \to T_{gh}(G)$. A tangent vector field X is called a **left invariant vector field**, if $l^h_{g*}(X_h) = X_{gh}$, that is, the tangent vector at h is pushed forward into the tangent vector at gh of the same tangent vector field. Right invariant vector fields are defined analogously.

Let $\mathfrak{g}$ *be the set of all left invariant vector fields on the Lie group* G. Then,

1. $\mathfrak{g}$ *is a real vector space isomorphic to* $T_e(G)$ *by the isomorphism* $\pi : X \mapsto X_e$. Consequently, $\dim \mathfrak{g} = \dim T_e(G) = \dim G$.
2. $X \in \mathfrak{g}$ *is smooth.*
3. $X, Y \in \mathfrak{g} \Rightarrow [X, Y] \in \mathfrak{g}$, *that is,* $\mathfrak{g}$ *is a Lie algebra* (p. 68).
4. *Let* $\{X_1, \ldots, X_n\}$ *be a base of the vector space* $\mathfrak{g}$, *then there are constants* c_{ij}^k *such that*

$$[X_i, X_j] = \sum_{k=1}^{n} c_{ij}^k X_k, \quad c_{ij}^k + c_{ji}^k = 0, \quad \sum_l (c_{ij}^l c_{lk}^m + c_{jk}^l c_{li}^m + c_{ki}^l c_{lj}^m) = 0. \qquad (6.1)$$

These constants are uniquely defined by G *and the base* $\{X_i\}$ *of* $\mathfrak{g}$.

Proof Linear combinations of left invariant vector fields are clearly left invariant vector fields, hence $\mathfrak{g}$ is a subspace of the real vector space $\mathcal{X}(G)$. π is injective, since $\pi(X) = \pi(Y) \Rightarrow X_e = Y_e \Rightarrow X_g = Y_g$ for all $g \in G$ due to left invariance. π is also surjective, since for every $X_e \in T_e(G)$ there is $X \in \mathcal{X}(G)$ with $X_g = l_{g*}^e(X_e)$. Hence, π is an isomorphism of vector spaces.

Smoothness of left invariant vector fields is traced back to smoothness of the group operations of Lie groups and properties of the push forward, analyzed in (3.29, 3.30) on the basis of the commutative diagram on p. 73. In the present case, $h \in U_\alpha$, $l_g(h) = gh \in U_\beta$ and local coordinates $\boldsymbol{x}_\alpha = \boldsymbol{\varphi}_\alpha(h) \in \boldsymbol{U}_\alpha$ and $\boldsymbol{y}_\beta = \boldsymbol{\varphi}_\beta(gh) \in \boldsymbol{U}_\beta$ are to be considered, where smoothness of the group operations means that the coordinates $\boldsymbol{y}_\beta = (\boldsymbol{l}_g)_{\beta\alpha}(\boldsymbol{x}_\alpha)$ are smooth functions of the coordinates $\boldsymbol{x}_\alpha$ for all admissible charts $(U_\alpha, \boldsymbol{\varphi}_\alpha)$ and $(U_\beta, \boldsymbol{\varphi}_\beta)$. Equation 3.30 with $F = l_g$ now reads $l_{g*}^h(\partial/\partial x_\alpha^i) = \sum_j (\partial (l_g)_{\beta\alpha}^j / \partial x_\alpha^i)(\partial/\partial y_\beta^j)$. The first factor of the last expression is the jth component of the vector field at $\boldsymbol{y}_\beta$ and, as the derivative of the smooth function $(\boldsymbol{l}_g)_{\beta\alpha}$, is a smooth function of the $\boldsymbol{x}_\alpha$.

If X and Y are l_g-related (p. 78), then $[X, Y]$ is l_g-related, and hence Lie products of left invariant vector fields are left invariant vector fields. From that, the existence and uniqueness of the constants c_{ij}^k follows. Their properties are a direct consequence of the properties (3.17) of the Lie product. □

Depending on context both isomorphic Lie algebras $\mathfrak{g}$ and $T_e(G)$ are called the **Lie algebra of the Lie group** G. The relevance of this Lie algebra lies in the fact that it locally, and in the important case of pathwise connected, simply connected Lie groups also globally, completely determines the Lie group. In physics one speaks of the elements of the Lie algebra as of the infinitesimal generators of the Lie group. The constants c_{ij}^k are called the **structure constants** of the Lie group G.

Let $\omega \in \mathcal{D}^r(G)$ be an r-form on G. At each point $h \in G$, ω_h is an element of $\Lambda_r(T_h^*)$ (cf. (4.32)). A left translation l_g induces as a pull back a mapping $l_{g,h}^* : \Lambda_r(T_{gh}^*(G)) \to \Lambda_r(T_h^*(G))$. The r-form ω is called a **left invariant *r*-form**, if

$l^*_{g,h}(\omega_{gh}) = \omega_h$, that is, the r-form at gh is pulled back to its own value at h. Left invariant r-forms form the vector space $\mathcal{D}^r_{\text{inv}}(G)$ and the exterior algebra

$$\mathcal{D}_{\text{inv}}(G) = \sum_{r=0}^{n} \mathcal{D}^r_{\text{inv}}(G), \quad n = \dim G. \tag{6.2}$$

Like the case of tangent vector fields, here due to a property of pull back, smoothness of r-forms is a consequence of left invariance. Right invariant r-forms are defined analogously.

The left invariant 1-forms, which in local coordinates are $\sum_i \omega_i(h)dx^i$, are called the **Maurer–Cartan forms**.

Left invariant r-forms have the following properties:

1. *They are smooth.*
2. $\mathcal{D}_{\text{inv}}(G)$ *is a subalgebra of* $\mathcal{D}(G)$. *By the isomorphism* $\pi^* : \omega \mapsto \omega_e, \mathcal{D}_{\text{inv}}(G)$ *is isomorphic to* $\Lambda(T^*_e(G))$; *in particular* $\mathcal{D}^1_{\text{inv}}(G)$ *is isomorphic to* $T^*_e(G)$ *and hence dual to* $\mathfrak{g}$.
3. $\omega \in \mathcal{D}^1_{\text{inv}}(G)$ *and* $X \in \mathfrak{g} \Rightarrow \langle \omega, X \rangle = \text{const.}$ *on* G.
4. $\omega \in \mathcal{D}^1_{\text{inv}}(G)$ *and* $X, Y \in \mathfrak{g} \Rightarrow \langle d\omega, X \wedge Y \rangle = -\langle \omega, [X, Y] \rangle$.
5. *If* $\vartheta^i \in \mathcal{D}^1_{\text{inv}}(G)$ *and* $\{\vartheta^1, \dots, \vartheta^n\}$ *is the base dual to* $\{X_1, \dots, X_n\}$, *then the* ***Maurer–Cartan equations*** *or* ***structure equations***

$$d\vartheta^i = -\sum_{1 \le j < k \le n} c^i_{jk} \vartheta^j \wedge \vartheta^k, \quad c^i_{jk} = \langle \vartheta^i, [X_j, X_k] \rangle \tag{6.3}$$

hold.

The isomorphism of 2 is proved analogously to the case of vector fields, and 3 is a direct consequence. 4 follows from (4.49) where the second line vanishes due to 3, and 5 follows from the duality $\langle \vartheta^i, X_j \rangle = \delta^i_j$ together with (4.22, 6.1) and 4. Let $\vartheta = \sum_i \vartheta^i X_i$, then $\langle \vartheta_e, \cdot \rangle$ maps $T_e(G)$ isomorphically onto $\mathfrak{g}$. The tangent-vector valued 1-form ϑ is called the **canonical Maurer–Cartan form**.

Depending on context, in the whole concept of Lie algebra of Lie groups sometimes right invariant vector fields and forms are used, mainly for the sake of convenience of notation. Since in both cases the Lie algebra $\mathfrak{g}$ is isomorphic to the same $T_e(G)$ and $\mathcal{D}_{\text{inv}}(G)$ is isomorphic to the same $\Lambda(T^*_e(G))$, the buildings in both cases are isomorphic to each other. However, the composition of two right translations is $r_{g'} r_g : h \mapsto hgg'$ and hence $r_{gg'} = r_{g'} r_g$. This contravariant behavior transfers to the push forward r^h_{g*}, and therefore *the Lie product* $[X, Y]$ *of left invariant vector field corresponds to the Lie product* $[Y, X]$ *of right invariant vector fields. Correspondingly, the structure constants of both cases differ by a sign while all the above given relations remain valid for both cases.*

For an **Abelian Lie group** (like for instance $\mathbb{R}^n$) left and right invariant vector fields coincide, hence all structure constants vanish and the corresponding **Lie algebra** is also **Abelian**: all Lie products are zero.

6.2 Lie Group Homomorphisms and Representations

A mapping $F : G \to H$ of a Lie group G into a Lie group H is a **Lie group homomorphism**, if it is both a smooth mapping of manifolds and a homomorphism of groups, that is, $F(gh^{-1}) = F(g)F(h^{-1})$. If it is a diffeomorphism of manifolds then it is an **isomorphism**, because in that case F is onto and F^{-1} exists and hence $F^{-1}(F(g))F^{-1}(F(h^{-1})) = gh^{-1} = F^{-1}(F(gh^{-1})) = F^{-1}(F(g)F(h^{-1}))$ which proves that F^{-1} is also a homomorphism. A (Lie group) isomorphism from G onto itself is a (Lie group) **automorphism**. Naturally, the automorphisms of G form a group with respect to composition as group operation (exercise). If H is the transformation group $\mathrm{Aut}\,(V)$ (automorphism group) of some vector space V (p. 100), for instance $H = Gl(n,\mathbb{R})$ or $H = Gl(n,\mathbb{C})$, then the homomorphism F is called a **representation** of the Lie group G.

A K-linear mapping $L : \mathfrak{g} \to \mathfrak{h}$ ($K = \mathbb{R}$ or $\mathbb{C}$) from a Lie algebra $\mathfrak{g}$ over K into a Lie algebra $\mathfrak{h}$ over K which preserves Lie products, $L([X,Y]) = [L(X), L(Y)]$, is a (Lie algebra) **homomorphism**. (It is an ordinary homomorphism of algebra.) If it is one–one and onto, then it is an **isomorphism**. An isomorphism from $\mathfrak{g}$ onto itself is an **automorphism**. If $\mathfrak{h}$ is the algebra $\mathrm{End}(V)$ of K-linear mappings of some vector space V over K into itself (endomorphisms, forming an algebra with respect to composition as multiplication, exercise), for instance $\mathfrak{h} = \mathfrak{gl}(n,\mathbb{R})$ or $\mathfrak{h} = \mathfrak{gl}(n,\mathbb{C})$ (all real or complex $n \times n$-matrices, respectively), then the homomorphism L is called a **representation** of the Lie algebra $\mathfrak{g}$.

Let G and H be Lie groups with Lie algebras $\mathfrak{g}$ and $\mathfrak{h}$, and let $F : G \to H$ be a Lie group homomorphism. Then, for every $X \in \mathfrak{g}$, X and $F_*(X)$ are F-related, and $F_* : \mathfrak{g} \to \mathfrak{h}$ is a Lie algebra homomorphism.

Proof By definition of the tangent map F_* of the mapping F (p. 71), $F_*(X)_{F(g)} = F_*^g(X_g)$, which also means that X and $F_*(X)$ are F-related (p. 78). It is to be proved that $F_*(X)$ is a left invariant vector field on H. Let e be the unit in G and $\tilde{e}$ the unit in H. Since F is a Lie group homomorphism, $l_{F(g)} \circ F = F \circ l_g$, and hence $F_*(X)_{F(g)} = F_*^g(X_g) = F_*^g(l_{g*}^e(X_e)) = (F \circ l_g)_*(X_e) = (l_{F(g)} \circ F)_*(X_e) = l_{F(g)*}^{\tilde{e}}(F_*(X)_{\tilde{e}})$ where in the third and fifth equality the covariance of the push forward (p. 73) was used. Hence, $F_*(X) \in \mathfrak{h}$. It remains to prove that $F_*([X,Y]) = [F_*(X), F_*(Y)]$. But this follows from the previous result and the statement on p. 78. □

Quite similarly it is shown that F^* pulls back left (right) invariant r-forms $\tilde{\omega}$ on H to left (right) invariant r-forms $\omega = F^*(\tilde{\omega})$ on G. In particular, since invariant 1-forms are dual to invariant vector fields, F^* on $\mathcal{D}^1_{\mathrm{inv}}(H) = \mathfrak{h}^*$ is transposed to F_* on $\mathfrak{g}$. Since the exterior differentiation d commutes with F^*, (4.43), it follows from (6.3) for the pulled back 1-forms

$$d(F^*(\tilde{\vartheta}^i)) = -\sum_{1 \le j < k \le n} \tilde{c}^i_{jk} F^*(\tilde{\vartheta}^j) \wedge F^*(\tilde{\vartheta}^k), \tag{6.4}$$

where the $\tilde{\vartheta}^i$ form a base of Maurer–Cartan forms of H and $\tilde{c}^i_{jk}$ are the structure constants of H.

There are intimate algebraic interrelations between Lie groups G and their Lie algebras $\mathfrak{g}$. Let a Lie group G of dimension m be given with its Lie algebra $\mathfrak{g}$. Let a collection of linearly independent left invariant 1-forms $\omega^1, \dots, \omega^{\tilde{m}}$ out of $\mathfrak{g}^* \approx \mathcal{D}^1_{\text{inv}}(G)$ be given, $\tilde{m}$ not necessarily related to m. The natural question arises, is there a Lie group H of dimension $\tilde{m}$ with Lie algebra $\mathfrak{h}$, with a base $\tilde{\omega}^1, \dots, \tilde{\omega}^{\tilde{m}}$ of $\mathfrak{h}^*$ and a Lie group homomorphism $F : G \to H$ so that $F^*(\tilde{\omega}^i) = \omega^i, i = 1, \dots, \tilde{m}$ holds.

Observe that, if F exists, its **graph** is an embedded submanifold (G, γ) of the product manifold $G \times H$ (which is also a Lie group with its Lie algebra $\mathfrak{g} \oplus \mathfrak{h}$) with $\gamma : G \ni g \mapsto (g, F(g)) \in G \times H$. Introduce the canonical projections $\pi_G : G \times H \to G$ and $\pi_H : G \times H \to H$ which both are Lie group homomorphisms. Hence, the 1-forms

$$\{\nu^i = \pi^*_G(\omega^i) - \pi^*_H(\tilde{\omega}^i) = \pi^*_G(F^*(\tilde{\omega}^i)) - \pi^*_H(\tilde{\omega}^i) \,|\, i = 1, \dots, \tilde{m}\} \tag{6.5}$$

are left invariant 1-forms on $G \times H$. π^*_G and π^*_H are pull backs from G and H, respectively, to $G \times H$, and F^* pulls back from H to G. Since the $\pi^*_G(\omega^i)$ are obviously linearly independent from the $\pi^*_H(\tilde{\omega}^i)$ (they belong to subspaces of $(\mathfrak{g} \oplus \mathfrak{h})^*$ linearly independent of each other) and the $\tilde{\omega}^i, i = 1, \dots, \tilde{m}$ were supposed to form a base of $\mathfrak{h}^*$ and hence to be linearly independent from each other, the forms (6.5) are also linearly independent. Consider the two-sided ideal $\mathcal{I}$ of $\mathcal{D}_{\text{inv}}(G \times H)$ generated by the forms (6.5), that is, the algebra

$$\mathcal{I} = \text{span}_K\{\mathcal{D}_{\text{inv}}(G \times H) \wedge \nu^i \wedge \mathcal{D}_{\text{inv}}(G \times H)\} \tag{6.6}$$

which is the span of all elements of the set on the right hand side. From (6.4),

$$\begin{aligned} d\big(\pi^*_G(F^*(\tilde{\omega}^i)) - \pi^*_H(\tilde{\omega}^i)\big) &= \sum c^i_{jk}\big(\pi^*_G(F^*(\tilde{\omega}^j)) \wedge \pi^*_G(F^*(\tilde{\omega}^k)) - \pi^*_H(\tilde{\omega}^j) \wedge \pi^*_H(\tilde{\omega}^k)\big) \\ &= \sum c^i_{jk}\big([\pi^*_G(F^*(\tilde{\omega}^j)) - \pi^*_H(\tilde{\omega}^j)] \wedge \pi^*_G(F^*(\tilde{\omega}^k)) \\ &\quad + \pi^*_H(\tilde{\omega}^j) \wedge [\pi^*_G(F^*(\tilde{\omega}^k)) - \pi^*_H(\tilde{\omega}^k)]\big), \end{aligned}$$

where the c^i_{jk} are the structure constants of $G \times H$. This result shows $d\mathcal{I} \subset \mathcal{I}$, which is expressed by saying that $\mathcal{I}$ is a **differential ideal** of $\mathcal{D}_{\text{inv}}(G \times H)$.

Now, pull back the 1-forms ν^i from $G \times H$ to the graph of F by the embedding mapping γ of the graph of F into $G \times H$. With $\pi_G \circ \gamma = \text{Id}_G$ and $\pi_H \circ \gamma = F$ one finds

$$\gamma^*(\nu^i) = (\pi_G \circ \gamma)^*(F^*(\tilde{\omega}^i)) - (\pi_H \circ \gamma)^*(\tilde{\omega}^i) = F^*(\tilde{\omega}^i) - F^*(\tilde{\omega}^i) = 0,$$

where also $\gamma^* \circ \pi^*_G = (\pi_G \circ \gamma)^*$ was used and the corresponding relation for π_H. Hence, on the graph of F there hold $\tilde{m}$ independent relations $\nu^i = 0$, and $d\nu^i = 0 \bmod \nu^1, \dots, \nu^{\tilde{m}}$. By the dual Frobenius theorem (p. 80) this means that the graph of F is the integral manifold of the completely integrable Pfaffian system $\nu^i = 0$, $i = 1, \dots, \tilde{m}$ on $G \times H$.

These considerations presupposed the existence of F. Now, suppose that only a homomorphism $f : \mathfrak{g} \to \mathfrak{h}$ is given (in the above case $f = F_*$). The transpose f^* of f maps 1-forms on H to 1-forms on G. In a way analogous to the above it is straightforwardly demonstrated that the 1-forms on $G \times H$,

$$\{v^i = \pi_G^*(f^*(\tilde{\omega}^i)) - \pi_H^*(\tilde{\omega}^i) \mid i = 1, \ldots, \tilde{m}\} \tag{6.7}$$

generate a differential ideal $\mathcal{I}$ of $\mathcal{D}_{\text{inv}}(G \times H)$ and hence define a graph of a homomorphism $F : G \to H$ as the unique integral manifold of the system $v^i = 0$, $i = 1, \ldots, \tilde{m} = \dim H$ through the point $(e, \tilde{e}) \in G \times H$, if G is pathwise connected. A first consequence is the following theorem:

Let the Lie group G be pathwise connected, and let F and F' be Lie group homomorphisms from G into the Lie group H such that the Lie algebra homomorphisms F_* and F'_* from $\mathfrak{g}$ into $\mathfrak{h}$ are identical. Then, $F \equiv F'$.

Proof As homomorphisms, F and F' agree at the unit $e \in G$. Moreover, F^* and F'^* agree as the transposes to F_* and F'_*. Hence, F and F' define identical differential ideals on $G \times H$ and hence have identical graphs. □

6.3 Lie Subgroups

Let G and H be Lie groups, and let H be a subset of G, not necessarily provided with the relative topology as a topological space, but such that

1. H is a subgroup of G,
2. (H, Id) is an embedded submanifold of G.

A Lie group which is isomorphic to H is called a **Lie subgroup** of G. If one speaks of uniqueness of a Lie subgroup, uniqueness of H as a subset of G is always meant. The topology of the embedding must be such that smoothness of the group operations is provided, the embedding need not be regular. H is called a **closed Lie subgroup** of G if in addition the subset H is closed in the topology of G. It can be shown [1] that the Lie subgroup H is a regular embedding, iff it is a closed Lie subgroup of G.

If $\mathfrak{g}$ is a Lie algebra and $\mathfrak{h} \subset \mathfrak{g}$ is a linear subspace of $\mathfrak{g}$ closed under the Lie product $[X, Y]$ of $\mathfrak{g}$, then $\mathfrak{h}$ is also a Lie algebra; it is called a **subalgebra** of $\mathfrak{g}$.

Let H be a Lie subgroup of the Lie group G, and let $\mathfrak{h}$ and $\mathfrak{g}$ be their Lie algebras. Then $\mathfrak{h}$ is a subalgebra of $\mathfrak{g}$.

This simply follows from $(\mathrm{Id}_H)_* = \mathrm{Id}_{\mathfrak{h}}$ where $(\mathrm{Id}_H)_*$ is a Lie algebra homomorphism (see p. 177).

Let G be a connected Lie group, and let U be a neighborhood of the unit e. Then, U generates G, which means

$$G = \bigcup_{n=1}^{\infty} U^n, \quad U^n = \{\underbrace{g_1 \cdots g_n}_{\text{group product}} \mid g_i \in U\}. \tag{6.8}$$

Proof Let $V = U \cap U^{-1}, U^{-1} = \{g^{-1} \mid g \in U\}$, and let $H = \cup_{n=1}^{\infty} V^n \subset \cup_{n=1}^{\infty} U^n$. It is easily seen that H is a subgroup of G. It is also an open subset of G since for every $g \in H$ the set gV is a neighborhood of g and $gV \subset H$. Thus, for every $g \in G$ the coset gH is an open subset of G. Since cosets are disjoint, either $gH = H$ or $gH \cap H = \emptyset$, which means that the open subset H of G as the complement of all cosets $gH \neq H$ is also closed in G. Since G is connected and H is not empty, $H = G$. □

Since a Lie group is a finite dimensional manifold, it has a neighborhood U of the unit e for which $\overline{U}$ is compact and hence contains a countable dense set. From that and the above theorem it follows easily that the connected component G_e is second countable. Hence,

A Lie group G is second countable, iff G/G_e is countable.

In this text the latter is always presupposed, that is, a Lie group is supposed to have at most countably many connected components and so to be second countable.

Let G be a Lie group with Lie algebra $\mathfrak{g}$, and let $\mathfrak{h}$ be a subalgebra of $\mathfrak{g}$. Then there is a unique connected Lie subgroup H of G which has $\mathfrak{h}$ as its Lie algebra.

Proof $\mathfrak{h}$ is a $(\dim \mathfrak{h} = \tilde{m})$-dimensional involutive distribution on G (Sect. 3.6). By the Frobenius theorem, there is a unique maximal connected integral manifold (H', F) through $e \in G$. Let $H = F(H')$. $\mathfrak{h}$ is left invariant, therefore for every $h \in H$, $(H', l_{h^{-1}} \circ F)$ is also an integral manifold of $\mathfrak{h}$ through e, and, because of the maximality of (H', F), $l_{h^{-1}} \circ F(H') \subset H$. Hence, if $h, k \in H$, then $h^{-1}k \in H$ and H is a subgroup of G. One must show that $(h, k) \mapsto h^{-1}k$ is smooth in the topology of H inherited from the embedding (H', F). This follows since F is smooth and one–one and $l_{h^{-1}}$ is a diffeomorphism: $h^{-1}k = (l_{h^{-1}} \circ F)(F^{-1}(k))$ is a smooth function of h for fixed k, in particular h^{-1} is a smooth function of h. Also, $k^{-1}h$ is a smooth function of k for fixed h and so is $(k^{-1}h)^{-1} = h^{-1}k$. Thus, H is a Lie subgroup of G.

Assume that there is another connected Lie subgroup K of G which has $\mathfrak{h}$ as its Lie algebra. Both must coincide in a neighborhood of e, and therefore they are identical due to the previous theorem. □

In summary, there is a one–one correspondence between the connected Lie subgroups of a Lie group and the subalgebras of its Lie algebra. It can be shown (Sect. 9.2) that for every subgroup of a Lie group there is at most one manifold structure which makes it into a Lie subgroup.

6.4 Simply Connected Covering Group

Universal coverings have deep consequences in physics, therefore they are considered here in some detail. Who is not so much interested in the technical details may just take notice of the theorems in italics and skip the proofs. The following analysis is essentially due to Pontrjagin [2, §50].

A continuous mapping $\pi : \tilde{M} \to M$ of a topological space $\tilde{M}$ *onto* a topological space M is called a **covering**, if every point $x \in M$ has a neighborhood U which is **evenly covered** by π, meaning that the preimage $\pi^{-1}(U)$ of U is a (possibly infinite) union of disjoint open sets V_α of $\tilde{M}$ each of which is homeomorphic to U. M is called the **base of the covering**, and $\tilde{M}$ is called the **covering space**. Two homeomorphic covering spaces $\tilde{M}$ and $\tilde{M}'$ with coverings π and π' onto M are considered **equivalent coverings** of M, if there exists a homeomorphism $F : \tilde{M}' \to \tilde{M}$ for which $\pi' = \pi \circ F$.

For example, $\pi : t \mapsto \phi = e^{it}$ is an ∞-fold covering of the unit circle S^1 in the complex plane (with $\operatorname{Arg}\phi$ as local coordinate) by the real line $\mathbb{R}$ (with global coordinate t). In general of course, only local coordinate relations are possible. More sophisticated familiar examples of coverings are Riemann surfaces with branch points and poles removed as coverings of domains of holomorphy of complex functions in the complex plane.

So far, nothing on the connectedness of M was presupposed. If, however, N is a connected topological space which is continuously mapped by F into $\tilde{M}$ and by $\pi \circ F$ into some $U \subset M$ so that the intersection $F(N) \cap V_\alpha$ with one of the sets V_α of an even covering of U is non-empty, then obviously $F(N) \subset V_\alpha$. $F(N)$ as the continuous image of a connected space is connected, and the V_α are mutually disconnected since they are disjoint and homeomorphic to the open set U and hence open.

In particular, if $F : I \to M, I = [0,1]$ is a path in M starting at $x = F(0)$ and $F^* : I \to \tilde{M}$ is a path in $\tilde{M}$ with $F = \pi \circ F^*$, then it is straightforward to demonstrate that the path F^* is uniquely defined by its starting point $F^*(0)$ and by F (Fig. 6.1). Moreover, a continuous deformation of F causes a continuous

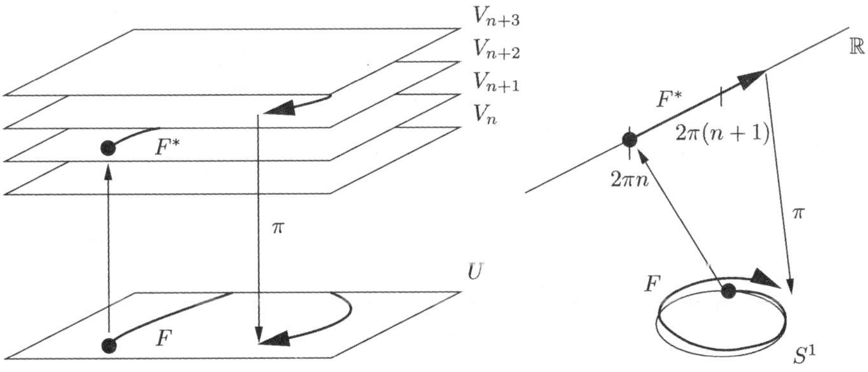

Fig. 6.1 Lifting of a path F from M to a path F^* in the covering $\tilde{M}$. *Left* an evenly covered open set $U \subset M$ and some of its covering sets $V_\alpha \subset \tilde{M}$ are shown. F^* is uniquely defined by F and $F^*(0)$ (*black dot*). If F leaves U and then returns, F^* need not return in the same V_α where it started, but its end point is still uniquely defined by F and $F^*(0)$. This is illustrated on the *right side* for the covering of S^1 by $\mathbb{R}$

deformation of F^*. This is the basis of the homotopy of coverings. In the following, as in Sect. 2.5, F_- means a path running against F (from the end point of F to the starting point of F through the same intermediate points in opposite direction) and $F'F$ means the **concatenation of paths**, first running through F and then through F'. For concatenation the end point of F must be the starting point of F'. A closed path starting and ending at point x is a **loop** with **base point** x.

Let $\pi : \tilde{M} \to M$ be a covering of a pathwise connected space M by a pathwise connected space $\tilde{M}$, let $x \in M$ be any point and let $\tilde{x}$ be some point of $\pi^{-1}(x)$. Then, the covering π generates a homomorphism from the homotopy group $\pi_1(\tilde{M}, \tilde{x})$ into the homotopy group $\pi_1(M, x)$. In fact this homomorphism is an isomorphism of $\pi_1(\tilde{M}, \tilde{x})$ onto some subgroup $\rho(\pi, \tilde{x})$ of $\pi_1(M, x)$, and, if $\tilde{x}'$ runs through all points of $\pi^{-1}(x)$, then $\rho(\pi, \tilde{x}')$ runs through all members $g^{-1}\rho(\pi, \tilde{x})g,\ g \in \pi_1(M, x)$, of the conjugacy class of the subgroup $\rho(\pi, \tilde{x})$ of $\pi_1(M, x)$.

Proof It is clear that a continuous mapping π maps loops into loops and continuous deformations of loops into continuous deformations of loops. Hence it induces a homomorphism from $\pi_1(\tilde{M}, \tilde{x})$ into $\pi_1(M, x)$. However, it was stated above that the homotopy classes of loops of $\pi_1(\tilde{M}, \tilde{x})$ are uniquely determined by $\tilde{x}$ and by the homotopy classes of loops of $\pi_1(M, x)$, which means that the homomorphism is injective and hence is an isomorphism into a subgroup of $\pi_1(M, x)$.

Now, let $\tilde{x}$ and $\tilde{x}'$ be two arbitrary points of $\pi^{-1}(x)$, let F^* and $F^{*\prime}$ be loops with base point $\tilde{x}$ and $\tilde{x}'$, respectively, and let $F^{*\prime\prime}$ be a path from $\tilde{x}$ to $\tilde{x}'$. All three paths are mapped by π into loops F, F', F'' in M with base point x. In $\tilde{M}$, the paths $F^*, F^{*\prime\prime}_-F^{*\prime}F^{*\prime\prime}$ are loops with the same base point $\tilde{x}$ and $F^{*\prime}, F^{*\prime\prime}F^*F^{*\prime\prime}_-$ are loops with the same base point $\tilde{x}'$, so that their images of the mapping π belong to the homotopy classes $[F], [F'']^{-1}[F'][F'']$ of $\rho(\pi, \tilde{x})$ and $[F'], [F''][F][F'']^{-1}$ of $\rho(\pi, \tilde{x}')$. Hence, $[F'']^{-1}\rho(\pi, \tilde{x}')[F''] \subset \rho(\pi, \tilde{x})$ and $[F'']\rho(\pi, \tilde{x})[F'']^{-1} \subset \rho(\pi, \tilde{x}')$, which means $\rho(\pi, \tilde{x}') = [F'']\rho(\pi, \tilde{x})[F'']^{-1}$.

Let now $[F'']$ be any element of $\pi_1(M, x)$, and let F'' be one of its loops. Choose $\tilde{x}$ as the starting point of a corresponding path $F^{*\prime\prime}$ in $\tilde{M}$ which latter is uniquely determined by $\tilde{x}$ and F''. It ends at some point $\tilde{x}' \in \pi^{-1}(x)$, and the relation at the end of the last paragraph holds. This proves the last statement of the theorem. □

Since in pathwise connected spaces the homotopy groups $\pi(M, x)$ are isomorphic for all $x \in M$ and conjugated subgroups are also isomorphic, up to isomorphisms the subgroup of the last theorem is independent of x and is denoted by $\sigma(\pi)$.

Let F^* be an arbitrary path in $\tilde{M}$ starting at $\tilde{x}$. It is closed, iff $F = \pi \circ F^*$ is closed and $[F] \in \rho(\pi, \tilde{x})$. Indeed, if F^* is closed, the condition follows. Let F be closed and $[F] \in \rho(\pi, \tilde{x})$. Then, there is a loop $F^{*\prime}$ in $\tilde{M}$, which starts in $\tilde{x}$ and which is mapped by π into $[F]$. Since $\pi \circ F^{*\prime} \cong F$, there is a continuous deformation of $F^{*\prime}$ into F^* without moving the end point. Then, F^* is closed together with $F^{*\prime}$.

More generally, let F^* and $F^{*\prime}$ be two arbitrary paths starting from $\tilde{x}$. They will have the same end point in $\tilde{M}$, iff F and F' have the same end point in M and $[F'_-F] \in \rho(\pi, \tilde{x})$. Indeed, if F^* and $F^{*\prime}$ have the same end point, then F and F' have the same end point, $F^{*\prime}_-F^*$ is closed and hence $[F'_-F] \in \rho(\pi, \tilde{x})$. Reversely, F'_-F is the image of a path $F^{*\prime\prime}_-F^*$, where $F^{*\prime\prime}_-$ starts at $F^*(1)$ and is mapped by π to F'_-. Since F'_-F is closed and $[F'_-F] \in \rho(\pi, \tilde{x})$, $F^{*\prime\prime}_-F^*$ is also closed and the end point of $F^{*\prime\prime}_-$ is $\tilde{x}$. Thus, both paths $F^{*\prime}$ and $F^{*\prime\prime}$ start at $\tilde{x}$ and are mapped to F', hence they are identical, and $F^{*\prime}$ ends at $F^*(1)$.

Now, let in the latter case in particular F and F' both be closed. Then, $[F]$ and $[F']$ both are elements of $\pi_1(M, x)$, and F^* and $F^{*\prime}$ both have end points in $\pi^{-1}(x)$. If $[F'_-F] = [F']^{-1}[F] \in \rho(\pi, \tilde{x})$, that is $[F] \in [F']\rho(\pi, \tilde{x})$, then these end points fall together. Clearly, the number of end points, that is the number of points of $\pi^{-1}(x)$, is equal to the number of (left) cosets $[F']\rho(\pi, \tilde{x})$ of $\rho(\pi, \tilde{x})$ as subgroup of $\pi_1(M, x)$ (which is called the index of the subgroup). If the cardinality of $\pi_1(M, x)$ is finite, then this index is the ratio of cardinalities of $\pi_1(M, x)$ and $\rho(\pi, \tilde{x})$. In pathwise connected spaces this cardinality α (finite or infinite) is independent of x; it is called the **multiplicity of covering**, the covering is called **α-fold**.

Now, the most important questions of existence and uniqueness of coverings can be answered. First, uniqueness is considered.

Let π and π' be two coverings of a pathwise connected and locally pathwise connected space M by covering spaces $\tilde{M}$ and $\tilde{M}'$, respectively. Let $x \in M$, $\tilde{x} \in \pi^{-1}(x)$, $\tilde{x}' \in \pi'^{-1}(x)$, and $\rho(\pi, \tilde{x}) \subset \rho(\pi', \tilde{x}')$. Then, there exists a covering $\tilde{\pi}: \tilde{M} \to \tilde{M}'$ such that $\pi' \circ \tilde{\pi} = \pi$. Moreover, π and π' are equivalent coverings of M, iff $\sigma(\pi) = \sigma(\pi')$.

Proof Let $F, F^*, F^{*\prime}$ be paths starting at $x, \tilde{x}, \tilde{x}'$ and ending at $y, \tilde{y}, \tilde{y}'$, respectively, and let F be the image of both paths F^* and $F^{*\prime}$ by π and π', respectively. Since $\rho(\pi, \tilde{x}) \subset \rho(\pi', \tilde{x}')$, a deformation of the path F which does not change the end point $\tilde{y}$ of F^* will not change the end point y' of $F^{*\prime}$ either, hence $\tilde{y}'$ is uniquely defined by $\tilde{y}$, and the just described construction defines a mapping $\tilde{\pi}: \tilde{M} \to \tilde{M}': \tilde{y} \mapsto \tilde{y}'$, for which $\pi' \circ \tilde{\pi} = \pi$. If $\rho(\pi, \tilde{x}) = \rho(\pi', \tilde{x}')$, then π and π' may be interchanged to prove that $\tilde{\pi}$ is one–one and onto.

It remains to show that $\tilde{\pi}$ is a covering. Let $F, F^*, F^{*\prime}$ be as above, and let U be a neighborhood of y such that $V_\alpha \ni \tilde{y}$ and $V'_{\alpha'} \ni \tilde{y}'$ are open sets of even coverings of U by π and π', respectively. Such an U exists since M is locally pathwise connected. Let $\tilde{F}^*$ be a path in V_α from $\tilde{y}$ to some point $\tilde{z}$, let $\tilde{F} = \pi(\tilde{F}^*)$, and let $\tilde{F}^{*\prime}$ be a path starting at $\tilde{y}'$ with $\pi'(\tilde{F}^{*\prime}) = \tilde{F}$. Because of the even coverings, $\tilde{F}^{*\prime}$ is in $V'_{\alpha'}$. Moreover, $\pi'(\tilde{F}^{*\prime}F^{*\prime}) = \tilde{F}F$, and hence by construction of $\tilde{\pi}$, $\tilde{\pi}(\tilde{z}) \in V'_{\alpha'}$. Since the restrictions of π and π' to V_α and $V'_{\alpha'}$, respectively, are homeomorphisms, $\tilde{\pi}|_{V_\alpha} = \pi'^{-1} \circ \pi|_{V_\alpha}$ is also a homeomorphism from V_α onto $V'_{\alpha'}$. By choosing $\tilde{y}'$ arbitrarily in M', the preimage $\tilde{\pi}^{-1}(V'_{\alpha'})$ with $\tilde{y}' \in V'_{\alpha'}$ consists of all V_α for which $\tilde{y} \in \tilde{\pi}^{-1}(\tilde{y}')$ which proves that $\tilde{\pi}$ is a covering.

If $\sigma(\pi) = \sigma(\pi')$, then according to the first theorem of this section a one–one correspondence between the points of $\pi^{-1}(x)$ and the points of $\pi'^{-1}(x)$ can be chosen so that $\rho(\pi, \tilde{x}) = \rho(\pi', \tilde{x}')$, and then $\tilde{\pi}$ is also one–one and hence a homeomorphism making π and π' to be equivalent coverings. $\square$

Existence of a covering is governed by the following theorem:

Let M be a pathwise connected, locally pathwise connected and semi-locally 1-connected topological space. (Every point $x \in M$ has a neighborhood U such that every loop in U with base point x is contractible in M into x.) Let ρ be a given subgroup of a given subgroup of $\pi_1(M, x)$. There exists a covering of M by a pathwise connected space $\tilde{M}$ such that $\rho(\pi, \tilde{x}) = \rho$ for $\tilde{x} \in \pi^{-1}(x)$. In particular, there exists a covering by a simply connected covering space $\bar{M}$ which is uniquely defined up to homeomorphisms.

$\bar{M}$ is called the **universal covering space** of M.

Proof of the theorem Step one is establishing $\tilde{M}$ as a set. Two paths F and F' in M starting at x are considered equivalent by ρ, if they have the same end point and $[F'\_F] \in \rho$. This subdivides the set of all paths in M starting at x into equivalence classes $\{F\}$. Now, $\tilde{M}$ is taken to be the set of these equivalence classes, and $\pi(\{F\}) = y$ is defined to be the end point y of F. Since M is pathwise connected, $\pi(\tilde{M}) = M$.

Next, a topology is introduced in $\tilde{M}$. Let F be any path in M from x to y, and let U be a neighborhood of y in M. Let $\tilde{U}$ be the set of all points $\{F_z F\}$ where F_z is a path in U from y to $z \in U$. Since M is locally pathwise connected and semi-locally 1-connected, for every point $y \in M$ there exists a neighborhood base $\mathcal{B}_y$ of pathwise connected sets U for which this construction is possible. $\{F_z F\}$ depends only on $\{F\}$ and on z. Indeed, let F', F'_z be any other paths for which $\{F'\} = \{F\}$ and F'_z is in U from y to z. Then, $(F_z F)\_F'_z F' = F\_F_z\_F'_z F' \cong F\_F'$ since $F_z\_F'_z$ is null-homotopic in M. Since $\{F'\} = \{F\}$ means that $[F\_F'] \in \rho$, also $[(F_z F)\_F'_z F'] \in \rho$ or $\{F'_z F'\} = \{F_z F\}$. Hence, the restriction of π to $\tilde{U}$ is a one–one mapping. Taking for every point $\{F\}$ of $\tilde{M}$ all sets $\tilde{U}$ for all $U \in \mathcal{B}_{\pi(\{F\})}$ to form a neighborhood base $\tilde{\mathcal{B}}_{\{F\}}$ of $\{F\}$ defines a topology on $\tilde{M}$ which makes π into a local homeomorphism from sets $\tilde{U}$ to sets U. Indeed, let any union of sets $\tilde{U}$ be an open set of $\tilde{M}$. Take any two sets $\tilde{U}$ and $\tilde{U}'$ which have a common point $\{F\}$. Then, there exists a common point $z = \pi(\{F\})$ of $U = \pi(\tilde{U})$ and $U' = \pi(\tilde{U}')$ in M and hence a neighborhood $V \ni z$ of the above type with $V \subset U \cap U'$. By construction of neighborhood bases in $\tilde{M}$, the set $\tilde{V}$ is a neighborhood of $\{F\}$, and $\tilde{V} \subset \tilde{U} \cap \tilde{U}'$. Hence, every intersection of two open sets of $\tilde{M}$ as just defined is an open set of $\tilde{m}$ and these sets form a topology of $\tilde{M}$. This topology is Hausdorff: Let $\{F\} \neq \{F'\}$. If $\pi(\{F\}) \neq \pi(\{F'\})$ then there are disjoint open sets U and U' of the Hausdorff space M with $\pi(\{F\}) \in U$ and $\pi(\{F'\}) \in U'$, and hence $\tilde{U}$ and $\tilde{U}'$ are two disjoint open sets with $\{F\} \in \tilde{U}$ and $\{F'\} \in \tilde{U}'$. If $\pi(\{F\}) = \pi(\{F'\})$, take any neighborhood U of $\pi(\{F\})$ of the above type and let $\tilde{U}$ and $\tilde{U}'$ be the corresponding neighborhoods of $\{F\}$ and $\{F'\}$. Since both are homeomorphic to U, they are either disjoint or identical. The latter case is excluded since it would imply $\{F'\} = \pi^{-1}(\pi(\{F'\})) = \pi^{-1}(\pi(\{F\})) = \{F\}$.

From the above it is already clear that π is a covering of M. Let F be any path in M starting at x. Let $F_s(t) = F(st), s \in I = [0,1]$, then F_s is a path continuously depending on s. F_0 is the one point path at x, and $\{F_0\} = \tilde{x}$ as well as $\pi(\{F_1\}) = F(1)$. To prove that $\tilde{M}$ is pathwise connected, it suffices to prove that $\{F_s\}$ is a continuous function of s in $\tilde{M}$. This is rather obvious, since for s_0 an $\epsilon > 0$ can be chosen so that the end points of F_s are in some of the above described neighborhoods U for $|s - s_0| < \epsilon$. These F_s can be represented as $F'F_{s_0}$ where F' is completely in U. Hence, $\tilde{U}$ is a neighborhood of $\{F_{s_0}\}$ in $\tilde{M}$ which contains all $\{F_s\} = \{F'F_{s_0}\}$ for $|s - s_0| < \epsilon$.

To prove that $\rho(\pi, \tilde{x}) = \rho$, consider the path $I \ni s \mapsto \{F_s\}$ in $\tilde{M}$ which is closed, iff F is closed in M and $[F] \in \rho(\pi, \tilde{x})$. Now $F \in \{F_1\}$, and $\{F_1\} = \{F_0\} = \tilde{x}$, iff F is closed and $[F] \in \rho$.

Finally, let $\rho = \{e\}$ be trivial. Then, $\rho(\pi, \tilde{x}) \approx \pi_1(\tilde{M}, \tilde{x})$ is also trivial, and hence M is simply connected. Since for every simply connected covering $\bar{\pi} : \bar{M} \to M$ the fundamental group $\pi_1(\bar{M})$ is trivial, it follows immediately from the previous uniqueness theorem that $\bar{M}$ and the latter $\tilde{M}$ are equivalent and hence homeomorphic. □

Now, let M be a second countable m-dimensional manifold. Since it is second countable and locally homeomorphic to $\mathbb{R}^m$, it can be covered by a countable number of open sets each of which is homeomorphic to an open ball in $\mathbb{R}^m$. Any loop in M runs through a countable sequence of these open sets, and loops running through the same sequence are obviously homotopy equivalent. Since there is at most a countable number of distinct such sequences, $\pi_1(M, x)$ is countable for every $x \in M$. Consequently, the multiplicity of any covering π of each component of M is at most countable. Hence, the covering space $\tilde{M}$ of any covering of M is second countable. Requiring that the local homeomorphisms of evenly covered open sets are diffeomorphisms defines uniquely a differentiable structure on $\tilde{M}$ which makes π into a smooth covering by a smooth manifold $\tilde{M}$ for which the linear mapping $\pi_*^{\tilde{x}}$ of the tangent spaces is nowhere singular.

If G is a connected Lie group, then, since G is locally homeomorphic to $\mathbb{R}^m$, it is locally pathwise connected and semi-locally 1-connected. From (6.8) it follows that it is also pathwise connected. Hence, it has a universal covering space $\bar{G}$ which has a uniquely defined differentiable structure for which the covering π is smooth and $\pi_*^{\tilde{x}}$ is nowhere singular. In fact, $\bar{G}$ can be provided with a group structure which makes it into the **universal covering group** of G. It remains to establish the group structure of $\bar{G}$.

Let G be further on a connected Lie group, and let D be a discrete subgroup of G, that is, the one point sets of D are mutually disconnected in the topology of G. Consider the quotient space G/D of the left cosets of G with respect to D, that is, of sets $\{dg \mid d \in D\}$ for all $g \in G$, provided with the quotient topology as the finest topology for which the canonical projection $\pi : G \to G/D$ is continuous. Its open sets are the sets U for which $\pi^{-1}(U)$ is open in G. The elements of D form a discrete grid in G, so that there is a neighborhood U of the unit e of G such that the sets $dU, d \in D$ are disjoint. Each of these sets is diffeomorphically projected onto U by π. Hence, π is a covering of the manifold G/D by the covering space G, and G/D and G have the same dimension.

Let F be any path in G from $e = F(0)$ to some element $d = F(1)$ of D, and let $[F]$ be its homotopy class. Let F' be another path from e to d'. Then, $F(1)F'$ is a path from d to dd' which is obtained by a left translation of F' by d. Introduce the product of homotopy classes as $[F'][F] = [F(1)F'F]$ where $F(1)F'F$ is the path F concatenated with the translated path $F(1)F'$. It will be seen that this makes the set of classes $[F]$ into a group. Since the end points of the paths F and F' are in D, the projections $\pi(F)$ and $\pi(F')$ in G/D are loops with base point $\pi(e)$, and the projection $\pi(F(1)F')$ of the translated path $F(1)F'$ is equal to $\pi(F')$. Since π is continuous, homotopy equivalent paths F are projected into homotopy equivalent loops $\pi(F)$ in G/D. It is obvious that the corresponding projections $\pi_*([F]) = [\pi(F)]$ of homotopy classes $[F]$ form the fundamental group $\pi_1(G/D, \pi(e))$. Moreover, $\pi_*([F'][F]) = [\pi(F(1)F'F)] = [\pi(F')][\pi(F)] = \pi_*([F'])\pi_*([F])$, and hence π_* is an isomorphism between the multiplicative set of homotopy classes of paths F from e to elements of D and the fundamental group $\pi_1(G/D, \pi(e))$ of the space G/D. As anticipated, the former set with the introduced multiplication is a group.

In G, the homotopy classes of loops based on e form the fundamental group $\pi_1(G, e)$ which is isomorphically mapped into the subgroup $\rho(\pi, e)$ of $\pi_1(G/D, \pi(e))$ by the covering π. Let $[F_0] \in \pi_1(G, e)$, and let $[F]$ be *any* homotopy class of paths from e into D. Then, since $F_0(1) = e = F_-(1)$, $[F][F_0][F]^{-1} = [FF_0][F_-] = [FF_0F_-] = [F_0]$. ($FF_0F_-$ just moves homotopically the base point of F_0 from e to the end point of F.) Hence, $[F][F_0][F]^{-1} = [F_0]$ for every $[F_0] \in \pi_1(G, e)$ and every $[F]$ with $\pi_*([F]) \in \pi_1(G/D, \pi(e))$. Inversely, if the last relation holds, then $[FF_0F_-] = [F_0]$ which is only possible, if F_0 is closed. In summary, $\rho(\pi, e)$ is the **central normal subgroup** of $\pi_1(G/D, \pi(e))$, that is, the subgroup of all elements $[F_0]$ with $[F][F_0][F]^{-1} = [F_0]$ for all $[F] \in \pi_1(G/D, \pi(e))$. If the end points of paths F from e run through D, then $\pi_*([F]\pi_1(G, e)) = \pi_*([F])\rho(\pi, e)$ runs through the quotient group $\pi_1(G/D, \pi(e))/\rho(\pi, e)$ which latter hence is isomorphic with D. If in particular G is simply connected and hence $\rho(\pi, e)$ is trivial, then $\pi_1(G/D, \pi(e))$ is isomorphic with D.

As a simple example, consider the n-torus group $\mathbb{T}^n$. Let $G = \mathbb{R}^n = \{(x_1, \ldots, x_n) \mid x_i \in \mathbb{R}\}$, and let $L = \mathbb{Z}^n = \{(k_1, \ldots, k_n) \mid k_i \in \mathbb{Z}\}$ be the n-dimensional unit lattice. Let $\mathbb{T}^n = \mathbb{R}^n/L = \{(t_1, \ldots, t_n) \mid t_i = x_i \bmod 1\}$. Since $\mathbb{R}^n$ is simply connected, $\pi_1(\mathbb{T}^n, 0)$ is isomorphic to L : Paths from the origin of $\mathbb{R}^n$ to one of the lattice points correspond to k_i-fold windings around the n non-homotopic circles of $\mathbb{T}^n$. This fact was heuristically already used in Sects. 2.5 and 5.9.

In a sense inverse to the above is the following theorem:

Each connected Lie group G has a simply connected covering space $\bar{G}$ which is again a Lie group and the covering $\pi : \bar{G} \to G$ is a Lie group homomorphism the kernel of which is a discrete subgroup of $\bar{G}$.

Proof It was already seen that G has a simply connected covering space $\bar{G}$. Choose an arbitrary element of $\pi^{-1}(e)$ (where e is the unit of G) to be the unit $\bar{e}$ of $\bar{G}$. Let $\bar{F}_{\bar{g}}$ and $\bar{F}_{\bar{h}}$ be two paths in $\bar{G}$ from $\bar{e}$ to arbitrarily chosen points $\bar{g}$ and $\bar{h}$, respectively. Let $F_g = \pi(\bar{F}_{\bar{g}})$, $F_h = \pi(\bar{F}_{\bar{h}})$, $g = \pi(\bar{g})$ and $h = \pi(\bar{h})$. Let $F' = gF_hF_g$ be the path in G obtained by concatenation of F_g and the g-translated image of F_h and let $\bar{F}'$ be a path in $\bar{G}$ starting at $\bar{e}$ and being projected by π onto F'. Its end point $\bar{k}$ depends only on $\bar{g}$ and $\bar{h}$ and not on the particular paths chosen. Indeed, let $\bar{F}'_{\bar{g}}$ and $\bar{F}'_{\bar{h}}$ alternatively chosen paths, let F'_g and F'_h be their projections, and let $\bar{F}''$ be a path starting at $\bar{e}$ and being projected onto $F'' = gF'_hF'_g$. Since $[F'_{g}\ F_g], [F'_{h-}F_h] \in \rho(\pi, \bar{e})$ and $(FF')_- = F'_-F_-$, it follows that $F''_-F' = (gF'_hF'_g)_-(gF_hF_g) = F'_{g-}gF'_{h-}gF_hF_g \cong (F'_{g-}F_g)(F'_{h-}F_h)$, and hence $[F''_-F'] \in \rho(\pi, \bar{e})$ with the consequence that F' and F'', both starting at $\bar{e}$, have the same end point $\bar{k}$. On this basis, the product $\bar{g}\bar{h} = \bar{k}$ in $\bar{G}$ is correctly defined, and by considering corresponding paths associativity of this product, unit property of $\bar{e}$ and the existence of $\bar{g}^{-1}$ is demonstrated. Furthermore, $\pi(\bar{g}\bar{h}) = \pi(\bar{g})\pi(\bar{h})$ was underlying the construction of the product. Hence, $\bar{g}$ is a group and the covering π is a group homomorphism.

It remains to show that the product $\bar{g}\bar{h}^{-1}$ is smooth in $\bar{G}$. This is straightforwardly demonstrated with the help of paths $F_s(t) = F(st)$ smoothly depending on s in G and using the fact that π is a local diffeomorphism.

Finally, since π is a covering, there is a neighborhood U of e in G the preimage of which consists of disjoint open sets of $\bar{G}$ homeomorphic with U. In particular, the preimage of e which is the kernel of the homomorphism π is discrete. □

Hence, for every connected Lie group G there exists a simply connected Lie group $\bar{G}$ which is a covering of G. The natural question arises, whether and in which sense $\bar{G}$ is unambiguously determined. It was already demonstrated that simply connected coverings of G are diffeomorphic as manifolds. That they are also isomorphic as groups follows from the connection between the Lie groups and their Lie algebras.

Let G and H be connected Lie groups, and let $F : G \to H$ be a Lie group homomorphism. Then F is a covering, iff $F_* : \mathfrak{g} \to \mathfrak{h}$ is a Lie algebra isomorphism.

Proof Suppose that F is a covering. Then F_* must be injective. Otherwise $F_* : T_g(G) \to T_{F(g)}(H)$ would have a non-trivial kernel at every point g. These kernels form an involutive distribution having an integral manifold (Frobenius theorem) which is mapped into a point of H by F, and F could not be a local homeomorphism. F_* must also be surjective, since otherwise F would define a proper submanifold of H. Being an injective and surjective homomorphism, F_* is a Lie algebra isomorphism.

Suppose now that F_* is a Lie algebra isomorphism. Then, by the inverse function theorem, p. 74 f, F is everywhere a local diffeomorphism, and, since $F(e) = e$, $F(G)$ contains a neighborhood of the unit $e \in H$ and hence, by (6.8), $F(G) = H$. It remains to show that a neighborhood of every point of H is evenly covered by G. Observe, since F is a local homeomorphism, that $F^{-1}(e) = \operatorname{Ker} F = K$ is a discrete normal subgroup of G. Therefore, there exists a small enough neighborhood U of $e \in G$ such that $(U^{-1}U) \cap K = \{e\}$. Using the continuous group operations it is not difficult to show that $F(U)$ is a neighborhood of $e \in H$ evenly covered by F. This even cover may be translated to every point of H. □

Let G and H be Lie groups, and let G be simply connected. Let $\tilde{F} : \mathfrak{g} \to \mathfrak{h}$ be a homomorphism. Then there exists a unique homomorphism $F : G \to H$ such that $F_* = \tilde{F}$. In particular, if simply connected Lie groups have isomorphic Lie algebras, then they are isomorphic.

Uniqueness was proved at the end of Sect. 6.2. The proof of existence which uses considerations similar to those of Sect. 6.3 is skipped, see for instance [1, p. 101].

Now, for any Lie group G the Lie algebra $\mathfrak{g}$ is isomorphic to the tangent space $T_e(G)$ (p. 175). If G and H are diffeomorphic (as manifolds) Lie groups and g and h are elements mapped to each other by the diffeomorphism, then the tangent spaces $T_g(G)$ and $T_h(H)$ are isomorphic (inverse function theorem, p. 74 f). Hence, from the above it follows that connected Lie groups have (up to isomorphism) uniquely defined simply connected covering groups, which are called **universal covering groups**. Just to mention an important simple example from physics: the group of transformations of spinors $SU(2)$ is the universal covering group of the group of rotations in 3-space $SO(3)$. (See Sect. 6.6 for details.)

6.5 The Exponential Mapping

Recall the formal Taylor expansion of a real function f of a real variable, analytic in a neighborhood of x:

$$f(x+t) = e^{t\frac{d}{dx}} f(x) = f(x) + t\frac{df}{dx}(x) + \frac{t^2}{2!}\frac{d^2f}{dx^2}(x) + \cdots \qquad (6.9)$$

The real line $G_1 = \mathbb{R}$ is a simple case of a simply connected Lie group with respect to addition as group operation. Its Lie algebra consists of the one-dimensional vector fields $t(d/dx)$, $t \in \mathbb{R}$, that is $\mathfrak{g}_1 = \mathbb{R}$.

Consider *any* Lie group G and its Lie algebra $\mathfrak{g}$. Let $X \in \mathfrak{g}$ be a left invariant vector field on G, then $\exp_*(X) : \mathfrak{g}_1 = \mathbb{R} \to \mathfrak{g} : t \mapsto tX$ is a Lie algebra homomorphism. (The notation exp will become evident soon.) According to the end of last section there is a unique Lie group homomorphism $\exp(X) : G_1 = \mathbb{R} \to G :$

$t \mapsto \exp(X)_t$ for every $X \in \mathfrak{g}$ from $\mathbb{R}$ onto a **1-parameter subgroup** of G. On the other hand, the tangent vector fields (d/dx) and X are $\exp(X)$-related (see p. 78) and hence, according to the Frobenius theorem, there is a unique integral curve of X in the manifold G for which $\exp(X)_0 = e$, the latter since $\exp(X)$ is a Lie group homomorphism. Moreover, since X is left invariant, there are unique integral curves of X for which $l_g \circ \exp(X)_0 = g$ for every $g \in G$, in particular for every $g \in \exp(X)_{\mathbb{R}} = \{\exp(X)_t \,|\, t \in \mathbb{R}\}$. Thus, *the 1-parameter subgroup* $\exp(X)_{\mathbb{R}}$ *consists of the integral curve of* X *through* $e \in G$, *and the left invariant tangent vector fields on a Lie group are always complete.* (See p. 81.)

Note that this is a global statement. Locally, one could introduce a local coordinate system in G like (3.33) and argue with (6.9). Now, with the help of left translations one easily finds globally, that is for *all* $t, t_1, t_2 \in \mathbb{R}$,

$$\begin{aligned} \exp(tX) &= \exp(tX)_1 = \exp(X)_t, \\ \exp((t_1 + t_2)X) &= \exp(t_1 X)\exp(t_2 X), \\ \exp(-tX) &= (\exp(tX))^{-1}. \end{aligned} \tag{6.10}$$

So far, $\exp(X)$ for every fixed $X \in \mathfrak{g}$ was a mapping $t \mapsto \exp(X)_t$ from $\mathbb{R}$ to G. With the relations (6.10) one may put $\exp(X)_1 = \exp(X)$ and consider exp as a mapping from $\mathfrak{g}$ to G.

As a mapping from $\mathfrak{g}$ to G, exp maps a sufficiently small neighborhood u of the origin of $\mathfrak{g}$ homeomorphic to a neighborhood U of e in G, and, according to (6.8) G^e as a whole is obtained by all kinds of products of factors out of U. However, the mapping $\exp : \mathfrak{g} \to G^e$ need not be onto (compare the exercise on p. 200) nor need it be one–one (compare the mapping $\exp : \mathbb{R} = \mathfrak{s}^1 \to S^1 : t \mapsto e^{i2\pi t}$). It can be shown that exp is a smooth map. Moreover, it can be shown for every Lie group that there exists a unique complete analytic atlas (that is, all transition functions between charts are analytic) so that the group operations and the mapping exp are analytic. (See for instance [2].)

With the help of the exponential mapping the interrelation between a Lie group and its Lie algebra can further be explored. If $F : G \to H$ is a Lie group homomorphism, then the following diagram is commutative:

$$\begin{array}{ccc} \mathfrak{g} & \xrightarrow{F_*} & \mathfrak{h} \\ {\scriptstyle \exp}\downarrow & & \downarrow{\scriptstyle \exp} \\ G & \xrightarrow{F} & H \end{array} \tag{6.11}$$

Indeed,

$$F(\exp(X)) = \exp(F_*(X)). \tag{6.12}$$

Since F is a homomorphism, $F(\exp(X)_{\mathbb{R}})$ is a 1-parameter subgroup of H. On the other hand, $\exp(F_*(X))_{\mathbb{R}}$ is an integral curve of $F_*(X)$ in H, and both are uniquely defined by $F_*(X)$ at $e \in H$. Left translation from e to $g \in H$ and left invariance of $F_*(X)$ prove their equality.

If, in particular, $F(G)$ is a Lie subgroup of H and $X \in F_*(\mathfrak{g})$, then $\exp(tX) \in F(G)$ for all $t \in \mathbb{R}$. If $\exp(tX) \in F(G)$ for some interval of values t, then $\exp(t_0X) = F(g)$ for an inner value t_0 of that interval and some $g \in G$, and $\exp((t-t_0)X) = \exp(tX)(F(g))^{-1} \in F(G)$ is a local 1-parameter group through e in $F(G)$, hence $(t-t_0)X \in F_*(\mathfrak{g})$ implying $X \in F_*(\mathfrak{g})$.

The following basic results can be proved with the help of the exponential mapping [1]:

Let H be a Lie group, and let G be an algebraic subgroup of H closed in the topology of H. Then there is a unique complete atlas of G making it into a Lie subgroup of H.

Let $F : G \to H$ be a Lie group homomorphism, and let $K = \operatorname{Ker} F, \mathfrak{k} = \operatorname{Ker} F_*$. Then K is a closed Lie subgroup of G with Lie algebra $\mathfrak{k}$.

6.6 The General Linear Group *Gl(n,K)*

A most important case of a Lie algebra is formed by the n^2-dimensional real vector space of all $n \times n$-matrices $A, B, \ldots$ with the multiplication (commutator)

$$[A, B] = AB - BA \tag{6.13}$$

where AB is the ordinary matrix multiplication. This Lie algebra is called the **general linear algebra** $\mathfrak{gl}(n, \mathbb{R})$.

A base $X_{(ij)}$, $i,j = 1, \ldots, n$ may be introduced consisting of matrices $X_{(ij)}$ having unity as matrix element of the ith row and jth column and zeros at all other entries. Then, obviously

$$[X_{(ij)}, X_{(kl)}] = \delta^j_k X_{(il)} - \delta^i_l X_{(kj)}, \tag{6.14}$$

and comparison with (6.1) yields the structure constants

$$c^{(pq)}_{(ij)(kl)} = \delta^p_i \delta^q_l \delta^j_k - \delta^p_k \delta^q_j \delta^i_l. \tag{6.15}$$

Any matrix A can be expanded in this base as $A = \sum_{i,j=1}^n A^i_j X_{(ij)}$ where the vector components of A in the vector space $\mathfrak{gl}(n, \mathbb{R})$ are the ordinary matrix elements A^i_j in row i and column j. As a topological vector space, $\mathfrak{gl}(n, \mathbb{R})$ is homeomorphic to $\mathbb{R}^{n^2}$.

The complex general linear algebra $\mathfrak{gl}(n, \mathbb{C})$ is obtained just by replacing the real components A^i_j with complex ones. Hence, it has the same base and structure constants as $\mathfrak{gl}(n, \mathbb{R})$, but is homeomorphic to $\mathbb{C}^{n^2} \sim \mathbb{R}^{2n^2}$. The following consideration is the same for both cases.

Let $|A| = \max_{ij} |A^i_j|$, then it is easily seen that $|A^m| \leq n^{m-1}|A|^m$ holds for the mth power of A. Denote the $n \times n$ unit matrix by $\mathbf{1}$ and consider the series

$$\exp(A) = \mathbf{1} + A + \frac{A^2}{2!} + \cdots + \frac{A^m}{m!} + \cdots = \sum_{m=0}^{\infty} \frac{A^m}{m!}. \tag{6.16}$$

Its partial sums are $n \times n$-matrices, and for all A with $|A| < c$ it converges uniformly for any fixed positive constant c, since the absolute value of each matrix element of the mth item is bounded by $n^{m-1}c^m/m!$ and $\sum (nc)^m/m! = e^{nc}$ converges. Since the matrix multiplication is continuous in $\mathfrak{gl}(n,K)$, $K = \mathbb{R}$ or $\mathbb{C}$,

$$B\left(\sum_{m=0}^{\infty} \frac{A^m}{m!}\right) = \sum_{m=0}^{\infty} B\frac{A^m}{m!} \tag{6.17}$$

and accordingly for the right multiplication, and hence

$$Be^A B^{-1} = e^{(BAB^{-1})} \tag{6.18}$$

for any $B \in \mathfrak{gl}(n,K)$. Now, for any matrix A there is a matrix B such that BAB^{-1} is upper-triangular, and the product of two upper-triangular matrices is again an upper-triangular matrix. Hence, the right hand side of (6.18) for such a B is an upper-triangular matrix, and, if $a_1, \ldots, a_n$ are the diagonal elements of BAB^{-1}, then $e^{a_1}, \ldots, e^{a_n}$ are the diagonal elements of that right hand side. In particular, no matter what the numbers a_i are,

$$\det e^A = \det(Be^A B^{-1}) = \det e^{(BAB^{-1})} = \prod_{i=1}^{n} e^{a_i} = e^{\mathrm{tr}(BAB^{-1})} = e^{\mathrm{tr}\,A} \neq 0. \tag{6.19}$$

Here, the simple matrix rules $\det(AB) = \det A \det B$ and $\mathrm{tr}(ABC) = \mathrm{tr}(CAB)$ were used.

At the beginning of this chapter it was already mentioned that the **general linear group**

$$Gl(n,K) = \{A \,|\, \det A \neq 0\} \tag{6.20}$$

is a Lie group. Consider for *any* matrix A' the 1-parameter subgroup $t \mapsto e^{tA'}, t \in \mathbb{R}$. Its tangent vector at $t = 0$, that is, for $e^{tA'} = \mathbf{1}$, is A'. It is simply obtained by term wise differentiating the uniformly converging power series for $e^{tA'}$ with respect to t. This proves

$$\exp(A') = e^{A'}, \quad e^{\mathfrak{gl}(n,K)} \text{ generates } (Gl(n,K))^{\mathbf{1}}. \tag{6.21}$$

The general linear algebra is the Lie algebra of the general linear group, the exponential mapping coincides with the ordinary matrix exponentiation and yields the component containing the unity of the general linear group. $Gl(n,K)$ is pathwise connected for $K = \mathbb{C}$. It is not difficult to see that a path from any non-singular matrix A_0 to any other non-singular matrix A_1 in $\mathbb{C}^n$ can always be infinitesimally deformed into a path avoiding $\det A = 0$ (for instance encircling $\det A = 0$ in the complex plane always in the positive sense). This does not hold for $K = \mathbb{R}$ for which there are two components with $\det A \gtrless 0$. $Gl(n, \mathbb{C})$ is not simply connected, which is most easily seen for $Gl(1, \mathbb{C}) = \mathbb{C} \setminus 0$. However, $(Gl(n, \mathbb{R}))^{\mathbf{1}}$ is simply connected: Let A_0 and A_1 be two arbitrary matrices both with positive (negative) determinant. There are paths $B_0(t), B_1(t), t \in [0,1], B_i(0) = \mathbf{1}, \det B_i(t) = 1$ which continuously transform $B_i(t)A_iB_i^{-1}(t)$ into upper triangular matrices without changing the determinant. Having determinants of the same sign, the signs of diagonal elements of both triangular matrices can only differ in an even number of them. Group neighboring diagonal elements not having the same signs into pairs and consider a (2×2)-matrix. Put $a_{ii}(t) = ta_{ii}$, $a_{12}(t)a_{21}(t) = (t^2 - 1)a_{11}a_{22}$, $a_{21}(1) = a_{21}(-1) = 0$ and let t run from 1 to -1. It reverses the signs of diagonal elements without changing the determinant on a path from upper triangular form to upper triangular form. It does also not change the determinant of the full matrix which is up to the considered (2×2)-block upper triangular. (This can directly be inferred from the Laplace expansion with respect to the two rows containing the (2×2)-block.) In this way successively all diagonal elements of $B_1(1)A_1B_1^{-1}(1)$ differing in sign from those of $B_0(1)A_0B_0^{-1}(1)$ can be moved together and then sign reverted. A further continuous path brings the absolute values of the diagonal elements into coincidence without changing signs. Concatenation of all changes completes the path from A_0 to A_1 in $Gl(n, \mathbb{R})$. Since any path from $\det A > 0$ to $\det A < 0$ must unavoidably cross $\det A = 0$, this proves that the polynomial condition $\det A = 0$ defines a smooth hypersurface in $\mathbb{R}^{n^2}$ dividing it into just two connected components. Consider *any* loop in the component with $\det A > 0$, and take the point on it with minimal $\det A$ as base point of the loop. (The minimum exists since a loop is compact.) Transform every point of the loop by the above transformation into the base point. This transformation may be chosen as a continuous function of the points of the loop and keeps $\det A$ above the value at the base point, contracting the loop into the base point. Hence, $(Gl(n, \mathbb{R}))^{\mathbf{1}}$ is simply connected.

Any $n \times n$-matrix may be considered as a linear mapping of the n-dimensional vector space K^n over the field K into itself (endomorphisms $\mathrm{End}(K^n)$). Likewise, a non-singular matrix may be considered as the transformation matrix of an automorphism of K^n. Hence, one has also

$$\exp : \mathrm{End}(K^n) \to \mathrm{Aut}(K^n). \tag{6.22}$$

Again, this mapping need not be surjective, for instance, if $\mathrm{Aut}(K^n)$ is not connected. If G is *any* Lie group and $F : G \to \mathrm{Aut}(K^n)$ is a representation of G, and if $X \in \mathfrak{g}$, then it follows immediately from (6.11) that

$$F(\exp X) = \exp(F_*(X)) = \mathbf{1} + F_*(X) + \frac{F_*(X)^2}{2!} + \cdots \tag{6.23}$$

where $F_*(X) \in \mathfrak{gl}(n, K)$ is a matrix obtained from $[dF(\exp(tX))/dt]_{t=0}$.

It has been shown (theorem by Ado) that every finite-dimensional Lie algebra is isomorphic to a subalgebra of $\mathfrak{gl}(n,\mathbb{R})$ for some n. (That means that also $\mathfrak{gl}(n',\mathbb{C})$ is isomorphic to such a subalgebra, of course with $n > n'$.) By exponentiation, every subalgebra of $\mathfrak{gl}(n,\mathbb{R})$ generates a connected Lie group to which this subalgebra is the corresponding Lie algebra. Every such Lie group has a uniquely (up to Lie group isomorphism) determined simply connected covering group. This provides a one–one correspondence between Lie subalgebras of $\mathfrak{gl}(n,\mathbb{R})$ and simply connected Lie groups.

(A complete classification of all Lie algebras has not yet been achieved, to say nothing about a complete classification of all Lie groups.)

Some important Lie subgroups of $Gl(n,K)$ are shortly considered:

The **special linear group**

$$Sl(n,K) = \{A \mid \det A = 1\}, \quad n > 1, \tag{6.24}$$

is a closed connected Lie subgroup of $Gl(n,K)$ and has its Lie algebra

$$\mathfrak{sl}(n,K) = \{A' \mid \operatorname{tr} A' = 0\}. \tag{6.25}$$

Indeed, if $\operatorname{tr} A' = 0$, then $\det\exp(A') = 1$ follows directly from (6.19). Conversely, $\det\exp(A') = 1$ implies $\operatorname{tr} A' = (2\pi i)k$ with some integer k, and only in the case $k = 0$ the A' may form a vector space over the field K. This trace condition reduces the number of independent diagonal elements of A by one, hence the dimension of $\mathfrak{sl}(n,K)$ and of $Sl(n,K)$ is equal to $n^2 - 1$, in the case of complex algebra for $K = \mathbb{C}$, and $\mathfrak{sl}(n,\mathbb{C})$ and $Sl(n,\mathbb{C})$ have the dimension $2n^2 - 2$ in real algebra. $Sl(n,K)$ is closed since $\det A = 1$ is a polynomial equation in K^n. Let A_0 and A_1 be two arbitrary elements of $Sl(n,\mathbb{C})$. Since $Gl(n,\mathbb{C})$ is connected, there is a path $B(t)$ in $Gl(n,\mathbb{C})$ connecting $A_0 = B(0)$ with $A_1 = B(1)$. For every t there is a path $D(u,t)B(t)$, where $D(u,t) = \lambda(u,t)\mathbf{1}$ and $\lambda(u,t)$ is a continuous non-zero complex scalar function with $\lambda(0,t) = 1$, $\lambda(1,t) = (\det B(t))^{-1}$ (It is continuous in *both* variables u and t and may for instance be chosen always to go around the origin of the complex plane in the positive sense). Now, $D(u,t)B(t)$ continuously deforms the path $B(t)$ into a path $A(t) \in Gl(n,\mathbb{C})$ (in the relative topology from $\mathbb{C}^{n^2}$) from A_0 to A_1, where now $A(t)$ is in $Sl(n,\mathbb{C})$. Hence, $Sl(n,\mathbb{C})$ is connected. In fact it is even simply connected. An analogous argument shows that $Sl(n,\mathbb{R}) = Sl(n,\mathbb{C}) \cap Gl(n,\mathbb{R})$ is a connected subspace of $(Gl(n,\mathbb{R}))^{\mathbf{1}}$. ($Sl(1,K) = \{1\}$ is trivial.)

The **unitary group**

$$U(n) = \{A \mid A^\dagger = A^{-1}\} \tag{6.26}$$

is a connected compact (closed bounded) Lie subgroup of $Gl(n,\mathbb{C})$. $A^\dagger$ means the Hermitian conjugate of the matrix A. Indeed, from $1 = (AA^\dagger)^i_i = \sum_k |A^i_k|^2$ it follows that $|A^i_k| \le 1$, and hence $U(n)$ is bounded in $\mathbb{C}^{n^2}$. It is closed since the condition $A^\dagger = A^{-1}$ implies $\det A \neq 0$, and hence this former condition can be expressed as a set of polynomial equations in the matrix elements. One further has

$$\exp(A'^{\dagger}) = \mathbf{1} + A'^{\dagger} + \frac{A'^{\dagger 2}}{2!} + \cdots = (\mathbf{1} + A' + \frac{A'^{2}}{2!} + \cdots)^{\dagger}$$
$$= (\exp(A'))^{\dagger} = (\exp(A'))^{-1} = \exp(-A')$$

from which chain it is seen that the commutator algebra of skew-Hermitian matrices

$$\mathfrak{u}(n) = \{A' \,|\, A'^{\dagger} + A' = 0\} \tag{6.27}$$

is the Lie algebra of $U(n)$. Since the matrices $A' \in \mathfrak{u}(n)$ necessarily have a vanishing real part of the diagonal matrix elements, it is an algebra over $K = \mathbb{R}$. Although the matrices themselves may be complex, $U(n)$ is a real Lie group and $\mathfrak{u}(n)$ is a real Lie algebra, both with real dimension n^2. $U(1)$ is the unit circle in the complex plane (which is not simply connected). The angle ϕ may be taken as its real coordinate.

Let $D_{p,q}$ be the diagonal matrix with the first p diagonal entries equal to 1 and the last q diagonal entries equal to -1, $p, q \geq 1$, $p + q = n$. The **generalized unitary group**

$$U(p,q) = \{A \,|\, D_{p,q} A^{\dagger} D_{p,q} = A^{-1}\} \tag{6.28}$$

is a real subgroup of $Gl(n, \mathbb{C})$ with the real Lie algebra

$$\mathfrak{u}(p,q) = \{A' \,|\, D_{p,q} A'^{\dagger} D_{p,q} + A' = 0\}. \tag{6.29}$$

$U(p,q)$ is not compact as the example

$$\begin{pmatrix} \cosh\theta & \sinh\theta \\ \sinh\theta & \cosh\theta \end{pmatrix} \in U(1,1), \quad \theta \in \mathbb{R} \tag{6.30}$$

shows. The real dimension is again n^2.

The **orthogonal group**

$$O(n,K) = \{A \,|\, A^t = A^{-1}\} \tag{6.31}$$

is a closed Lie subgroup of $Gl(n,K)$. Here, A^t means the transposed of the matrix A. For $K = \mathbb{R}$, it is compact by the same argument as in the unitary case. However, since $1 = \det(AA^{-1}) = \det A \det A^t = (\det A)^2$, it consists of the two components with $\det A = \pm 1$ and is not connected. A chain of equations analogous to the unitary case shows that the commutator algebra of skew-symmetric matrices

$$\mathfrak{o}(n,K) = \{A' \,|\, A'^t + A' = 0\} \tag{6.32}$$

is the Lie algebra of $O(n,K)$. Since all matrices of $\mathfrak{o}(n,\mathbb{C})$ have zero diagonal elements, complex coefficients will not violate the skew-symmetry condition; $\mathfrak{o}(n,\mathbb{C})$ is a complex Lie algebra and $O(n,\mathbb{C})$ a complex Lie group, both with complex dimension $n(n-1)/2$ because of the vanishing diagonal of $A' \in \mathfrak{o}(n,\mathbb{C})$

and the skew-symmetry of the off-diagonal elements. The corresponding real dimension is $n(n-1)$. $O(n,\mathbb{R}) = O(n) = U(n) \cap Gl(n,\mathbb{R})$ and $\mathfrak{o}(n,\mathbb{R}) = \mathfrak{o}(n) = \mathfrak{u}(n) \cap \mathfrak{gl}(n,\mathbb{R})$ consist of real matrices and are of real dimension $n(n-1)/2$. ($O(1,K) = O(1)$ is discrete and consists of the two elements 1 and -1; hence its Lie algebra is trivial, $\mathfrak{o}(1) = \{0\}$.)

The **generalized orthogonal group** (again $p,q \geq 1,\ p+q=n$)

$$O(p,q) = \{A \mid D_{p,q}A^t D_{p,q} = A^{-1}\} \tag{6.33}$$

is a non-compact non-connected Lie subgroup of $Gl(n,\mathbb{R})$ with the Lie algebra

$$\mathfrak{o}(p,q) = \{A' \mid D_{p,q}A'^t D_{p,q} + A' = 0\}. \tag{6.34}$$

It is not difficult to see that $O(p,q)$ has the four components $(O(p,q)^{\mathbf{1}} = O^+(p,q),\ -\mathbf{1}O^+(p,q),\ D_{p,q}O^+(p,q)$ and $-D_{p,q}O^+(p,q))$. The matrix (6.30) is obviously also an element of $O(1,1)$. The real dimension of $O(p,q)$ is again $n(n-1)/2$.

The **special unitary group**

$$SU(n) = U(n) \cap Sl(n,\mathbb{C}) = \{A \mid A^\dagger = A^{-1}, \det A = 1\}, \quad n > 1, \tag{6.35}$$

is a simply connected compact real Lie subgroup of both $U(n)$ and $Sl(n,\mathbb{C})$ with the Lie algebra

$$\mathfrak{su}(n) = \mathfrak{u}(n) \cap \mathfrak{sl}(n,\mathbb{C}) = \{A' \mid A'^\dagger + A' = 0,\ \mathrm{tr} A' = 0\}. \tag{6.36}$$

Its (real) dimension is $n^2 - 1$.

The generalized special unitary group

$$SU(p,q) = U(p,q) \cap Sl(n\mathbb{C}), \quad p,q \geq 1,\ p+q=n, \tag{6.37}$$

has also dimension $n^2 - 1$, but is not compact. Again, (6.30) is also an element of $SU(1,1)$.

The **special orthogonal group**

$$SO(n,K) = O(n,K) \cap Sl(n,K) = \{A \mid A^t = A^{-1}, \det A = 1\}, \quad n > 1, \tag{6.38}$$

is a connected Lie subgroup of both $O(n,K)$ and $Sl(n,K)$ with the Lie algebra

$$\mathfrak{so}(n,K) = \mathfrak{o}(n,K), \tag{6.39}$$

since the skew-symmetry implies a vanishing diagonal and hence tracelessness. $SO(n,\mathbb{C})$ is not compact and has complex dimension $n(n-1)/2$ and real dimension $n(n-1)$. $SO(n,\mathbb{R}) = SO(n)$ is compact and has real dimension $n(n-1)/2$.

The generalized special orthogonal group

$$SO(p,q) = O(p,q) \cap Sl(n,\mathbb{R}), \quad p,q \geq 1,\ p+q=n, \tag{6.40}$$

has also real dimension $n(n-1)/2$. It is again not compact, and (6.30) is also an element of $SO(1,1)$.

Finally, let

$$J_n = \begin{pmatrix} 0 & \mathbf{1}_n \\ -\mathbf{1}_n & 0 \end{pmatrix} \tag{6.41}$$

be the matrix which replaces the first n coordinates of K^{2n} with the second n coordinates and the second ones with the negative first ones. (For $\mathbb{R}^2$ it just rotates by $-\pi/2$.) The **symplectic group**

$$Sp(2n) = \{A \,|\, J_n A^t J_n = A^{-1}, A^\dagger = A^{-1}\}, \quad n > 1, \tag{6.42}$$

is a simply connected compact real Lie subgroup of $U(2n)$ with the Lie algebra

$$\mathfrak{sp}(2n) = \{A' \,|\, J_n A'^t J_n + A' = 0, A'^\dagger + A' = 0\}. \tag{6.43}$$

Like $U(n)$ it is a real group and algebra, although the elements are complex. It is a simple exercise to see that its (real) dimension is $n(2n+1)$. The elements A' of the algebra $\mathfrak{sp}(2n)$ have two skew-Hermitian $n \times n$ diagonal blocks being the negative transposed of each other and two symmetric $n \times n$ off-diagonal blocks being the skew-Hermitian conjugate of each other. $Sp(2) = SU(2)$.

The **symplectic *K*-group**

$$Sp(2n, K) = \{A \,|\, J_n A^t J_n = A^{-1}\}, \quad n > 1, \tag{6.44}$$

is a connected (but non-compact and not simply connected) Lie subgroup of $Gl(2n, K)$ with the Lie algebra

$$\mathfrak{sp}(2n, K) = \{A' \,|\, J_n A'^t J_n + A' = 0\}. \tag{6.45}$$

The K-matrices A' consist of two $n \times n$ diagonal blocks being the negative transposed of each other and two independent symmetric off-diagonal $n \times n$ blocks. The K-groups and K-algebras have the K-dimension $n(2n+1)$. $Sp(2, K) = Sl(2, K)$.

The Lie groups $SU(n), n \geq 2, Sp(2n), n \geq 2, SO(n), n \geq 7$, are compact simply connected and as such universal covering groups corresponding to their respective Lie algebras. If the Lie algebra $\mathfrak{g}$ of a Lie group G is isomorphic to one of the algebras $\mathfrak{su}(n), \mathfrak{sp}(2n), \mathfrak{o}(n)$ with integers n from above, then the corresponding group $SU(n), Sp(2n)$ or $SO(n)$, respectively, is the universal covering group of G. ($\mathfrak{so}(3) \approx \mathfrak{sp}(2) \approx \mathfrak{su}(2), \mathfrak{so}(4) \approx \mathfrak{su}(2) \oplus \mathfrak{su}(2), \mathfrak{so}(5) \approx \mathfrak{sp}(4), \mathfrak{so}(6) \approx \mathfrak{su}(4)$.) All *compact, simply connected, simple* Lie groups (Lie groups having simple Lie algebras, see Compendium C.4) were classified by W. Killing and H. Cartan (which leads in addition to the just mentioned classical groups to the five so-called exceptional groups E_6, E_7, E_8, F_4, G_2, see again Compendium C.4).

The relevance of the general linear group and its subgroups lies in the fact that they may be understood as transformation groups of an n-dimensional vector pace K^n over the field K. If $\boldsymbol{x} \in K^n$ with coordinates x^i, $i = 1, \ldots, n$, then $A\boldsymbol{x} \in K^n$ is the transformed vector with coordinates $A^i_j x^j$. The composition of two transformations, that is their subsequent performance, corresponds to matrix multiplication and hence to the group operation. Hence, any set of transformations closed with respect to composition is a group. For instance, $U(n)$ is the group of unitary transformations in the n-dimensional unitary space leaving the scalar product invariant. These are the unitary 'rotations' of the group $SU(n)$ as well as reflections from coordinate hyperplanes and their combinations. Accordingly, $O(n,\mathbb{R})$ are the transformations of the n-dimensional Euclidean space leaving the scalar product invariant. The group of affine motions in the Euclidean space is the **semi-direct product** $E(n) = O(n,\mathbb{R}) \rtimes \mathbb{R}^n \subset Gl(n+1,\mathbb{R})$ consisting of the matrices $(A, \boldsymbol{x}), A \in O(n,\mathbb{R}), \boldsymbol{x} \in \mathbb{R}^n$ as mentioned in the introduction to this chapter. The Lorentz group is the group $O(1,3)$ consisting of four components obtained by time inversion, spatial reflection and their composition, and $O^+(1,3)$ is the proper orthochronous Lorentz group while $O(1,3) \rtimes \mathbb{R}^4 \subset Gl(5,\mathbb{R})$ is the Poincaré group of time and space translations and Lorentz transformations. Finally, the group $Sp(2n, K)$ leaves a symplectic form, like (4.52) for $K = \mathbb{R}$, on the space K^{2n} invariant. Hence, $\mathfrak{sp}(2n,\mathbb{R})$ contains the Jacobi matrices of canonical transformations (in phase space) of Hamilton mechanics. For more details see for instance [3].

6.7 Example from Physics: The Lorentz Group

Two points in flat space–time (absence of a gravitational field) which may be connected by a light signal obey the condition

$$(ct)^2 - \boldsymbol{x}^2 = (x^0)^2 - (x^1)^2 - (x^2)^2 - (x^3)^2 = x^\mu (D_{1,3})_{\mu\nu} x^\nu = 0, \tag{6.46}$$

where c is the velocity of light. A transformation $x^\mu \to x'^\mu = L^\mu_\nu x^\nu$, which leaves the velocity of light invariant, must obey the condition $x'^\mu (D_{1,3})_{\mu\nu} x'^\nu = 0$, while

$$0 = x'^\mu (D_{1,3})_{\mu\nu} x'^\nu = L^\mu_\kappa x^\kappa (D_{1,3})_{\mu\nu} L^\nu_\lambda x^\lambda = x^\kappa (L^t D_{1,3} L)_{\kappa\lambda} x^\lambda. \tag{6.47}$$

In order that for all x obeying (6.46) also (6.47) follows and vice versa, $L^t D_{1,3} L = D_{1,3}$ or equivalently $D_{1,3} L^t D_{1,3} = L^{-1}$ must hold, because a real quadratic form that has zeros is uniquely determined by all its zeros. Hence, $L \in O(1,3)$, and the classical Lorentz group is the generalized orthogonal group $O(1,3)$.

The group $O(1,3)$ obviously contains the element

$$B(\theta, \boldsymbol{e}_1) = \begin{pmatrix} U(1,1) & 0 \\ 0 & \mathbf{1}_2 \end{pmatrix}, \quad U(1,1) = \begin{pmatrix} \cosh\theta & \sinh\theta \\ \sinh\theta & \cosh\theta \end{pmatrix}, \tag{6.48}$$

with $U(1,1)$ as in (6.30). In order to reveal the physical meaning of θ, consider first the limit $\theta \to 0$. In lowest order, $x'^0 = ct' = ct + \theta x^1$, $x'^1 = \theta ct + x^1$. In the original reference system, the origin of the primed x'^1-axis is described by $x^1 = -\theta ct$, hence it moves with the velocity $v = -\theta c$, measured in the original system in its e_1-direction, while $t' = t + vx^1/c^2 \approx t$. In the limit $|\theta| \to |v/c| \to 0$ the Galilei transformation is obtained. For a general θ, the origin of the primed system $0 = x'^1 = ct \sinh\theta + x^1 \cosh\theta$ moves with the velocity

$$\tanh\theta = -v/c \tag{6.49}$$

along the e_1-axis, and this relation implies $\cosh\theta = 1/\sqrt{1-(v^2/c^2)}$, $\sinh\theta = -(v/c)/\sqrt{1-(v^2/c^2)}$. From $|\tanh\theta| < 1$ the restriction $|v/c| < 1$ follows. θ is called the *rapidity parameter*.

Experimentally, the speed of light c is in all reference systems moving with constant speed relative to each other the same. Hence, the Lorentz transformations $L \in O(1,3)$ describe physically correctly the transformation of space and time from one reference system to another one moving relative to the first with a constant speed v.

A Lorentz transformation to a system moving in any direction e in 3-space is obtained as

$$B(\theta, e) = \begin{pmatrix} 1 & 0 \\ 0 & R_3 \end{pmatrix} B(\theta, e_1) \begin{pmatrix} 1 & 0 \\ 0 & R_3^t \end{pmatrix} = \begin{pmatrix} \cosh\theta & -e^t \sinh\theta - e \\ -e \sinh\theta & \mathbf{1}_3 + ee^t(\cosh\theta - 1) \end{pmatrix}, \tag{6.50}$$

where $R_3 = (efg)$ with three mutually orthogonal unit (column) vectors e, f, g in $\mathbb{R}^3$. A general rotation of the reference system in 3-space,

$$R(\alpha, \beta, \gamma) = \begin{pmatrix} 1 & 0 \\ 0 & R_3(\alpha, \beta, \gamma) \end{pmatrix}, \quad R_3(\alpha, \beta, \gamma) \in SO(3), \tag{6.51}$$

which can be uniquely characterized by the Euler angles α, β, γ is another particular Lorentz transformation. Both particular transformations (6.50) and (6.51) have the properties $L^0_0 > 0, \det L = 1$. Since θ may be any real number and $SO(3)$ is connected, both transformations belong to $O^+(1,3)$ which is called the proper orthochronous Lorentz group (orthochronous because it preserves the direction of time flow).

Every element of $O^+(1,3)$ may uniquely be written as $L = B(\theta, e)R(\alpha, \beta, \gamma)$.

Proof Let $x'^\mu = L^\mu{}_\nu x^\nu$ be any element of $O^+(1,3)$. If $x'^0 = x^0$, then necessarily $B(\theta, e) = B(0) = \mathbf{1}_4$. Otherwise $x'^0 \neq \pm x^0$ and one may choose the unit vector e perpendicular to $\mathbf{e}_0$ in the plane spanned by e_0 and e'_0, so that $x'^0 = x^0 \cosh\theta + e \cdot x \sinh\theta$ with some value θ. Put $R(\alpha, \beta, \gamma) = (B(\theta, \mathbf{e}))^{-1} L$, then $(R(\alpha, \beta, \gamma))^0_0 = 1$ and L has the demanded form. Let $B(\theta', \mathbf{e}')R(\alpha', \beta', \gamma') = L = B(\theta, \mathbf{e})R(\alpha, \beta, \gamma)$ be another such decomposition of L. Then, $(B(\theta', \mathbf{e}'))^{-1}B(\theta, \mathbf{e}) = R(\alpha', \beta', \gamma')(R(\alpha, \beta, \gamma))^{-1}$ is a product of two rotations and hence a rotation, which implies $((B(\theta', e'))^{-1}B(\theta, e))^0_\mu = \delta^0_\mu$ and hence $\theta' = \theta, e' = e$. □

The transformation (6.50) is called a *boost*, and any element of $O^+(1,3)$ may be uniquely decomposed into a 3-rotation followed by a boost (or alternatively into an in general different boost followed by a 3-rotation). In the last section the simple fact was already stated that $O(1,3)$ consists of four connected components. Two of them are orthochronous and two are proper in the sense that their elements do not imply an odd number of spatial reflections.

So far, the Lorentz transformation was interpreted in the *passive* sense of the description of the same point in space–time seen from different reference systems. Consider a particle with rest mass m_0 placed at the origin of the reference system and hence with zero momentum $\boldsymbol{p}$. From another primed reference system the origin of which is moving with velocity v in direction $-\boldsymbol{e}$ as measured in the unprimed system, this particle is seen as moving with velocity v in the direction $\boldsymbol{e}$. Hence, in this system it has energy $m_0c^2/\sqrt{1-(v^2/c^2)} = m_0c^2\cosh\theta$ and momentum $\boldsymbol{e}m_0v/\sqrt{1-(v^2/c^2)} = -\boldsymbol{e}m_0c\sinh\theta$. The four-momentum $(p^\mu) = (E/c,\boldsymbol{p}^t)^t$ of the particle at rest in the unprimed reference system (as a column vector) is $(m_0c,\boldsymbol{0}^t)^t$ while that in the primed reference system is $(m_0c\cosh\theta, -\boldsymbol{e}^t m_0c\sinh\theta)^t$. This may be written as $(E'/c,\boldsymbol{p}'^t)^t = B(\theta,\boldsymbol{e})\ (m_0c,\boldsymbol{0}^t)^t$ and may also be interpreted in the active sense that the particle at rest in the fixed unprimed reference system is boosted to velocity $\boldsymbol{v} = v\boldsymbol{e}$ by the transformation $B(\theta,\boldsymbol{e})$. However, boosts alone do not form a group: the composition of two boosts is a Lorentz transformation, but in general not again a boost (check it). Conversely, the transition from one boost to another boost is also a Lorentz transformation, but in general not a boost. Hence, the generalization of the just considered relation is

$$\begin{pmatrix} E'/c \\ \boldsymbol{p}' \end{pmatrix} = L\begin{pmatrix} E/c \\ \boldsymbol{p} \end{pmatrix}, \quad L \in O(1,3) \tag{6.52}$$

with an interpretation again in the active sense that the physical content of the fixed unprimed reference system is first rotated and then boosted by the unique rotation and boost content of L according to the above theorem.

As is well known from physics, the infinitesimal generators of a 3-rotation (generators of the Lie algebra $\mathfrak{o}(3,\mathbb{R})$) are the three components of Schrödinger's angular momentum operator multiplied by the imaginary unit and the infinitesimal generators of boosts (infinitesimal boosts may be described by infinitesimal Galilei transformations as seen above after (6.48)) are the three components of the momentum operator. Hence, the dimension of the Lorentz group is $6 = 4(4-1)/2$ in accord with the general dimension of $O(p,q)$.

Instead of describing a point of space–time as a four-vector in Minkowski space $\mathbb{R}^4$ provided with a pseudo-metric $g_{\mu\nu} = (D_{1,3})_{\mu\nu}$ as in (6.46), it may likewise be characterized by a complex Hermitian 2×2 matrix which also has four real entries, by the correspondence

$$(x^\mu) \mapsto X = x^\mu \sigma_\mu = \begin{pmatrix} x^0 + x^3 & x^1 - ix^2 \\ x^1 + ix^2 & x^0 - x^3 \end{pmatrix} \mapsto \left(\frac{1}{2}\mathrm{tr}(X\sigma_\mu)\right) = (x^\mu), \qquad (6.53)$$

with

$$\sigma_0 = \mathbf{1}_2, \quad \sigma_1 = \begin{pmatrix} 0 & 1 \\ 1 & 0 \end{pmatrix}, \quad \sigma_2 = \begin{pmatrix} 0 & -i \\ i & 0 \end{pmatrix}, \quad \sigma_3 = \begin{pmatrix} 1 & 0 \\ 0 & -1 \end{pmatrix}. \qquad (6.54)$$

$\sigma_i, i = 1, 2, 3$ are the Pauli matrices, while all real quadruples with arithmetic operations according to $(a, b, c, d) = a\sigma_0 + bi\sigma_3 + ci\sigma_2 + di\sigma_1$ form a realization of the field of Hamilton's **quaternions**. Obviously, (6.53) provides a one–one mapping between the points (x^μ) and X. Moreover,

$$\det X = (x^0)^2 - (x^1)^2 - (x^2)^2 - (x^3)^2 \qquad (6.55)$$

defines Minkowski's pseudo-metric on the space of the complex Hermitian matrices X. A Lorentz transformation must now be a linear transformation of matrices X which preserves Hermiticity and keeps the determinant of X constant. A simple such transformation is

$$X' = AXA^\dagger, \quad \det A = 1, \quad \text{that is, } A \in Sl(2, \mathbb{C}). \qquad (6.56)$$

In fact it would suffice to demand $|\det A| = 1$, but A and $e^{i\lambda}A$, λ real, provide the same transformation (6.56) with $\det(e^{i\lambda}A) = e^{i2\lambda}\det A$. Hence, for every A' with $|\det A'| = 1$ one may choose $A = (\det A')^{-1/2}A'$ with $\det A = 1$.

Every transformation (6.56) is via (6.53) mapped to a Lorentz transformation of (x^μ), and this mapping is obviously both smooth and a group homomorphism. Hence, there is a Lie group homomorphism $Sl(2, \mathbb{C}) \to O(1, 3)$. Since $Sl(2, \mathbb{C})$ is simply connected, it is smoothly mapped into the connected component of $O^+(1, 3)$. This latter mapping is even onto. Consider first a rotation by an angle $-\varphi$ around the $\boldsymbol{e}_3$-axis, $x'^1 = x^1 \cos\varphi - x^2 \sin\varphi$, $x'^2 = x^1 \sin\varphi + x^2 \cos\varphi$ while x^0 and x^3 do not change. It is easy to check by direct calculation that it is provided via (6.53) and (6.56) by

$$\begin{aligned} \exp((i\varphi/2)\sigma_3) &= \mathbf{1}_2 + i\frac{\varphi}{2}\sigma_3 - \frac{1}{2!}\left(\frac{\varphi}{2}\right)^2 - \frac{i}{3!}\left(\frac{\varphi}{2}\right)^3 \sigma_3 + \cdots \\ &= \begin{pmatrix} \cos(\varphi/2) + i\sin(\varphi/2) & 0 \\ 0 & \cos(\varphi/2) - i\sin(\varphi/2) \end{pmatrix} \end{aligned} \qquad (6.57)$$

which is obviously an element of $Sl(2, \mathbb{C})$. Similar expressions hold for rotations around the other spatial axes. Since any rotation of $SO(3)$ can be performed by rotating around $\boldsymbol{e}_3$ by the first Euler angle, then rotating around the new $\boldsymbol{e}_2$-axis by the second Euler angle and finally rotating around the thus obtained $\boldsymbol{e}_3$-axis by the third Euler angle, every $SO(3)$-rotation corresponds to a product of three $Sl(2, \mathbb{C})$-transformations. Similarly it is seen that a boost along the $\boldsymbol{e}_3$-axis is provided by

$$\exp((\theta/2)\sigma_3) = \mathbf{1}_2 + \frac{\theta}{2}\sigma_3 + \frac{1}{2!}\left(\frac{\theta}{2}\right)^2 + \frac{1}{3!}\left(\frac{\theta}{2}\right)^3 \sigma_3 + \cdots$$
$$= \begin{pmatrix} \cosh(\theta/2) + \sinh(\theta/2) & 0 \\ 0 & \cosh(\theta/2) - \sinh(\theta/2) \end{pmatrix}. \quad (6.58)$$

a general boost along any $\boldsymbol{e}$-direction is then obtained from (6.58) by replacing σ_3 with $\boldsymbol{e}\cdot\boldsymbol{\sigma}$ where $\boldsymbol{\sigma}$ means the 3-vector of the three Pauli matrices. Finally, since any proper orthochronous Lorentz transformation can be written as an $SO(3)$-rotation followed by a boost, it may be likewise written as a transformation (6.56) with A generated by expressions $\exp(\boldsymbol{\lambda}\cdot\boldsymbol{\sigma})$ where $\boldsymbol{\lambda}$ is a general complex 3-vector, and $\det\exp(\boldsymbol{\lambda}\cdot\boldsymbol{\sigma}) = \exp\operatorname{tr}(\boldsymbol{\lambda}\cdot\boldsymbol{\sigma}) = \exp(0) = 1$ since the Pauli matrices are traceless. Exercise: Show that a rotation by π around the y-axis followed by a boost in z-direction cannot be given by a single exponent $\exp(\boldsymbol{\lambda}\cdot\boldsymbol{\sigma})$.

A traceless complex 2×2-matrix has three independent complex entries and can be expressed as a complex linear combination of the three Pauli matrices. From the result of the last paragraph it follows, that the Pauli matrices generate the Lie algebra $\mathfrak{sl}(2,\mathbb{C})$ with six real dimensions. The three matrices $i\sigma_k, k = 1,2,3$ generate infinitesimal rotations, and the matrices $\sigma_k, k = 1,2,3$ generate infinitesimal boosts. On physical grounds it is clear, and it can of course be shown technically, that the Lie algebras $\mathfrak{o}(1,3)$ and $\mathfrak{sl}(2,\mathbb{C})$ are isomorphic.

If one, however, replaces the matrices (6.57) or (6.58) by their negative (which does not change $\det A$ because of even rank), the transformation (6.56) is not affected. Two elements of $Sl(2,\mathbb{C})$ differing in a sign only lead to the same Lorentz transformation:

$$O^+(1,3) \approx Sl(2,\mathbb{C})/\{\mathbf{1}_2, -\mathbf{1}_2\}. \quad (6.59)$$

While $Sl(2,\mathbb{C})$ is simply connected, $O^+(1,3)$ having a Lie algebra isomorphic to that of $Sl(2,\mathbb{C})$ cannot be simply connected. *$Sl(2,\mathbb{C})$ is the universal covering group of $O^+(1,3)$.*

The matrices $\exp(i\boldsymbol{\lambda}\cdot\boldsymbol{\sigma})$ with real λ^k are in fact unitary as any exponentiation of a skew-Hermitian matrix. Hence, the rotations belong to the subgroup $SU(2)$ of $Sl(2,\mathbb{C})$, and it holds that

$$SO(3) \approx SU(2)/\{\mathbf{1}_2, -\mathbf{1}_2\}. \quad (6.60)$$

The Lie algebra $\mathfrak{su}(2)$ is the real Lie algebra generated by $i\sigma_k, k = 1,2,3$ and is isomorphic to the Lie algebra of angular momenta. *$SU(2)$ is the universal covering group of $SO(3)$.*

The representation theory of the groups $SU(2)$ and $Sl(2,\mathbb{C})$ can be found in textbooks of quantum mechanics and is not considered here. Only a few final remarks are in due place. In the last section, $Sl(2,\mathbb{C}) = Sp(2,\mathbb{C})$ was mentioned. As a consequence, every $Sl(2,\mathbb{C})$-transformation leaves a skew-symmetric bilinear form on the representation space invariant the skew-symmetric matrix providing it being often called the 'metric spinor'. With respect to the $SU(2)$ being a subgroup of $Sl(2,\mathbb{C})$, due to unitarity there is additionally a unitary bilinear form left invariant. This brings it about that $Sl(2,\mathbb{C})$ has two unitarily inequivalent two-dimensional irreducible representations (undotted and dotted spinors) while $SU(2)$ has only one (up to unitary equivalence). Moreover, $Sl(2,\mathbb{C}) \times \{\mathbf{1}_2, P, T, TP\}$ where $A = P$ and $A = T$, respectively, provide space and time inversion in (6.56) is the cover of the complete Lie group $O(1,3)$ relevant in quantum theory, which has no two-dimensional faithful (see next section) representation. These facts were pointed out by van der Waerden immediately after Dirac's formulation of the relativistic theory of the electron.

6.8 The Adjoint Representation

As stated in Sect. 6.2, a homomorphism from a Lie group G into the Lie group $Gl(n,K)$ is called a representation of the Lie group G. Even for a finite dimensional Lie group there may be infinite dimensional irreducible representations in the Lie group $Gl(V)$ of automorphisms of an infinite dimensional vector space V. A representation is called a **faithful representation**, if the homomorphism is injective which means that its kernel is trivial. The group may then be identified with its image of this representation. An important faithful representation of a special class of Lie groups is the adjoint representation.

Consider any Lie group G and fix one element $x \in G$. Then,

$$R_x(g) = xgx^{-1} = l_x \circ r_{x^{-1}}(g) = r_{x^{-1}} \circ l_x(g) \tag{6.61}$$

is an automorphism of G called an **inner automorphism**. Indeed, $R_x(gh^{-1}) = R_x(g)R_x(h^{-1})$ for all $g, h \in G$ and both translations l_x and r_x were shown in Sect. 6.1 to be injective. Since Lie group homomorphisms are pushed forward to Lie algebra homomorphisms (Sect. 6.2), R_x is pushed forward to the Lie algebra automorphism

$$\mathrm{Ad}(x) = (R_x)_* : \mathfrak{g} \to \mathfrak{g}. \tag{6.62}$$

As a Lie algebra automorphism, $\mathrm{Ad}(x)$ is a non-singular linear transformation of the vector space $\mathfrak{g}$ of dimension $n = \dim G$ and hence is an element of the Lie group $Gl(n,K)$.

$$\mathrm{Ad} : G \to \mathrm{Aut}(\mathfrak{g}) \subset Gl(n,K). \tag{6.63}$$

This mapping is a Lie group homomorphism, because it is smooth as a composition of smooth mappings leading to (6.63), and first $R_{xy^{-1}}(g) = xy^{-1}g(xy^{-1})^{-1} = xy^{-1}gyx^{-1} = R_x(R_{y^{-1}}(g))$ and hence $R_{xy^{-1}} = R_x \circ R_{y^{-1}}$, and then finally $\mathrm{Ad}(xy^{-1}) = (R_{xy^{-1}})_* = (R_x \circ R_{y^{-1}})_* = (R_x)_* \circ (R_{y^{-1}})_* = \mathrm{Ad}(x)\mathrm{Ad}(y^{-1})$ where the last expression means matrix multiplication. As a Lie group homomorphism into $Gl(n,K)$, Ad is an n-dimensional Lie group representation. It is called the adjoint representation of G. Again invoking the push forward from Lie group homomorphisms to Lie algebra homomorphisms,

$$\mathrm{ad} = \mathrm{Ad}_* : \mathfrak{g} \to \mathrm{End}(\mathfrak{g}) \subset \mathfrak{gl}(n, K) \tag{6.64}$$

is a Lie algebra representation of $\mathfrak{g}$ as an algebra of linear transformations of the vector space $\mathfrak{g}$ itself.

The diagram (6.11) applied to $F = R_x$ and $F = \mathrm{Ad}$ yields the following two commutative diagrams

$$\begin{array}{ccc} \mathfrak{g} & \xrightarrow{\mathrm{Ad(x)}} & \mathfrak{g} \\ {\scriptstyle \exp}\downarrow & & \downarrow{\scriptstyle \exp} \\ G & \xrightarrow[R_x]{} & G \end{array} \qquad \begin{array}{ccc} \mathfrak{g} & \xrightarrow{\mathrm{ad}} & \mathrm{End}(\mathfrak{g}) \\ {\scriptstyle \exp}\downarrow & & \downarrow{\scriptstyle \exp} \\ G & \xrightarrow[\mathrm{Ad}]{} & \mathrm{Aut}(\mathfrak{g}) \end{array} \tag{6.65}$$

meaning

$$\exp(t\mathrm{Ad}(x)(Y)) = x \exp(tY) x^{-1} \quad \text{and} \quad \exp(t\mathrm{ad}(X)) = \mathrm{Ad}(\exp(tX)). \tag{6.66}$$

Differentiation of the last relation with respect to t at $t = 0$ yields $\mathrm{ad}(X) = [(d/dt)\mathrm{Ad}(\exp(tX))]_{t=0}$. Now, $\mathrm{ad}(X) \subset \mathrm{End}(\mathfrak{g})$, and elements of Lie algebras are left invariant vector fields of their respective Lie groups. Hence,

$$\begin{aligned} \mathrm{ad}(X)(Y_e) &= \frac{d}{dt}[\mathrm{Ad}(\exp(tX))Y_e]_{t=0} = \frac{d}{dt}\left[(R_{\exp(tX)})_* Y_e\right]_{t=0} \\ &= \frac{d}{dt}\left[(r_{\exp(-tX)} \circ l_{\exp(tX)})_* Y_e\right]_{t=0} \\ &= \frac{d}{dt}\left[(r_{\exp(-tX)})_* \circ (l_{\exp(tX)})_* Y_e\right]_{t=0} \\ &= \frac{d}{dt}\left[(r_{\exp(-tX)})_* Y_{\exp(tX)}\right]_{t=0} = \lim_{t \to 0} \frac{(r_{\exp(-tX)})_* Y_{\exp(tX)} - Y}{t}. \end{aligned}$$

(The last but one equality is due to the left invariance of Y.) Recall that $\exp(tX)$ describes the integral curve $\phi_t(e)$ of the vector field X on G through $e \in G$. Since at $e \in G$ left and right invariant vector fields coincide, $(r_{\exp(-tX)})_*$ is just Φ_t of (4.36), and (cf. (3.37))

$$\mathrm{ad}(X)Y = L_X Y = [X, Y]. \tag{6.67}$$

This is called the adjoint representation of the Lie algebra.

The center ZG of a Lie group G is defined as the subgroup consisting of the elements of G commuting with all elements of G separately (the latter as distinct from a mere invariant subgroup):

$$ZG = \{z \in G \mid gzg^{-1} = z \text{ for all } g \in G\}. \tag{6.68}$$

Accordingly, the center $\mathfrak{zg}$ of a Lie algebra $\mathfrak{g}$ is

$$\mathfrak{zg} = \{Z \in \mathfrak{g} \mid [X, Z] = 0 \text{ for all } X \in \mathfrak{g}\}. \tag{6.69}$$

The center of a connected Lie group G is the kernel of the adjoint representation.

Proof Let $z \in ZG$. Then, for every $X \in \mathfrak{g}$ and all $t \in \mathbb{R}$, $\exp(tX) = z(\exp(tX))z^{-1} = \exp(t\mathrm{Ad}(z)(X))$. Hence, from the left diagram (6.65), $X = \mathrm{Ad}(z)(X)$ which means $z \in \mathrm{Ker\,Ad}$. Conversely, let $z \in \mathrm{Ker\,Ad}$. Then, $z(\exp(tX))z^{-1} = \exp(t\mathrm{Ad}(z)(X)) = \exp(tX)$, and z commutes with all $\exp(tX)$ forming a neighborhood of $e \in G$. Since G is connected, z commutes with all G and hence is in its center. □

If G' is a Lie group and ZG' is its center, then $G = G'/ZG'$ is a Lie group with trivial center which can be identified with its adjoint representation and hence with a Lie subgroup of some $Gl(n, K)$, $n = \dim G$. Comparing (6.1) with (6.67), this identification of G with $\mathrm{Ad}(G)$ and of $\mathfrak{g}$ with $\mathrm{ad}(\mathfrak{g})$ yields

$$(\mathrm{ad}(X_i))^k_j = (X_i)^k_j = c^k_{ij}, \quad \mathrm{Ad}(\exp(tX_i)) = \exp(tX_i) = \exp(t(c^k_{ij})) \tag{6.70}$$

where (c^k_{ij}) in the last exponent means the matrix with matrix elements c^k_{ij} for i fixed.

References

1. Warner, F.W.: Foundations of Differentiable Manifolds and Lie Groups. Springer, New York (1983)
2. Pontrjagin, L.S.: Topological Groups, 2nd edn. Gordon and Breach, New York (1966)
3. Scheck, F.: Mechanics, 2nd Corrected and Enlarged edn. Springer, Berlin (1994)

Chapter 7
Bundles and Connections

In Chap. 3, manifolds were introduced as a special category of topological spaces having locally a topology like a (finite-dimensional) metric vector space. In order to glue together these quite simple local patches, in addition to the topology a differentiable structure (pseudo-group, complete atlas) of transition functions $\psi_{\beta\alpha}$ was introduced which allowed to develop an analysis on manifolds. Globally, however, manifolds may be very complex. Fiber bundles form a special category of manifolds which locally behave like a topological product of manifolds, but which again are provided with additional specific structure to allow for a rich content of theory. Their global topology may be as complex as that of any manifold; in fact, fiber bundles are build on any type of manifold.

In order to have an illustrative introductory example, consider a smooth real function F defined on a manifold M, the circle $M = S^1$ say. The graph of F is a set $\mathcal{F} = \{(x, F(x)) | x \in M\}$ of pairs $(x, F(x))$ which can be viewed as points of the product manifold $M \times \mathbb{R}$ which in the considered example would be an (infinite) cylinder. Considered merely as a product manifold, arbitrary coordinate patches of that cylinder (charts) may be used to describe it by means of homeomorphic mappings onto open domains $\boldsymbol{U} \in \mathbb{R}^2$. However, this way the important feature of a function is lost, namely that $\mathcal{F}$ has precisely one point $(x, F(x))$ for every $x \in M$. Moreover, often pointwise algebraic operations with functions on M are of interest, that is, the algebraic structure of $\mathbb{R}$ as an Abelian group (of additions) or even as a vector space matters. A simple but in physics particularly important case is that of complex functions on M of absolute value equal to unity (phases). Then, instead of $\mathbb{R}$, the Abelian Lie group $G = U(1)$ forms the space of values of F. In this case, the graph of F is a subset of $M \times G$. If M is a manifold of quantum states, then G may be an Abelian gauge group. It is clear immediately that also non-Abelian groups are of great relevance.

Again, only **smooth bundles** are considered in this volume, and for the sake of brevity the adjective smooth is omitted throughout (but recalled now and then).

7.1 Principal Fiber Bundles

A **principal fiber bundle** (P, M, π, G), or in short notation P, consists of

1. a manifold P,
2. a Lie group G which acts freely on P from the right, that is, there is a smooth mapping $R_g : P \times G \to P : (p, g) \mapsto pg = R_g p$ with $R_{gh^{-1}} = R_{h^{-1}} R_g$ and so that $R_g p \neq p$ unless $g = e$, the unit in G,
3. M is the quotient space P/G with respect to the action of G in P, and the canonical projection $\pi : P \to M$ is smooth,
4. P is **locally trivial**, that is, for every $x \in M$ there is a neighborhood $U \subset M$ of x so that $\pi^{-1}(U)$ is diffeomorphic to $U \times G$ in the sense that there is a smooth bijection $\psi : \pi^{-1}(U) \to U \times G$ so that $\psi(p) = (\pi(p), \phi(p))$ for all $p \in \pi^{-1}(U)$ with $\phi(pg) = \phi(p)g$ for all $g \in G$.

The points 2 and 3 together mean that $\pi^{-1}(x) \approx G$ for every $x \in M$ and hence that G acts transitively on $\pi^{-1}(x)$, that is, for every pair (p, p') of points of $\pi^{-1}(x)$ there is a $g \in G$ with $p' = pg$.

M is called the **base space** of the bundle and P is called the **bundle space** while π is the **bundle projection,** $\pi^{-1}(x)$ is the **fiber** over $x \in M$, and in a principal fiber bundle all fibers are isomorphic to the Lie group G, the **structure group** of the bundle.

The simplest principal fiber bundle is the **trivial bundle** or product bundle $M \times G$ (Fig. 7.1). If, moreover, $G = \mathbb{R}$ as the Abelian Lie group with respect to addition of numbers, then $P = \{(x, r) | \, x \in M, r \in \mathbb{R}\}$ as a manifold is the infinite cylinder over M as base (viewed for instance by fixing $r = 0$) and π is the projection onto the base of that cylinder.

Note that in general *the only connection between P and M is the mapping* π. Despite the simplified sketches in Fig. 7.1 and in some of the following figures *there is no reason to think of M as a subset of P. Subsets of P homeomorphic to M (which even need not exist in general) are the images of global sections of P defined below.*

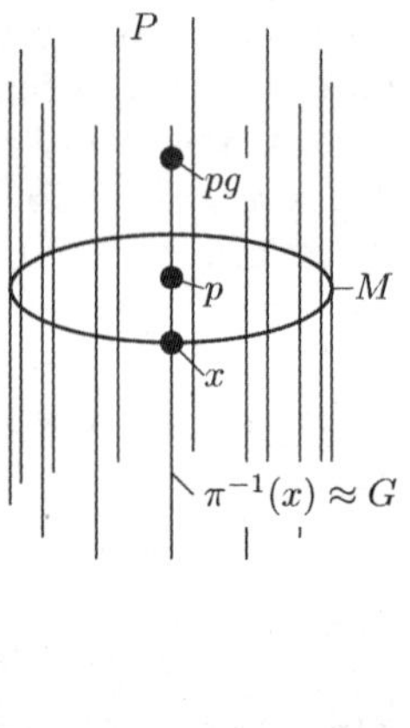

Fig. 7.1 A trivial principal bundle $P = M \times G$ with $M = S^1$ and $G = \mathbb{R}$. See also emphasized text above

The trivial bundle $M \times G$ is, however, not the only principal fiber bundle P with M as base space and G as structure group. To see this, consider an open cover $\{U_\alpha\}$ of M fine enough so that for every U_α the restriction $\pi^{-1}(U_\alpha)$ of P to U_α is trivial according to point 4 of the above definition, that is, on the trivial bundle $\pi^{-1}(U_\alpha)$ there is a diffeomorphism $p \mapsto \psi_\alpha(p) = (\pi(p), \phi_\alpha(p))$. Let $U_\alpha \cap U_\beta$ be non-empty and let $p \in \pi^{-1}(U_\alpha \cap U_\beta)$. Then, $\phi_\beta(pg)(\phi_\alpha(pg))^{-1} = \phi_\beta(p)gg^{-1}(\phi_\alpha(p))^{-1} = \phi_\beta(p)(\phi_\alpha(p))^{-1}$, where $(\phi_\alpha(pg))^{-1}$ is the inverse group element to $\phi_\alpha(pg)$ in G. Hence, $\phi_\beta(pg)(\phi_\alpha(pg))^{-1}$ does not depend on g. Moreover, since G acts freely and transitively on $\pi^{-1}(x)$, for every $p, p' \in \pi^{-1}(x)$ there is $g \in G$ so that $p' = pg$. Consequently, $\phi_\beta(pg)(\phi_\alpha(pg))^{-1}$ depends on p only through $\pi(p) \in U_\alpha \cap U_\beta \subset M$. As a result, for every pair (U_α, U_β) with $U_\alpha \cap U_\beta \neq \emptyset$ there is a **transition function** $\psi_{\beta\alpha}(\pi(p)) = \phi_\beta(p)(\phi_\alpha(p))^{-1}$, $\pi(p) \in U_\alpha \cap U_\beta$, with the obvious properties

$$\psi_{\alpha\beta}(x) = (\psi_{\beta\alpha}(x))^{-1}, \quad \psi_{\gamma\alpha}(x) = \psi_{\gamma\beta}(x)\psi_{\beta\alpha}(x) \quad \text{for all } x \in U_\alpha \cap U_\beta \cap U_\gamma. \tag{7.1}$$

The function values of these transition functions are elements of the Lie group G, and on the right hand sides of (7.1) the group operations are meant. Of course, for $\gamma = \beta = \alpha$ the second relation implies $\psi_{\alpha\alpha} = e$ and hence, for $\gamma = \alpha$, it also implies the first relation. (If G is Abelian and additively written, all group multiplications used so far in this section are to be replaced by additions.)

Recall that a (principal) fiber bundle is a special manifold, and hence it has transition functions of its coordinate neighborhoods as a manifold. Because of the more complex structure of a fiber bundle as a manifold, its transition functions $\tilde{\psi}_{\beta\alpha}(p) = (\boldsymbol{\psi}_{\beta\alpha}(\boldsymbol{x}_\alpha), \psi_{\beta\alpha}(x))$ (here marked with a tilde) have also a more complex structure: $\boldsymbol{\psi}_{\beta\alpha}$ is the transition function on M as in Chap. 3, and the G-group valued function $\psi_{\beta\alpha}$ was analyzed above. See Fig. 7.2 on the next page.

Take as an example again $M = S^1 \ni e^{i\alpha} = z$ with coordinate α on M, consider the open cover $\{U_1, U_2\}, U_1 = S^1 \setminus \{1\}, U_2 = S^1 \setminus \{-1\}$ of M. Let $G = \mathbb{R} \cup \{I\}$ be the Lie group of all translations by $g \in \mathbb{R}$ and of the inversion I of the real line, in (somewhat non-standard) multiplicative writing $gh = hg = g + h, gI = Ig = -g, I^2 = e$. Consider the case $\psi_{21}(z) = \psi_{12}(z) = e$. Then, $P = S^1 \times G$ which can again be visualized as the cylinder of Fig. 7.1. Now, consider the possibility $\psi_{21}(z) = \psi_{12}(z) = I$. It fulfils the conditions (7.1): e.g. $\psi_{11}(z) = \psi_{12}(z)\psi_{21}(z) = I^2 = e$. It is easily seen that P for this case is the **Möbius band** infinitely extended perpendicular to S^1 (cf. Fig. 1.2 for a finite version). One may consider it either as the tape $U_1 \times G$ turned around and glued together at $z = 1$ or as the tape $U_2 \times G$ turned around and glued together at $z = -1$. Clearly it is distinct from the cylinder already as a manifold, both cases are not homeomorphic.

Let M be any manifold and let G be any Lie group. It is not difficult to show (e.g. [1, vol. I]) that if there is an open cover $\{U_\alpha\}$ of M and if there are

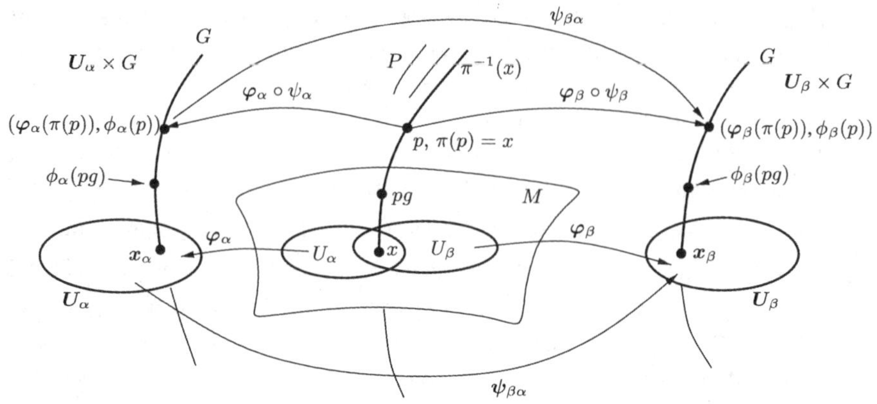

Fig. 7.2 The interrelations between a principal fiber bundle P and its local trivializations $U_\alpha \times G$ as well as the transition functions between the latter. Note that the fiber above x, here drawn as a line, can have any dimension. (Strictly speaking, instead of $\varphi_\alpha \circ \psi_\alpha$ it should be written $(\varphi_\alpha \times \mathrm{Id}_G) \circ \psi_\alpha$)

transition functions $\psi_{\beta\alpha} : U_\alpha \cap U_\beta \to G$ for all non-empty $U_\alpha \cap U_\beta$ which fulfil the conditions (7.1), then there exists a corresponding principal fiber bundle (P, M, π, G).

The principal fiber bundle (P, M, π, G) may be constructed as follows: Take the trivial bundles $Q_\alpha = U_\alpha \times G$ and form their disjoint union $Q = \sqcup_\alpha Q_\alpha$. A point in Q is a triple (α, x, g), $x \in U_\alpha$, $g \in G$. Introduce the equivalence relation $R : (\alpha, x, g) \equiv (\beta, x', g')$, if $x = x' \in U_\alpha \cap U_\beta$ and $g' = \psi_{\beta\alpha} g$. Take the quotient space $P = Q/R$. It is easy to see that P is a principal fiber bundle with structure group G, base space $M = P/G$, open cover $\{U_\alpha\}$ of M and transition functions $\psi_{\beta\alpha}$.

This rises the question of the morphisms of the category of bundles. Leaving aside general bundle morphisms, a **bundle homomorphism** of principal fiber bundles is a triple $(F, \bar{F}, \bar{\bar{F}})$ of smooth mappings from a bundle (P', M', π', G') into a bundle (P, M, π, G) where $F : P' \to P, \bar{F} : M' \to M$ so that the diagram

$$\begin{array}{ccc} P' & \xrightarrow{F} & P \\ {\scriptstyle \pi'}\downarrow & & \downarrow{\scriptstyle \pi} \\ M' & \xrightarrow[\bar{F}]{} & M \end{array} \tag{7.2}$$

is commutative, and $\bar{\bar{F}} : G' \to G$ is a Lie group homomorphism so that $F(p'g') = F(p')\bar{\bar{F}}(g')$ for every $p' \in P'$ and every $g' \in G'$. Because of (7.2), F maps fibers of P' into fibers of P. Indeed, $(\pi \circ F)(p') = x \in M$ equals $(\bar{F} \circ \pi')(p')$ which only depends on $\pi'(p') = x' \in M'$. Hence, for all p' in the fiber $\pi'^{-1}(x')$ above x' the image $F(p')$ is in the fiber $\pi^{-1}(x)$ above x. In the following, $(F, \bar{F}, \bar{\bar{F}})$ is often in short denoted simply by F or by $F : P' \to P$, where in the last notation P' and P are the short notations of (P', M', π', G') and (P, M, π, G).

$(F, \bar{F}, \bar{\bar{F}}) : (P', M', \pi', G') \to (P, M, \pi, G)$ is a **bundle embedding**, if the mapping $\bar{F} : M' \to M$ is an embedding of manifolds and $\bar{\bar{F}} : G' \to G$ is injective. Identifying (P', M', π', G') with its image by a bundle embedding F, it is called a **subbundle** of (P, M, π, G).

If moreover $M' = M$ and $\bar{F} = \mathrm{Id}_M$, but $\bar{\bar{F}}(G') \neq G$, then F is called a **reduction of the structure group** G of P to G' and P' is called a **reduced fiber bundle**. A principal fiber bundle P is called **reducible**, if there exists a reduction $F : P' \to P$. It can straightforwardly be shown [1, vol. I], that

a principal fiber bundle with structure group G is reducible to the structure group G', iff there is an open cover of M with transition functions obeying (7.1) *and having values only in G'.*

Of particular interest in physics is the case of bundle isomorphisms with $M' = M$, $\bar{F} = \mathrm{Id}_M$. If F is an isomorphism, then $\bar{\bar{F}}$ must also be an isomorphism which means that G' and G are isomorphic, and hence $\bar{\bar{F}}$ may be viewed as an automorphism of P onto itself which maps fibers onto fibers: $\pi^{-1}(x) = \pi'^{-1}(x)$ for all $x \in M$. It is therefore often called a **vertical automorphism** of a principal fiber bundle. As in general for automorphisms, these vertical automorphisms form a group ${}^v\mathrm{Aut}(P)$ which is called the **group of gauge transformations** of P with the symmetry group G. It will be discussed in more detail in the next chapter. In fact, modern gauge theory in physics and the theory of principal fiber bundles were developed in parallel in the second half of 20th century.

Let $\mathfrak{g}$ be the Lie algebra of *right* invariant vector fields on the Lie group G of the principal fiber bundle P and let $\mathcal{X}(P)$ be the Lie algebra of (smooth) tangent vector fields on the manifold P. For every $X \in \mathfrak{g}$, the 1-parameter subgroup $\exp(tX)$ of G induces a local 1-parameter group (p. 81) $\phi_t(p) = p\exp(tX)$ through every point $p \in P$ which is tangent to the fiber containing p because the action of G maps fibers of P onto themselves, and, by differentiation with respect to t, it induces a (smooth) tangent vector field $X^* \in \mathcal{X}(P)$ which is everywhere tangent to fibers of P. How are X and X^* related algebraically? Recall that a tangent vector on a manifold is defined by its action on smooth real functions on that manifold. Let $f : P \to \mathbb{R}$ be a smooth function understood as a differential 0-form on P. Pull the right action of G on P, $R_g p = pg$, $R_{gh^{-1}} = R_{h^{-1}} R_g$ back by f (p. 72): $R_g^* f(R_g p) = f(p)$. From

$$R_g^* R_{h^{-1}}^* f(pgh^{-1}) = R_g^* f(pg) = f(p) = R_{gh^{-1}}^* f(pgh^{-1})$$

for every $f \in \mathcal{C}(P) = \mathcal{D}^0(P)$ it follows that $R_g^* R_{h^{-1}}^* = R_{gh^{-1}}^*$. (Observe how the contravariance of the right action of G is neutralized by the contravariance of the pull back; that is why principal fiber bundles are defined with a right action of the structure group.) The result is, that the restriction of R^* to any fiber $\pi^{-1}(x)$, $x \in M$, of P is a representation of the Lie group G in the infinite-dimensional functional space $\mathcal{C}(\pi^{-1}(x))$ as representation space. It is called the **regular representation** of G. In other words, there is a Lie group homomorphism from G into

R^*. According to the theorem on p. 177, this homomorphism is pushed forward to a Lie algebra homomorphism $R^*_* : \mathfrak{g} \to \mathcal{X}(\pi^{-1}(x))$, and $R^*_*(X) = X^*$ (cf. (6.11)). Suppose that $X^*_p = 0$ on some point p. This would imply $p\exp(tX) = p$. Since G acts freely on P, this means $\exp(tX) = e$ for all t and hence $X = 0$.

$X^* = R^*_*(X)$ is called the **fundamental vector field** corresponding to X. For $X \neq 0$ it is nowhere zero on P. From that and the fact that $\dim T_p(\pi^{-1}(x)) = \dim(\pi^{-1}(x)) = \dim G = \dim \mathfrak{g}$ it follows that R^*_* is an isomorphism of vector spaces. (The infinite-dimensional regular representation R^* of G is obviously reducible.) Moreover, from the content of Sect. 6.8 it is easily obtained that

if $X^* = R^*_*(X)$, *then for every* $g \in G$ *there is a fundamental vector field* $(R_g)_*(X^*)$ *corresponding to* $(\mathrm{Ad}(g^{-1}))X \in \mathfrak{g}$.

Here, $(R_g)_*$ is the push forward by the right action R_g of G on P to the corresponding action on the Lie algebra $\mathcal{X}(P)$ of tangent vector fields on P.

Finally, the functions on M anticipated in the introduction to this chapter are treated by the notion of bundle sections. A **local section** of a fiber bundle (P, M, π, G) is a smooth function $s : M \supset U \to P$ for which $\pi \circ s = \mathrm{Id}_U$, that is, $\pi(s(x)) = x$ for every $x \in U$. If s is defined on all M, it is called a **global section** or simply a **section**. In Sect. 7.6 below, vector bundles are considered which always have (global) sections. For a principal fiber bundle this is not the case in general.

A principal fiber bundle has a (global) section, iff it is trivial.

Proof Let $P = M \times G$. Then, $s : x \mapsto (x, e)$ is a section. Conversely, let $s : M \to P$ be a section of (P, M, π, G). The sets $\{s(x)g | g \in G\} \approx G$ for each fixed $x \in M$ are the fibers of P yielding a global trivialization $P = M \times G$. □

Take for instance a Möbius band as M and the (discrete multiplicative) Lie group $G = \{1, -1\}$ locally describing orientation on M. There is no global section $s : M \to G$ smooth on M, not even a continuous one.

This section is closed with a number of examples of principal fiber bundles.

Let G be a Lie group and let H be a closed Lie subgroup of G. The quotient space G/H of left cosets gH of H in G is a **homogeneous manifold** or **homogeneous space** with respect to the action of G, that is, G acts transitively (by group multiplication) on G/H. Let $\pi : g \mapsto gH$ be the canonical projection, it is a surjective Lie group homomorphism with kernel H. Then, $(G, G/H, \pi, H)$ is a principal fiber bundle. Principal fiber bundles of this type form a subcategory of principal fiber bundles characterized as those for which the bundle space is a Lie group and the base space is a homogeneous space of that Lie group. For more details see Sect. 9.2.

Let M be a pathwise connected manifold and let $\pi_1(M)$ be its fundamental group. A manifold is locally homeomorphic to some $\mathbb{R}^n$, hence a pathwise connected manifold is locally pathwise connected and semi-locally 1-connected (Sect. 6.4). Let $\tilde{M}$ be its universal covering manifold, and let $\pi : \tilde{M} \to M$ be the canonical projection. Then, $(\tilde{M}, M, \pi, \pi_1(M))$ is a principal fiber bundle. If, for instance, M is the unit cell of an infinite crystal (three-dimensional torus $\mathbb{T}^3$), then $\pi_1(\mathbb{T}^3) \approx \mathbb{Z}^3 \ni \boldsymbol{n} = (n_1, n_2, n_3)$ and $\tilde{M} = \mathbb{R}^3 = \cup_{\boldsymbol{n}\in\mathbb{Z}^3}(M + \boldsymbol{n})$ is the infinite repetition of M. The fiber over a point $\boldsymbol{x}$ of M (the unit cell) is the lattice $\{\boldsymbol{x} + \boldsymbol{n}\}$ of points equivalent to $\boldsymbol{x}$ by the discrete translational symmetry.

Let $\mathbb{C}^{n+1}_\circ = \mathbb{C}^{n+1} \setminus \{0\}$ be the punctured complex vector space (with the topology from $\mathbb{R}^{2n+2}$), and let $G = Gl(1, \mathbb{C})$ be the multiplicative group of non-zero complex numbers. Then, $\mathbb{C}P^n = \mathbb{C}^{n+1}_\circ / Gl(1, \mathbb{C})$ is the n-dimensional projective complex space, and, with the canonical projection $\pi : \mathbb{C}^{n+1}_\circ \to \mathbb{C}P^n$, $(\mathbb{C}^{n+1}_\circ, \mathbb{C}P^n, \pi, Gl(1, \mathbb{C}))$ is a principal fiber bundle. Recall that $U(1)$ is a (closed) subgroup of $Gl(1, \mathbb{C})$. Let $S^{2n+1} \subset \mathbb{C}^{n+1}_\circ$ be the unit sphere. Then, $(S^{2n+1}, \mathbb{C}P^n, \pi, U(1))$ is a reduced fiber bundle of the principal fiber bundle $(\mathbb{C}^{n+1}_\circ, \mathbb{C}P^n, \pi, Gl(1, \mathbb{C}))$. Here, π is just the restriction of the above projection π to S^{2n+1}. This latter case is extremely relevant in physics for $n = \infty$ with the topology from the norm of the complex Hilbert space l^2. The projective Hilbert space is the space of quantum states, its unit sphere that of normalized states, and $U(1)$ is the gauge group for particle conservation. (See textbooks on quantum theory.)

For the n-sphere S^n in $\mathbb{R}^{n+1}$ and for $G = \{e, I\}$ with the inversion I of space (G is a discrete Lie group), $\mathbb{R}P^n = S^n/G$ is the real projective space, and, again with the canonical projection π, $(S^n, \mathbb{R}P^n, \pi, G)$ is a principal fiber bundle.

The most important special category of principal fiber bundles is considered now.

7.2 Frame Bundles

Let M be an m-dimensional K-manifold, $K = \mathbb{R}$ or $\mathbb{C}$. A **linear frame** at point $x \in M$ consists of point x and an ordered base $(X_1, \ldots, X_m)$ in the tangent space $T_x(M)$ on M at point x. Denote a linear frame as $p = (x, X_1, \ldots, X_m)$, and denote the set of all linear frames at all points of M by $L(M)$. It is easily seen that the Lie group $Gl(m, K)$ acts freely from the right on $L(M)$ and maps linear frames at x into linear frames at the same point x. Indeed, let $g = (g^j_i) \in Gl(m, K)(g^j_i \in K)$, then $p' = pg = (x, \sum_{j=1}^m X_j g^j_i,\ i = 1, \ldots, m)$, and $p' = p$ implies $g = e = \delta^j_i$. (Matrix convention is used throughout this book understanding an upper index as row index and a lower one as column index.) It is also clear that $Gl(m, K)$ acts transitively on any set of linear frames at any fixed point $x \in M$. Let $\pi : p = (x, X_1, \ldots, X_m) \mapsto x$ be the projection from $L(M)$ onto M. In order to see that $(L(M), M, \pi, Gl(m, K))$ is a principal fiber bundle, a differentiable structure must be defined on $L(M)$ so that π is smooth.

The differentiable structure on $L(M)$ is obtained in a straightforward way: Take an atlas $\mathcal{A}_M$ of M and choose a coordinate neighborhood U of $x \in M$ where $X_i = X^k_i(\partial/\partial x^k)$ (Einstein summation over k). For a base of $T_x(M)$, the matrix X^k_i of the coefficients of the tangent vectors X_i in this coordinate neighborhood (which are smooth functions of the coordinates in U) is not degenerate, that is, its determinant is non-zero. This yields a diffeomorphism between $\pi^{-1}(U) \subset L(M)$ and $U \times Gl(m,K)$. Taking $\{\pi^{-1}(U)|U \in \mathcal{A}_M\}$ as an atlas of $L(M)$ and $(x^k, X_i^k, i = 1,\dots,m)$ as local coordinates in $\pi^{-1}(U)$ makes $L(M)$ into an $m(m+1)$-dimensional manifold, for which obviously $\pi : L(M) \to M$ is smooth. The principal fiber bundle $L(M)$ is called the (linear) **frame bundle** over M.

A technical possibility to obtain the points of $L(M)$ is the following: Take the base $e_1 = (1,0,\dots,0),\dots,e_m = (0,\dots,0,1)$ of K^m. Then, any point p of the fiber over $x = \pi(p)$ in $L(M)$ can be obtained from a non-degenerate linear mapping $u(p) : K^m \to T_{\pi(p)}(M) : e_i \mapsto u(p)e_i = X_i$. In local coordinates one has $X_i^k = \sum u_j^k e_i^j = u_i^k$, and with $g = (g_j^k) \in Gl(m,K)$ and $pg = (x, (Xg)_i)$ one finds $(Xg)_i^k = \sum u_j^k g_{j'}^j e_i^{j'} = \sum u_j^k g_i^j$, that is, $u(pg) = u(p)g$. This shows again that, as for every principal fiber bundle, the typical fiber is isomorphic to the structure group, $Gl(m,K)$ in the considered case. With this convention, which is amply used later, for every fiber over some point x there is a one–one correspondence between $p \in \pi^{-1}(x)$ and linear mappings $u(p) : p = (\pi(p), u(p)e_i)$.

Figure 7.3 shows a number of frames of $L(S^2)$ as an example. (Moving frames (repère mobile) as a central technical tool in the theory of Lie groups were introduced by E. Cartan.) Below it will be seen that for every (paracompact) K-manifold M the structure group of $L(M)$ may be reduced from $Gl(m,K)$ to the unitary group $U(m)$ for $K = \mathbb{C}$ and to the orthogonal group $O(m)$ for $K = \mathbb{R}$. From Fig. 7.3 it is intuitively clear that orthogonal frame bundles can be treated as (smooth) principal fiber bundles.

Instead of taking the tangent space $T_x(M)$ on M at x to be the (linear) vector space of the frame bundle, the affine-linear space $A_x(M)$ may be considered with the group of affine-linear transformations introduced in Sect. 6.1 as transformation group. This group is described there explicitly and is denoted $A(m,\mathbb{R}) = Gl(m,\mathbb{R}) \rtimes \mathbb{R}^m$ (semi-direct product). There is a short exact sequence

$$0 \to \mathbb{R}^m \xrightarrow{\alpha} A(m,\mathbb{R}) \xrightarrow{\beta} Gl(m,\mathbb{R}) \to e,$$

(where $\mathbb{R}^m$ is considered as the Abelian group of vector addition) and a homomorphism

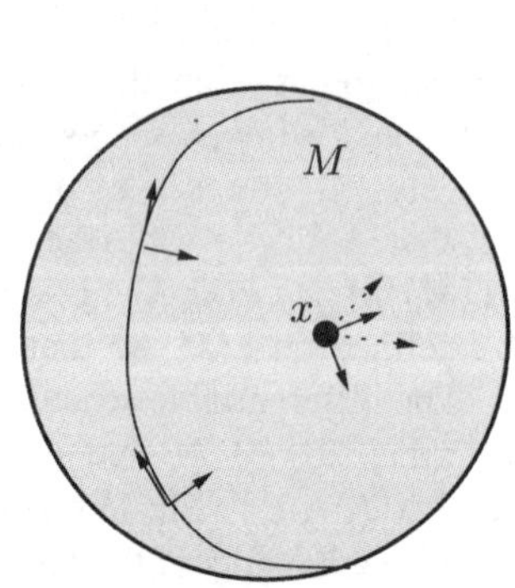

Fig. 7.3 The manifold $M = S^2$ with some examples of frames. At point x the frames of the full and dotted arrow lines both belong to $L(S^2)$. (All arrows are understood tangent to S^2)

$$\gamma : Gl(m,\mathbb{R}) \to A(m,\mathbb{R}) : g \mapsto \gamma(g) = \begin{pmatrix} g & 0 \\ 0 & 1, \end{pmatrix}, \quad g \in Gl(m,\mathbb{R}),$$

so that $\beta \circ \gamma = \mathrm{Id}_{Gl(m,\mathbb{R})}$. There is also a short exact sequence

$$0 \to \mathbb{R}^m \xrightarrow{\alpha_*} \mathfrak{a}(m,\mathbb{R}) \xrightarrow{\beta_*} \mathfrak{gl}(m,\mathbb{R}) \to 0$$

and a homomorphism $\gamma_* : \mathfrak{gl}(m,\mathbb{R}) \to \mathfrak{a}(m,\mathbb{R})$ for the corresponding Lie algebras, so that $\mathfrak{a}(m,\mathbb{R}) = \mathfrak{gl}(m,\mathbb{R}) \oplus \mathbb{R}^m$ (semi-direct sum). In the same matrix notation as used for $A(m,\mathbb{R})$ the elements of $\mathfrak{a}(m,\mathbb{R})$ are

$$\begin{pmatrix} A & X \\ 0 & 0 \end{pmatrix} = \begin{pmatrix} A & 0 \\ 0 & 0 \end{pmatrix} + \begin{pmatrix} 0 & X \\ 0 & 0 \end{pmatrix}, \qquad A : (m \times m)\text{-matrix}, \quad X : m\text{-column}.$$

If $p = (x, X_1, \ldots, X_m)$ is a linear frame at $x \in M$, then $\tilde{p} = (x, X_1, \ldots, X_m, X_{m+1})$ is an **affine frame** at that point, where X_{m+1} stands for the affine shift vector. As in the case of linear frames, let $\tilde{g} \in A(m,\mathbb{R})$ act from the right on an affine frame as $\tilde{p}' = \tilde{p}\tilde{g} = (x, \sum_{j=1}^{m+1} X_j \tilde{g}^j_i, i = 1, \ldots, m+1)$. Denote the set of all affine frames $\tilde{p}$ on M by $A(M)$, and the projection $(x, X_1, \ldots, X_{m+1}) \mapsto x$ by $\tilde{\pi}$, then $(A(M), M, \tilde{\pi}, A(m,\mathbb{R}))$ is a principal fiber bundle. It is the **affine frame bundle** over M. Like in the case of linear frame bundles, by introducing the same natural base in $\mathbb{R}^{m+1}$ as above in K^m, a linear mapping $\tilde{u}(\tilde{p}) : \mathbb{R}^{m+1} \to T_{\tilde{\pi}(\tilde{p})}(M)$ generates every frame out of the canonical frame of the fixed natural base. (Show that the base vector $e_{m+1} = (0, \ldots, 0, 1)$ corresponds to a zero shift in the transformation on p. 174 since the $m+1$st coordinate of the vectors in $\mathbb{R}^m$ is fictitious.)

7.3 Connections on Principle Fiber Bundles

Now, manifolds are again treated as $\mathbb{R}$-manifolds. Let (P, M, π, G) be a principal fiber bundle, let $T_p(P)$ be the tangent space on P at point p, and let G_p be the linear subspace of $T_p(P)$ which is tangent to the fiber of P containing p. A **connection** Γ on P specifies a subspace Q_p of $T_p(P)$ at every point $p \in P$ so that

1. $T_p(P) = G_p \oplus Q_p$,
2. $Q_{pg} = (R_g)_* Q_p$ for every $p \in P$ and every $g \in G$ (see below),
3. Q_p depends smoothly on $p \in P$.

Here, G_p and Q_p are again treated just as topological vector spaces, not as Euclidean spaces. Scalar products and angles between vectors are not defined. For the direct sum of vector spaces see p. 16. Orthogonality also is not defined and not demanded between G_p and Q_p. (Orthogonality between vectors of $T^*_p(P)$ and $T_p(P)$, however, is always defined by $\langle \omega, X \rangle = 0$ as usual.) Nevertheless, any vector $X \in T_p(P)$ has a unique decomposition $X = {}^vX + {}^hX, {}^vX \in G_p$, ${}^hX \in Q_p$. To give these two components a name, ${}^vX$ is called the **vertical component** and ${}^hX$ is called the **horizontal component**; to say it again, no angle between vertical and horizontal components matters. These names are suggested by Fig. 7.1 where fibers are 'vertical' on the 'horizontal' base manifold M, although also Fig. 7.1 is just some visualization, and angles between the base manifold and fibers do not matter. Strictly speaking, what is denoted M in that figure is rather some section $s : M \to P$, which, as orientation on the Möbius band showed, even does not always exist globally for a principal fiber bundle. Nevertheless, given any point $p \in P$, $T_p(P)$ and G_p always exist, since P and the fiber are manifolds, the latter as a space isomorphic to a Lie group. Hence, Q_p as a complement to G_p in the vector space $T_p(P)$ may always be defined, although not uniquely: there is freedom in choosing a connection. G_p is called the **vertical space** and Q_p is called the **horizontal space**. The structure group G of a fiber bundle allows to transform distinct points on a fiber into each other, to compare them or to combine them in pointwise manipulations of functions on M. The connection is the general tool to transform distinct fibers into one another by 'parallel' transport, and thus to compare functions on M at distinct points and to obtain derivatives.

For a fixed $g \in G$, the right action $R_g : P \to P : p \mapsto pg$ is a smooth mapping of the manifold P onto itself. For every $p \in P$, it is pushed forward to a linear mapping $(R_g)_* : T_p(P) \to T_{pg}(P)$ (see p. 71 and the transformation of fundamental vector fields by $g \in G$ in Sect. 7.1). While the fundamental vector fields are vertical in the new nomenclature, $(R_p)_*$ of course yields also a linear mapping of horizontal vectors at p to vectors at pg. The condition 2 says that the image of this mapping must again be a horizontal vector at pg and the mapping of Q_p must be onto Q_{pg}. Since by condition 1 $\dim Q_p = \dim T_p(P) - \dim G_p$ and the latter two spaces have dimensions independent of p (as tangent spaces of manifolds), the dimension of Q_p must also be independent of p, and $(R_g)_*$ must be a regular linear mapping (isomorphism of vector spaces).

In Sect. 7.1, the isomorphism of vector spaces R^*_* was considered which exists for every principal fiber bundle and which maps every $X \in \mathfrak{g}$ to a fundamental vector field X^* on P which is vertical at every point $p \in P$, that is $X^*_p \in G_p$. Conversely, consider a covector ω_p with $\mathfrak{g}$-valued components and a linear mapping $\langle \omega_p, \cdot \rangle$ from $T_p(P)$ into $\mathfrak{g} \approx T_e(G)$ which maps *any* tangent vector $X^*_p \in T_p(P)$ to the uniquely defined vector $\langle \omega_p, X^*_p \rangle = X \in \mathfrak{g}$ for which $R^*_*(X) = {}^vX^*_p$. (For the sake of distinction, again vectors of $\mathfrak{g}$ are denoted by X here and tangent vectors to P by X^*.) X is indeed uniquely defined by X^*_p, since ${}^vX^*_p$ is uniquely defined for every X^*_p and R^*_* is an isomorphism between $\mathfrak{g}$ and the space of fundamental vector fields on P and hence provides a bijection between $\mathfrak{g}$ and the vertical space G_p. Clearly, $\langle \omega_p, X^*_p \rangle = 0$, iff X^*_p is horizontal. The mapping ω_p is a $\mathfrak{g}$-vector-valued linear function on $T_p(P)$ for every $p \in P$. Since fibers of a principal fiber bundle depend smoothly on p and because of condition 3 of the definition of Q_p, for every (smooth) tangent vector field X^* on P, $X^* \in \mathcal{X}(P)$, the mapping ω equal to ω_p for all p may be considered as a smooth mapping from $\mathcal{X}(P)$ to $\mathfrak{g}$-valued functions on P. Introduce a (fixed) base $\{E_i | i = 1, \ldots, \dim G\}$ in

$\mathfrak{g}$, so that $X_g = \sum_i X^i(g)E_i$ for every $X \in \mathfrak{g}$ with real components $X^i(g)$. Then, ω induces $\dim G$ real functions $\omega^i : \mathcal{X}(P) \to \mathcal{C}(P) : X^* \mapsto \langle \omega, X^* \rangle^i = \langle \omega^i, X^* \rangle$, with $\langle \omega^i, X^* \rangle(p) = \langle \omega^i_p, X^*_p \rangle$, which in fact are 1-forms on P. For that reason, ω is considered as a vector-valued or $\mathfrak{g}$-valued 1-form on P, it is called the **connection form** of the connection Γ.

The connection form ω has the following two decisive properties:

1. $\langle \omega, R^*_*(X) \rangle = X$ for every $X \in \mathfrak{g}$,
2. $\langle (R_g)^*\omega, X^* \rangle = \langle \mathrm{Ad}(g^{-1})\omega, X^* \rangle$ for every $g \in G$ and every $X^* \in \mathcal{X}(P)$.

Property 1 follows directly from the definition of the connection form. Consider as a vertical vector field (vertical X^*_p at every $p \in P$) a fundamental vector field X^*. One has

$$\begin{aligned}\langle ((R_g)^*\omega)_p, X^*_p \rangle &= \langle \omega_{pg}, (R_g)_*(X^*_p) \rangle = \langle \omega_{pg}, (R^*_*(\mathrm{Ad}(g^{-1})X)_{pg}) \rangle = \mathrm{Ad}(g^{-1})X \\ &= \mathrm{Ad}(g^{-1})\langle \omega_p, X^*_p \rangle = \langle \mathrm{Ad}(g^{-1})\omega_p, X^*_p \rangle.\end{aligned}$$

The first equality expresses just the general duality between pulling back a form and pushing forward a vector field $(((R_g)^*\omega)_p = (R_g)^*\omega_{pg})$. The second equality is an application of the rule for pushing forward a fundamental vector field by $(R_g)_*$ given on p. 210. The third and fourth equalities use property 1 forth and back, in the last step with $R^*_*(X) = X^*$. The last expression follows since $\mathrm{Ad}(g^{-1})$ acts on $\mathfrak{g}$ and hence on the $\mathfrak{g}$-valued covector ω_p of the last two expressions. On the other hand, for a horizontal vector ${}^hX^*_p$, $(R_g)_*({}^hX^*_p)$ is also horizontal by the condition 2 of the definition of a connection Γ. Hence, the second expression of the above chain of equations is already zero. (Recall from the text above, that $\langle \omega, X^* \rangle = 0$, if X^* is horizontal.) Hence, by linearity, in the first and last expressions of the above chain of equations the vertical vector X^*_p may be replaced by *any* vector $X^*_p \in T_p(P)$. In particular, since $p \in P$ is arbitrary, X^*_p may belong to any vector field $X^* \in \mathcal{X}(P)$, and property 2 holds. Now, given a $\mathfrak{g}$-valued 1-form ω with properties 1 and 2, define

$$Q_p = \{X^*_p \in T_p(P) | \langle \omega_p, X^*_p \rangle = 0\}. \tag{7.3}$$

It is easily seen that this defines a connection Γ.

There is a one–one correspondence between connections Γ *and* $\mathfrak{g}$*-valued* 1*-forms* ω *having properties* 1 and 2. *The correspondence is expressed by* (7.3).

Consider now the bundle projection $\pi : P \to M$ of the principal fiber bundle (P, M, π, G). It is a smooth mapping between manifolds and hence is pushed forward to a linear mapping $\pi_* : T_p(P) \to T_{\pi(p)}(M)$. In a neighborhood U of $x = \pi(p) \in M$ there is a local trivialization $P \supset \pi^{-1}(U) \approx U \times G$ and hence

$\dim P = \dim M + \dim G$. Pushing this trivialization forward to the tangent spaces on $\pi^{-1}(U)$ and considering a connection Γ on P, it is easily seen that G_p is mapped by the push forward to $\mathfrak{g}$ and Q_p is mapped *isomorphically* to $T_x(U)$. (The tangent space on $U \times G$ at p may be realized as $T_{\pi(p)}(U) \oplus \mathfrak{g}$, with $T_{\pi(p)}(U) = T_{\pi(p)}(M)$.) Since the trivialization is a local bundle isomorphism, the same statement can be made for the connection on P itself:

For every connection on a principle fiber bundle the bundle projection π is pushed forward to a linear bijection (isomorphism) π_* of Q_p onto $T_{\pi(p)}(M)$.

A *horizontal* tangent vector field $X^* \in \mathcal{X}(P)$ is called a (horizontal) **lift of a tangent vector field** $X \in \mathcal{X}(M)$, if $\pi_*(X^*_p) = X_{\pi(p)}$ for every $p \in P$. (Now, tangent vector fields on M are denoted by X.) X^* is invariant under the action of $(R_g)_*$, since the horizontal space Q_p is invariant and, since $\pi(R_g(p)) = \pi(p)$, it must hold that $\pi_*((R_g)_* X^*_p) = \pi_*(X^*_p)$.

Given a connection on a principal fiber bundle (P, M, π, G), there is a one–one correspondence between tangent vector fields X on M and horizontal tangent vector fields on P invariant under $(R_g)_*$; the latter being the lifts X^* of X. This correspondence observes addition and Lie products of tangent vector fields as well as multiplication by real functions.

It is readily seen that a horizontal vector field X^* on P which is invariant under G is the lift of $X = \pi_*(X^*)$. That the lift of every $X \in \mathcal{X}(M)$ is smooth can easily be checked in a local trivialization of P. The rest is obvious.

So far, two ways are obtained to define a connection on a principal fiber bundle, by specifying a family Γ of horizontal tangent spaces Q_p obeying conditions 1 to 3 or by specifying a $\mathfrak{g}$-valued 1-form ω having properties 1 and 2 Instead of specifying a global 1-form ω on P, a third way is to specify a family of local $\mathfrak{g}$-valued 1-forms on M as considered below. All three ways are of equal practical importance.

As on p. 207, let $\{U_\alpha\}$ be an open cover of M so that a family of diffeomorphisms $\psi_\alpha : \pi^{-1}(U_\alpha) \to U_\alpha \times G : p \mapsto (\pi(p), \phi_\alpha(p))$ is a local trivialization of P. Let $\psi_{\alpha\beta}$, $\psi_{\alpha\beta}(x) \in G$, $x \in U_\alpha \cap U_\beta \subset M$ be the corresponding transition functions. Let $s_\alpha : U_\alpha \to \pi^{-1}(U_\alpha) : x \mapsto s_\alpha(x) = \psi_\alpha^{-1}(x, e)$ be the **canonical local section,** where e is the unit in G. In fact, any local section on U_α may be expressed as $s(x) = s_\alpha(x) g(x)$ through the canonical local section and a function $U_\alpha \ni x \mapsto g(x) \in G$. In particular, on $U_\alpha \cap U_\beta$ the canonical local sections s_α and s_β are linked by the transition function $\psi_{\alpha\beta} : U_\alpha \cap U_\beta \to G$, indeed

$$s_\beta(x) = \psi_\beta^{-1}(x, e) = (\psi_\alpha^{-1} \circ \psi_\alpha \circ \psi_\beta^{-1})(x, e) = s_\alpha(x)\psi_{\alpha\beta}(x).$$

ψ_β^{-1} maps $(x, e) \in U_\beta \times G$ to the point $s_\beta(x) = p \in P$ with $\pi(p) = x$ and $\phi_\beta(p) = e$. Then, ψ_α maps this point p to $(x, \phi_\alpha(p)) = (x, \phi_\alpha(p)(\phi_\beta(p))^{-1}) = (x, e)\phi_\alpha(p)$ $(\phi_\beta(p))^{-1} = (x, e)\psi_{\alpha\beta}(x)$, where in the first equality use was made that $\phi_\beta(p) =$

$e = (\phi_\beta(p))^{-1}$, and in the second equality the action of G on a principal fiber bundle was employed. Finally, $\psi_\alpha^{-1}((x,e)\psi_{\alpha\beta}(x)) = \psi_\alpha^{-1}(x,e)\psi_{\alpha\beta}(x) = s_\alpha(x)\psi_{\alpha\beta}(x)$.

The canonical section s_α is a mapping of the manifold U_α into the manifold P, hence it may be pushed forward to a linear mapping $s_{\alpha *}$ of the tangent spaces $T_x(M)$ into the tangent spaces $T_{s_\alpha(x)}(P)$. Likewise, the mapping $\psi_{\alpha\beta}$ of the manifold $U_\alpha \cap U_\beta$ into G may be pushed forward to a linear mapping $\psi_{\alpha\beta *}$ from the spaces $T_x(M)$ into the spaces $T_{\psi_{\alpha\beta}(x)}(G)$. Since these push forwards are differentials (Sect. 3.5), the Leibniz rule applies to the above displayed relation: For every tangent vector $X_x \in T_x(M)$, $x \in U_\alpha \cap U_\beta$,

$$s_{\beta *}(X_x) = s_{\alpha *}(X_x)R(\psi_{\alpha\beta}) + (s_\alpha(x))_* \psi_{\alpha\beta *}(X_x),$$

where R is the representation of G by right action onto the vector space $T_{s_\alpha(x)}(P)$, $(R_g)_*(Y) = YR(g)$, and $(s_\alpha(x))_*$ is the push forward of $s_\alpha(x)$, for fixed x considered as a mapping $G \ni g \mapsto s_\alpha(x)g \in P$, to a linear mapping from $T_{\psi_{\alpha\beta}(x)}(G)$ into $T_{s_\beta(x)}(P)$ with $s_\beta(x) = s_\alpha(x)\psi_{\alpha\beta}(x)$, cf. Fig. 7.4.

Let $\{E_i | i = 1, \ldots, \dim G\}$ be a fixed base in $\mathfrak{g}$ and let $\omega = \sum_i \omega^i E_i$ be a connection form on P. Then, as $\omega_p^i \in T_p^*(P)$, $\langle \omega^i_{s_\beta(x)}, s_{\beta *}(X_x)\rangle$ is a real number and, if x is varied through U_β, it is a smooth real function on U_β. Hence, $\sum_i \langle \omega^i_{s_\beta(x)}, s_{\beta *}(X_x)\rangle E_i$ is a smooth vector valued function on U_β with values in $\mathfrak{g}$. It is denoted by ω_β and is the pull back of $\omega^i \in \mathcal{D}^1(P)$ to $\mathcal{D}^1(U_\beta)$ by s_β^* : $\langle \omega^i_\beta, X_x\rangle = \langle \omega^i_{s_\beta(x)}, s_{\beta *}(X_x)\rangle = \langle s_\beta^*(\omega^i_{s_\beta(x)}), X_x\rangle$, that is, $\omega_\beta = s_\beta^*(\omega)$. Applying ω on both sides of the above displayed Leibniz rule one obtains for $X \in \mathcal{X}(U_\alpha \cap U_\beta)$

$$\langle \omega_\beta, X\rangle = \langle (\mathrm{Ad}(\psi_{\alpha\beta}^{-1})\omega_\alpha), X\rangle + \langle \vartheta_{\alpha\beta}, X\rangle, \qquad \omega_\alpha = s_\alpha^*(\omega), \quad \vartheta_{\alpha\beta} = \psi_{\alpha\beta}^*(\vartheta), \tag{7.4}$$

where ϑ is the canonical Maurer–Cartan 1-form (p. 176) of G and the $\mathfrak{g}$-valued 1-forms $\omega_\alpha = s_\alpha^*(\omega)$ are called **local connection forms**. They are pull backs of

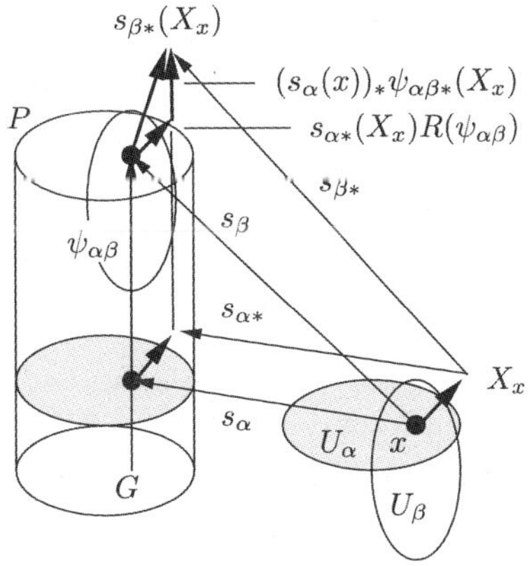

Fig. 7.4 Local sections s_α and s_β with transition function $\psi_{\alpha\beta}$ and push forwards

the connection form ω from the canonical local section $s_\alpha(U_\alpha) \subset P$ to $U_\alpha \subset M$. The first term on the right hand side was transformed with the property 2 of connection forms: $\langle \omega, s_{\alpha *}(X)R(\psi_{\alpha\beta})\rangle = \langle \omega, (R_{\psi_{\alpha\beta}})_*(s_{\alpha *}(X))\rangle = \langle ((R_{\psi_{\alpha\beta}})^*\omega), s_{\alpha *}(X)\rangle = \langle ((R_{\psi_{\alpha\beta}})^*\omega)_\alpha, X\rangle = \langle (\mathrm{Ad}(\psi_{\alpha\beta}^{-1})\omega)_\alpha, X\rangle = \langle (\mathrm{Ad}(\psi_{\alpha\beta}^{-1})\omega_\alpha), X\rangle$. The last step realizes the independent linear action of $\mathrm{Ad}(g)$ and of s_α^* on X : $\left(\sum_j \mathrm{Ad}(\psi_{\alpha\beta}^{-1})^i_j\omega^j\right)_\alpha = \sum_j \mathrm{Ad}(\psi_{\alpha\beta}^{-1})^i_j\omega^j_\alpha$. As regards the second term on the right hand side of (7.4), consider the left invariant vector field Y on G which for $g = \psi_{\alpha\beta}(x)$ equals $\psi_{\alpha\beta *}(X_x) \in T_{(x,\psi_{\alpha\beta}(x))}((U_\alpha \cap U_\beta) \times G)$ and apply the canonical Maurer–Cartan 1-form: $\langle \vartheta, \psi_{\alpha\beta *}(X_x)\rangle = \langle \vartheta, Y\rangle = Y_e$. The isomorphism $G \approx \pi^{-1}(x)$ translates Y into a fundamental vector field $Y_e^* = R_*^*(Y_e)$ on P, the value of which at $p = s_\alpha(x)\psi_{\alpha\beta}(x)$ is $(s_\alpha(x))_*\psi_{\alpha\beta *}(X_x)$. Now, $\langle \omega, (s_\alpha(x))_*\psi_{\alpha\beta *}(X_x)\rangle = \langle \omega, R_*^*(Y_e)\rangle = Y_e = \langle \vartheta, \psi_{\alpha\beta *}(X_x)\rangle = \langle (\psi_{\alpha\beta}^*(\vartheta)), X_x\rangle$.

The transition formula (7.4) from U_α to U_β for the local connection forms of a connection form ω, that is, from ω_α to ω_β looks quite involved. Consider the important special case where G is $Gl(n,K)$ or a subgroup thereof, that is, where both G and $\mathfrak{g} \approx T_e(G)$ consist of $n \times n$-matrices. Recall that $\psi_{\alpha\beta}(x) \in G$ and $\langle \omega, X\rangle \in \mathfrak{g}$. The relation (7.4) becomes a matrix equation and reads

$$\langle \omega_\beta, X\rangle = \psi_{\alpha\beta}^{-1}\langle \omega_\alpha, X\rangle \psi_{\alpha\beta} + \psi_{\alpha\beta}^{-1}\psi_{\alpha\beta *}(X).$$

The first expression is due to the definition of the adjoint representation of G in this case, and in the second expression $\psi_{\alpha\beta}^{-1}$ pulls back the vertical vector $\psi_{\alpha\beta *}(X) \in T_{(x,\psi_{\alpha\beta}(x))}((U_\alpha \cap U_\beta) \times G)$ to a vertical vector of $T_{(x,e)}((U_\alpha \cap U_\beta) \times G) \approx \mathfrak{g}$. Recall, that $\psi_{\alpha\beta *}$ is the differential of $\psi_{\alpha\beta}$.

There is a one–one correspondence of connection forms ω on P and families of local connection forms ω_α on M obeying (7.4). *The correspondence is expressed by the second relation* (7.4).

A local connection form ω_α is a connection form on the trivial bundle $U_\alpha \times G$. It is easily seen that on a trivial bundle $M \times G$ the lifts $Q_{(x,g)}$ of the tangent spaces $Q_{(x,e)}$ on the reduced bundle $M \times \{e\}$, that is, all tangent spaces on all submanifolds $M \times \{g\}$, form a connection. It is called the **canonical flat connection**. (Later it becomes clear why it is called flat.) Since all manifolds are supposed to be paracompact, the technique of partitioning of unity (Sect. 2.4) can be used to show that on every principal fiber bundle any local connection may be continued to a global connection [1, vol. I, Sect. II.2].

A connection exists on every principal fiber bundle.

Let $(L(M), M, \pi, Gl(m,\mathbb{R}))$, $m = \dim M$ be the frame bundle over M. A connection form ω on $L(M)$ is called a **linear connection**. (There is a modification compared to the general case which is explained in more detail at the end of Sect. 7.7.) A linear connection is a $\mathfrak{gl}(m,\mathbb{R})$-valued 1-form $\omega = \sum_{ij}\omega^i_j E^j_i$ with properties 1 and 2 on p. 215, where $\{E_i^j | i,j = 1,\ldots,m\}$ is a fixed base in $\mathfrak{gl}(m,\mathbb{R})$, for instance given by the real $m \times m$-matrices E_i^j having a unit entry in the ith row and jth column and zeros otherwise, $(E_i^j)^l_m = \delta^l_i \delta^j_m$. Recall from Sect. 7.2 that a linear frame is an ordered base $(X_1,\ldots,X_m)$ of $T_x(M)$, and $L(M) \ni p = (x, X_1,\ldots,X_m)$.

Consider a local trivialization of $L(M)$ by an open cover $\{U_\alpha\}$ of M and introduce local coordinates $\boldsymbol{\varphi}_\alpha : U_\alpha \to \boldsymbol{U}_\alpha \subset \mathbb{R}^m : x \mapsto \sum_k x^k e_k$, where $\{e_k\}$ is the base of $\mathbb{R}^m$ introduced in Sect. 7.2. As was done there, consider again the linear bijection $u(p) : \mathbb{R}^m \to T_{\pi(p)}(M)$, $u(pg) = u(p)g$ and find local coordinates $\psi_\alpha(p) = (x^k(p), u_i^k(p))$ on $U_\alpha \times Gl(m,\mathbb{R}) \subset L(M)$ and $\psi_\alpha(pg) = (x^k(p), u_j^k(p)g_i^j)$, where $u_i^k(p)$ is a real non-degenerate $m \times m$-matrix. Therefore, the coordinate expression of a tangent vector is

$$T_p(L(M)) \ni X_p^* = \sum_k X^k(p)\frac{\partial}{\partial x^k} + \sum_{ik} X_i^k(p)\frac{\partial}{\partial u_i^k} = {}^hX_p^* + {}^vX_p^*.$$

While the first coordinate expression has no component tangent to the fiber and hence belongs to the horizontal space ${}^hX_p^*$ only, the second one may, depending on the connection, belong partially to both ${}^hX_p^*$ and ${}^vX_p^*$. Nevertheless, the horizontal space must be m-dimensional since it is isomorphic to $T_x(M)$ and the vertical space must be m^2-dimensional since it is isomorphic to $\mathfrak{gl}(m,\mathbb{R})$. The canonical local section is $s_\alpha : x \mapsto \psi_\alpha^{-1}(x_\alpha^k(x), \delta_i^k)$ and $s_\beta(x) = \psi_\alpha^{-1}(x_\alpha^k(x), \delta_i^k)\psi_{\alpha\beta} = \psi_\alpha^{-1}(x_\alpha^k(x), (\psi_{\alpha\beta})_i^k)$.

Let θ be the $\mathbb{R}^m$-(vector)-valued 1-form on $L(M)$, defined as

$$\langle \theta_p, X_p^* \rangle = u^{-1}(\pi_*(X_p^*)), \quad X_p^* \in T_p(L(M)), \tag{7.5}$$

where $\pi_* : T_p(L(M)) \to T_{\pi(p)}(M)$ is the push forward of π as previously which projects any tangent vector X^* on the bundle space $L(M)$ to the tangent space on the base space M, and $u = u(p) : \mathbb{R}^m \to T_{\pi(p)}(M)$ is the linear bijection as above and in Sect. 7.2 which transforms the orthonormal standard base of the $\mathbb{R}^m$ into the frame p. u^{-1} then represents the vector $X_x = \pi_*(X_p^*)$ in the frame p. θ is called the **canonical form on** $L(M)$ (sometimes called the soldering form which 'solders' structural objects of the points of $L(M)$ like tangent vectors to the base space M). If X_p^* is vertical, then $\pi_*(X_p^*) = 0$ ($\pi(p(t))$ has zero derivative at t where the tangent vector $X_{p(t)}^*$ to the curve $p(t)$ is vertical) and hence $\langle \theta_p, X_p^* \rangle = 0$ for vertical X_p^*. In the case of a general X_p^* a group action yields $\langle (R_g^*(\theta_{pg})), X_p^* \rangle = \langle \theta_{pg}, R_{g*}(X_p^*) \rangle = (ug)^{-1}(\pi_*(R_{g*}(X_p^*))) = g^{-1}u^{-1}(\pi_*(X_p^*)) = g^{-1}\langle \theta_p, X_p^* \rangle$. (Since $R_{g*}(X_p^*)$ is in the same fiber as X_p^*, $\pi_*(R_{g*}(X_p^*)) = \pi_*(X_p^*)$.) Hence, $R_g^*(\theta) = g^{-1}\theta$.

Now, since $Q_p \approx \mathbb{R}^m$, let B be a linear mapping of $\mathbb{R}^m \ni \boldsymbol{X}$ into the space $\mathcal{H}(L(M)) \ni B(\boldsymbol{X})$ of horizontal vector fields on $L(M)$, $B(\boldsymbol{X})_p \in Q_p$, defined by

$$\langle \omega, B(\boldsymbol{X}) \rangle = 0, \quad \langle \theta, B(\boldsymbol{X}) \rangle = \boldsymbol{X}. \tag{7.6}$$

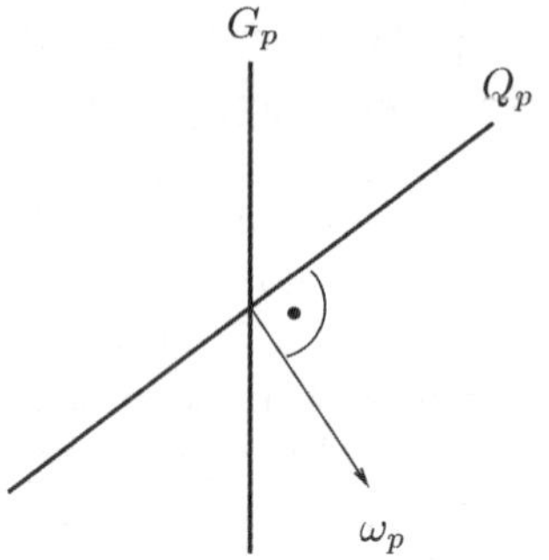

Fig. 7.5 Example of vertical space G_p and horizontal space Q_p of a two-dimensional tangent space $T_p(P)$ (drawing plane), showing how the connection form $\omega, \omega_p \in T_p^*(P)$, determines Q_p with G_p independently given

The first relation ensures that ${}^vB(X) = 0$ while the second relation spells out as $\pi_*(B(X)_p) = (u(p))(X)$, and, since $Q_p \approx T_{\pi(p)}(L(M))$, $B(X)$ is uniquely defined by (7.6). It is called the **standard horizontal vector field** corresponding to X. There are m^2 linearly independent fundamental vertical vector fields, which are independent of the connection ω, and m linearly independent standard horizontal vector fields, which depend on the choice of the connection ω by the first relation of (7.6) (Fig. 7.5).

7.4 Parallel Transport and Holonomy

The connection Γ on a principal fiber bundle (P, M, π, G) is used to define the parallel transport of fibers on the base space M. Let $F : I \to M$, $I = [0, 1] \subset \mathbb{R}$, be a path in M from $x_0 = F(0)$ to $x_1 = F(1)$. A (horizontal) **lift** F^* **of the path** F is a path $F^* : I \to P$ which is projected to F so that $\pi \circ F^* = F$ and which has a horizontal tangent vector in every of its points $F^*(t)$, $t \in I$. If $X \in \mathcal{X}(M)$ is a tangent vector field on M and if F is an integral curve of X, then F^* is obviously an integral curve in P of the lift X^* of X. Since there is a one–one correspondence of tangent vector fields X on M and their lifts X^* on P, which was stated on p. 216, and since there is a unique maximal integral curve of X^* through every point $p \in P$ by Frobenius' theorem, there is precisely one lift F^* of the path F starting at a given point $p_0 \in \pi^{-1}(x_0)$. In other words, for every $p_0 \in \pi^{-1}(x_0)$ there is a uniquely defined lift F^* which transports p_0 to a point $p_1 \in \pi^{-1}(x_1)$, for given x_1 on F uniquely defined by F and p_0. This is written as $p_1 = \tilde{F}(p_0)$. Obviously, $\tilde{F}(p_0 g) = \tilde{F}(p_0)g$, since $(R_g)_*(X^*) = X^*$ for every horizontal vector field. Hence, $\tilde{F} : \pi^{-1}(x_0) \to \pi^{-1}(x_1)$ is a Lie group isomorphism. It is called the **parallel transport** of the fiber along the path F from x_0 to x_1.

If F is a path in M, then F_-, $F_-(t) = F(1-t)$ is the inverse path, and $\tilde{F}_- = \tilde{F}^{-1}$ is the inverse isomorphic mapping of fibers. If F is a path from x_0 to x_1 and F' is a path from x_1 to x_2, then the concatenation (p. 182) $F'' = F'F$ is a path from x_0

to x_2 (not necessarily smooth at x_1, but this does not pose a problem in the present context, piecewise smooth paths may be allowed). Obviously, $\tilde{F}'' = \tilde{F}' \circ \tilde{F}$.

If F is a loop with base point x, then $\tilde{F}$ is an automorphism of $\pi^{-1}(x)$. Every loop F yields such an automorphism. Let $\mathcal{L}_x$ be the family of all loops in M with base point x. From the last paragraph it follows that all automorphisms due to the loops of $\mathcal{L}_x$ form a group, the **holonomy group** H_x of the connection Γ with base point x. If $\mathcal{L}_x^0$ is the family of all null-homotopic loops with base point x, then the corresponding subgroup of the holonomy group is the **restricted holonomy group** H_x^0.

Take a loop F based on x, and take any point $p \in \pi^{-1}(x)$. It is parallel transported by the loop to $p' = \tilde{F}(p) \in \pi^{-1}(x)$, and, since G acts transitively from the right on $\pi^{-1}(x)$, there is $g_F \in G$ so that $\tilde{F}(p) = pg_F$. Clearly, $\widetilde{F'F}(p) = pg_F g_{F'} = R_{g_{F'} g_F}(p)$. This provides a homomorphism from the holonomy group H_x of automorphisms of $\pi^{-1}(x)$ into the right action R of the structure group G of P, and, since G acts freely on $\pi^{-1}(x)$, into G itself. The image of this homomorphism in G is a subgroup of G, it is called the **holonomy group** H_p **with reference point** p. The restricted holonomy group H_p^0 with reference point p is likewise defined. If the reference point is changed within a fiber from p to pg, then $\widetilde{F'F}(pg) = pgg_F g_{F'} = p(gg_F g^{-1} gg_{F'} g^{-1})g = \widetilde{F'F}(p)g$. Hence, the holonomy group H_p with reference point p is changed into $H_{pg} = gH_p g^{-1}$ (and H_p^0 is changed into $gH_p^0 g^{-1}$).

Observe, that by the above definitions the holonomy group H_x is a subgroup of $\mathrm{Aut}(\pi^{-1}(x)) \approx \mathrm{Aut}(G)$, while H_p is a subgroup of G itself. Let F and F' be two loops with base point $x = \pi(p)$ and so that $p' g_F \neq p' g_{F'}$ for some $p' \in \pi^{-1}(x)$, that is the automorphisms corresponding to F and F' are not the same. Then, since G acts freely on $\pi^{-1}(x)$, $p' g_F \neq p' g_{F'}$ for all $p' \in \pi^{-1}(x)$. Hence, F and F' yield two different elements in every $H_{p'}$, which means that the homomorphism from H_x to $H_{p'} \subset G$ is injective. H_p and H_x for $x = \pi(p)$ are isomorphic.

More generally, let p and p' be two points (not necessarily of the same fiber) which may be parallel transported into each other by a lift of some path F from $\pi(p)$ to $\pi(p')$, $p' = \tilde{F}(p)$. Then, for every loop $F_x \in \mathcal{L}_x(\mathcal{L}_x^0)$ with base point $x = \pi(p)$ there is a loop $F_{x'} = FF_x F_- \in \mathcal{L}_{x'}(\mathcal{L}_{x'}^0)$ with base point $x' = \pi(p')$. Let $p_{F_x} = \tilde{F}_x(p) = pg_{F_x}$, that is, $g_{F_x} \in H_p$. Then, $p'_{F_{x'}} = (\tilde{F} \circ \tilde{F}_x \circ \tilde{F}^{-1})(p') = \tilde{F}(\tilde{F}_x(p)) = \tilde{F}(pg_{F_x}) = \tilde{F}(p) g_{F_x} = p' g_{F_x}$. In the last but one equality, it was used that $\tilde{F}$ is a Lie group isomorphism from $\pi^{-1}(x)$ to $\pi^{-1}(x')$. Hence, $g_{F_x} \in H_{p'}$, too:

If p can be parallel transported to p', then $H_p = H_{p'}$ and $H_p^0 = H_{p'}^0$.

It can be proved [1, vol. I, Sect. II.3] that

if M is pathwise connected (and paracompact), then for every $p \in (P, M, \pi, G)$ the holonomy group H_p is a Lie subgroup of G whose connected component of unity is H_p^0, while H_p / H_p^0 is countable.

As a very simple example reconsider the universal covering of S^1 by $\mathbb{R}$ of Fig. 6.1 on p. 181. At the end of Sect. 7.1 the universal covering of a connected manifold was considered as a principal fiber bundle, in the present case $(\mathbb{R}, S^1, \pi, \pi_1(S^1))$ where the bundle projection is $\pi : \mathbb{R} \ni t \mapsto \phi = e^{it} \in S^1$, and $\pi_1(S^1) \approx \mathbb{Z}$ is the fundamental group of the circle S^1. Since this is a discrete Lie group, its Lie algebra is trivial, and there are no vertical vector fields. Like the whole bundle $P = \mathbb{R}$, the horizontal space is one-dimensional and coincides with $\mathbb{R}$ at every point $p = t$, which is likewise the tangent space on S^1 at every point $x = e^{it}$. A lift of the loop based on $\phi = 1$ and running once around S^1 is an interval $[2\pi n, 2\pi(n+1)] \in \mathbb{R}$, $n \in \mathbb{Z}$. Hence, the holonomy group $H_t = \mathbb{Z}$ for every $t \in \mathbb{R} = P$, while $H_t^0 = 0$ (both groups in additive writing). If a loop F from $\phi = 1$ returns to $\phi = 1$ without running around S^1, then F^* from $2\pi n$ returns to $2\pi n$. $H_t = H_t/H_t^0 = \mathbb{Z}$ is a countable discrete Lie subgroup of $G = \pi_1(S^1)$, which in this case coincides with G itself.

The reader easily verifies that the holonomy group H_p for every point p of the Möbius band is $\{e, I\}$, while H_p^0 is again trivial.

Less trivial examples of holonomy groups will be considered later.

7.5 Exterior Covariant Derivative and Curvature Form

Like the $\mathfrak{g}$-valued 1-form ω, the connection form with property 2 on p. 215, consider more generally $\mathfrak{g}$-valued r-forms $\sigma = (\sigma^1, \dots, \sigma^{\dim G})$, so that $\langle \sigma^i, X_1 \wedge \cdots \wedge X_r \rangle$, $X_j \in \mathcal{X}(P)$, are real functions on P and

$$(R_g)^* \sigma = \mathrm{Ad}(g^{-1}) \sigma \quad \text{for every } g \in G. \tag{7.7}$$

Such a form is called a **pseudo-tensorial r-form of type** $(\mathrm{Ad}, \mathfrak{g})$. It is said to be horizontal, if $\langle \sigma_p^i, (X_1)_p \wedge \cdots \wedge (X_r)_p \rangle = 0$ whenever at least one of the tangent vectors $(X_j)_p$ at $p \in P$ is vertical (tangent to the fiber). A horizontal pseudo-tensorial r-form is called a **tensorial r-form**. Note that a connection form ω is vertical in this sense, it is a pseudo-tensorial 1-form of type $(\mathrm{Ad}, \mathfrak{g})$, but not a tensorial 1-form.

For every pseudo-tensorial r-form σ, a tensorial r-form ${}^h\sigma$ may be uniquely defined by

$$\langle {}^h\sigma, X_1 \wedge \cdots \wedge X_r \rangle = \langle \sigma, {}^hX_1 \wedge \cdots \wedge {}^hX_r \rangle. \tag{7.8}$$

Indeed, because of the r-linearity of σ, ${}^h\sigma$ is uniquely defined by the above relation, and together with σ it is of type $(\mathrm{Ad}, \mathfrak{g})$. Furthermore, it vanishes whenever at least one of the vectors X_j is vertical, which means that ${}^hX_j$ vanishes. For a connection form ω always ${}^h\omega = 0$ holds.

The **exterior covariant derivative** D of a pseudo-tensorial r-form σ is defined as

$$D\sigma = {}^h(d\sigma). \tag{7.9}$$

It is a linear mapping from pseudo-tensorial r-forms to tensorial $(r+1)$-forms. Indeed, by the linearity of the exterior derivative, together with σ the exterior derivative $d\sigma$ is a pseudo-tensorial form.

The tensorial 2-form

$$\Omega = D\omega \tag{7.10}$$

is called the **curvature form** of the connection Γ given by the connection form ω. This name derives from the geometric meaning in the case of the Riemannian geometry (Chap. 9). Recall that there is no angle between a vector and a covector, both living in different spaces, nevertheless one often speaks of orthogonality, if a covector annihilates a vector, $\langle\omega, X\rangle = 0$. Likewise, there is no radius of curvature of a manifold not having gotten a metric. Nevertheless, the curvature form measures the deviation of parallel transport between two points along distinct paths, and the manifold is said to be flat (see below), if the curvature form vanishes.

Let $X = {}^hX$ and $Y = {}^hY$ be two horizontal tangent vectors at $p \in P$. Then, (7.8) yields $\langle {}^h\sigma, X \wedge Y\rangle = \langle\sigma, X \wedge Y\rangle$ for any pseudo-tensorial 2-form σ. Hence, $\langle d\omega, X \wedge Y\rangle = \langle\Omega, X \wedge Y\rangle$ in this case. Now, let $X = {}^hX$ further be horizontal and $Y' = {}^vY'$ be vertical. Continue X to a horizontal vector field on P and Y' to the uniquely defined (vertical) fundamental vector field $Y^* = R^*_*(Y)$, equal to Y' at p and corresponding to $Y \in \mathfrak{g}$. First of all, according to (3.37), $[X, Y^*] = -[Y^*, X] = -\lim_{t\to 0}((\phi_{-t})_*(X) - X)/t$ where the 1-parameter group ϕ_t created by Y^* is a subgroup of G and therefore it leaves the horizontal vector field X horizontal. Hence, $[X, Y^*]$ is a horizontal vector field. Now, (4.49) yields $\langle d\omega, X \wedge Y^*\rangle = -L_X\langle\omega, Y^*\rangle + L_{Y^*}\langle\omega, X\rangle - \langle\omega, [X, Y^*]\rangle = 0$. The first Lie derivative vanishes since $\langle\omega, Y^*\rangle = \langle\omega, R^*_*(Y)\rangle = Y$ is constant, in the second and third $\langle\omega, \ldots\rangle = 0$ since the argument is horizontal. Finally, if both X' and Y' are vertical and X^* and Y^* are the corresponding fundamental vector fields, then $\langle d\omega, X^* \wedge Y^*\rangle = -\langle\omega, [X^*, Y^*]\rangle = -[X, Y] = -[\langle\omega, X^*\rangle, \langle\omega, Y^*\rangle]$. Again the two Lie derivatives vanish as derivatives of a constant, and in the remaining term $\langle\omega, X^*\rangle = X$ was used twice.

Let $X, Y \in T_p(P)$ be two arbitrary tangent vectors, decompose them into their horizontal and vertical components and continue them into tangent vector fields as above. By virtue of the bilinearity of the 2-form $d\omega$, E. Cartan's **structure equations** for a connection ω on a principal fiber bundle,

$$\langle d\omega, X \wedge Y\rangle = -[\langle\omega, X\rangle, \langle\omega, Y\rangle] + \langle\Omega, X \wedge Y\rangle, \tag{7.11}$$

are obtained. In symbolic writing they are often expressed as $d\omega = -[\omega, \omega] + \Omega$. Eq. 7.11 is a $\mathfrak{g}$-valued equation consisting of $\dim G$ real equations. They may be obtained by introducing a base $\{E_i | i = 1, \ldots, \dim G\}$ in $\mathfrak{g}$ with corresponding structure constants c^k_{ij}. Then, $\omega = \sum_i \omega^i E_i$, $\Omega = \sum_i \Omega^i E_i$ and from the left $\mathrm{Ad}(g)$ invariance of ω and (6.3) one has

$$d\omega^i = -\frac{1}{2}\sum_{jk} c^i_{jk}\omega^j \wedge \omega^k + \Omega^i. \tag{7.12}$$

(In addition the obvious relation $\sum c^i_{jk}\omega^j\omega^k = \sum(1/2)c^i_{jk}\omega^j \wedge \omega^k$ following from the properties of the structure constants was used, observe that each vector component $d\omega^i$ of the $\mathfrak{g}$-vector is a 2-form in $\Lambda_2(T^*_p(P))$, and the wedge-product of 1-forms is such a 2-form.)

A word on notation. In exterior calculus the convention

$$[\omega, \sigma]_{i_1\ldots i_{r+s}} = (\omega_{i_1\ldots i_r}\sigma_{i_{r+1}\ldots i_{r+s}} - \sigma_{i_{r+1}\ldots i_{r+s}}\omega_{i_1\ldots i_r}) \tag{7.13}$$

is used. If ω and σ are matrices, then the matrix element $(\omega_{i_1\ldots i_r}\sigma_{i_{r+1}\ldots i_{r+s}})^k_l$ may not be the same as $(\sigma_{i_{r+1}\ldots i_{r+s}}\omega_{i_1\ldots i_r})^k_l$, and one of them may even not be defined according to the concatenation rule for matrices. Then, $[\omega, \sigma]$ would not exist. However, in general $[\omega, \omega] = \omega \wedge \omega$ for a 1-form, while $([\sigma, \omega] + [\omega, \sigma])_{ij} = (\sigma_i\omega_j - \omega_j\sigma_i + \omega_i\sigma_j - \sigma_j\omega_i)$ need not vanish for general 1-forms, and hence generally it may be that $[\sigma, \omega] \neq -[\omega, \sigma]$. In analogy to the derivation of (7.11), the exterior covariant derivative of a tensorial 1-form σ may be obtained as $D\sigma = d\sigma + ([\sigma, \omega] + [\omega, \sigma])/2$.

Like the local connection forms ω_α of a connection form ω, **local curvature forms** $\langle\Omega^i_\alpha, X_x \wedge Y_x\rangle = \langle\Omega^i_{s_\alpha(x)}, s_{\alpha*}(X_x) \wedge s_{\alpha*}(Y_x)\rangle = \langle s^*_\alpha(\Omega^i_{s_\alpha(x)}), X_x \wedge Y_x\rangle$ on open sets $U_\alpha \subset M$ of local bundle trivializations may be introduced with the help of the canonical local sections s_α, that is, $\Omega_\alpha = s^*_\alpha(\Omega)$ are pull backs of the curvature form on P to $U_\alpha \subset M$. However, since Ω is a tensorial form with the property (7.7) and since $\psi_{\alpha\beta*}({}^hX)$ vanishes, the transition relations are simply

$$\Omega_\beta = \mathrm{Ad}(\psi^{-1}_{\alpha\beta})\Omega_\alpha \quad \text{or} \quad \Omega_\beta = \psi^{-1}_{\alpha\beta}\Omega_\alpha\psi_{\alpha\beta}, \tag{7.14}$$

where the second relation again holds, if G is a subgroup of $Gl(n, K)$ in matrix notation. Since a pull back is a homomorphism of exterior algebras commuting with the exterior differentiation, one immediately has $d\omega_\alpha = -[\omega_\alpha, \omega_\alpha] + \Omega_\alpha$.

Taking the exterior derivative of (7.12), one finds $0 = dd\omega^i = -\sum c^i_{jk}d\omega^j \wedge \omega^k + d\Omega^i$ as an equation of 3-forms. Let X, Y, Z be three horizontal tangent vectors at $p \in P$. Since ω^k annihilates horizontal vectors, it follows that $\langle d\Omega^i, X \wedge Y \wedge Z\rangle = 0$. In view of (7.8, 7.9), this may be expressed as

$$D\Omega = 0. \tag{7.15}$$

These are the **Bianchi identities** for the curvature form. Alternatively, for any pseudo-tensorial r-form $DD\sigma = D^h(d\sigma) = {}^h(dd\sigma)$, and $D^2 = 0$ is inherited from $d^2 = 0$; hence (7.15) immediately follows from (7.10).

On p. 218 the canonical flat connection of a trivial principal fiber bundle $P = M \times G$ was introduced. Consider the canonical Maurer–Cartan form ϑ of G and the projection $\pi_2 : M \times G \to G$. Then,

$$\omega = \pi_2^*(\vartheta) \tag{7.16}$$

is the canonical flat connection form. Indeed, the ϑ_g^i form a dual base to a base in $T_g(G)$ and hence $\pi_2^*(\vartheta) = (\pi_2 \circ \vartheta)^* = \vartheta^* \circ \pi_2^*$ pulls any vector $X \in T_{(x,g)}(P)$ first back to $T_g(G)$ and then isomorphically to $T_e(G)$. Hence, $\langle \omega, X \rangle = 0$, iff the pull back of X to $T_g(G)$ vanishes, that is, iff X is tangent to $M \times \{g\}$ (cf. Fig. 7.5).

Now, $d\omega = d(\pi_2^*(\vartheta)) = \pi_2^*(d\vartheta) = \pi_2^*(-[\vartheta, \vartheta]) = -[\pi_2^*(\vartheta), \pi_2^*(\vartheta)] = -[\omega, \omega]$, and hence $\Omega = 0$. In the third equality the Maurer–Cartan equations of a Lie group where used.

A connection in a general principal fiber bundle (P, M, π, G) is called a **flat connection**, if every point $x \in M$ has a neighborhood U for which there exists an isomorphism $F : \pi^{-1}(U) \to U \times G$ mapping horizontal spaces on $\pi^{-1}(U)$ to tangent spaces on $U \times \{g\}$. Since the above considerations were local ones, it is clear that $\Omega = 0$ for a flat connection. However, the reverse is also true, which is the result of three theorems presented here without proof (see for instance [1, vol. I, Chap. II]).

Reduction theorem: *Let (P, M, π, G) be a principal fiber bundle, let M be pathwise connected (and paracompact), and let Γ be a connection on P with connection form ω and curvature form Ω. For every $p \in P$, denote $P(p)$ the set of all points $p' \in P$ which may be parallel transported to p. Then, $P(p)$ is a reduced fiber bundle with the reduction of the structure group from G to H_p. Let $F : P(p) \to P$ be the corresponding bundle homomorphism with the push forward $F_* : \mathfrak{h}_p \to \mathfrak{g}$, and let Γ' be a connection on $P(p)$ with connection form ω' and curvature form Ω'. Then, $F_*(\omega') = F^*(\omega)$, $F_*(\Omega') = F^*(\Omega)$, where F^* pulls ω^i and Ω^i back from $\mathcal{D}(P)$ to $\mathcal{D}(P(p))$, however, still forming vectors of $\mathfrak{g}$.*

Ambrose–Singer theorem on holonomy: *In the settings and notation of the previous theorem, the Lie algebra $\mathfrak{h}_p$ is generated by all those elements of $\mathfrak{g}$ which may be expressed as $\langle \Omega_{p'}, X_{p'} \wedge Y_{p'} \rangle$, where $X_{p'}$ and $Y_{p'}$ are arbitrary horizontal vectors in $T_{p'}(P)$.*

Theorem on flat connections: *A connection on a principal fiber bundle is a flat connection, iff the corresponding curvature form vanishes.*

Let Γ be a flat connection on (P, M, π, G), $\Omega = 0$, and let M be connected. Let $p \in P$ be arbitrary and consider the holonomy bundle through p. Denote it $\tilde{M} = P(p)$. In view of the Ambrose–Singer theorem, $\mathfrak{h}_p$ is trivial. Hence, $\mathfrak{h}_p^0$ is also trivial, and, since H_p^0 is a connected Lie subgroup of G and hence uniquely defined by $\mathfrak{h}_p^0$, it is also trivial. Consequently, $H_p = H_p/H_p^0$ is countable and therefore discrete. It follows that $\tilde{M}$ is a covering space of M. In particular, if M is simply connected, then P is isomorphic to the trivial bundle $M \times G$ and Γ is isomorphic to the canonical flat connection of the latter.

The theory of principal fiber bundles forms the base of the theory of more special and important vector bundles considered in the following sections. However, it also yields immediately the mathematics of gauge field theories and, more generally, of geometric phases (Berry phases) in quantum physics, which will be considered in the next chapter.

7.6 Fiber Bundles

Before more general covariant derivatives of parallel transport of vector and tensor fields are considered with the help of a connection, as a further step more special structure is introduced into fiber bundles.

A general bundle over M is a triple (E, M, π) of two topological spaces, E and M and a smooth surjective mapping $\pi : E \to M$. In a fiber bundle (E, M, π_E, F, G), M is a manifold (locally homeomorphic to $\mathbb{R}^m$ for some $m = \dim M$), and all spaces $\pi_E^{-1}(x)$, $x \in M$ are isomorphic to each other and isomorphic to a manifold F, the typical fiber. Moreover, there is a Lie group G of transformations of F which introduces more structure into F (for instance the group $Gl(n, K)$, $n = \dim F$ introduces the structure of a K-vector space into F) and which in physics often has the meaning of a symmetry group. In a principle fiber bundle the typical fiber is the group G itself which acts on itself from the right. In order to adjust the action of G to the fiber bundle (E, M, π_E, F, G), it is incorporated by a principle fiber bundle (P, M, π, G).

A **fiber bundle** (E, M, π_E, F, G), or in short E, consists of

1. a principal fiber bundle (P, M, π, G),
2. G acts on F from the left, that is, $G \times F \to F : (g,f) = gf,\ g \in G,\ f \in F$, is a linear mapping and hence a representation of the Lie group G,
3. $E = P \times_G F$, that is, $(p,f) = (pg, g^{-1}f)$ is an equivalence relation R in $P \times F$, and $E = (P \times F)/R$, the elements of E are denoted $p(f)$,
4. $\pi_E : E \to M : p(f) \mapsto \pi(p)$,
5. every local diffeomorphism $\pi^{-1}(U) \sim U \times G, U \subset M$, induces a local diffeomorphism $\pi_E^{-1}(U) \sim U \times F$.

Item 3 may be understood as a mapping $p : F \to \pi_E^{-1}(x) \subset E, x \in M : f \mapsto p(f)$ of the typical fiber F into E. In this respect, an **isomorphism of fibers** is a mapping $p \circ p'^{-1} : \pi_E^{-1}(x') \to \pi_E^{-1}(x)$ where $x' = \pi(p'), x = \pi(p)$. Since $x' = x$ implies $p' = pg^{-1}$ for some $g \in G$, $p \circ p'^{-1} = p \circ g \circ p^{-1}$ in this case, the group of automorphisms of a fiber $\pi_E^{-1}(x)$ is isomorphic to the structure group itself. Item 5 fixes the topology in E in such a way that for every local trivialization of $\pi^{-1}(U)$ in P there is a local trivialization of $\pi_E^{-1}(U)$ in E. Of course, this is only possible, if there exists a bijection between $\pi_E^{-1}(U)$ defined by the previous items and $U \times F$. Consider $(U \times G \times F)/R = \{\{(x, gg', g'^{-1}f) | g' \in G\}\}$. Choosing $g' = g^{-1}$, any

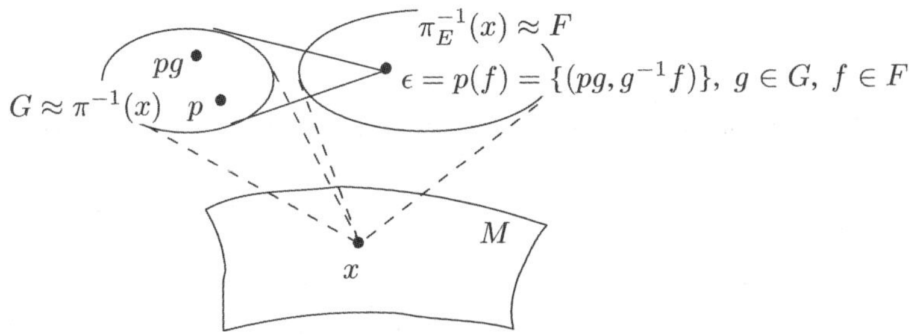

Fig. 7.6 Sketch of a fiber bundle (E, M, π_E, F, G) associated with a principal fiber bundle (P, M, π, G). A point ϵ of a fiber over $x \in M$ is an equivalence class of pairs $(pg, g^{-1}f)$

point of this set may be represented as (x, e, gf), and since $F = eF \subset GF \subset F$ for a representation of G in F, the distinct points of this type are in one–one and onto correspondence with the points of $U \times F$. This also shows that the fibers of E are isomorphic to the **typical fiber** F.

M is again the **base space** of the bundle and E is called the **bundle space,** π_E is called the **bundle projection,** $\pi_E^{-1}(x)$ is the **fiber** over $x \in M$, and G is the **structure group** of the fiber bundle (E, M, π_E, F, G) associated with the principal fiber bundle (P, M, π, G).

This appears to be a quite complex definition, nevertheless the structure of a fiber bundle (Fig. 7.6) is very common in analysis and physics as seen from the examples below. By definition, every fiber bundle E is based on a principle fiber bundle P. In this respect, a fiber bundle is more special than its principle fiber bundle, it has additional structure, introduced by an additional typical fiber F. On the other hand, taking F as the primary structure as in many applications, a principal fiber bundle may appear as a special case of a fiber bundle, in which the typical fiber F and the structure group G (the typical fiber of P) coincide. Many texts treat the principal fiber bundle in this sense as a special case after having introduced into the theory of (general) fiber bundles.

In the latter sense, a **local section** of a fiber bundle (E, M, π_E, F, G) assigns a point ϵ of the fiber $\pi_E^{-1}(x)$ over x to every point $x \in U \subset M$. Thus it is defined as a smooth function $s : M \supset U \to E$ for which $\pi_E \circ s = \mathrm{Id}_U$, and if this holds for all M, then s is called a (global) **section** of E.

Before continuing with the general theory, for illustration a number of important examples are now considered which will be treated in more detail subsequently.

Let $V \approx K^n$ be an n-dimensional K-vector space, $K = \mathbb{R}$ or $\mathbb{C}$, so that $\mathrm{Aut}(V) \approx Gl(n, K)$ is the Lie group of general linear transformations of V. Let (P, M, π, G) be some principal fiber bundle, and fix a representation R of G in $Gl(n, K)$. The fiber bundle (E, M, π_E, V, G) with the left action R of G on V is called a (real or complex) **vector bundle** over the manifold M with the structure group G. Sections s on M are (smooth) **vector fields** on M of the type V. (Consider electromagnetic fields on space–time as an example.) As physicists are well aware of, a vector is not just a column of numbers with respect to the fixed canonical base of the typical space K^n. Instead it is a physical entity which has a meaning independent of any base. If G is the group $Gl(n,K)$, then pg can be understood as transformation from a base p to another equivalent base pg by applying g from the right to p. (Compare the frame bundles of Sect. 7.2.) If the vector with respect to the base p is represented by the column f, then *the same vector* is represented with respect to the base pg by the transformed column $g^{-1}f$. Precisely in this sense a vector bundle associated with a principal fiber bundle is needed to give a general vector field on M (not just a tangent vector field) a meaning independent of a reference base at each point x of M (compare (3.11) with (3.14)).

The set $\mathcal{S}(M)$ of all sections on M forms an infinite-dimensional vector space (functional space of vector fields) with respect to pointwise addition or multiplication by a constant $k \in K$. Pointwise means at points x of M, or within fibers $\pi_E^{-1}(x)$ of E. Addition and multiplication means, if $\epsilon_1 = p(f_1)$ and $\epsilon_2 = p(f_2)$ where $p \in \pi^{-1}(x)$ and $\epsilon_i \in \pi_E^{-1}(x)$, then $\epsilon_1 + \epsilon_2 = p(f_1 + f_2)$ and $k\epsilon_1 = p(kf_1)$. If the product of a smooth function $F \in \mathcal{C}(M)$ with a vector field $s \in \mathcal{S}(M)$ is pointwise taken, $(Fs)(x) = F(x)s(x)$, then $\mathcal{S}(M)$ may also be considered as a module over the ring $\mathcal{C}(M)$ of smooth functions. Every vector bundle has trivially the global section $x \mapsto 0$. It can be shown with the partition of unity technique, that for paracompact M every local section of a vector bundle and more generally of a fiber bundle the typical fiber F of which is contractible, given on a closed subset of M, can be continued into a global section; what does not always exist as will be shown in Sect. 8.2 below is a vector field *without nodes*.

Let (E, M, π_E, V, G) and $(E', M, \pi_{E'}, V', G)$ be two vector bundles over the same manifold M. The **sum of vector bundles** which is also called the **Whitney sum**, $(E \oplus E', M, \pi_{E\oplus E'}, V \oplus V', G)$, or in short $E \oplus E'$, is a vector bundle over M the typical fiber of which is the direct sum $V \oplus V'$ of vector spaces V and V' with the obvious bundle projection ($\pi_{E\oplus E'}^{-1}(x) = \pi_E^{-1}(x) \oplus \pi_{E'}^{-1}(x)$). The left action of the (common) structure group G on $V \oplus V'$ is the direct sum of representations $R \oplus R'$ from E and E'. The sum of more than two items is defined analogously. Likewise, the **tensor product of vector bundles**, $(E \otimes E', M, \pi_{E\otimes E'}, V \otimes V', G)$, or in short $E \otimes E'$, is a vector bundle over M the typical fiber of which is the tensor product $V \otimes V'$ of vector spaces V and V' again with the obvious bundle projection. The left action of the structure group G on $V \otimes V'$ is the tensor product $R \otimes R'$ of representations (in the obvious meaning of the tensor product of transformation matrices, cf. (4.7)). Again, the tensor product of more than two factors is defined analogously. Likewise, the **exterior product of vector bundles** is obtained.

Let V^* be the dual space to V, that is, $\langle \omega, X \rangle \in K$, $\omega \in V^*$, $X \in V$ is bilinear. The **dual bundle**, $(E^*, M, \pi_{E^*}, V^*, G)$, or in short E^*, is a vector bundle over M the typical fiber of which is V^* and the representation of G in V^* is the dual R^* of the representation R of G in V, that is, $\langle R^*(g)\omega, R(g)X \rangle = \langle \omega, X \rangle$ for all $g \in G$. Hence, $\langle p(\omega), p(X) \rangle = \langle pg(\omega), pg(X) \rangle$ for $p \in P$, $p(\omega) \in E^*$, $p(X) \in E$, is a bilinear scalar invariant under the action of G. (Think for instance of an electric field as an element of V and an electric dipole density as an element of V^* under the group of rotation, both on a spatial manifold M.)

In particular, the **tangent bundle** $T(M) = (T(M), M, \pi_T, K^m, Gl(m, K))$, $m = \dim M$ (p. 106) is an m-dimensional K-vector bundle associated with the frame bundle $L(M)$ as principal fiber bundle over M. It is easily seen that $\pi_T^{-1}(x) \approx T_x(M)$ is the tangent space on M at x and $\mathcal{S}(T(M)) = \mathcal{X}(M)$ is the space of tangent vector fields. The structure group $Gl(m, K)$ ensures that tangent vector fields have an unambiguous meaning independent of local coordinate systems and independent of the choice of a local frame. The dual of the tangent bundle is the **cotangent bundle** $T^*(M) = (T^*(M), M, \pi_{T^*}, K^m, Gl(m, K))$. Its fibers $\pi_{T^*}^{-1}(x) \approx T_x^*(M)$ are the cotangent spaces on M at x and its sections form the space $\mathcal{S}(T^*(M)) = \mathcal{D}^1(M)$ of differential 1-forms. Finally, by taking the tensor product of r factors $T(M)$ and s factors $T^*(M)$ one obtains the **tensor bundle** $T_{r,s}(M)$ of type (r, s) over M, and by taking the exterior product of r factors $T^*(M)$ one obtains the **exterior** r **bundle** $\Lambda_r^*(M)$ over M.

To a physicist, tensor bundles associated with frame bundles elucidate the usefulness of the definition of fiber bundles: In order to express a tensor in numbers, a frame is needed. Transforming the frame into another equivalent one demands to transform the tensor components inversely.

Now, the question of reducibility (p. 209) of a principal fiber bundle can be reconsidered. Let (P, M, π, G) be a principal fiber bundle, and let H be a closed Lie subgroup of G (Fig. 7.7). It was already shown that $(G, G/H, \pi_G, H)$ is a principal fiber bundle with base space G/H, bundle projection $\pi_G : g \mapsto gH$ and structure group H. The left cosets $gH, g \in G$ form the quotient space G/H on which G acts from the left. Since G acts on P from the right, H as its subgroup acts also on P from the right. The orbits $pH \subset P$ of this action form the quotient space P/H (in which p and $ph, h \in H$ form the same point pH). Hence, the fiber bundle

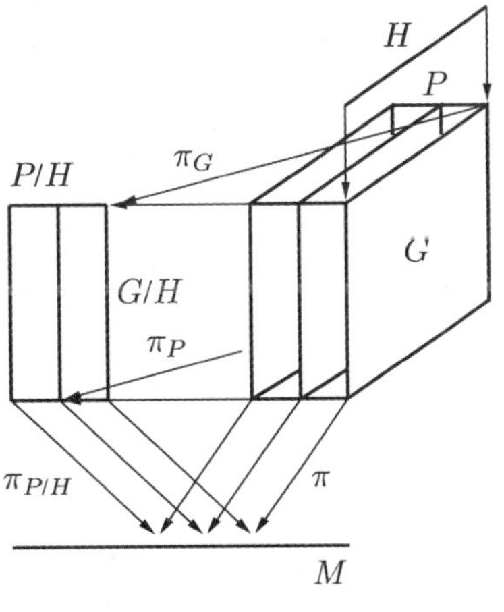

Fig. 7.7 A sketch of the interrelations between the bundles (P, M, π, G), $(P/H, M, \pi_{P/H}, G/H, P)$, $(G, G/H, \pi_G, H)$, $(P, P/H, \pi_P, H)$

$(P/H, M, \pi_{P/H}, G/H, G)$ associated with (P, M, π, G) may be considered with the typical fiber G/H and the bundle projection $\pi_{P/H} : P/H \to M$ induced by $\pi : P \to M$ in an obvious manner. It is not difficult to see that $(P, P/H, \pi_P, H)$ is also a principal fiber bundle with base space P/H, bundle projection $\pi_P : p \mapsto pH$ and structure group H. Indeed, let $U \in M$ yield a local trivialization $\pi_{P/H}^{-1}(U) \approx U \times G/H$ of the fiber bundle $(P/H, M, \pi_{P/H}, G/H, G)$ and let $V \in G/H$ be so that $\pi_G^{-1}(V) \approx V \times H \subset G$. Then $U \times V \subset U \times G/H \approx \pi_{P/H}^{-1}(U)$. There is $W \subset \pi_{P/H}^{-1}(U)$ which corresponds to $U \times V$ by the latter isomorphism, and $\pi_P^{-1}(W) \approx W \times H$. Hence, $(P, P/H, \pi_P, H)$ is locally trivial.

The structure group G of the principal fiber bundle (P, M, π, G) can be reduced to the closed Lie subgroup H, iff the associated fiber bundle $(P/H, M, \pi_{P/H}, G/H, G)$ has a section $s : M \to P/H$.

Proof Let G be reducible to H and let (P', M, π', H) be the reduced principal fiber bundle with the corresponding bundle embedding $F : P' \to P$. Let π_P be the projection from P to P/H in the principal fiber bundle $(P, P/H, \pi_P, H)$. If p' and p'' lie in the same fiber of P', then $p'' = p'h$ with some $h \in H$. Therefore, $\pi_P(F(p'')) = \pi_P(F(p')h) = \pi_P(F(p'))$ does not depend on $p' \in \pi'^{-1}(x)$ but depends only on $x \in M$. Hence, $s = \pi_P \circ F : M \to P/H$ is a section on $(P/H, M, \pi_{P/H}, G/H, G)$.

Conversely, let $s : M \to P/H$ be a section on $(P/H, M, \pi_{P/H}, G/H, G)$. For every $x \in M$, $\pi_P^{-1}(s(x)) \subset P$ is non-empty. Let p' and p'' belong to this set which implies $p'' = p'h$ for some $h \in H$. Since G acts freely on P and H is a subgroup of G, H acts also freely on P, that is, $\pi_P^{-1}(s(x)) \approx H$ is a fiber over $x \in M$. Let $P' = \pi_P^{-1}(s(M)) \subset P$, it is not difficult to see that (P', M, π', H) with $\pi' = \pi|_{P'}$ is a principal fiber bundle, reduced from (P, M, π, G) by reduction of G to H. □

As was already mentioned, every fiber bundle (E, M, π_E, F, G) with a contractible typical fiber F has a section. Since the elements of $Gl(m, \mathbb{R})$ may be expressed by matrices e^A with general real $m \times m$-matrices A, and the elements of $O(m)$ may in the same manner be expressed with skew-symmetric matrices A, the quotient space $Gl(m, \mathbb{R})/O(m)$, the space of linear deformations of the $\mathbb{R}^m$, is given by matrices e^A with A symmetric. Hence, $Gl(m, \mathbb{R})/O(m)$ is diffeomorphic to the $m(m+1)/2$-dimensional real space of symmetric $m \times m$-matrices A, which is a vector space. Hence, the typical fiber of $(L(M)/O(m), M, \pi_{L(M)/O(m)}, Gl(m, \mathbb{R})/O(m), Gl(m, \mathbb{R}))$ is contractible and the bundle has a section, which means that the frame bundle $(L(M), M, \pi, Gl(m.\mathbb{R}))$ can be reduced to $(L_O(M), M, \pi', O(m))$, where $L_O(M)$ consists of orthonormalized frames of orthonormal base vectors only, and π' is the corresponding restriction of π. (Here, normalization of the orthogonal frames is just an admissible convention, since $O(m)$ preserves norm of vectors.)

Analogously, the complex frame bundle $(L(M), M, \pi, G(m, \mathbb{C}))$ can be reduced to $(L_U(M), M, \pi', U(m))$, again consisting of frames of orthonormalized base vectors, but this time unitarily related over the field of complex numbers.

7.7 Linear and Affine Connections

Linear and affine connections are special connections on vector bundles. Before considering them, the parallel transport is generalized from principal fiber bundles to general fiber bundles.

Let (E, M, π_E, F, G) be a fiber bundle associated with the principal fiber bundle (P, M, π, G), let a connection Γ on P be given, and let $\epsilon = p(f))$, $f \in F$, be any point of E (see 3 of the definition of a fiber bundle on p. 226). The point $\epsilon = \{(pg, g^{-1}f) | g \in G\}$ can be represented (for $g = e$) by the point p of the principal fiber bundle P and the point f of the typical fiber F. The tangent space $T_\epsilon(E)$ on E at point ϵ is split into the direct sum of the **vertical and horizontal spaces**, $T_\epsilon(E) = F_\epsilon \oplus Q_\epsilon$. The vertical space F_ϵ is by definition tangent to the fiber $p(F) = \pi_E^{-1}(\pi(p)) \subset E$. Since $p(F) \approx F$, it holds that $F_\epsilon \approx T_f(F)$, $\dim F_\epsilon = \dim F$. Now, consider the projection $P \times F \to E : (p,f) \mapsto \epsilon$. Fixing f yields the restriction $\pi_f : P \times \{f\} \to E$. The image of Q_p of the connection Γ by its push forward, $\pi_{f*} : T_p(P) \to T_\epsilon(E)$, is by definition the horizontal space $Q_\epsilon = \pi_{f*}(Q_p)$. Represent ϵ by $(pg, g^{-1}f)$ instead and consider $Q_{pg} = (R_g)_* Q_p$ and $\pi_{g^{-1}f} : P \times \{g^{-1}f\} \to E$. Now, $\pi_{g^{-1}f*}(Q_{pg}) = \pi_{g^{-1}f*} \circ (R_g)_*(Q_p) = \pi_{f*}(Q_p) = Q_\epsilon$, and, as it should be, the definition of Q_ϵ does not depend on the chosen representative of ϵ from $P \times F$. The projection π_f induces a local mapping $\pi_f|_{U \times G} : U \times G \times \{f\} \to U \times F : ((x,g),f) \mapsto (x,(e,g^{-1}f))$ or $(x,g) \mapsto (x, g^{-1}f)$ which maps fibers of P into fibers of E over the same point x and thus implies a mapping Id_U. Hence, $Q_p \approx T_{\pi(p)}(M) = T_{\pi_E(\epsilon)}(M) \approx Q_\epsilon$, and $\dim Q_p = \dim M = \dim Q_\epsilon$ with the consequence $\dim F_\epsilon + \dim Q_\epsilon = \dim T_\epsilon(E)$. Moreover, F_ϵ and Q_ϵ are obviously linearly independent and thus indeed $T_\epsilon(E) = F_\epsilon \oplus Q_\epsilon$.

A (horizontal) **lift Φ^* of the path** $\Phi : I \to M, I = [0,1] \subset \mathbb{R}$, in E is a path $\Phi^* : I \to E$ which is projected to Φ so that $\pi_E \circ \Phi^* = \Phi$ and which has a horizontal tangent vector in every of its points $\Phi^*(t)$, $t \in I$. (In this section a path is denoted by Φ because F is reserved for the typical fiber here.) Like in the case of a principal fiber bundle (p. 220 f), if Φ is a path from x_0 to x_1, then for every $\epsilon_0 \in \pi_E^{-1}(x_0)$ there is a uniquely defined lift Φ^* which transports ϵ_0 to a uniquely defined point $\epsilon_1 \in \pi_E^{-1}(x_1)$. Indeed, if (p_0, f) is a representation of $\epsilon_0 = p_0(f)$ and $\Phi_P^* : t \mapsto p_t$ is the unique lift of Φ in P starting at p_0, then $\epsilon_t = p_t(f)$ is the lift Φ^*. It is the **parallel transport** along the path Φ from ϵ_0 to ϵ_1. A local section $s : U \to E$ is called parallel, if $s_*(T_x(M)) = Q_{s(x)}$ at every $x \in U$. A **parallel section** s (local or global) is parallel transported into itself.

Now, the considerations are specialized to vector bundles (E, M, π_e, V, G), where $V \approx K^n$, a representation R of G in $Gl(n, K)$ is operative as the left action of G on V, and a connection Γ on the principal fiber bundle (P, M, π, G) is fixed. It is this situation for which the **covariant derivative** of vector fields is introduced (Fig. 7.8, next page).

Let $s : M \supset U \to E$ be a local section (smooth V-vector field) on U, let $\Phi : I \to U$ be a path in U and let $X = \Phi_*^t(\partial/\partial t) \in T_{x_t}(M)$ be a tangent vector on M at

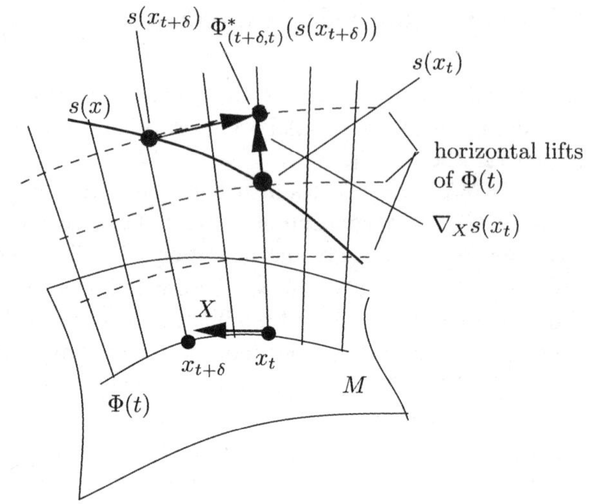

Fig. 7.8 A sketch of the covariant derivative of a vector field $s(x)$. (One single vector component of $s(x)$ is drawn)

$x_t = \Phi(t) \in U$ for some $t \in]0,1[$ which is tangent to Φ (in local coordinates $X = \sum_{i=1}^{\dim M}(\partial x_t^i/\partial t)(\partial/\partial x^i)$) and pushed forward by Φ^t_* from $\partial/\partial t \in T_t(I)$. Then, the covariant derivative of s at x_t in the direction of X is defined as

$$\nabla_X s(x_t) = \lim_{\delta\to 0}\frac{\Phi^*_{(t+\delta,t)}(s(x_{t+\delta})) - s(x_t)}{\delta}, \tag{7.17}$$

where $\Phi^*_{(t+\delta,t)}$ means the parallel (or horizontal) transport from $x_{t+\delta}$ to x_t along the (inverted) path Φ. It is intuitively clear and not difficult but tedious to show that the right hand side expression depends on X but not on the actual path Φ to which X is tangent at x_t. The same notation as on the left hand side above is used, if $X \in \mathcal{X}(M)$ is a tangent vector field (that is, $\nabla_X s(x) = \nabla_{X_x} s(x)$). For a (local) section (V-vector field) s in E, $\nabla_X s$ is again a (local) section (V-vector field) in E. For a parallel section s, the numerator of the right hand side expression vanishes, since the parallel transport brings $s(x_{t+\delta})$ back to $s(x_t)$. Hence, $\nabla_X s = 0$ for all X for a parallel section s.

It is easy to convince oneself of the additivity of the covariant derivative with respect to X and s:

$$\nabla_{X_1+X_2} s = \nabla_{X_1} s + \nabla_{X_2} s, \quad \nabla_X(s_1+s_2) = \nabla_X s_1 + \nabla_X s_2. \tag{7.18}$$

The second relation is obvious and the first can be obtained by using vector fields defined on U and their families of integral curves with smoothness arguments (Fig. 7.9, the analysis is again straightforward but tedious). It is also clear that a rescaling of δ only in the numerator of (7.17), which is equivalent to an inverse rescaling of the denominator only, amounts to the same as a rescaling of X. Moreover, if λ is a smooth K-valued function on M, then one has $\Phi^*_{(t+\delta,t)}$

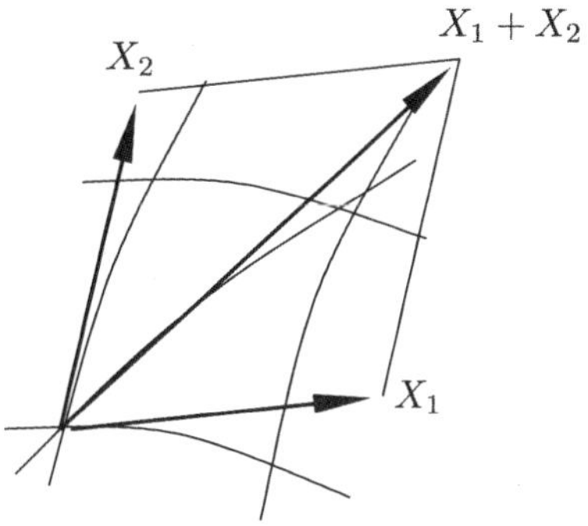

Fig. 7.9 Families of integral curves of tangent vector fields

$(\lambda(x_{t+\delta})s(x_{t+\delta})) = \lambda(x_{t+\delta})\Phi^*_{(t+\delta,t)}(s(x_{t+\delta}))$ and $\lim_{\delta\to 0}(\lambda(x_{t+\delta}) - \lambda(x_t))/\delta = X\lambda$. Hence,

$$\nabla_{\lambda X}s = \lambda\nabla_X s, \quad \nabla_X(\lambda s) = \lambda\nabla_X s + (X\lambda)s. \tag{7.19}$$

If the X are tangent vector fields on M (or on $U \subset M$), then all relations (7.18, 7.19) are relations between sections in E (V-vector fields).

By the very definition of a fiber bundle, it is associated with a principal fiber bundle. A connection, defined on the principal fiber bundle determines the parallel transport also on the associated fiber bundle. If the latter is a vector bundle, covariant derivatives are defined on the basis of the parallel transport. There are ample examples of vector bundles in physics. For instance matter fields are described by vectors of representations of abstract groups of 'inner' symmetry ($SU(2) \times U(1)$ in electroweak theory, or $SU(3) \times SU(3) \times U(1)$ in quantum chromodymanics) which are functions of position in the base manifold M being space–time in these cases. The structure of M itself determines the 'outer' four-tensor symmetry of each of the above vector components. This latter structure is the subject of tangent, cotangent and general tensor bundles, and is now considered.

Recall, that tangent, cotangent and tensor bundles are associated with the frame bundle $(L(M), M, \pi, Gl(m,\mathbb{R}))$, $m = \dim M$ as principal fiber bundle. (Here, the real case is considered.) Connections on $L(M)$ are called **linear connections** and were considered at the end of Sect. 7.3. There, m standard horizontal vector fields X_i were defined by (7.6), the values of which at any point $p \in L(M)$ span the horizontal space: $Q_p = \mathrm{span}_{\mathbb{R}}\{X_{ip} = B(\boldsymbol{X}_i)_p | i = 1,\dots,m\}$ where the $\boldsymbol{X}_i$ are taken to be any base of $\mathbb{R}^m$.

The standard horizontal vector fields were uniquely defined via (7.6) by two 1-forms: the connection form ω, in the present case of type $(\mathrm{Ad}, \mathrm{gl}(m,\mathbb{R}))$, that is, being a $\mathrm{gl}(m,\mathbb{R})$-valued pseudo-tensorial 1-form which transforms under the action of $G = Gl(m,\mathbb{R})$ according to the adjoint representation of G (cf. (7.7)) and whose exterior covariant derivative is the (tensorial) curvature form Ω, and by the soldering canonical $\mathbb{R}^m$-valued 1-form θ of (7.5). On p. 219 it was found that $\langle (R^*_g(\theta_{pg})), X_p\rangle = \langle g^{-1}\theta_p, X_p\rangle$, and hence, by the defining property (7.7), θ is a tensorial 1-form of type $(Gl(m,\mathbb{R}), \mathbb{R}^m)$. (It is tensorial, that is, horizontal, since

$\langle \theta_p, X_p \rangle = 0$ for every vertical vector X_p.) Since for the m standard horizontal vector fields X_i defined above $\langle \theta, X_i \rangle - \boldsymbol{X}_i$,

the tensorial 1*-form* θ *consists of* m 1*-forms* θ^i *which are dual to the standard horizontal vector fields* $X_i : \langle \theta^i, X_j \rangle = \delta^i_j$.

The tensorial 2-form of type $(Gl(m, \mathbb{R}), \mathbb{R}^m)$

$$\Theta = D\theta \tag{7.20}$$

is called the **torsion form** of the linear connection Γ which latter defines θ and ω.

Let $X, Y \in T_p(L(M))$. By definition (7.8, 7.9), if X and Y are two horizontal tangent vectors, then $\langle \Theta, X \wedge Y \rangle = \langle d\theta, X \wedge Y \rangle$. If X' and Y' both are vertical, then fundamental vector fields X^* and Y^* may be chosen whose values at p are X' and Y'. Since Θ as defined by (7.20) is horizontal, $\langle \Theta, X' \wedge Y' \rangle = 0$. On the other hand (cf. (4.49)), $\langle d\theta, X^* \wedge Y^* \rangle = L_{X^*}\langle \theta, Y^* \rangle - L_{Y^*}\langle \theta, X^* \rangle - \langle \theta, [X^*, Y^*] \rangle$. Since $R^*_* : \mathfrak{g} \to \mathcal{X}(\pi^{-1}(x))$ is an isomorphism of vector spaces, $[X^*, Y^*] = [R^*_*(X), R^*_*(Y)] = R^*_*([X, Y])$, and hence $[X^*, Y^*]$ is vertical. Thus, all three of the above right hand expressions for $\langle d\theta, X^* \wedge Y^* \rangle$ vanish because θ is horizontal. Hence, at p again $\langle d\theta, X' \wedge Y' \rangle = 0 = \langle \Theta, X' \wedge Y' \rangle$. It remains to consider the case where X is horizontal and (without loss of generality) equal to the value at p of the standard horizontal vector field $B(\boldsymbol{X}), \boldsymbol{X} \in \mathbb{R}^m$, and Y' is vertical and as above represented by the fundamental vector field Y^*. In this case, still $\langle \Theta, X' \wedge Y' \rangle = 0$ since Θ is horizontal. Moreover, $\langle d\theta, B(\boldsymbol{X}) \wedge Y^* \rangle = L_{B(\boldsymbol{X})}\langle \theta, Y^* \rangle - L_{Y^*}\langle \theta, B(\boldsymbol{X}) \rangle - \langle \theta, [B(\boldsymbol{X}), Y^*] \rangle$. The first expression on the right hand side vanishes again since Y^* is vertical. The second expression vanishes since $\langle \theta, B(\boldsymbol{X}) \rangle = \boldsymbol{X}$ is constant. It remains to analyze the last term. First of all (compare p. 223), $[B(\boldsymbol{X}), Y^*] = -[Y^*, B(\boldsymbol{X})] = -\lim_{t\to 0}((\phi_{-t})_*(B(\boldsymbol{X})) - B(\boldsymbol{X}))/t = -\lim_{t\to 0}(B(\tilde{\phi}_t \boldsymbol{X}) - B(\boldsymbol{X}))/t = -B(\lim_{t\to 0}(\tilde{\phi}_t \boldsymbol{X} - \boldsymbol{X})/t) = -B(Y\boldsymbol{X})$. In the present case, ϕ_t created by $Y^* = R^*_*(Y)$, $Y \in \mathfrak{g}$, is a 1-parameter subgroup of $Gl(m, \mathbb{R})$ which corresponds via R^*_* to $\tilde{\phi}_t = \exp(tY)$. In the last but one equality of the above chain of equations the linearity of the mapping $B : \mathbb{R}^m \to Q_p$ was used. Now recall that $Y = \langle \omega, Y^* \rangle$ and summarize $-\langle \theta, [B(\boldsymbol{X}), Y^*] \rangle = \langle \theta, B(Y\boldsymbol{X}) \rangle = Y\boldsymbol{X} = \langle \omega, Y^* \rangle \langle \theta, B(\boldsymbol{X}) \rangle$ or $\langle d\theta, X \wedge Y' \rangle = \langle \omega, Y' \rangle \langle \theta, X \rangle$. The order of terms in the last product matters since the first factor is $\mathfrak{gl}(m, \mathbb{R})$-valued and the second is $\mathbb{R}^m$-valued, the product (like $Y\boldsymbol{X}$ above) is a matrix product of an $(m \times m)$-matrix with an m-column vector.

By decomposing tangent vectors in their horizontal and vertical components and using the multi-linearity of forms, the **first structure equation** of a linear connection on a manifold M (that is, on its frame bundle $L(M)$)

$$\langle d\theta, X \wedge Y \rangle = -(\langle \omega, X \rangle \langle \theta, Y \rangle - \langle \omega, Y \rangle \langle \theta, X \rangle) + \langle \Theta, X \wedge Y \rangle, \tag{7.21}$$

is readily obtained. The **second structure equation**,

$$\langle d\omega, X \wedge Y \rangle = -[\langle \omega, X \rangle, \langle \omega, Y \rangle] + \langle \Omega, X \wedge Y \rangle, \qquad (7.22)$$

which is of course the same as in the general case, is repeated here for comparison.

By fixing a base $\{e_1, \ldots, e_m\}$ of $\mathbb{R}^m$ and a base $\{E_1^1, \ldots, E_m^1, E_1^2, \ldots, E_m^m\}$ of $\mathrm{gl}(m, \mathbb{R})$, with $\theta = \sum \theta^i e_i$, $\Theta = m\Theta^i e_i$, $\omega = \sum \omega_j^i E_i^j$, $\Omega = \sum \Omega_j^i E_i^j$ the structure equations may be written in components as

$$d\theta^i = -\sum_j \omega_j^i \wedge \theta^j + \Theta^i, \quad d\omega_j^i = -\sum_k \omega_k^i \wedge \omega_j^k + \Omega_j^i. \qquad (7.23)$$

The second equation compares to (7.12) with the structure constants (6.15) of the general linear group. These equations are symbolically often written as $d\theta = -\omega \wedge \theta + \Theta$, $d\omega = -\omega \wedge \omega + \Omega$. Besides the mnemonic power of such a writing, it demonstrates the algebraic power of E. Cartan's exterior calculus by focussing onto the exterior algebraic structure of the expressions and not diverting by the maybe quite complex inner structure (hence the name exterior calculus). Of course, using it needs a certain routine. In particular, like in operator calculus it is strongly recommended never to change the order of factors in expressions obtained. (Compare the product $\omega\theta$ above.)

There is a choice of standard horizontal vector fields B_i and of fundamental vector fields E_i^{j*} determined by

$$\langle \theta^k, B_i \rangle = \delta_i^k, \quad \langle \theta^k, E_i^{j*} \rangle = 0, \quad \langle \omega_l^k, B_i \rangle = 0, \quad \langle \omega_l^k, E_i^{j*} \rangle = \delta_i^k \delta_l^j, \qquad (7.24)$$

which form an absolutely parallel base of $T_p(L(M))$ of horizontal and vertical vectors at every point p and thus provide the decomposition of any X_p which could not explicitly be given by the displayed expressions before (7.5). (It is not difficult to see that the tangent vectors B_i and E_i^{j*} are nowhere zero and everywhere linearly independent.)

Taking the exterior covariant derivative of $d\theta$ and using the first structure equation yields $0 = -d\omega \wedge \theta + \omega \wedge d\theta + d\Theta$. Therefore, $\langle D\Theta, X \wedge Y \wedge Z \rangle = \langle d\Theta, {}^hX \wedge {}^hY \wedge {}^hZ \rangle = \langle d\omega \wedge \theta, {}^hX \wedge {}^hY \wedge {}^hZ \rangle - \langle \omega \wedge d\theta, {}^hX \wedge {}^hY \wedge {}^hZ \rangle$. The last term vanishes because ω vanishes on horizontal vector fields. The first term is equal to $\langle \Omega \wedge \theta, {}^hX \wedge {}^hY \wedge {}^hZ \rangle$ which on its part is equal to $\langle \Omega \wedge \theta, X \wedge Y \wedge Z \rangle$, since $\Omega \wedge \theta$ as the (wedge) product of two horizontal forms is horizontal. Summarizing, the first **Bianchi identity**

$$D\Theta = \Omega \wedge \theta \qquad (7.25)$$

is obtained while the **second Bianchi identity** is as previously $D\Omega = 0$, (7.15). As an example of the rule not to change the order of factors in exterior calculus (here the order of the forms Ω and θ), the application of (7.25) to three vectors is presented:

$$\langle D\Theta, X \wedge Y \wedge Z\rangle = \langle \Omega, X \wedge Y\rangle\langle \theta, Z\rangle + \langle \Omega, Y \wedge Z\rangle\langle \theta, X\rangle + \langle \Omega, Z \wedge X\rangle\langle \theta, Y\rangle.$$

The left hand side is an alternating 3-form applied to an alternating product of three vectors (trilinear mapping to real numbers). It is invariant under common alternation of the components of the form and the vector product. This invariance is used to keep the order of the form components fixed. (The three anti-cyclic permutations of the vectors are absorbed into the application of the alternating 2-form Ω to two of the vectors.)

With a linear connection on a manifold M defined, **covariant derivatives** of tensor fields on M can be formed. If $t \in \mathcal{T}_{r,s}(M)$ is a tensor field of type (r,s) and $X \in \mathcal{X}(M)$ is a tangent vector field, then, since the tensor bundle $T_{r,s}(M) = (T_{r,s}, M, \pi_{T_{r,s}}, \mathbb{R}^{m^{r+s}}, Gl(m,\mathbb{R}))$ is a special vector bundle associated with the frame bundle $L(M)$ as its principal fiber bundle ($Gl(m,\mathbb{R})$ acts on the typical fiber $\mathbb{R}^{m^{r+s}}$ by a tensor product of r factors of the representation in R^m and s factors of its transposed) and t is a section on $T_{r,s}(M)$, the general approach (7.17) applies. (To consider the covariant derivative of t at a given point $x \in M$, it is enough that $X = X_p$ is given at that point and t is given in a neighborhood of x or even on a curve through x only to which X is tangent.) It is readily seen, that $\nabla_X : \mathcal{T}(M) \to \mathcal{T}(M)$ is a derivation D in the sense of (4.13). By the theorem proved on p. 109,

∇_X is uniquely determined by its action on $\mathcal{C}(M)$ and on $\mathcal{X}(M)$.

In analogy to that proof it can be shown that

any derivation $D : \mathcal{T}(M) \to \mathcal{T}(M)$ has the form $D = \nabla_X + S'$ with a uniquely determined tangent vector field X and a uniquely determined endomorphism S' given by a tensor field s' of type $(1,1)$.

The covariant derivative of a smooth function $F \in \mathcal{C}(M)$ is simply

$$\nabla_X F = XF. \tag{7.26}$$

This was shown before (7.19). For the application of ∇_X on tangent vectors $s \in \mathcal{X}(M)$, the rules (7.18, 7.19) hold.

Recall (p. 100), that a homogeneous tensor t of type (r,s) at $x \in M$ may be considered as an s-linear mapping of $T_x(M) \times \cdots \times T_x(M)$ (s factors) into $(T_x(M))_{r,0}$ by the expression $t(X_1,\ldots,X_s) = C_{1,1}\cdots C_{s,s}(t \otimes X_1 \otimes \cdots \otimes X_s)$. With t, $\nabla_X t$ is of the same type (r,s). Considering $\nabla_X t$ as such an s-linear mapping into $(T_x(M))_{r,0}$, one may write $(\nabla_X t)(X_1,\ldots,X_s) = (\nabla t)(X_1,\ldots,X_s;X)$ and hence consider the homogeneous tensor ∇t of type $(r, s+1)$. The tensor ∇t is called the **covariant differential** of the tensor t. In this sense, ∇ is a mapping from $\mathcal{T}_{r,s}(M)$ to $\mathcal{T}_{r,s+1}(M)$. One has

$$(\nabla t)(X_1,\ldots,X_s;X) = \nabla_X(t(X_1,\ldots,X_s)) - \sum_{i=1}^{s} t(X_1,\ldots,\nabla_X X_i,\ldots,X_s). \tag{7.27}$$

Proof Apply ∇_X to $t(X_1, \ldots, X_s) = C_{1,1} \cdots C_{s,s}(t \otimes X_1 \otimes \cdots \otimes X_s)$ and observe (4.13) for $D = \nabla_X$. □

One now may apply ∇ a second time and obtain $(\nabla^2 t)(X_1, \ldots, X_s; X; Y) = (\nabla_Y(\nabla t))(X_1, \ldots, X_s; X)$, or recursively more generally

$$(\nabla^n t)(\ldots; X_1; \ldots; X_n) = (\nabla_{X_n}(\nabla^{n-1} t))(\ldots; X_1; \ldots; X_{n-1}). \tag{7.28}$$

Like in the general case of sections in a vector bundle, the tensor field t is a **parallel tensor field**, if $\nabla_X t = 0$ for all $X \in T_x(M)$ at all $x \in M$, that is, $\nabla t = 0$.

The alert reader might be intrigued by the question why there are two structure equations in the case of a linear connection on M while there is in general only one (the second). Some insight into this situation is obtained by considering **generalized affine connections** as introduced by Kobayashi and Nomizu. These are connections on the affine frame bundle considered in Sect. 7.2. Take a connection form $\tilde{\omega}$ defining a connection $\tilde{\Gamma}$ on the affine frame bundle $A(M)$. It is a pseudo-tensorial 1-form of type $(\mathrm{Ad}, \mathfrak{a}(m, \mathbb{R}))$. Pull it back to the linear frame bundle $L(M)$ by the homomorphism γ considered in Sect. 7.2. According to the semi-direct sum $\mathfrak{a}(m, \mathbb{R}) = \mathfrak{gl}(m, \mathbb{R}) \oplus \mathbb{R}^m$ one obtains

$$\gamma^*(\tilde{\omega}) = \omega + \varphi,$$

where ω is a pseudo-tensorial 1-form of type $(\mathrm{Ad}, \mathfrak{gl}(m, \mathbb{R}))$ and φ is of type $(Gl(m, \mathbb{R}), \mathbb{R}^m)$. It acts linearly on $\mathbb{R}^m$ (on the last column of the $(m+1) \times (m+1)$-matrix representation given in Sect. 7.2) and produces $\mathbb{R}^m$-vectors, hence it can be represented by an $\mathbb{R}^m$-tensor t_φ of type $(1, 1)$. On $L(M)$, the vertical spaces are isomorphic to $\mathfrak{gl}(m, \mathbb{R})$ which does not have the $m+1$st column, hence φ is horizontal on $L(M)$ and constitutes a tensorial 1-form of type $(Gl(m, \mathbb{R}), \mathbb{R}^m)$ there.

As a pseudo-tensorial 1-form of type $(\mathrm{Ad}, \mathfrak{gl}(m, \mathbb{R}))$, ω defines a linear connection Γ on $L(M)$. The mapping between connections, $\tilde{\Gamma} \mapsto (\Gamma, t_\varphi)$, where t_φ is any tensor field of type $(1, 1)$ on M turns out to be one–one, it comprises a pushed forward homomorphism $\beta_* : \tilde{\Gamma} \mapsto \Gamma$ (from $\beta : A(m, \mathbb{R}) \to GL(m, \mathbb{R})$). Take the exterior derivative of the above displayed relation (it commutes with the homomorphism γ^*, see (4.43)) and obtain $\gamma^*(d\tilde{\omega}) = d\omega + d\varphi$. Let X, Y be two horizontal vector fields on $L(M)$, then the right hand side of the last equation yields $\langle (d\omega + d\varphi), X \wedge Y \rangle = \langle (\Omega + D\varphi), X \wedge Y \rangle$. Its left hand side yields, with the structure equation of $\tilde{\Gamma}$, $\langle d\tilde{\omega}, \gamma_*(X) \wedge \gamma_*(Y) \rangle = -[\langle \tilde{\omega}, \gamma_*(X) \rangle, \langle \tilde{\omega}, \gamma_*(Y) \rangle] + \langle \tilde{\Omega}, \gamma_*(X) \wedge \gamma_*(Y) \rangle$. Since X, Y are horizontal for Γ, $\langle \omega, X \rangle = \langle \omega, Y \rangle = 0$ and $\langle \tilde{\omega}, X \rangle = \langle \varphi, X \rangle, \langle \tilde{\omega}, Y \rangle = \langle \varphi, Y \rangle$. However, $\mathbb{R}^m$ is Abelian and hence $[\langle \varphi, X \rangle, \ \langle \varphi, Y \rangle] = 0$ and $\langle \gamma^*(d\tilde{\omega}), X \wedge Y \rangle = \langle d\tilde{\omega}, \gamma_*(X) \wedge \gamma_*(Y) \rangle = \langle \tilde{\Omega}, \gamma_*(X) \wedge \gamma_*(Y) \rangle = \langle \gamma^*(\tilde{\Omega}), X \wedge Y \rangle$. In total,

$$\gamma^*(\tilde{\Omega}) = \Omega + D\varphi.$$

Use again the structure equation of $\tilde{\Gamma}$ on $A(M)$, $d\tilde{\omega} = -\tilde{\omega} \wedge \tilde{\omega} + \tilde{\Omega}$, pull it back to $L(M)$ and insert $\omega + \varphi$ for $\gamma^*(\tilde{\omega})$. Split the resulting equation $d(\omega + \varphi) = -\omega \wedge \omega - \omega \wedge \varphi + \Omega + D\varphi$ into the $\mathfrak{gl}(m, \mathbb{R})$-components and the $\mathbb{R}^m$-components and obtain finally

$$d\varphi = -\omega \wedge \varphi + D\varphi, \quad d\omega = -\omega \wedge \omega + \Omega.$$

In view of this result, a generalized affine connection $\tilde{\Gamma}$ on M is called an **affine connection**, if the $\mathbb{R}^m$-valued 1-form φ is the canonical form θ on $L(M)$. In this case the above relations are just the structure equations of a linear connection Γ on M. The canonical form θ as introduced by (7.5) maps the horizontal space identical into the horizontal space, hence the corresponding tensor t_φ is the unit tensor, and one is left with a one–one correspondence between affine connections $\tilde{\Gamma}$ and linear connections Γ on M. Therefore these two names are used synonymously in the literature. However, if one uses the principal fiber bundle $A(M)$ instead of $L(M)$ to define a linear connection, the two structure equations are again merged into a single one like in the general case.

7.8 Curvature and Torsion Tensors

Let a linear connection Γ be given on a manifold M, and let X, Y be two tangent vector fields on M. Let X^*, Y^* be lifts of X, Y into $L(M)$, and consider $\langle\Theta, X^* \wedge Y^*\rangle$ with the torsion form Θ of the connection Γ. Since Θ is a tensorial 2-form of type $(Gl(m,\mathbb{R}), \mathbb{R}^m)$, this expression defines a vector $\boldsymbol{T}_p \in \mathbb{R}^m$ at every point $p \in L(M)$. With the linear mapping $u(p)$ introduced in Sect. 7.2, $\boldsymbol{T}_p$ is mapped to a tangent vector of $T_{\pi(p)}(M)$, and the whole result depends linearly on $X \wedge Y$. Hence it may be expressed by a tensor field T of type (1,2) on M as

$$\langle T, X \wedge Y\rangle = u\langle\Theta, X^* \wedge Y^*\rangle, \tag{7.29}$$

which is to be understood that the value of the left hand side at $x = \pi(p) \in M$ is given by $u(p)$ applied to the value at p of the argument of u on the right hand side. It is easily seen, that the result at x does not depend on the actual point $p \in \pi^{-1}(x)$. Indeed, let $p' = pg$, $g \in Gl(m,\mathbb{R})$. Then as lifts, $X^*_{p'} = (R_g)_* X^*_p$, $Y^*_{p'} = (R_g)_* Y^*_p$. The right hand side of (7.29) at p' is $u(p')\langle\Theta_{p'}, X^*_{p'} \wedge Y^*_{p'}\rangle = u(p')\langle\Theta_{p'}, (R_g)_* X^*_p \wedge (R_g)_* Y^*_p\rangle = u(p')\langle R^*_g(\Theta_{p'}), X^*_p \wedge Y^*_p\rangle = u(p')\langle g^{-1}\Theta_p, X^*_p \wedge Y^*_p\rangle = u(p')g^{-1}\langle\Theta_p, X^*_p \wedge Y^*_p\rangle$. In the last but one equality it was used that Θ is a tensorial 2-form of type $(Gl(m,\mathbb{R}), \mathbb{R}^m)$ (compare (7.7) with $Gl(m,\mathbb{R})$ instead of Ad). Now, since $u(p')g^{-1} = u(p'g^{-1}) = u(p)$, the result is the same as that at p. Hence, for every $x \in M$, (7.29) is uniquely defined by the right hand side and is a tangent vector of $T_x(M)$ for every pair of tangent vectors X, Y, which means that $T \in \mathcal{T}_{1,2}(M)$ is a tensor field of type $(1,2)$ alternating in the lower indices (in coordinate representation). It is called the **torsion tensor field** or simply also the torsion of the linear connection Γ on M. Equation 7.29 is called the **torsion operation** on the pair X, Y.

In an analogous manner the **curvature operation** on a pair X, Y is defined. Since the curvature form Ω is a tensorial 2-form of type $(\mathrm{Ad}, \mathfrak{gl}(m,\mathbb{R}))$, the curvature operation at x is an element of the Lie algebra $\mathfrak{gl}(m,\mathbb{R})$ and hence a (not necessarily regular) linear transformation of tangent vectors. This transformation of a tangent vector field Z is defined as

$$C(\langle R, X \wedge Y\rangle \otimes Z) = u(\langle\Omega, X^* \wedge Y^*\rangle(u^{-1}Z)), \tag{7.30}$$

where C means the contraction (4.9) of the tensor product (in local coordinates of the lower index of the tensor $\langle R, X \wedge Y\rangle$ of type $(1,1)$ with the upper index of Z). On the right hand side, $\langle\Omega, X^* \wedge Y^*\rangle \in \mathfrak{gl}(m,\mathbb{R})$ and $u^{-1}Z \in \mathbb{R}^m$. Hence, the argument of u is again in $\mathbb{R}^m$ which is mapped by u into $\mathcal{T}(M)$. The independence of the right hand side on $p \in \pi^{-1}(x)$ to which X and Y are lifted is seen in the same manner as above, only now $R_g^*(\Omega_{p'}) = \mathrm{Ad}(g^{-1})\Omega_p = g^{-1}\Omega_p g$ and $u(p')^{-1} = (u(p)g^{-1})^{-1} = gu(p)^{-1}$. The result depends multi-linearly on X, Y, Z and is a tangent vector field on M. Hence, it may be represented by a tensor field of type $(1,3)$ which is alternating in its first two lower indices in a coordinate representation. It is called the **curvature tensor field** or simply also the curvature of the linear connection Γ on M.

Of course, a neighborhood of x suffices to define the curvature and torsion tensors at x.

The torsion and curvature operations can be expressed in terms of covariant derivatives as

$$\langle T, X \wedge Y\rangle = \nabla_X Y - \nabla_Y X - [X,Y], \quad \langle R, X \wedge Y\rangle = [\nabla_X, \nabla_Y] - \nabla_{[X,Y]}. \quad (7.31)$$

Proof Start with $\langle T_x, X_x \wedge Y_x\rangle = u(p)\langle \Theta_p, X_p^* \wedge Y_p^*\rangle$, $\pi(p) - x$. With (7.21), since X^* and Y^* are horizontal lifts and hence $\langle \omega, X^*\rangle = 0 = \langle \omega, Y^*\rangle$, $\langle \Theta_p, X_p^* \wedge Y_p^*\rangle = \langle d\theta_p, X_p^* \wedge Y_p^*\rangle = L_{X_p^*}\langle \theta, Y^*\rangle - L_{Y_p^*}\langle \theta, X^*\rangle - \langle \theta_p, [X^*, Y^*]_p\rangle = X_p^*\langle \theta, Y^*\rangle - Y_p^*\langle \theta, X^*\rangle - \langle \theta_p, [X^*, Y^*]_p\rangle$. With (7.5), the last term yields $u(p)\langle \theta_p, [X^*, Y^*]_p\rangle = \pi_*([X^*, Y^*]_p) = [X,Y]_x$. It remains to consider $u(p)(X_p^*\langle \theta, Y^*\rangle)$. Take first $\langle \theta_p, Y_p^*\rangle = u(p)^{-1}(Y_x)$ which is a vector in $\mathbb{R}^m$ whose components are functions of p. Let $\phi_t(x)$, $\phi_0(x) = x$, be a curve in M through x to which X_x is tangent, and let $\phi_t^*(p)$ be its lift through p which is a curve in the frame bundle $L(M)$ to which X_p^* is tangent. Hence,

$$X_p^*\langle \theta, Y^*\rangle = \lim_{\delta \to 0} \frac{u(\phi_t^*(p))^{-1}(Y_{\phi_t(x)}) - u(p)^{-1}(Y_x)}{t}$$

and

$$u(p)(X_p^*\langle \theta, Y^*\rangle) = \lim_{\delta \to 0} \frac{u(p) \circ u(\phi_t^*(p))^{-1}(Y_{\phi_t(x)}) - Y_x}{t}.$$

Here, $u(\phi_t^*(p))^{-1}$ maps $Y_{\phi_t(x)}$ into $\mathbb{R}^m$ and $u(p)$ maps this image into $T_x(M)$. Since $\phi_t^*(p)$ and $p = \phi_0^*(p)$ are connected by a horizontal path in $L(M)$, the two mappings realize a horizontal transport of $Y_{\phi_t(x)}$ from $\phi_t(x)$ to $x = \phi_0(x)$. Call this transport $\Phi_{(t,0)}^*$ and compare to (7.17) to see that the result is $(\nabla_X Y)_x$. Putting together these findings proves the first relation (7.31).

Now, start with $C(\langle R_x, X_x \wedge Y_x\rangle \otimes Z_x) = u(p)(\langle \Omega_p, X_p^* \wedge Y_p^*\rangle(u(p)^{-1}Z_x))$. Since X^* and Y^* are again horizontal lifts, $\langle \Omega, X^* \wedge Y^*\rangle = \langle d\omega, X^* \wedge Y^*\rangle = L_{X^*}\langle \omega, Y^*\rangle - L_{Y^*}\langle \omega, X^*\rangle - \langle \omega, [X^*, Y^*]\rangle = -\langle \omega, {}^v[X^*, Y^*]\rangle$. (Recall that ω annihilates horizontal vectors.) Let A^* be the fundamental vector field on $L(M)$ which at p equals $A_p^* = {}^v[X^*, Y^*]_p$, so that $A = \langle \omega_p, A_p^*\rangle$ is an element of $\mathrm{gl}(m, \mathbb{R})$. So far, $C(\langle R_x, X_x \wedge Y_x\rangle \otimes Z_x) = u(p)(-\langle \omega, {}^v[X^*, Y^*]_p\rangle\langle \theta_p, Z_p^*\rangle)$.

$F = \langle \theta, Z^* \rangle$ is an $\mathbb{R}^m$-valued function on $L(M)$ which is tensorial of type $(Gl(m,\mathbb{R}),\mathbb{R}^m)$. Hence,

$$A_p^* F = \lim_{t \to 0} \frac{F(p\exp(tA)) - F(p)}{t} = \lim_{t \to 0} \frac{\exp(-tA)F(p) - F(p)}{t} = -AF(p).$$

Therefore, $-\langle \omega_p, {}^v[X^*, Y^*]_p \rangle \langle \theta_p, Z_p^* \rangle) = {}^v[X^*, Y^*]_p \langle \theta, Z^* \rangle = ([X^*, Y^*]_p - {}^h[X^*, Y^*]_p)$ $\langle \theta, Z^* \rangle$. From the first part of the proof above, $u(p)({}^h[X^*, Y^*]_p \langle \theta, Z^* \rangle) = (\nabla_{[X,Y]} Z)_x$ for the horizontal vector ${}^h[X^*, Y^*]_p$. On the other hand, again with the first part of the proof and using the horizontality of X^*, Y^*, one obtains $u(p)(X_p^*(Y_p^* \theta(Z^*))) = u(p)(X_p^*(u(p)^{-1} \circ u(p)(Y_p^* \langle \theta, Z^* \rangle))) = u(p)(X_p^*(u(p)^{-1}(\nabla_Y Z)_x)) = u(p)(X_p^* \langle \theta_p, (\nabla_Y Z)_p^* \rangle) = (\nabla_X \nabla_Y Z)_x$. Putting everything together and observing that one may formally write $(\nabla_X Z)_x = C(\nabla_X \otimes Z))_x$ (∇_X acts like a transformation tensor of type $(1,1)$), the proof of the second relation (7.31) is completed, since Z was chosen completely arbitrarily. □

7.9 Expressions in Local Coordinates on *M*

In this section, finally local coordinate expressions are derived for the forms, the covariant derivative and the torsion and curvature tensors of a linear connection on M. For the sake of simplicity of notation, the same letter is used for points in manifolds and in corresponding coordinate spaces. As in Sect. 7.2, local coordinates $p = (x^k, X_i^k, i = 1, \ldots, m)$, so that $x = (x^1, \ldots, x^m)$ and $X_i = \sum_k X_i^k (\partial/\partial x^k)$, $\det(X_i^k) \neq 0$, are introduced in $\pi^{-1}(U) \subset L(M)$ where U is a coordinate neighborhood of x in M. The inverse of the $(m \times m)$-matrix (X_i^k) is denoted by $(\tilde{X}_i^k)$, so that

$$\sum_j X_j^k \tilde{X}_i^j = \sum_j \tilde{X}_j^k X_i^j = \delta_i^k. \tag{7.32}$$

In addition, in $\mathbb{R}^m$ the natural base $\{e_1, \ldots, e_m\}$ is introduced, for which

$$\theta = \sum_i \theta^i e_i. \tag{7.33}$$

For any $p = (x^k, X_i^k) \in U$, the mapping $u(p) : \mathbb{R}^m \to T_x(M)$, $x = \pi(p)$, maps e_i to $\sum_j X_i^j (\partial/\partial x^j)_x$ (see Sect. 7.2). Let

$$Y_p^* = \sum_j Y^j(p) \left(\frac{\partial}{\partial x^j} \right)_x + \sum_{jk} Y_k^j(p) \left(\frac{\partial}{\partial X_k^j} \right)_p$$

be any tangent vector in $T_p(L(M))$. Its projection to the tangent space on M, $T_x(M)$, is $\pi_*(Y_p^*) = \sum_j Y^j(p)(\partial/\partial x^j)_x$, hence

$$\langle\theta_p, Y_p^*\rangle = u(p)^{-1}(\pi_*(Y_p^*)) = \sum_{ij}(\tilde{X}_j^i(p)Y^j(p))e_i,$$

since $u(p)^{-1}(\partial/\partial x^j)_x = \sum_i \tilde{X}_j^i(p)e_i$. Comparison with the representation (7.33) yields $\langle\theta_p^i, Y_p^*\rangle = \sum_j \tilde{X}_j^i(p)Y^j(p)$ or, with $\langle dx^j, \partial/\partial x^k\rangle = \delta_k^j$,

$$\theta_p^i = \sum_j \tilde{X}_j^i(p)dx_x^j \quad \text{or in short} \quad \theta^i = \sum_j \tilde{X}_j^i dx^j \tag{7.34}$$

as the local coordinate expression of the **canonical form** θ on $L(M)$. As an $\mathbb{R}^m$-valued 1-form, the local coordinate expression $\tilde{X}_j^i$ of θ has an upper index i as an $\mathbb{R}^m$-vector and a lower index j as a 1-form according to the general local coordinate representation (3.24) of an exterior form.

Consider the transition properties of θ between two overlapping coordinate neighborhoods $U_\alpha \cap U_\beta \ni x$. According to (3.14), the tangent vector X_i on M transforms as $X_{\beta i}^j = \sum_k (\psi_{\beta\alpha})_k^j X_{\alpha i}^k$ where $\psi_{\beta\alpha}$ is the Jacobian matrix of the coordinate transformation given by (3.6). Hence, $\tilde{X}^i$ transforms like a cotangent vector, $\tilde{X}_{\beta j}^i = \sum_k \tilde{X}_{\alpha k}^i (\psi_{\beta\alpha}^{-1})_j^k$, in order that $\sum_j \tilde{X}_{\beta i}^j X_{\beta j}^k = \delta_i^k = \sum_j \tilde{X}_{\alpha i}^j X_{\alpha j}^k$. With the second relation (3.11) this ensures that $\sum_j \tilde{X}_{\beta j}^i dx_\beta^j = \sum_j \tilde{X}_{\alpha j}^i dx_\alpha^j$. As seen, θ^i behaves indeed like a tensor of type $(0,1)$ on M, which yields another justification to call it a tensorial 1-form. (Recall that originally θ was introduced in Sect. 7.3 as a 1-form on $L(M)$.)

The connection form ω of a linear connection Γ on M is a $\mathfrak{gl}(m,\mathbb{R})$-valued 1-form. For its corresponding local coordinate expression the natural base $\{E_j^i\}$ in $\mathfrak{gl}(m,\mathbb{R})$ is needed, which consists of matrices with a unity in the ith column and jth row and zeros otherwise (p. 219). The analogue of (7.33) is

$$\omega = \sum_{ik} \omega_k^i E_i^k. \tag{7.35}$$

On a coordinate neighborhood $U_\alpha \subset M$, the analogue of (7.34) is first considered for the local connection forms ω_α only, which are pull-backs from the canonical local section $s_\alpha \subset L(M)$ to U_α:

$$\omega_{\alpha k}^i = \sum_j \Gamma_{jk}^i(x)dx_x^j \quad \text{or in short} \quad \omega_{\alpha k}^i = \sum_j \Gamma_{jk}^i dx^j.$$

The components Γ_{jk}^i of the local connection form are called **Christoffel symbols**. Unlike the components of the canonical form θ, the Christoffel symbols to not form a tensor on M. The local connection forms ω_α must have the transition properties (7.4), that is, the pairing with a tangent vector field X on M, $X_\beta^j = \sum_n (\psi_{\beta\alpha})_n^j X_\alpha^n$, must obey the relation

$$\langle \omega^i_{\beta k}, X^j_\beta \rangle = \sum_{lmn} (\psi^{-1}_{\alpha\beta})^i_l \langle \omega^l_{\alpha m}, X^n_\alpha \rangle (\psi_{\alpha\beta})^m_k (\psi_{\beta\alpha})^j_n + \langle (\vartheta_{\alpha\beta})^i_k, X^n_\alpha \rangle (\psi_{\beta\alpha})^j_n$$

where for X_x the local connection forms ω_α and ω_β were the pulled back connection forms from $p_\alpha = s_\alpha(x)$ and $p_\beta = s_\beta(x)$ given by the canonical local sections. (Recall also that $\psi_{\alpha\beta} = \psi^{-1}_{\beta\alpha}$.)

In order to find the coordinate transition expression $\vartheta_{\alpha\beta} = \psi^*_{\alpha\beta}(\vartheta)$ of the Maurer–Cartan 1-form ϑ of $Gl(m,\mathbb{R})$, the way (7.4) was obtained has to be reconsidered. In the coordinate neighborhood $U_\alpha \ni x$, let s_α be the canonical local section which maps x to the frame $(x, (\partial/\partial x^1_\alpha), \dots, (\partial/\partial x^m_\alpha))$. Clearly, $(x, (\partial/\partial x^1_\beta), \dots, (\partial/\partial x^m_\beta)) = (x, \sum_j (\psi^{-1}_{\beta\alpha})^j_1 (\partial/\partial x^j_\alpha), \dots, (\psi^{-1}_{\beta\alpha})^j_m (\partial/\partial x^j_\alpha)) = (x, (\partial/\partial x^1_\alpha), \dots, (\partial/\partial x^m_\alpha))\psi_{\alpha\beta}$ as was used in (7.4). Any frame p of $\pi^{-1}(x)$ is obtained by acting from the right on $s_\alpha(x)$ with an element g of the Lie group $Gl(m,\mathbb{R})$, $p = s_\alpha(x)g$. (Recall that according to property 2 on p. 215 the same group acts on ω as an element of $\mathrm{gl}(m,\mathbb{R})$ by the adjoint representation.) Use again the natural base E^i_k in $\mathrm{gl}(m,\mathbb{R})$ as above. An element g in natural coordinates of the Lie group $Gl(m,\mathbb{R})$ is represented by g^i_k, $\det(g^i_k) \neq 0$. Let $(\tilde{g}^i_k)$ be the matrix inverse to (g^i_k). The Maurer–Cartan form maps left invariant tangent vector fields on $Gl(m,\mathbb{R})$ into $\mathrm{gl}(m,\mathbb{R})$ and maps $(\partial/\partial g^i_j)_e \in T_e(Gl(m,\mathbb{R}))$ to E^j_i: $\langle \vartheta_e, (\partial/\partial g^i_j)_e \rangle = E^j_i$. Let G^j_i be the left invariant tangent vector field which at e is $G^j_{ei} = (\partial/\partial g^i_j)_e$, that is, $G^j_{gi} = \sum_{kl} \tilde{g}^k_i g^j_l (\partial/\partial g^k_l)_g$. It must also hold that $\langle \vartheta_g, G^i_{gj} \rangle = E^j_i$. As a $\mathrm{gl}(m,\mathbb{R})$-valued 1-form, write $\vartheta = \sum_{klmn} \vartheta^{lm}_{kn}\, dg^k_l\, E^n_m$ and use $\langle dg^k_l, (\partial/\partial g^i_j) \rangle = \delta^k_i \delta^j_l$. There must be $\sum_{klmn} \vartheta^{lm}_{gkn}\, \tilde{g}^k_i\, g^j_l\, E^n_m = E^j_i$ which finally results in $\vartheta^{lm}_{gkn} = g^m_k \tilde{g}^l_n$ or $\vartheta_g = \sum_{klmn} g^m_k \tilde{g}^l_n\, dg^k_l\, E^n_m$. Now,

$$\vartheta_{\alpha\beta} = \psi^*_{\alpha\beta}(\vartheta) = \vartheta_{\psi_{\alpha\beta}} = \sum_{klmn} (\psi_{\alpha\beta})^m_k (\psi^{-1}_{\alpha\beta})^l_n\, d(\psi_{\alpha\beta})^k_l E^n_m, \tag{7.36}$$

where

$$d(\psi_{\alpha\beta})^i_j = \sum_k \frac{\partial x^i_\alpha}{\partial x^j_\beta \partial x^k_\beta} dx^k_\beta = \sum_k (d\psi_{\alpha\beta})^i_{jk} dx^k_\beta. \tag{7.37}$$

Putting everything together results in

$$\Gamma^i_{\beta jk} = \sum_{lmn} \Gamma^l_{\alpha mn} (\psi_{\beta\alpha})^i_l (\psi^{-1}_{\beta\alpha})^m_j (\psi^{-1}_{\beta\alpha})^n_k + \sum_l (\psi_{\beta\alpha})^i_l (d\psi^{-1}_{\beta\alpha})^l_{jk}, \tag{7.38}$$

which shows that Γ^i_{jk} is indeed not a tensor on M (whence ω was called a *pseudo-tensorial 1-form*).

In local coordinates $p = (x^k, X^k_i)$, a vertical tangent vector on $L(M)$ has the form $Y^* = \sum_{jk} Y^j_k (\partial/\partial X^j_k)$, and since $\langle dx^i, (\partial/\partial X^j_k) \rangle = 0$, the pairings of local connection forms with vertical tangent vectors vanish. The connection form ω itself must, however, have the properties 1 and 2 given on p. 215. Hence, ω must consist of a term which, if paired with vertical tangent vectors, restores these properties 1 and 2 and of another term which pulls back to ω_α. It is easily found that the right expression is

$$\omega = \sum_{ijk} \tilde{X}^i_k \left(dX^k_j + \sum_{lm} \Gamma^k_{lm} X^m_j dx^l \right) E^j_i. \tag{7.39}$$

Indeed, consider the fundamental vector field $(X^j_i)^* = R^*_*(E^j_i)$ which on the canonical local section s_α with local coordinates $s_\alpha(x) = (x^k, \delta^k_i)$ is $(X^j_i)^*_{s_\alpha(x)} = (\partial/\partial X^i_j)_{s_\alpha(x)}$. A general point p on the fiber over x is $p = (x^k, X^k_i) = s_\alpha(x)g, g = X^k_i$ in local coordinates. A fundamental vector field is a left invariant vector field on $Gl(m, \mathbb{R})$, hence $(X^j_i)^*_p = g(X^j_i)^*_{s_\alpha(x)} = \sum_k X^k_i (\partial/\partial X^k_j)_p = \sum_{kl} X^k_i \delta^j_l (\partial/\partial X^k_l)_p$. Now, $\langle \omega, (X^j_i)^* \rangle = \sum_{klmrs} \tilde{X}^m_s \langle dX^s_r, (\partial/\partial X^k_l) \rangle X^k_i \delta^j_l E^r_m = \sum_{klm} \tilde{X}^m_k X^k_i \delta^j_l E^l_m = E^j_i$ and property 1 is fulfilled. Property 2 is directly read off the factor at E^j_i in (7.39), since the second term of (7.38) vanishes for $\alpha = \beta$. Moreover, since s^*_α pulls back from X^k_j to δ^k_j, it pulls the first term of (7.39) back to zero and the second term to the expression for ω_α introduced after (7.35). Hence, (7.39) is the final local coordinate expression for the linear connection form.

Every transformation step from (7.4) to (7.38) was one–one. Hence, symbols Γ^i_{jk} transforming according to (7.38) yield local connection forms ω_α which obey (7.4). On the other hand, as it was seen there, local connection forms ω_α obeying (7.4) define uniquely a connection form ω on $L(M)$ and hence a linear connection Γ on M.

Symbols Γ^i_{jk} in local coordinates having the transition properties (7.38) *define uniquely a connection form through* (7.39) *and thus a linear connection Γ on M.*

Next, let $X_l = (\partial/\partial x^l)$ be a tangent vector on U_α and let $X^*_{lp} = (\partial/\partial x^l)_x + \sum_{mn} (X^*_l)^m_n (\partial/\partial X^m_n)_p$ be its horizontal lift through $p = (x^k, X^k_i)$. Then, $0 = \langle \omega^i_j, X^*_l \rangle = \sum_k \tilde{X}^i_k (X^*_l)^k_j + \sum_{km} \tilde{X}^i_k \Gamma^k_{lm} X^m_j$. Multiplication with X^n_i and summation over i yields $(X^*_l)^n_j = -\sum_m \Gamma^n_{lm} X^m_j$ and hence

$$X^*_l = \frac{\partial}{\partial x^l} - \sum_{kmn} \Gamma^m_{lk} X^k_n \frac{\partial}{\partial X^m_n} = {}^h X^*_l. \tag{7.40}$$

This expression only now determines the splitting of a general tangent vector on $L(M)$ into its horizontal and vertical parts, first mentioned on p. 219, by giving the structure of horizontal vectors in terms of local coordinates.

Now, the tangent bundle $(T(M), M, \pi_T, \mathbb{R}^m, Gl(m, \mathbb{R})) = T(M)$ associated with the frame bundle $L(M)$ is considered where $Gl(m, \mathbb{R})$ acts on the vector space $\mathbb{R}^m$ by the identical representation, and $L(M)$ is provided with a linear connection Γ. Let $f : L(M) \to \mathbb{R}^m$ be a $\mathbb{R}^m$-valued function with the property $f(pg) = g^{-1}f(p)$. It may be understood to be a tensorial 0-form of type $(Gl(m, \mathbb{R}), \mathbb{R}^m)$. In combination with the canonical local section s_α in $L(M)$ it defines a local section $f_\alpha : U_\alpha \to T(M) : x \mapsto (s_\alpha(x)g, g^{-1}f(s_\alpha(x)))$ on U_α in the tangent bundle $T(M)$, where the section point for x corresponds to the tangent vector with components $f^l(s_\alpha(x))$ in the frame δ^l_i given by the natural base $\{e_1, \ldots, e_m\}$ at x. In particular, the functions $f_{\alpha i} : p = (x^k, X^k_j) \mapsto \delta^k_i$ for given i, that is, $f^l_{\alpha i} = \tilde{X}^l_i$ by matrix multiplication to the frame coordinates, provide this property. The section $f_{\alpha i}$ consists of the tangent vector $(\partial/\partial x^i)$ at every $x \in U_\alpha$. According to (7.17), its directional derivative along a path with horizontal tangent vector X is $\nabla_X(\partial/\partial x^i)$. Apply the horizontal vector X^*_j from (7.40) to $f_{\alpha i}$ and obtain $\nabla_{\partial/\partial x^j}(\partial/\partial x^i) = -\sum_{klmn} \Gamma^{m}_{jk} X^k_n (\partial/\partial X^m_n) \tilde{X}^l_i e_l = \sum_{klmnrs} \Gamma^{m}_{jk} X^k_n \tilde{X}^r_i (\delta^n_r \delta^s_m) \tilde{X}^l_s e_l = \sum_{lm} \Gamma^m_{ji} \tilde{X}^l_m e_l$ where in the second equality (2.26) was used. By reinserting $(\partial/\partial x^m) = \sum_l \tilde{X}^l_m e_l$ in the last expression, the final result

$$\nabla_{\partial/\partial x^j}(\partial/\partial x^i) = \sum_k \Gamma^k_{ji}(\partial/\partial x^k) \tag{7.41}$$

is obtained. Replacing in these considerations the tangent bundle $T(M)$ by the cotangent bundle $T^*(M)$ replaces $f^l_{\alpha i} = \tilde{X}^l_i$ by $f^i_{\alpha l} = X^i_l$ only and results in

$$\nabla_{\partial/\partial x^j} dx^i = -\sum_k \Gamma^i_{jk} dx^k. \tag{7.42}$$

This analysis underlines the role of the frame bundle as principal fiber bundle with which tangent, cotangent and general tensor bundles over M are associated. Would one try to define connections directly on those bundles, the definition would unavoidably depend on the used local coordinate systems in a quite involved way. The use of the frame bundle makes it possible to define linear connections independently of local coordinate systems, and in addition leads to general forms of coordinate expressions. (It was invented by E. Cartan.)

The properties (7.18, 7.19) of **covariant derivatives** can now be used to obtain the local coordinate expressions of the derivatives of general tensor fields as sections of tensor bundles, from (7.41, 7.42). With the general local coordinate expression (4.33) one gets

$$(\nabla_X t(x))^{i_1\ldots i_r}_{j_1\ldots j_s} = X^k \left(\frac{\partial t^{i_1\ldots i_r}_{j_1\ldots j_s}}{\partial x^k} + \sum_{\mu=1}^r \Gamma^{i_\mu}_{kl} t^{i_1\ldots i_{\mu-1} l i_{\mu+1}\ldots i_r}_{j_1\ldots j_s} - \sum_{\mu=1}^s \Gamma^l_{k j_\mu} t^{i_1\ldots i_r}_{j_1\ldots j_{\mu-1} l j_{\mu+1}\ldots j_s} \right). \tag{7.43}$$

The notation

$$\left(\nabla_{\partial/\partial x^k} t(x)^{i_1\ldots i_r}_{j_1\ldots j_s}\right) = (\nabla t(x))^{i_1\ldots i_r}_{j_1\ldots j_s k} = t^{i_1\ldots i_r}_{j_1\ldots j_s;k}(x) \tag{7.44}$$

is generally used. The tensor contraction of the right hand side with any tangent vector X gives (7.43) back, which shows that the expression in parentheses on the right hand side of (7.43) forms the components of a tensor of type $(r, s+1)$, the

covariant differential of t. Higher order covariant differentials are recursively obtained:

$$(\nabla^n t))^{i_1\dots i_r}_{j_1\dots j_s k_1\dots k_n} = t^{i_1\dots i_r}_{j_1\dots j_s;k_1;\dots;k_n}. \tag{7.45}$$

Compare (7.27, 7.28). These are the generalizations of tensor gradients from trivial connections for which the Christoffel symbols vanish. (They do not vanish even in flat connections, if general non-linear coordinates are used.)

The local coordinate expressions of the torsion and curvature tensors are now straightforwardly obtained from (7.31). First write the left hand sides as

$$\langle T, X\wedge Y\rangle = T^i_{jk}X^jY^k\frac{\partial}{\partial x^i}, \quad \langle R, X\wedge Y\rangle = R^i_{jkl}X^kY^l dx^j\frac{\partial}{\partial x^i}, \tag{7.46}$$

and then use (7.41) and (7.18, 7.19) on the the right hand sides to find (exercise)

$$T^i_{jk} = \Gamma^i_{jk} \quad \Gamma^i_{kj} \tag{7.47}$$

and

$$R^i_{jkl} = \frac{\partial\Gamma^i_{lj}}{\partial x^k} - \frac{\partial\Gamma^i_{kj}}{\partial x^l} + \Gamma^m_{lj}\Gamma^i_{km} - \Gamma^m_{kj}\Gamma^i_{lm}. \tag{7.48}$$

For a smooth function F on M one also directly finds that

$$F_{;j;k} - F_{;k;j} = T^i_{jk}F_{;i} \tag{7.49}$$

and for a tangent vector field X that

$$X^i_{;l;k} - X^i_{;k;l} = R^i_{jkl}X^j - T^j_{kl}X^i_{;j}. \tag{7.50}$$

Thus, the covariant derivatives of functions commute, if the torsion of the linear connection vanishes, and the covariant derivatives of vector fields commute, if the linear connection is flat and torsion free.

A smooth curve $x(t)$ in M which locally solves the equations

$$\frac{d^2x^i}{dt^2} + \sum_{jk}\Gamma^i_{jk}\frac{dx^j}{dt}\frac{dx^k}{dt} = 0, \quad i = 1,\dots,m \tag{7.51}$$

is called a **geodesic**. The vector $X = (dx^i/dt)(\partial/\partial x^i)$ is tangent to this curve. From (7.43) it follows that $\nabla_X X = 0$.

The tangent vector to a geodesic is parallel transported to itself on the geodesic.

Finally, the Lie derivative (4.36) is compared to the covariant derivative (7.17). A direct comparison of (3.37) with the first relation (7.31) yields immediately

$$L_X Y = \nabla_X Y - \nabla_Y X - \langle T, X\wedge Y\rangle \tag{7.52}$$

for any tangent vector field Y, while for any 1-form σ

$$(L_X\sigma)_j = (\nabla_X\sigma)_j + \sigma_i X^i_{;j} - T^k_{ij} X^i \sigma_k \tag{7.53}$$

is an exercise. While in Fig. 4.1 for the Lie derivative the transport is along the flow of X (local 1-parameter group) without a twist, in Fig. 7.8 for the covariant derivative it is *horizontally* in the direction of X.

Reference

1. Kobayashi, S., Nomizu, K.: Foundations of Differential Geometry, Interscience, vols. I and II. New York (1963, 1969)

Chapter 8
Parallelism, Holonomy, Homotopy and (Co)homology

This chapter is devoted to most topical and important applications of topology and geometry in physics: gauge field theory and the physics of geometric phases which vastly emerges from the notion of the Aharonov–Bohm phase and later more generally from the notion of a Berry phase (see [1, 2]) and even penetrates chemistry and nuclear chemistry. The central notion in these applications is holonomy. Since holonomy is based on lifts of integral curves of tangent vector fields on the base manifold M of a bundle, and maximal integral curves may end in singular points of tangent vector fields, non-singularity of tangent vector fields plays its role. Non-zero tangent vector fields can be expressed as sections of the 'punctured tangent bundle' on M. This is a subject of the interrelation of holonomy with homotopy of fiber bundles, an important issue by itself. Therefore the chapter starts with two sections on homotopy of fiber bundles before gauge fields and finally geometric phases in general are considered. All these issues fall also into the vast realm of characteristic classes and index theory. In a first reading the first two sections may be skipped.

8.1 The Exact Homotopy Sequence

Let a fiber bundle (E, M, π_E, F, G) associated with a principal fiber bundle (P, M, π, G) be given. (In particular E may be P itself.) So far (horizontal) lifts of paths in M to E were considered. Now the goal is to lift homotopy classes of mappings of n-dimensional spheres into M. Recall, that a special intermediate bijective mapping $\boldsymbol{P}$ of n-spheres S^n to one point compactified cubes $\overline{I^n}$ was needed in order to define a group structure on the sets of homotopy classes (Sect. 2.5, in particular Fig. 2.8). In the following, I^n, I denote the n-cube, unit interval closed in the ordinary Euclidean topology, $\mathring{I}^n.\mathring{I}$ denote their interior, and $\overline{I^n}$ as previously denotes the one-point compactification of $\mathring{I}^n$.

Let $\widetilde{\Phi} : \overline{I^n} \to M$ be a mapping with $\widetilde{\Phi}(\partial I^n) = x_0 \in M$, where $\partial I^n = \boldsymbol{x}_\infty$ is the boundary of the unit cube I^n which is defined to be the point $\boldsymbol{x}_\infty$ of the one-point compactification of the open cube $\mathring{I}^n$ to $\overline{I^n}$. Then, $\Phi = \widetilde{\Phi} \circ \boldsymbol{P}$ is the mapping of the n-sphere S^n into M which in Sect. 2.5 was considered for a general topological space X instead of a manifold M, and for which $\Phi(s_0) = x_0$. The part of the homotopy class of Φ with the mapping $s_0 \mapsto x_0$ fixed is $[\Phi] = \{\Phi_{\mathrm{H}} : I \times S^n \to M \mid \Phi_{\mathrm{H}}(0, \cdot) = \Phi, \Phi_{\mathrm{H}}(t, s_0) = x_0 \text{ for } t \in I\}$. It corresponds to $[\widetilde{\Phi}] = \{\widetilde{\Phi}_{\mathrm{H}} : I \times I^n \to M \mid \widetilde{\Phi}_{\mathrm{H}}(0, \cdot) = \widetilde{\Phi}, \widetilde{\Phi}_{\mathrm{H}}(I \times \partial I^n) = \{x_0\}\}$.

As in Sect. 5.5, all continuous mappings from a closed simplex (or a cube) of $\mathbb{R}^k$ into M may be arbitrarily closely uniformly approximated (in the metrics of $\mathbb{R}^k$ and of coordinate neighborhoods $\boldsymbol{U}_\alpha$ of M) by smooth mappings of some neighborhood of the simplex (cube) into M. In this sense all mappings are again supposed to be smooth in the following.

Let D^k be a (sufficiently smoothly bounded) domain in $\mathbb{R}^k$. A **general lift** of $\Phi : D^k \to M$ to E is a (smooth) function $\Phi^* : D^k \to E$ with $\pi_E \circ \Phi^* = \Phi$.

Let $\Phi : I \times I^n \to M$ be given and let $Q = (\{0\} \times I^n) \cup (I \times \partial I^n)$. Let $\Phi^*_Q : Q \to E$ be a general lift of the restriction $\Phi|_Q$ of Φ to Q. Then, there exists a general lift $\Phi^* : I \times I^n \to E$ of Φ with $\Phi^*|_Q = \Phi^*_Q$.

Proof Consider first the case that $\Phi(I \times I^n)$ lies in a trivializing coordinate neighborhood U of the base space M of E, so that $\pi_E^{-1}(U) \approx U \times F$. Then, $\Phi^*_Q|_{\{0\}\times I^n}$ maps $\boldsymbol{x} \in I^n$ to $(\Phi(\boldsymbol{x}), \phi(\boldsymbol{x}))$, where ϕ is some mapping from I^n to the typical fiber F. Consider Φ as a homotopy of $\Phi|_{I^n}$ and take *any* homotopy of ϕ in F to obtain a general lift of Φ.

Next, take a homeomorphism q of $I \times I^n$ onto itself which maps $\{0\} \times I^n$ onto Q. Such a homeomorphism exists, it can explicitly be constructed in the following way (Fig. 8.1). Map I^n homeomorphically into an n-ball B^n and hence $I \times I^n$ into a spherical cylinder $I \times B^n$. Then, in a first step (a), embed the domain $I \times B^n$ ($n = 1$ in the figure) into a large enough ball B^{n+1}, and then stretch it homogeneously along the drawn arrows from some inner point in such a way that the domain $\{0\} \times B^n$ (thick line of the figure) is mapped onto the lower half-sphere of the boundary of the $(n+1)$-ball (thick arc). This is a homeomorphic map of $I \times B^n$ onto the ball B^{n+1}. Next, shift the $(n+1)$-ball upwards as shown in part (b) of the figure and shrink it homogeneously along the drawn arrows to $I \times B^n$ which after going back to $I \times I^n$ maps the lower half-sphere to Q. The composition of all homeomorphisms yields the sought map q from $I \times I^n$ to $I \times I^n$ mapping $\{0\} \times I^n$ to Q. (Is the mapping from I^n to B^n necessary in this construction?) The mapping $\widetilde{\Phi} = \Phi \circ q$ maps $I \times I^n$ to U, and $\widetilde{\Phi}|_{\{0\}\times I^n} = \Phi|_Q \circ q$. $\widetilde{\Phi}^*_Q = \Phi^*_Q \circ q$ is a general lift of $\widetilde{\Phi}|_{\{0\}\times I^n}$. It was seen in the previous paragraph that it has an extension to a lift $\widetilde{\Phi}^*$ of $\widetilde{\Phi}$. $\Phi^* = \widetilde{\Phi}^* \circ q^{-1}$ is now the wanted general lift of Φ. On the trivializing neighborhood U the statement holds in an elementary way also for $n = 0$.

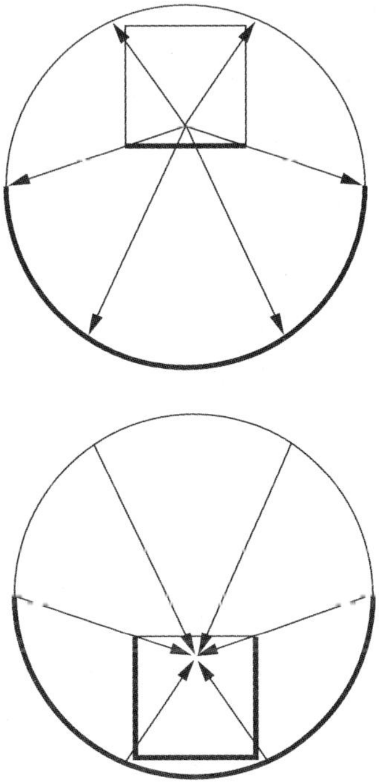

Fig. 8.1 The mapping q in two steps for $n = 1$. See text for explanation

Now, the general case is reduced to the just considered by subdividing $I \times I^n$ into N^{n+1} small cubes of equal size, small enough that each of them is mapped by Φ into a trivializing coordinate neighborhood in M. In a first step, I of the homotopy is replaced by the interval $[0, 1/N]$, and then, step by step the continuation is performed to all I. Start at the small cube with corner $(0,\ldots,0)$ and scan lexicographically with respect to the coordinates of the corner closest to the origin through the cubes. At any stage, Φ^* is determined on the n-face of the cube with $x^1 =$ const. and on some of the n-faces sharing an $(n-1)$-face with the latter. An n'-face is an n'-cube, and the above extension procedure can be applied to it, provided Φ^* is already defined on all its $(n'-1)$-faces sharing $(n'-2)$-faces with the n-face $x^1 =$ const.. In case of necessity one has further to go down with n' at most to $n' = 0$ in which case the above continuation of Φ^* (from a point to an interval) is always possible. Stepping from there upwards again with n' finally extents Φ^* to the whole $(n+1)$-cube, and one can proceed to the next. Φ^* is finally established on $I \times I^n$. □

As seen from the given proof, in the above statement the cube I^n may be replaced by any domain which is homeomorphic to a ball, in particular also by an n-simplex. Even more generally, it may even be replaced by any polyhedron $|c|$ (see p. 141 f), when ∂I^n is replaced by any subpolyhedron $|c'|$ of $|c|$. By definition, a subpolyhedron of $|c|$ is a polyhedron which is also a subset of the skeleton of some complex realizing the polyhedron $|c|$.

Let M be contractible, and let $\Phi : I \times M \to M$ be given with $\Phi(0,x) = x_0$ fixed for all $x \in M$ and $\Phi(1,x) = x$. Let $\epsilon_0 \in E$ be some point. The just mentioned generalization of the above lifting proposition says that there is a general lift Φ^* of Φ, and for $\Phi(1,\cdot)$ this is a (global) section of E. Hence, every fiber bundle with a contractible base space has a section. This is also true for principal fiber bundles, for which it was shown in Sect. 7.1 that a principal fiber bundle P which has a section is trivial (that is, $P \approx M \times G$). By the very definition of fiber bundles associated with principal fiber bundles, triviality transfers also to the former:

A fiber bundle with contractible base space is trivial.

With the help of the lifting proposition a group homomorphism $\delta : \pi_n(M,x_0) \to \pi_{n-1}(F,f_0)$, $n > 1$, between homotopy groups may be constructed. Given a group element of $\pi_n(M,x_0)$, a representing mapping $\Phi : I^n \to M$, $\Phi(\partial I^n) = \{x_0\}$ is chosen. Lift x_0 to any point $\epsilon_0 = \{(p_0 g, g^{-1} f_0)\} \in \pi_E^{-1}(x_0)$. This can be extended to a general lift of Φ to $\Phi^* : I^n \to E$ which lifts $\Phi|_Q$, $Q = \partial I^n \setminus \{1\} \times (I^{n-1})^\circ$ to ϵ_0. Since $\Phi(\{1\} \times I^{n-1}) = \{x_0\}$, it holds that $\Phi^*(\{1\} \times I^{n-1}) \subset \pi_E^{-1}(x_0)$. Moreover, $\partial(\{1\} \times I^{n-1}) \subset Q$, and hence $\Phi^*|_{\{1\}\times I^{n-1}}$ represents a group element of $\pi_{n-1}(F,f_0)$. Let $\Phi' : I^n \to M$, $\Phi'(\partial I^n) = \{x_0\}$ be a mapping homotopic with Φ. That means that there is a mapping $\widetilde{\Phi} : I \times I^n \to M$ with $\widetilde{\Phi}|_{\{0\}\times I^n} = \Phi$, $\widetilde{\Phi}|_{\{1\}\times I^n} = \Phi'$. This can be lifted to $\widetilde{\Phi}^*$ with $\widetilde{\Phi}^*|_{\{0\}\times I^n} = \Phi^*$ and hence yields a homotopy in F between Φ^* and $\widetilde{\Phi}^*|_{\{1\}\times I^n}$. Hence, the constructed correspondence between representatives of group elements of $\pi_n(M,x_0)$ and $\pi_{n-1}(F,f_0)$ yields a correspondence between the group elements themselves independent of the chosen representatives (Fig. 8.2). Taking the construction (2.35), it is easy to see that the just constructed mapping δ from $\pi_n(M,x_0)$ to $\pi_{n-1}(F,f_0)$ is a group homomorphism for $n > 1$. For $n = 1$ it is still a well defined mapping from the fundamental group $\pi_1(M,x_0)$ to $\pi_0(F,f_0)$. Only $\pi_0(F,f_0)$ is not in general a group. It is a set in bijective relation to the pathwise connected components of F.

The **exact homotopy sequence** is

$$\cdots \xrightarrow{\delta} \pi_n(F,f_0) \xrightarrow{i_*} \pi_n(E,\epsilon_0) \xrightarrow{\pi_{E*}} \pi_n(M,x_0) \xrightarrow{\delta} \pi_{n-1}(F,f_0) \xrightarrow{i_*} \cdots \tag{8.1}$$

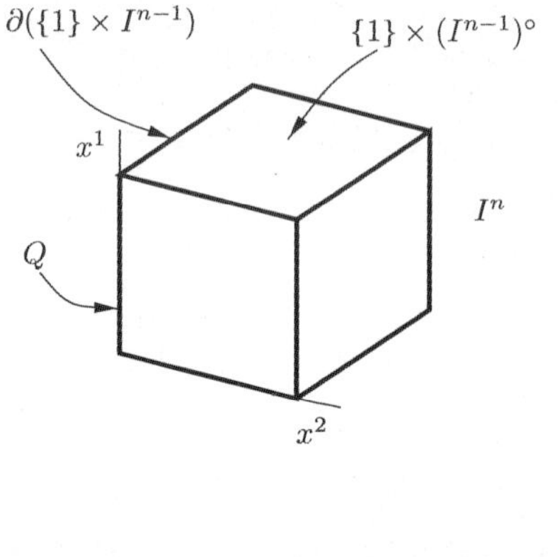

Fig. 8.2 A cube I^n, $n = 3$. Q is its surface without the interior of its upper face (open box). $\partial(\{1\} \times I^{n-1})$ consists of the edges of the upper face

where i is the inclusion mapping of F onto the fiber through $\epsilon_0 = \{(p_0 g, g^{-1} f_0)\} \in \pi_E^{-1}(x_0)$ in E, π_E is the bundle projection, and both mappings are pushed forward to the corresponding homotopy groups.

Proof of the exactness of the sequence The elements of $\pi_n(F, f_0)$ are homotopy classes of mappings from the cube I^n into F with ∂I^n mapped to f_0. The inclusion mapping i maps these mappings to mappings of the same cube into the fiber of E containing ϵ_0, and hence the images of these mappings are projected by π_{E*} to constant mappings of the cube I^n to the point $x_0 = \pi_E(\epsilon_0)$, which represent the zero element of $\pi_n(M, x_0)$. Hence Im $i_* \subset$ Ker π_{E*}.

The elements of $\pi_n(E, \epsilon_0)$ are homotopy classes of mappings Φ^* of the cube I^n into E with ∂I^n mapped to ϵ_0. By π_{E*} they are projected to mappings $\pi_E \circ \Phi^*$ of I^n into M with ∂I^n mapped to x_0. Then, δ maps the latter to the restriction to $\{1\} \times I^{n-1}$ of *any* general lift $(\pi_E \circ \Phi^*)^*$ which maps $\{0\} \times I^{n-1}$ to f_0. If one takes Φ^* as such a general lift, it maps $\{1\} \times I^{n-1}$ to $\{f_0\}$ which belongs to the zero of $\pi_{n-1}(F, f_0)$. Hence Im $\pi_{E*} \subset$ Ker δ.

Finally, the elements of $\pi_n(M, x_0)$ are homotopy classes of mappings Φ of the cube I^n into M with ∂I^n mapped to x_0. δ puts them to the homotopy classes of the restriction $\Phi^*|_{\{1\}\times I^{n-1}}$ of a general lift Φ^* with $\Phi^*|_{\{0\}\times I^{n-1}} = \{\epsilon_0\}$, which means that $\Phi^*|_{\{1\}\times I^{n-1}}$ is null-homotopic. Hence Im $\delta \subset$ Ker i_*.

To prove exactness, the reverse inclusions have also to be demonstrated.

Suppose that the homotopy class of $\Phi^* : I^n \to E$ belongs to Ker π_{E*}, that is, that $\pi_E \circ \Phi^*$ is null-homotopic in M. Lift the homotopy between $\Phi_0 \equiv x_0$ and $\pi_E \circ \Phi^*$ to $\widetilde{\Phi}^* : I^{n+1} \to E$. It shows that Φ^* is homotopic to a mapping of I^n into $\pi_E^{-1}(x_0)$ and thus belongs to Im i_*. Hence Im $i_* \supset$ Ker π_{E*}.

Suppose that the homotopy class of $\Phi : I^n \to M$ belongs to Ker δ. That means that Φ can be chosen so that $\Phi(\{1\} \times I^{n-1}) \subset \pi_E^{-1}(x_0)$ and $\Phi(\{0\} \times I^{n-1}) = \{\epsilon_0\}$. This can be lifted to Φ^* belonging to $\pi_n(E, \epsilon_0)$. Hence Im $\pi_{E*} \supset$ Ker δ.

The last step is a little bit more elaborate. Let $\Phi^* : I^{n-1} \to F$, $\Phi^*(\partial I^{n-1}) = \{f_0\}$ belong to Ker i_*. This means that there exists a mapping (homotopy) $\widetilde{\Phi}^* \circ \boldsymbol{P} : I \times S^{n-1} \to E$ with $\widetilde{\Phi}^* \circ \boldsymbol{P}|_{\{1\}\times S^{n-1}} = i \circ \Phi^* \circ \boldsymbol{P}$, $\widetilde{\Phi}^* \circ \boldsymbol{P}(\{0\} \times S^{n-1}) = \{\epsilon_0\}$ and $\widetilde{\Phi}^* \circ \boldsymbol{P}(I \times \{s_0\}) = \{\epsilon_0\}$. This implies $\widetilde{\Phi}^* : I \times I^{n-1} \to E$ with $\widetilde{\Phi}^*|_{\{1\}\times I^{n-1}} = i \circ \Phi^*$, $\widetilde{\Phi}^*(\{0\} \times I^{n-1}) = \{\epsilon_0\}$ and $\widetilde{\Phi}^*(I \times \partial I^{n-1}) = \{\epsilon_0\}$. Now it is easily seen that $\Phi = \pi_E \circ \widetilde{\Phi}^*$ represents an element of $\pi_n(M, x_0)$ which by δ is mapped to the element of $\pi_{n-1}(F, f_0)$ represented by Φ^*. Hence Im $\delta \supset$ Ker i_* □

If one defines the 'zero' of the set $\pi_0(X, x_0)$ to correspond to the pathwise connected component of X containing x_0, then it is easy to see that (8.1) extends to $n = 0$. If one further defines $\pi_n(F, f_0) = \pi_n(E, \epsilon_0) = \pi_n(M, x_0) = 0$ for $n < 0$ and $\delta(\pi_0(M, x_0)) = \{0\}$, then the exact homotopy sequence extends infinitely in both directions as an exact sequence.

The exact homotopy sequence can amply be used to compute homotopy groups. Let, for instance, $\pi_n(M, x_0) = 0$ for some n. This implies Ker $\pi_{E*} = \pi_n(E, \epsilon_0)$ and hence, by exactness of the homotopy sequence, that i_* is surjective. If now additionally $\pi_{n+1}(M, x_0) = 0$, then Im $\delta = 0$ in $\pi_n(F, f_0)$ and hence i_* is also injective, which means that

if $\pi_n(M, x_0) = \pi_{n+1}(M, x_0) = 0$, *then* $\pi_n(F, f_0) \approx \pi_n(E, \epsilon_0)$.

In the same way one obtains

if $\pi_{n-1}(F, f_0) = \pi_n(F, f_0) = 0$, *then* $\pi_n(E, \epsilon) \approx \pi_n(M, x_0)$,

in particular, for $n = 0$,

If F and M are (pathwise) connected, so is E,

and

if $\pi_{n-1}(E, \epsilon_0) = \pi_n(E, \epsilon_0) = 0$, *then* $\pi_n(M, x_0) \approx \pi_{n-1}(F, f_0)$.

If X is a discrete topological space, then $\pi_n(X, x_0) = 0$ for all $n > 0$. Hence it follows from the second of the above conclusions that

if the fiber F is discrete, then $\pi_n(E, \epsilon_0) \approx \pi_n(M, x_0)$ *for all* $n > 1$.

For instance, since $SU(2)$ is a twofold cover of $SO(3)$, it can be considered as a principal fiber bundle with base space $SO(3)$ and the discrete structure group $G = \mathbb{F}_2 = (\mathbb{Z} \bmod 2)$ consisting of two elements. On the other hand, according to (6.35), the elements of $SU(2)$ are represented by matrices

$$A = \begin{pmatrix} x_1 + ix_2 & x_3 + ix_4 \\ -x_3 + ix_4 & x_1 - ix_2 \end{pmatrix}, \quad \det A = x_1^2 + x_2^2 + x_3^2 + x_4^2 = 1, \tag{8.2}$$

and therefore $SU(2)$ is homeomorphic to S^3. Hence, $\pi_n(SO(3)) \approx \pi_n(SU(2)) \approx \pi_n(S^3)$ for all $n > 1$. (See also p. 226 f for more details.)

As another example (by taking $E = P$ and $F = G$), consider the principal fiber bundle $(\mathbb{R}, S^1, \pi, 2\pi\mathbb{Z})$ with $\pi(t \in \mathbb{R}) = \exp(it) \in S^1$, already discussed previously. It follows that $\pi_n(S^1) \approx \pi_n(\mathbb{R}) = 0$ for all $n > 1$, which is intuitively clear since an n-sphere with $n > 1$ cannot continuously be wound around S^1. Since $SO(2)$ is homeomorphic to S^1, also $\pi_n(SO(2)) = 0$ holds for $n > 1$.

As yet another example, consider the principal fiber bundle $(SO(3), S^2, \pi, SO(2))$ with bundle space $SO(3)$, base space S^2 and structure group $SO(2)$. (Any $SO(3)$-transformation corresponds bijectively to a directed rotation axis, that is, a point of S^2 and a rotation in the mathematically positive sense around this axis, which rotations are in bijective correspondence to $SO(2)$-transformations. More generally, it can be shown that $(SO(n), S^{n-1}, \pi, SO(n-1))$ is a principal fiber bundle, [3, Section 9.3]). Now, since $\pi_n(SO(2)) = 0$ for $n > 1$, the second of the above conclusions from the exact homotopy sequence yields $\pi_n(SO(3)) \approx \pi_n(S^2))$ and hence also $\pi_n(S^3) \approx \pi_n(S^2)$ for $n > 2$. This implies the Hopf theorem $\pi_3(S^2) \approx \pi_3(S^3) \approx \mathbb{Z}$ as a special case (cf. the end of Sect. 2.5).

Suppose that there exists a section $s : M \to E$ in the fiber bundle E. Then, $\pi_E \circ s = \mathrm{Id}_M$ and hence $\pi_{E*} \circ s_* = \mathrm{Id}_{\pi_n(M,x_0)}$, where the pushes forward from the spaces to their homotopy groups are considered. Therefore, every $[\Phi] \in \pi_n(M, x_0)$

is the image of some $[\Phi^*] \in \pi_n(E, \epsilon_0)$: Im $\pi_{E*} = \pi_n(M, x_0)$. Because of the exactness of (8.1) this means $\pi_n(M, x_0) = \text{Ker } \delta$, which is the same as $\delta \equiv 0$.

If E has a section, then

$$0 \to \pi_n(F, f_0) \xrightarrow{i_*} \pi_n(E, \epsilon_0) \xrightarrow{\pi_{E*}} \pi_n(M, x_0) \to 0 \quad \text{for all } n, \tag{8.3}$$

that is, $\pi_n(M, x_0) = \pi_n(E, \epsilon_0)/\pi_n(F, f_0)$.

8.2 Homotopy of Sections

The construction of (global) sections in a fiber bundle is a case of interrelation between homotopy and homology. In this section it will be presupposed that the base space M is a compact manifold (of dimension m as always in this text), which is homeomorphic to a polyhedron $|c_M|$ embedded into some $\mathbb{R}^n$, $n \geq m$. (Section 5.6; recall that every polyhedron of dimension m may be embedded into the $\mathbb{R}^{2m+1}$, hence, besides M being compact, the presupposition is not really restrictive.)

Let an abstract complex c_M corresponding to the polyhedron $|c_M|$ be fixed, and let c^r be the rth skeleton of $c_M = c^m$. For simplicity it will now further be assumed that $\pi_0(F) = 0$, that is, the typical fiber F is assumed to be pathwise connected. By smoothness, a section through $\epsilon_0 = p_0(f_0) = \{(p_0 g, g^{-1} f_0)\}$ consists of points $\epsilon \in E$, represented (for $g = e$) by (p, f), where f stays in the pathwise connected component of f_0 for all $x \in M$, if M is pathwise connected. Hence, for pathwise connected M, instead of a bundle $(E', M, \pi_{E'}, F', P)$ the subbundle (E, M, π_E, F, G) may be considered in the general case with F being the pathwise connected component of $f_0 \in F'$.

Next, assume that a section on $|c^l|$ is given for some $l < m$ and that $\pi_k(F) = 0$ for all $k < l$. (Recall that for a pathwise connected space F, $\pi_k(F, f_0) = \pi_k(F)$ does not depend on the point f_0.) Try to extend the section to $|c^{l+1}|$. If this is done, induction in l can be used, since $\pi_0(F, f_0) = 0$, and $|c^0|$ consists of discrete points (vertices) only for which the existence of a section is trivial.

Consider first a trivial fiber bundle $E = M \times F$.

Take a (regular) simplex given by $c_i^{l+1} \in c^{l+1}$. As any regular $(l+1)$-simplex, $|c_i^{l+1}|$ is homeomorphic to the $(l+1)$-ball. Its boundary belongs to the lth skeleton, $\partial c_i^{l+1} \in c^l$, and $|\partial c_i^{l+1}|$ is homeomorphic to the l-sphere S^l. A section s on $|\partial c_i^{l+1}|$ is homotopic to a mapping $s : S^l \to F$ and hence defines an element $g_s(c_i^{l+1})$ of the homotopy group $\pi_l(F)$. It is easily seen that the section s can be extended to all of $|c_i^{l+1}|$, iff $g_s(c_i^{l+1}) = 0$. (For instance by contracting the values of s on S^l to one point when contracting S^l to its center.)

Consider the $(l+1)$-chain module $C_{l+1}(c_M, \mathbb{Z})$ generated by the $(l+1)$-simplices of c^{l+1}. Any section s on $|\partial c^{l+1}|$ gives rise to a mapping of each simplex c_i^{l+1} of c^{l+1} to some element $g_s(c_i^{l+1})$ of $\pi_l(F)$. Since the homotopy group $\pi_l(F)$ is Abelian, this mapping may be extended by linearity to a mapping $\langle s, c_v^{l+1} \rangle$:

$C_{l+1}(c_M, \mathbb{Z}) \to \pi_l(F)$, where $c_v^{l+1} = \sum_i v_i c_i^{l+1}$, v_i integer. All these mappings for all sections form a cochain module $C^{l+1}(c_M, \pi_l(F))$ with coefficients in the homotopy group $\pi_l(F)$. If the cochain $\langle s, \cdot \rangle$ is trivial (zero-dimensional), then every section on $|c^l|$ may be extended to a section on $|c^{l+1}|$. Therefore, the cochain $\langle s, \cdot \rangle$ is called the **obstruction cochain** to the extension of the section s to $|c^{l+1}|$. It is obviously constant under continuous (homotopic) deformations of s.

With the help of the lifting proposition on p. 248, it can be demonstrated that all sections on $|c^{l-1}|$ are mutually homotopic as long as all homotopy groups $\pi_k(F)$ are trivial for all $k < l$. To see this, take two arbitrary sections s and s' and construct a homotopy of their restrictions to $|c^0|$, which is always possible since $|c^0|$ consists of isolated points. By means of the lifting proposition, extend this homotopy to a homotopy of $s|_{c^1}$ with some s_1 on $|c^1|$ which coincides with s' on $|\partial c^1|$. If $\pi_1(F) = 0$, all sections on $|c^1|$ coinciding on $|\partial c^1|$ are homotopic, and hence s_1 may be homotopically deformed into $s'|_{c^1}$. These two steps may be repeated until $|c^{l-1}|$ is reached.

Given two sections s and s' on $|c^l|$ which coincide on $|\partial c^l|$, for each simplex $|c_i^l|$ the mapping $\phi_{s,s'} : S^l \to F$ is considered, which maps the upper hemisphere of S^l homeomorphically to the simplex and composes this mapping with s, maps the lower hemisphere of S^l again homeomorphically to the simplex and composes with s', in such a way that both mappings coincide on the equator of S^l which is mapped onto $|\partial c_i^l|$. The homotopy class of this mapping is denoted by $\langle \phi_{s,s'}, c_i^l \rangle$ and forms by linear extension a cochain of the module $C^l(c_M, \pi_l(F))$. It is called the **difference cochain** between s and s'. Clearly $\langle \phi_{s,s'}, \cdot \rangle = 0$, iff s and s' are homotopic.

From the construction of both cochains it is clear that

$$\langle d\phi_{s,s'}, \cdot \rangle = \langle s, \cdot \rangle - \langle s', \cdot \rangle, \quad \langle d\phi_{s,s'}, c_v^{l+1} \rangle = \langle \phi_{s,s'}, \partial c_v^{l+1} \rangle. \tag{8.4}$$

Indeed, two arbitrary sections s and s' on $|c^l|$ are homotopic to sections s and $\tilde{s}'$ which coincide on $|c^{l-1}|$ since on $|c^{l-1}|$ all sections are homotopic. Putting $\phi_{s,s'} = \phi_{s,\tilde{s}'}$, $\phi_{s,s'}$ is defined for all sections s and s' on $|c^l|$. Moreover, (2.35) in additive writing for the group operation implies the left relation (8.4), if $\langle d\phi_{s,s'}, \cdot \rangle$ is defined by the right relation. (Note that by the above construction the coordinate x^1 of (2.35) runs in the opposite direction for s', hence the minus sign.)

Since $\partial^2 = 0$, the second relation (8.4) implies $d^2 = 0$. Take s' to be the constant section s_0 for which $\langle s_0, \cdot \rangle = 0$, and obtain $\langle s, \cdot \rangle = \langle d\phi_{s,s_0}, \cdot \rangle$, that is, the obstruction cochain is a coboundary (and also a cocycle, since $d^2 = 0$).

If $l = m$, there are no $(l+1)$-simplices in $|c_M| \sim M$. By extension of the second relation (8.4) to this case, $\langle \phi_{s,s'}, \cdot \rangle$ may be considered to be a cocycle on $c^m \setminus c^{m-1}$ ($\langle d\phi_{s,s'}, \cdot \rangle = 0$), and there are no non-trivial m-boundaries. It follows that, if $\pi_k(F) = 0$ for all $k < m$, then the set of homotopy classes of sections of $M \times F$ is in bijective correspondence with the cohomology group $H^m(c, \pi_m(F))$.

So much for a trivial bundle $M \times F$. If E is not a trivial fiber bundle, then an abstract complex for M is to be considered which corresponds to a subdivision of the polyhedron $|c_M| \sim M$ into simplices fine enough so that each simplex lies within a trivializing coordinate neighborhood U of M. Instead of F, now a fiber above some point $x_i \in U_i$ is to be taken, which is isomorphic to F as a fiber. However, instead of f_0 each m-simplex has now its own reference point ϵ_0, and instead of $\pi_k(F, f_0)$, now the set of isomorphic groups $\pi_k(\pi_E^{-1}(x_i), \epsilon_i)$ is to be considered, which leads to cochains with values in local groups. The technicalities are considered in [3, Section 11.4]. The definition of corresponding (co)homology groups of M is straightforwardly transfered to the new situation.

The obstruction cochain and the difference cochain are now defined to have coefficients in the local homotopy groups, which are all isomorphic and connected by group isomorphisms as transition functions. The relations (8.4) as local relations remain valid. In particular, from the very definition of difference cochains it is clear, that for a given section s of E, the mapping $\langle \phi_{s,s'}, \cdot \rangle$ into the cochain module is surjective. Indeed, given l and a section s on the l-skeleton of c_M, take any simplex c_i^l of the l-skeleton. By the definition of the homotopy group $\pi_l(\pi_c^{-1}(x_i))$, $x_i \in |c_i^l|$), for any predefined group element g_i there is a mapping $\widetilde{\Phi}_i^* : |c_i^l| \to \pi_E^{-1}(x_i), \Phi_i^*|_{|\partial c_i^l|} = \epsilon_i$ representing g_i. It is homotopic to a mapping $\Phi_i^* : |c_i^l| \to \pi_E^{-1}(x_i), \Phi_i^*|_{|\partial c_i^l|} = s|_{|\partial c_i^l|}$ since on the skeleton c^{l-1} all sections are homotopic. Let s' be the section on the l-skeleton which on $|c_i^l|$ is equal to Φ_i^*. It is a section because on $|\partial c_i^l|$ it coincides with the section s and hence it is continuous (and thus homotopic to a smooth section). Consequently, for every predefined cochain there exists a section s' of $|c^l|$ so that $\langle \phi_{s,s'}, \cdot \rangle$ is mapped to that cochain.

However, since on a non-trivial bundle a constant section does not necessarily exist, obstruction cochains are not necessarily coboundaries any more. Only, for any obstruction cochain $\langle s', \cdot \rangle$ and any coboundary $\langle d\phi, \cdot \rangle$, the cochain $\langle s, \cdot \rangle = \langle s', \cdot \rangle + \langle d\phi, \cdot \rangle$ is again an obstruction cochain. Moreover, as long as $\pi_k(F) = 0$ for $k < l$, every $\langle s, \cdot \rangle$ is a cocycle: $\langle ds, c_v^{l+1} \rangle = \langle s, \partial c_v^{l+1} \rangle = 0$. ($\langle s, \partial c_v^{l+1} \rangle$ maps to $\pi_{l-1}(\pi_E^{-1}(x_i), \epsilon_i) \approx \pi_{l-1}(F) = 0$ by assumption.) Hence, the obstruction cochains form a certain cohomology class corresponding to an element $h^{l+1}(E)$ of the cohomology group $H^{l+1}(c, \tilde{\pi}_l(F)) = Z^{l+1}(c, \tilde{\pi}_l(F))/B^{l+1}(c, \tilde{\pi}_l(F))$, where $\tilde{\pi}_l(F)$ means the set of local homotopy groups connected by transition isomorphisms. This cohomology class $h^{l+1}(E)$ is called a **characteristic class** of the fiber bundle E.

The fiber bundle (E, M, π_E, F, G) admits a section over the $(l+1)$-skeleton, iff the characteristic class is $h^{l+1}(E) = 0$; $h^{l+1}(E)$ is defined, iff all $h^k(E) = 0$ for $k \leq l$.

Characteristic classes will be considered in more detail later. As seen from above, they provide a measure of 'non-triviality' of a fiber bundle.

As a simple application, consider the problem of **singular points of tangent vector fields** on compact manifolds M. Consider the punctured tangent bundle $T_\circ(M) = (T_\circ(M), M, \pi_T, \mathbb{R}^m \setminus \{0\}, Gl(m, \mathbb{R}))$. Its typical fiber $F = \mathbb{R}^m \setminus \{0\}$ is homotopy equivalent to the sphere S^{m-1}. It was mentioned at the end of Sect. 2.5 that $\pi_k(S^{m-1}) = 0$ for $k < m-1$ and $\pi_{m-1}(S^{m-1}) = \mathbb{Z}$. (See [3, Section 7.1] for an outline of a proof.) Therefore, there is always a non-zero tangent vector field on the $(m-1)$-skeleton of any polyhedron $|c_M| \sim M$. The attempt to extend it to M will run into an obstruction, if $h^m(T_\circ(M))$ is non-zero.

$h^m(T_\circ(M))$ is an element of the cohomology group $H^m(M,\mathbb{Z})$. If M is connected and orientable, then, since Poincaré's duality (5.61) holds also for $K=\mathbb{Z}$ instead of $\mathbb{R}$, one has $H^m(M,\mathbb{Z}) \approx H_0(M,\mathbb{Z}) = \mathbb{Z}$, and $h^m(T_\circ(M))$ is an integer depending on M (since the typical fiber F was fixed).

Isolated nodes of a tangent vector field are called singularities. Let $x \in M$ be a singular point of a tangent vector field X. Put a small sphere S^{m-1} around that point, so that no other singularity is enclosed and the enclosed ball is inside a single coordinate neighborhood of M. Then, $X|_{S^{m-1}}$ can be considered as a mapping from S^{m-1} to $\mathbb{R}^m$. Composing it with the central projection $\mathbb{R}^m \setminus \{0\} \to S^{m-1}$, a mapping of S^{m-1} to itself is obtained. The degree of such a mapping was defined on p. 47. This degree is called the **index of the singular point of the tangent vector field**. Replacing the enclosed ball by a homeomorphic simplex, it is easily seen that the index of the singularity is the obstruction to continue the non-zero tangent vector field X from the boundary of the ball (sphere) into the whole ball. The obstruction in the given case is an element of the homotopy group $\pi_{m-1}(S^{m-1}) \approx \mathbb{Z}$.

Next, consider a decomposition of M into simplices small enough that each simplex contains at most one singularity. Take the non-zero vector field X on the $(m-1)$-skeleton of this decomposition, and let n define the characteristic class of $T_\circ(M)$, which is the obstruction to continue the non-zero X to all M. It is clear that n is the sum of all indices of all singularities of X.

The sum of the indices of all singularities of a tangent vector field does not depend on X, it only depends on M and is a topological invariant of M.

This is a simple case of an index theorem. By taking a sufficiently simple tangent vector field for which the index sum is easy to compute, one can show that n is Euler's characteristic in that case. (For instance the gradient vector field of a real function on M provided with a metric can be analyzed by means of Morse theory (Sect. 5.8).) Since the only non-zero Betti numbers of a sphere S^m are $\beta^0(S^m) = \beta^m(S^m) = 1$ (cf. (5.60)), Euler's characteristic of an even dimensional sphere is 2 and of an odd dimensional sphere is 0. Hence, an even dimensional sphere cannot have a non-zero tangent vector field without singularities. (In two dimensions: 'every hedgehog has a vortex'; in fact it has at least one vortex of index 2 or two vortices of index 1.) Odd dimensional spheres have non-singular tangent vector fields. In fact, again by Poincaré's duality, this is true for any odd-dimensional compact orientable manifold.

8.3 Gauge Fields and Connections on $\mathbb{R}^4$

The theory of connections on principal fiber bundles and gauge field theory describe the same situation in different languages; for several decades they were developed in parallel and largely independently.

First gauge field theories on $\mathbb{R}^m$ are considered which every physicist is familiar with, and then the general case is treated. In particular, $\mathbb{R}^4$ may describe Minkowski's space–time. (As will be seen, the Minkowski metric is needed only to specify the dynamics of the theory.)

As the prototype of a gauge field theory, reconsider **Maxwell's electrodynamics** (Sect. 5.9). Since $H^2_{dR}(\mathbb{R}^4) = 0$, one may start from (5.99) with the 1-form

$$A, \quad \text{in coordinates} \quad A = \sum_\mu A_\mu dx^\mu, \tag{8.5}$$

of **gauge potentials** from which the **gauge fields** derive as

$$\begin{aligned} &F = dA, \quad \text{in coordinates} \quad F = \sum_{\mu<\nu} F_{\mu\nu} dx^\mu \wedge dx^\nu, \\ &F_{\mu\nu} = \partial_\mu A_\nu - \partial_\nu A_\mu, \quad \partial_\mu \equiv \partial/\partial x^\mu. \end{aligned} \tag{8.6}$$

(Since $\mathbb{R}^4$ may be covered by a single chart, natural coordinates as a single local and global coordinate system are used.) As a consequence of (8.6), the homogeneous set of Maxwell's equations,

$$dF = 0, \tag{8.7}$$

immediately follows as identities, while in order to get the dynamics of the fields, an action integral is needed. The simplest action is the Maxwell action

$$\mathcal{S}[A] = \frac{1}{2} \int_{\mathbb{R}^4} F \wedge *F, \qquad (*F)_{\sigma\tau} = \frac{\delta^{1234}_{\mu\nu\sigma\tau}}{2} g^{\mu\kappa} g^{\nu\lambda} F_{\kappa\lambda}, \tag{8.8}$$

where the prefactor sets the energy scale and hence is convention, and Hodge's star operator (5.87) in the present case results in the second relation. Note that the star operator makes use of the Minkowski metric, so that in tensor notation

$$\frac{1}{2} \int_{\mathbb{R}^4} F \wedge *F = -\frac{1}{4} \int_{\mathbb{R}^4} F_{\mu\nu} F^{\mu\nu} d^4x = \frac{1}{2} \int_{\mathbb{R}^4} (\boldsymbol{E}^2 - \boldsymbol{B}^2), \tag{8.9}$$

where $\boldsymbol{E}$ and $\boldsymbol{B}$ are the electric and magnetic fields. The dynamics derives from the stationarity of the action which in view of (5.93) and (8.7) implies the second set of Maxwell's equations,

$$\delta F = 0 = d * F, \tag{8.10}$$

which coincides with (5.96) in the absence of matter. Clearly, because of (8.6), a gauge transformation

$$A \to A' = A + d\chi, \quad d\chi = \sum_\mu \partial_\mu \chi dx^\mu \tag{8.11}$$

with a smooth single valued function χ on $\mathbb{R}^4$ does not change Maxwell's equations.

To include the interaction with a **matter field** Ψ, the integral over the Lagrangian density $\mathcal{L}(\Psi, \partial_\mu \Psi)$ of the matter field has to be added to the action (8.8) (which must be Hermitian, e.g. $\mathcal{L}(\Psi, \partial_\mu \Psi) = i\overline{\Psi}\gamma^\mu \partial_\mu \Psi - m\overline{\Psi}\Psi$ in the case of the electron–positron field of mass m, where γ^μ are the Dirac matrices and $\overline{\Psi} = \Psi^\dagger \gamma^0$), *and then all partial derivatives* ∂_μ *are to be replaced by the gauge-covariant derivative*

$$\mathcal{D}_\mu = \partial_\mu - ieA_\mu \quad \text{or} \quad \mathcal{D} = d - ieA \tag{8.12}$$

in a minimal interaction, where e is the charge of the matter field. (Depending on units used in which the vacuum speed of light is $c \neq 1$ and the action unit is $\hbar \neq 1$, e is sometimes to be replaced by $(e/\hbar c)$ in (8.12); in this text the above choice $c = \hbar = 1$ is always made.) In the case of electrodynamics, (8.6) remains unaltered since the potential components A_μ commute among one another. However, the gauge transformation (8.11) has now to be supplemented with

$$\Psi \to \Psi' = e^{ie\chi}\Psi, \tag{8.13}$$

so that $\mathcal{D}'_\mu \Psi' = e^{ie\chi}\mathcal{D}_\mu \Psi$, and the action remains invariant. Equations 8.12 and 8.11 may also be combined into

$$\mathcal{D} \to \mathcal{D}' = e^{ie\chi}\mathcal{D}e^{-ie\chi}, \tag{8.14}$$

from which together with (8.13) one directly infers that the Lie group $U(1)$ is the *local* symmetry group of the gauge symmetry. (This is the so-called Weyl rotation in the charge space; H. Weyl found it in 1929 in a (failed) attempt to unify electromagnetism with gravity and called it the 'relativity principle in the charge space'.) Note that the local value of $i\chi$ may be taken as an element of the Lie algebra $\mathfrak{u}(1) = i\mathbb{R}$ which transforms according to the adjoint representation Ad of the group $U(1)$, compare (6.66). The second set of Maxwell's equations (8.10) is now completed to become

$$\delta F = J = *d*F, \quad J^\mu = e\overline{\Psi}\gamma^\mu\Psi. \tag{8.15}$$

The theory is simple because $U(1)$ is an Abelian group.

In 1954, Yang and Mills found a non-Abelian generalization, which however had to wait for two decades as a seeming formal curiosity until it finally celebrated its triumph in particle physics not only by saving field theory from agony. Replace $U(1)$ with any appropriate compact Lie group G of dimension n, under which the **matter fields** Ψ (N components) transform according to some N-dimensional unitary representation ψ of G:

$$\Psi'^i = \psi^i_j(g)\Psi^j, \quad \overline{\Psi}'_i = \overline{\Psi}_j(\psi^{-1})^j_i, \qquad g \in G. \tag{8.16}$$

Define the 'derivative'

$$(\mathcal{D}_\mu\Psi)^i = \partial_\mu\Psi^i + \mathcal{A}^i_{\mu j}\Psi^j = \left(\partial_\mu\delta^i_j + \mathcal{A}^i_{\mu j}\right)\Psi^j, \quad \text{in short} \quad \mathcal{D}_\mu = \partial_\mu + \mathcal{A}_\mu, \tag{8.17}$$

where the $(N \times N)$-matrix valued 1-forms $\mathcal{A}_\mu$ are subject to the adjoint representation Ad of G in the Lie algebra $\mathfrak{g}$ of G, which is an n-dimensional vector space (spanned by the 'infinitesimal generators' of G), so that there are n linearly independent 1-forms $\mathcal{A}_\mu$, which transform according to the transformation group Aut $(\mathfrak{g})$ for every 'outer' (spatial) index μ, compare (6.65, 6.66). The group G is called the **inner symmetry group of the gauge field theory** (isospin, color, ...), while the 1-form derives from the **outer symmetry of space–time** (scalar, vector, spinor, ...). As indicated by the writing in (8.17), the **gauge potentials** $\mathcal{A}_\mu$ are taken in the representation of the matter fields Ψ, that is, by $N \times N$-matrices. The form (8.17) should be covariant under G-transformations in the sense

$$\mathcal{D}'\Psi' = \psi\mathcal{D}\psi^{-1}\psi\Psi. \tag{8.18}$$

This readily implies

$$\mathcal{A}'_\mu = \psi\mathcal{A}_\mu\psi^{-1} + \psi(\partial_\mu\psi^{-1}) = \psi\mathcal{A}_\mu\psi^{-1} - (\partial_\mu\psi)\psi^{-1}. \tag{8.19}$$

Note that while $\mathcal{D}_\mu$ is understood as a differential operator, that is, ∂_μ of its first term of (8.12) operates on everything written right of this operator, the derivative in (8.19) is taken of ψ^{-1} and ψ, respectively, only. (Compare the end of Sect. 2.3 for the last rewriting of (8.19).) Note also that, if $G = U(1)$ and the one-dimensional representation $\psi = e^{i\chi}$ is operative, then $\mathcal{A}'_\mu = \mathcal{A}_\mu - i\partial_\mu\chi$; compare to (8.11) with $\mathcal{A} = -iA$ and $e = 1$. Introduce **gauge fields** (matrix valued 2-forms)

$$\mathcal{F}_{\mu\nu} = \mathcal{D}_\mu\mathcal{A}_\nu - \mathcal{D}_\nu\mathcal{A}_\mu = \partial_\mu\mathcal{A}_\nu - \partial_\nu\mathcal{A}_\mu + [\mathcal{A}_\mu, \mathcal{A}_\nu] = [\mathcal{D}_\mu, \mathcal{D}_\nu], \tag{8.20}$$

for which from the last expression and (8.18) the transformation property

$$\mathcal{F}'_{\mu\nu} = \psi\mathcal{F}_{\mu\nu}\psi^{-1} \tag{8.21}$$

derives. Now, as for any commutator product, $[\mathcal{D}_\lambda, [\mathcal{D}_\mu, \mathcal{D}_\nu]] + [\mathcal{D}_\mu, [\mathcal{D}_\nu, \mathcal{D}_\lambda]] + [\mathcal{D}_\nu, [\mathcal{D}_\lambda, \mathcal{D}_\mu]] = 0$ and $[\mathcal{D}_\lambda, [\mathcal{D}_\mu, \mathcal{D}_\nu]] = [\mathcal{D}_\lambda, \mathcal{F}_{\mu\nu}] = (\partial_\lambda\mathcal{F}_{\mu\nu}) + [\mathcal{A}_\lambda, \mathcal{F}_{\mu\nu}]$, where in the first term again the derivative extends to $\mathcal{F}_{\mu\nu}$ only. Therefore, the fields must obey the kinematic equations

$$[\mathcal{D}_\lambda, \mathcal{F}_{\mu\nu}] + [\mathcal{D}_\mu, \mathcal{F}_{\nu\lambda}] + [\mathcal{D}_\nu, \mathcal{F}_{\lambda\mu}] = 0, \tag{8.22}$$

which replace in Yang–Mills theory the first group of Maxwell's equations.

The Yang–Mills action integral

$$\mathcal{S}[\mathcal{A}, \Psi] = -\sum_a \frac{1}{2\lambda_a^2}\int_{\mathbb{R}^4} \mathrm{tr}\,(\mathcal{F}^a \wedge *\mathcal{F}^a) + \int_{\mathbb{R}^4} (i\overline{\Psi}\gamma^\mu\mathcal{D}_\mu\Psi - \overline{\Psi}m\Psi)d^4x \tag{8.23}$$

is invariant under the G-transformations. Distinct from electrodynamics, the fields themselves now carry charges which were (together with the imaginary factor, hence the minus sign in (8.23) compared to (8.8)) included into the potentials, compare (8.17) with (8.12). This is consequent since now the gauge fields directly interact with each other as seen from the last term of the last but one expression of (8.20). Therefore, the squares of the coupling constants λ_a (instead of e in Maxwell's theory) now appear in the denominator of the Lagrangian of the fields. In (8.23) it is assumed that the symmetry group G is semi-simple, and one coupling constant enters for each simple component a. The trace is the matrix trace over the product of representation matrices $(\mathcal{F}^a_{\mu\nu})^i_j$ of each simple component of the group. While the Yang–Mills action itself is invariant, the dynamical field equations derived from that action,

$$\begin{aligned}&[\mathcal{D}_\mu, *\mathcal{F}^{\mu\nu}] = -\lambda J^\nu, \quad \lambda_j J^{\mu i}_j = i\lambda_j^2 \overline{\Psi}_j \gamma^\mu \Psi^i,\\ &i\gamma^\mu \mathcal{D}_\mu \Psi - m\Psi = 0,\end{aligned} \tag{8.24}$$

are covariant under the **gauge transformations** $\psi(g(x^\mu))$ with the gauge function $g(x^\mu)$.

Note a significant difference between Abelian and non-Abelian gauge field theories. Due to (8.11), in the Abelian case, the gauge field strength (8.6) is gauge invariant and hence measurable. Due to (8.21), the non-Abelian gauge fields transform covariantly under gauge transformations $\psi^i_j(g)$ and, like the phase of the wave function (8.13), they themselves are not measurable. A simple example of a measurable quantity is $\operatorname{tr}\mathcal{F}_{\mu\nu}$, where the trace is taken with respect to the inner symmetry group G, that is with respect to the vector indices i and j of the matter field.

All these considerations regard the classical wave equations. Quantization of gauge field theories [4] has its own problems, which are not considered here.

Consider the vector bundle $(E, \mathbb{R}^4, \pi_E, \mathbb{C}^N, G)$ and the corresponding Hermitian conjugate bundle $E^\dagger$ associated with the trivial principal fiber bundle $(P, \mathbb{R}^4, \pi, G) = \mathbb{R}^4 \times G$, where G is the symmetry group of a local gauge field theory and the inner product space $\mathbb{C}^N$ is the unitary representation space for an N-dimensional unitary representation ψ of G corresponding to **matter fields**. Let Ψ be a (global) section of E (which always exists in a vector bundle), and let $\Psi^\dagger$ be the corresponding hermitian conjugate section in $E^\dagger$. Then,

$$\Psi'(x^\mu) = \psi(g(x^\mu))^{-1}\Psi(x^\mu), \quad x^\mu \in \mathbb{R}^4 \tag{8.25}$$

with a (global) section $g : \mathbb{R}^4 \to G$ of P (which exists since P is trivial) is a gauge transformation of the matter field Ψ. (By convention, in comparison to (8.16) ψ^{-1} instead of ψ is used here in view of what follows.)

Introduce the $\mathfrak{g}$-valued 1-form A as a connection form on P. Since P is trivial, a single global coordinate neighborhood $U = \mathbb{R}^4$ can be used. Let $p \mapsto (\pi(p), \phi_\alpha(p))$ be coordinates on P and introduce the canonical section $s_\alpha : x \mapsto (x, e)$ of P.

Denote the local connection form in these coordinates by A_α. According to Sect. 7.3 it is the pull back of the connection form A from s_α to $\mathbb{R}^4$. With $A_{\alpha\mu} = A_\alpha(\partial/\partial x^\mu)$, the local connection form expresses as $A_\alpha = \sum_\mu A_{\alpha\mu}(x)dx^\mu$. A coordinate transformation on the fibers of P by $\phi_\beta(p) = \phi_\alpha(p)g(\pi(p))$ provided by the gauge transformation $g(x)$ leads to a canonical section $s_\beta(x) = s_\alpha(x)g(x)$ corresponding to the new coordinates. According to (7.4) on p. 217, the connection form must transform according to

$$A_{\beta\mu} = (\mathrm{Ad}(g^{-1})A_\alpha)_\mu + g^*(\vartheta)_\mu \tag{8.26}$$

where ϑ is the Maurer–Cartan 1-form of G at $p = (x, g)$ and g^* pulls it back to a 1-form at x on $\mathbb{R}^4$.

Moving over to the representation of the Lie group G by a subgroup of $Gl(N, \mathbb{C})$ of complex $(N \times N)$-matrices $\psi(g)$ acting on $\mathbb{C}^N$, the elements $A_{\alpha\mu}$ of the Lie algebra $\mathfrak{g}$ are likewise represented by $(N \times N)$-matrices $\mathcal{A}_{\alpha\mu}$ which are elements of the Lie algebra $\mathfrak{gl}(N, \mathbb{C})$. The transformation low (8.26) now reads

$$\mathcal{A}_{\beta\mu} = \psi^{-1}\mathcal{A}_{\alpha\mu}\psi + \psi^{-1}(\partial\psi/\partial x^\mu), \tag{8.27}$$

where $\psi = \psi(g(x^\mu))$, and $\psi^*(\vartheta)_\mu = \psi^{-1}\psi_*(\partial/\partial x^\mu) = \psi^{-1}(\partial\psi/\partial x^\mu)$: the difference between the lifts of the tangent vector $\partial/\partial x^\mu$ on $\mathbb{R}^4$ to $p_\beta = (x, g)$ and to $p_\alpha = (x, e)$, respectively, in $T_g(G)$ is $\partial g/\partial x^\mu$, and its $(N \times N)$-representation $\partial\psi/\partial x^\mu$ is pulled back to $\mathfrak{gl}(N, \mathbb{C})$ by ψ^{-1}. Comparison of (8.16, 8.19) with (8.25, 8.27) (with ψ replaced by ψ^{-1}) reveals that

*the **gauge potentials** of a local gauge field theory yield a local connection form, represented in the space $\mathbb{C}^N$ of matter fields, of the principal fiber bundle $(P, \mathbb{R}^4, \pi, G)$ with the inner symmetry group G of the gauge field theory.*

Now, by putting $\langle\Omega, (\partial/\partial x^\mu) \wedge (\partial/\partial x^\nu)\rangle = \Omega_{\mu\nu}$ and $\langle d\mathcal{A}, (\partial/\partial x^\mu) \wedge (\partial/\partial x^\nu)\rangle = \partial\mathcal{A}_\nu/\partial x^\mu - \partial\mathcal{A}_\mu/\partial x^\nu$ one immediately infers from (7.11) on p. 223 that

*the **gauge fields***

$$\mathcal{F}_{\mu\nu} = (d\mathcal{A})_{\mu\nu} + [\mathcal{A}_\mu, \mathcal{A}_\nu] \quad \text{or} \quad \mathcal{F} = \mathcal{D}\mathcal{A} \tag{8.28}$$

form the local curvature form of the connection given by the local connection form $\mathcal{A}$ on $(P, \mathbb{R}^4, \pi, G)$, both represented in the space $\mathbb{C}^N$ of matter fields. The exterior covariant derivative in this representation is

$$\mathcal{D} = [d + \mathcal{A}, \cdot] \tag{8.29}$$

yielding the right version of (8.28).

Fixing a local gauge $\mathcal{A}(x^\mu)$ links the position space with the 'charge space' $\mathbb{C}^N$ and thus defines a parallel transport of vector fields $\Psi(x^\mu)$, which are sections of $(E, \mathbb{R}^4, \pi_E, \mathbb{C}^N, G)$. The **Bianchi identities** $\mathcal{D}\mathcal{F} = 0$ for the fields read $\langle\mathcal{D}\mathcal{F}, (\partial/\partial x^\lambda) \wedge (\partial/\partial x^\mu) \wedge (\partial/\partial x^\nu)\rangle = 0$ or

$$[\mathcal{D}_\lambda, \mathcal{F}_{\mu\nu}] + [\mathcal{D}_\mu, \mathcal{F}_{\nu\lambda}] + [\mathcal{D}_\nu, \mathcal{F}_{\lambda\mu}] = 0,$$

which is (8.22) and which forms the first group (8.7) of Maxwell's equations in Maxwell's electrodynamics, where $G = U(1)$ is Abelian and all forms commute. With respect to (8.29) compare also the text after (7.15).

Pure gauge potentials are gauge potentials $\mathcal{A}^{\mathrm{p}}_\mu$ which 'can be gauged away', that is, for which there exists a gauge fixing in which

$$\mathcal{A}^{\mathrm{p}}_\mu = \psi^{-1}(\partial\psi/\partial x^\mu) \tag{8.30}$$

and for which hence according to (8.27) for every trivializing coordinate neighborhood there exists a gauge transformation $\psi(g(x^\mu))$ for which $\mathcal{A}_\mu$ vanishes. This means that A is a flat connection in this case, and hence, by virtue of the theorem on flat connections,

A gauge potential is a pure gauge potential, iff the corresponding gauge fields $\mathcal{F}_{\mu\nu}$ *vanish.*

Pure gauge potentials may reflect topological properties of the base manifold on which the fields live and which may have consequences without direct gauge field interactions of matter fields as considered in the next section.

8.4 Gauge Fields and Connections on Manifolds

Instead of having $\mathbb{R}^4$ as the base space of a local gauge field theory, the latter may be considered on any manifold M. In many examples, M is just an open subset of $\mathbb{R}^m$, for instance with cut-outs where the gauge field diverges (point charges, monopoles, dipoles, ... More generally, M may be any curved space–time in the presence of a gravitational field. Even more generally, M may be a high-dimensional manifold of which space–time is a submanifold, and $M/\mathbb{R}^4$ is compact. This is the situation in string theory.

As a connection on the principal fiber bundle (P, M, π, G), the local gauge field theory readily transfers. The important difference is that P is in general not globally trivial any more. This enhances topological aspects strongly. It was already seen in Sect. 7.1 that a non-trivial principal fiber bundle does not have a global section. Hence, $g : x \mapsto g(x) \in G$ cannot be given globally, and (8.26, 8.27) cannot hold globally any more. However, if A and A' are two locally given sets of local connection forms with the same sets of transition functions $g(x)$, then their difference may be a globally given 1-form, for which it locally holds that

$$(A - A')_\beta = \mathrm{Ad}(g^{-1})(A - A')_\alpha, \quad \text{that is,} \quad (\mathcal{A}_{\beta\mu} - \mathcal{A}'_{\beta\mu}) = \psi^{-1}(\mathcal{A}_{\alpha\mu} - \mathcal{A}'_{\alpha\mu})\psi. \tag{8.31}$$

All, that follows (8.27) in the last section, transfers locally to the general case.

A simple example is **Dirac's monopole**. It is a case of magnetostatics as one time-independent part of Maxwell's electrodynamics. The symmetry group is

Fig. 8.3 The manifold $\mathbb{R}^3_\circ$ with spherical coordinates

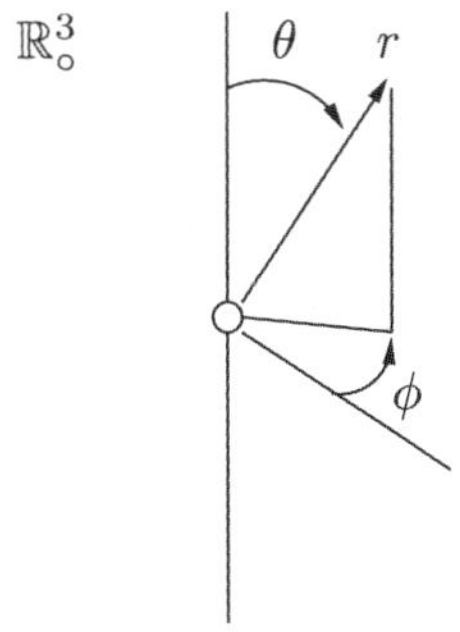

$U(1)$. Consider the punctured three-space $\mathbb{R}^3_\circ = \mathbb{R}^3 \setminus \{0\}$ as the base manifold of the principal fiber bundle $(P_\mathrm{D}, \mathbb{R}^3_\circ, \pi, U(1))$. Introduce polar coordinates (Fig. 8.3) $(r, \theta, \phi), r \neq 0$ in $\mathbb{R}^3_\circ$, and cover it by the two open sets $U_+ = \{\boldsymbol{r} \,|\, \theta \neq \pi\}$ and $U_- = \{\boldsymbol{r} \,|\, \theta \neq 0\}$. Define the local connection forms ($\mathfrak{u}(1)$-valued form, $\mathfrak{u}(1) = i\mathbb{R}$)

$$\mathcal{A}_+ = i\lambda(1 - \cos\theta)d\phi, \quad \mathcal{A}_- = -i\lambda(1 + \cos\theta)d\phi. \tag{8.32}$$

Write the $U(1)$-valued transition function as

$$\psi_{+-} = e^{i\chi}, \tag{8.33}$$

where χ is a real function on $U_+ \cap U_-$, that is, a real function of (r, θ, ϕ). Then, (8.27) reads

$$(\mathcal{A}_-)_\phi = \psi_{+-}^{-1}(\mathcal{A}_+)_\phi \psi_{+-} + \psi_{+-}^{-1}(\partial\psi_{+-}/\partial\phi) = (\mathcal{A}_+)_\phi + i(\partial\chi/\partial\phi). \tag{8.34}$$

Comparison with (8.32) yields $\chi = -2\lambda\phi$. The transition function must be uniquely defined on the intersection $U_+ \cap U_- = \{\boldsymbol{r} \,|\, \theta \neq 0, \pi\}$, which finally demands

$$\chi = -2\lambda\phi, \quad 2\lambda \in \mathbb{Z}, \tag{8.35}$$

since ϕ and $\phi + 2\pi$ describe the same point of $\mathbb{R}^3_\circ$, and hence it must be $e^{-i2\lambda 2\pi} = 1$. According to the theorem following (7.4), there is a connection form ω on P_D corresponding to the local connection forms (8.32) on $U_\pm \subset \mathbb{R}^3_\circ$, provided (8.35) is fulfilled. Since $U(1)$ is Abelian, the local curvature forms are

$$\mathcal{F}_\pm = d\mathcal{A}_\pm = i\lambda \sin\theta \, d\theta \wedge d\phi = i\lambda \frac{do}{r^2}. \tag{8.36}$$

They have a common analytic expression on both open sets $U_\pm$ and are proportional to the directed surface element do of spheres centered at the origin of $\mathbb{R}^3_\circ$, with an r-dependent coefficient.

Translating this result into physics means that $\mathcal{A}_\pm = -ieA_\pm$, where $A_\pm$ is the vector potential of the magnetic field $B^i = (i/2e)\delta^{ijk}_{123}\mathcal{F}_{jk}$ in Cartesian coordinates

x^i, in spherical coordinates the magnetic field has only a radial component $B^r = -\lambda/(er^2)$. The total magnctic flux through a sphere S^2 centered at the origin is independent of r and equal to

$$\Phi = \int_{S^2} \boldsymbol{B} \cdot do = \int_{S^2_+} dA_+ + \int_{S^2_-} dA_- = \int_{S^1} (A_+ - A_-) = -4\pi \frac{\lambda}{e} = 4\pi\mu, \quad (8.37)$$

where μ is the strength of the magnetic monopole sitting at the origin of $\mathbb{R}^3_\circ$. Here, $S^2_\pm$ mean the upper and lower half-sphere $\theta \lessgtr \pi/2, S^1$ is the equator $\theta = \pi/2$, and the trivial first integral has been rewritten and then treated with Stokes' theorem for later discussion. The result is Gauss' law for a magnetic monopole μ. Dirac's interest was attracted by the fact, that already in classical electrodynamics $\lambda = -e\mu$ (in ordinary units $\lambda = -e\mu/(\hbar c)$) is *topologically* quantized (!) to be half-integer. If somewhere in the universe there exists a magnetic monopole of strength $|\mu_0| = 1/(2e) = \mu_{\mathrm{Bohr}}/a_{\mathrm{Bohr}}$, then this would *explain* why all observed charges are multiples of e (a phenomenologically hard fact, with 22 orders of magnitude of relative experimental accuracy, for which otherwise there is no explanation). Here, μ_{Bohr} is Bohr's magneton and a_{Bohr} is Bohr's radius. After the surprising topological conclusion on p. 160 that a closed universe must be exactly electrically neutral, this is one more global topological conclusion of an intertwining of local magnetic and electric properties of the universe, resulting from the topological structure of **Maxwell's electrodynamics**. It does not mean that it is the correct explanation in physics since quantization of the fields themselves and linkage to other fields was not yet considered. Nevertheless, it reveals an important feature of the internal structure of Maxwell's theory. For a review on the actual theoretical and experimental status of magnetic monopoles see [5]. The example also shows that in gauge field theories the gauge potentials need exist only on open patches of the base space, in our case on $\mathbb{R}^3_\circ \setminus$ (some 'string' from the origin to infinity). The gauge fields may still be defined and smooth as tensor fields on all base space $\mathbb{R}^3_\circ$. (Gauge potentials correspond to the pseudo-tensorial connection form while gauge fields correspond to tensorial curvature forms.)

Returning to the principal fiber bundle $(P_{\mathrm{D}}, \mathbb{R}^3_\circ, \pi, U(1))$, it is easily seen that the quantization of λ is a case of a topological charge (Sect. 2.6). Consider the homotopy equivalence $U_+ \cap U_- \cong S^1$. Hence, the transition function ψ_{+-}, which takes on the role of an order parameter for the field, is homotopic to a function $F : S^1 \to U(1) \cong S^1$, for which the homotopy group $\pi_1(U(1)) = \pi_1(S^1) = \mathbb{Z}$ is relevant, resulting in a topological charge $2\lambda \in \pi_1(U(1))$. The above result is hence general and not related to the particular gauge fixing (8.32).

Another simple example is the **Aharonov–Bohm effect**. It refers to an electron moving outside of a confined magnetostatic field (Fig. 8.4). Here, the base space is $M = \mathbb{R}^3 \setminus S$, where S is a cylinder infinitely extending in x^3-direction, which contains a solenoid penetrated by a magnetic flux Φ and which keeps the electron outside by means of a potential wall. Outside of S there is no magnetic field. The electron is

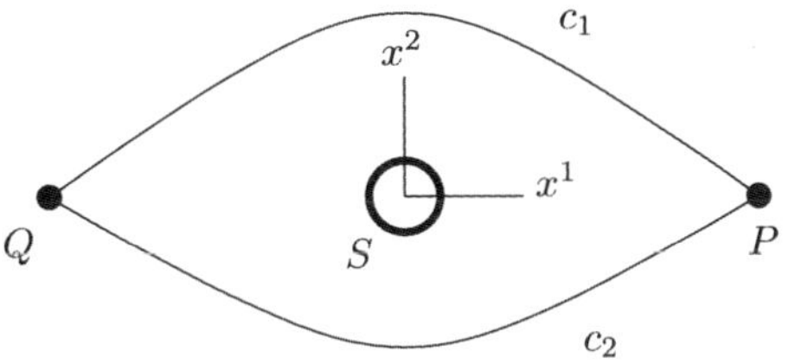

Fig. 8.4 The Aharonov–Bohm setup: S is the solenoid confining the magnetic field and excluding the electron by means of a potential wall, P and Q are considered possible positions of the electron, connected by typical paths c_1 and c_2. Cartesian coordinates are indicated, the solenoid extends infinitely in x^3-direction

injected at point P (for instance from an aperture before a cathode) and then after quantum propagation observed at point Q (for instance on a screen). In this case there is a globally defined local connection form $\mathcal{A}_i(\boldsymbol{r}) = -ieA_i(\boldsymbol{r})$, in cylinder coordinates (ρ, ϕ, x^3), $\rho^2 = (x^1)^2 + (x^2)^2$, $\phi = \arctan(x^2/x^1)$ outside of S given by

$$A = \frac{\Phi}{2\pi} d\phi, \tag{8.38}$$

for which it is directly seen that

$$F = dA = 0. \tag{8.39}$$

The connection ω on the manifold $(P_{\text{AB}}, M, \pi, U(1))$ given by A on M is flat (outside of S), there is no magnetic field outside of S. The formal reason for (8.39) is that (8.38) is a pure gauge potential, $A = d\chi$, $\chi = \phi\Phi/(2\pi)$. On the other hand, taking a circular area B^2 in the (x^1, x^2)-plane centered at the origin and containing the cross section of S, one finds trivially by means of Stokes' theorem

$$\int_{B^2} F = \int_{S^1} A = \Phi, \tag{8.40}$$

where S^1 is the oriented boundary of B^2. The magnetic flux in S is indeed Φ.

Consider for the sake of simplicity a non-relativistic electron with the Lagrangian $\hat{L} = -(\mathcal{D}_i)^2/2m = \hat{H}$ equal to the Hamiltonian. Let Ψ be a stationary wavefunction normalized according to an emission of one electron per unit time from the source P for $\Phi = 0$. For $\Phi \neq 0$, the wave function is $\Psi' = e^{ie\chi}\Psi = e^{ie\phi\Phi/(2\pi)}\Psi$. The geometry was chosen such that at point $P(\phi = 0)$ there is $\Psi' = \Psi$, since this value was fixed by normalization. However, at point Q, ϕ is not uniquely given. There are pairs of distinct homotopy classes of paths from P to Q, the pair (c_1, c_2) of Fig. 8.4 and similar pairs winding additionally a certain number of times around S in mutually opposite directions. Hence, for symmetry reasons, at point Q,

$$\Psi' = \sum_n a_n \Big(e^{ie\Phi(2n+1)/2} + e^{-ie\Phi(2n+1)/2} \Big) \Psi. \tag{8.41}$$

The absolute value of this function is periodic in Φ with the period $e\Phi_0 = 2\pi$ (or in ordinary units $\Phi_0 = hc/e$). $|\Psi'|^2$ has minima for $e\Phi = (2k+1)\pi$ and maxima halfway in-between. Although the electron wavefunction does not seem to directly see the flux Φ, it is equal to zero in S due to infinite potential walls, and although hence there is no Lorentz force from the magnetic field inside S onto the electron, it nevertheless reacts on the flux. It is, as if the electron sees directly the gauge potential and not only the gauge field like in classical physics. However, in truth it sees only the integral over the gauge potential over a closed loop, which is, as will be seen in the next but one section, a Berry phase. The Aharonov–Bohm effect has brilliantly been demonstrated in experiment.

As regards physics, another truth is that there are no infinite potential walls in nature, and hence the electron does see the field at the boundary of S by proximity (tunneling), and this boundary condition continues as a topological constraint via Stokes' theorem into all the outer space. The topological treatment relieves one from any detailed consideration of the proximity situation. This is a very general case with boundary conditions in physics. See also the discussion of polarization at the end of Sect. 8.6.

The wavefunction of the Aharonov–Bohm situation is a section of the complex line bundle (one-dimensional complex vector bundle) $(E, M, \pi_E, \mathbb{C}, U(1))$ associated with the principal fiber bundle P_{AB}. The paths contributing to (8.41) correspond to the elements of the holonomy group H_Q of the connection A with base point Q.

Relativistic field theory is conveniently first developed in Euclidean space $\mathbb{R}^4$ (with imaginary time) and then analytically continued (Wick rotated) to real time in the Minkowski space. As an example with a non-Abelian symmetry group, the **Belavin–Polyakov instanton**, is considered. Choose a Yang–Mills theory on $\mathbb{R}^4$ for which the field part of the action (first term of (8.23)) is finite. This demands that the gauge fields vanish for the spacial radius $r \to \infty$ and hence the gauge becomes pure. Technically this can be realized by compactifying the $\mathbb{R}^4$ to the sphere S^4 and demanding that the gauge is pure in the vicinity of the infinite point (south pole). Hence, the principal fiber bundle $(P_{\mathrm{BP}}, S^4, \pi, G)$ is operative.

As a simple case, take $G = SU(2)$ for the symmetry group. A general element of $SU(2)$ is

$$g = \exp\left(i\sum_{i=1}^{3} t^i\sigma_i\right), \quad \sigma_i\sigma_j = \delta_{ij}\mathbf{1}_2 + i\sum_k \delta_{ijk}^{123}\sigma_k, \quad t^i \in \mathbb{R}, \tag{8.42}$$

since the Lie algebra $\mathfrak{su}(2)$ is spanned by the Pauli matrices σ_i. Expanding the exponential function one gets $g = \mathbf{1}_2 + i\sum t^i\sigma_i - (1/2!)\sum t^i t^j(\delta_{ij}\mathbf{1}_2 + i\delta_{ijk}^{123}\sigma_k) + \cdots$. On summation over i and j the last spelled out term vanishes as a summation over a product of a symmetric factor $t^i t^j$ with an alternating factor δ_{ijk}^{123}. Hence, g may be cast into the form (compare (8.2))

$$g = u^0\mathbf{1}_2 + \sum_i u^i i\sigma_i, \quad u^\mu \in \mathbb{R}, \quad u^0 = \cos\sqrt{\sum (t^i)^2}. \tag{8.43}$$

Unitarity means $\mathbf{1}_2 = g^\dagger g = (\sum_{\mu=0}^{3}(u^\mu)^2)\mathbf{1}_2 + i\sum_{ijk} u^i u^j \delta_{ijk}^{123}\sigma_k$, where the last sum again vanishes, leaving $\sum_\mu (u^\mu)^2 = 1$ as the unitarity condition. The constraint $\det g = 1$ yields again the same condition. (It was already plugged into (8.42) by the traceless Pauli matrices, recall $\det(\exp A) = \exp(\operatorname{tr} A)$.) Hence, $SU(2)$ is homotopic to the sphere S^3. Put $\sigma_0 = i\mathbf{1}_2$, then the inverse relations are

$$u^\mu(g) = \operatorname{tr}(g\sigma_\mu)/(2i), \tag{8.44}$$

that is, the mapping $SU(2) \to S^3$ is even a bijection. (Distinct from (6.53, 6.54), here the sphere S^3 in the Euclidean space $\mathbb{R}^4$ figures.) The parameter space $S^3 \ni u^\mu$ is of course to be distinguished from the base space $\mathbb{R}^4 \ni x^\mu$ of the principal fiber bundle, in which the gauge fields live.

Returning to the base space $\mathbb{R}^4$, take the two patches $U_0 = \{(x^\mu) \in \mathbb{R}^4 \mid |x| < 2R\}$ and $U_\infty = \{(x^\mu) \in \mathbb{R}^4 \mid |x| > R/2\}$ of $\mathbb{R}^4$ for some fixed radius R, and gauge away the pure gauge outside $R/2$, that is, fix the local gauge potential

$$\mathcal{A}_{\infty\mu} = 0 \quad \Rightarrow \quad \mathcal{A}_{0\mu} = \psi^{-1}(\partial\psi/\partial x^\mu), \tag{8.45}$$

where $\psi = \psi_{0\infty}(g(x^\mu))$ is the transition function from U_∞ to U_0. It suffices to consider the transition function for $|x| = R$, that is, on another sphere S^3. In order to classify possible field configurations one has to classify the mappings $S^3 \ni (x^\mu) \mapsto g(x^\mu) \in SU(2) \sim S^3$. This is provided by the homotopy group $\pi_3(SU(2)) = \pi_3(S^3) = \mathbb{Z}$.

Use homogeneous coordinates $(w^\mu) = (x^\mu)/R$, $|w| = 1$, on the sphere S^3 of the base space and consider the mappings $S^3 \to S^3 \sim SU(2) : (w^\mu) \mapsto (u^\mu) = (\operatorname{tr}(g_n(w^\nu)\sigma_\mu)/(2i)) = (\operatorname{tr}((w^0\mathbf{1}_2 + w^i i\sigma_i)^n)\sigma_\mu)/(2i))$, $n \in \mathbb{Z}_+$. For $n = 1$ this is the identity mapping as seen from (8.44). For $n > 1$, the sphere S^3 is 'wrapped n times' by its preimage S^3 as can be seen from the last relation (8.43) since $g \mapsto g^n$ corresponds to $t^i \mapsto nt^i$, and the mapping $g \mapsto (u^\mu)$ is bijective. As is likewise easily seen for small t^i, the mappings preserve orientation of the manifold S^3 in the vicinity of its north pole ($u^0 = 1$), and hence everywhere due to smoothness. Let $j : S^3 \to S^3$ be the mapping which interchanges the coordinates w^1 and w^2 and hence reverses orientation of S^3. Replacing above $g_n(w^\mu)$ by $g_n(j(w^\mu))$ yields representatives for negative integer homotopy classes. All mappings for non-zero n are not homotopic to the trivial mapping $g_0(w^\mu) = e = \mathbf{1}_2$.

Belavin and Polyakov considered (anti)self-dual solutions $\mathcal{F} = \pm * \mathcal{F}$ of the Yang–Mills equations. Under this condition the field part of the Yang–Mills action becomes $\mp(2\lambda^2)^{-1}\int_{\mathbb{R}^4} \operatorname{tr}(\mathcal{F} \wedge \mathcal{F})$, where the integrand is a 4-form in a four-dimensional space and hence is closed, $d\operatorname{tr}(\mathcal{F} \wedge \mathcal{F}) = 0$. Since the patch U_0 is contractible, $\operatorname{tr}(\mathcal{F} \wedge \mathcal{F})$ is also exact (end of Sect. 5.5), that is, there exists a 3-form K so that $\operatorname{tr}(\mathcal{F} \wedge \mathcal{F}) = dK$. (Recall that $\mathcal{F} = 0$ on U_∞ and hence this relation is trivially fulfilled on U_∞ with any constant K.)

In the given case,

$$K = \mathrm{tr}\left(\mathcal{A}\wedge d\mathcal{A} + \frac{2}{3}\mathcal{A}\wedge\mathcal{A}\wedge\mathcal{A}\right), \quad dK = \mathrm{tr}\,(\mathcal{F}\wedge\mathcal{F}). \tag{8.46}$$

and

$$\int_{\mathbb{R}^4} \mathrm{tr}\,(\mathcal{F}\wedge\mathcal{F}) = \int_{U_0} dK = \int_{S^3} K = -\frac{1}{3}\int_{S^3} \mathrm{tr}\,(\mathcal{A}\wedge\mathcal{A}\wedge\mathcal{A}). \tag{8.47}$$

Proof Straightforwardly, with $d\mathcal{A} = \mathcal{F} - \mathcal{A}\wedge\mathcal{A}$,

$$\begin{aligned} dK &= \mathrm{tr}\left(d\mathcal{A}\wedge d\mathcal{A} + \frac{2}{3}(d\mathcal{A}\wedge\mathcal{A}\wedge\mathcal{A} - \mathcal{A}\wedge d\mathcal{A}\wedge\mathcal{A} + \mathcal{A}\wedge\mathcal{A}\wedge d\mathcal{A}\right) \\ &= \mathrm{tr}\,((\mathcal{F}-\mathcal{A}\wedge\mathcal{A})\wedge(\mathcal{F}-\mathcal{A}\wedge\mathcal{A}) \\ &\quad + \frac{2}{3}((\mathcal{F}-\mathcal{A}\wedge\mathcal{A})\wedge\mathcal{A}\wedge\mathcal{A} - \mathcal{A}\wedge(\mathcal{F}-\mathcal{A}\wedge\mathcal{A})\wedge\mathcal{A} + \mathcal{A}\wedge\mathcal{A}\wedge(\mathcal{F}-\mathcal{A}\wedge\mathcal{A}))\Big) \\ &= \mathrm{tr}\,(\mathcal{F}\wedge\mathcal{F} - \mathcal{F}\wedge\mathcal{A}\wedge\mathcal{A} - \mathcal{A}\wedge\mathcal{A}\wedge\mathcal{F} + \mathcal{A}\wedge\mathcal{A}\wedge\mathcal{A}\wedge\mathcal{A} \\ &\quad + \frac{2}{3}(\mathcal{F}\wedge\mathcal{A}\wedge\mathcal{A} - \mathcal{A}\wedge\mathcal{F}\wedge\mathcal{A} + \mathcal{A}\wedge\mathcal{A}\wedge\mathcal{F} - \mathcal{A}\wedge\mathcal{A}\wedge\mathcal{A}\wedge\mathcal{A})\Big). \end{aligned}$$

Now, using the alternating property of the wedge product and the cyclicity of the trace of matrices,

$$\begin{aligned} \mathrm{tr}\,(\mathcal{A}\wedge\mathcal{A}\wedge\mathcal{A}\wedge\mathcal{A}) &= \frac{1}{4!}\sum_{\kappa\lambda\mu\nu}\mathrm{tr}\,(\mathcal{A}_\kappa\mathcal{A}_\lambda\mathcal{A}_\mu\mathcal{A}_\nu)dx^\kappa\wedge dx^\lambda\wedge dx^\mu\wedge dx^\nu \\ &= -\frac{1}{4!}\sum_{\kappa\lambda\mu\nu}\mathrm{tr}\,(\mathcal{A}_\nu\mathcal{A}_\kappa\mathcal{A}_\lambda\mathcal{A}_\mu)dx^\nu\wedge dx^\kappa\wedge dx^\lambda\wedge dx^\mu = -\mathrm{tr}\,(\mathcal{A}\wedge\mathcal{A}\wedge\mathcal{A}\wedge\mathcal{A}). \end{aligned}$$

Hence, $\mathrm{tr}\,(\mathcal{A}\wedge\mathcal{A}\wedge\mathcal{A}\wedge\mathcal{A}) = 0$. Likewise, $\mathrm{tr}\,(\mathcal{F}\wedge\mathcal{A}\wedge\mathcal{A}) = -\mathrm{tr}\,(\mathcal{A}\wedge\mathcal{F}\wedge\mathcal{A}) = \mathrm{tr}\,(\mathcal{A}\wedge\mathcal{A}\wedge\mathcal{F})$ is found. This reduces the above result for dK to (8.46) and thus proves the latter. In the next section this will become a special case of a very general result.

In (8.47), S^3 is the sphere of radius R in $\mathbb{R}^4$, and Stokes' theorem was used in the second equality. Again using $d\mathcal{A} = \mathcal{F} - \mathcal{A}\wedge\mathcal{A}$, K may be cast into $K = \mathrm{tr}\,(\mathcal{A}\wedge\mathcal{F}) - (1/3)\mathrm{tr}\,(\mathcal{A}\wedge\mathcal{A}\wedge\mathcal{A})$. The first term vanishes on S^3, since $\mathcal{F} = 0$ there. □

Now, consider the strength of the instanton,

$$q = \int_{\mathbb{R}^4} \mathrm{tr}\,(\mathcal{F}\wedge\mathcal{F}). \tag{8.48}$$

For the sake of simplicity the identical (sometimes called fundamental) two-dimensional representation of the symmetry group is considered, $\psi(g) = g$, where g is the (2×2)-matrix (8.42). The results are qualitatively general, only the topological charge may get an additional dimension factor from the trace due to a more general representation. For $g_0(x^\mu) = e$, from (8.45) one has $\mathcal{A}_{0\mu} = 0$ and hence $q = q_0 = 0$. For $g_1(x^\mu) = w^0\mathbf{1}_2 + \sum_i w^i i\sigma_i$, one has $\mathcal{A}_{0\mu} = (w^0\mathbf{1}_2 + \sum_i w^i i\sigma_i)^{-1}(\partial(w^0\mathbf{1}_2 + \sum_i w^i i\sigma_i)/\partial x^\mu) = (w^0\mathbf{1}_2 + \sum_i w^i i\sigma_i)^{-1}(\partial(w^0\mathbf{1}_2 + \sum_i w^i i\sigma_i)/\partial w^\mu)/R$. Since this corresponds to the identity mapping $S^3 \to S^3$: $w^\mu \mapsto u^\mu(w^\nu) = w^\mu$, the gauge potential may be expressed as $\mathcal{A}_{0\mu}(u^\nu) = g^{-1}(\partial g/\partial u^\mu)(1/R) = (1/R)(\partial \ln g/\partial u^\mu) = (i/R)\sum_i \sigma_i(\partial t^i/\partial u^\mu)$ or $\mathcal{A}_0 = \sum_\mu \mathcal{A}_{0\mu}dx^\mu = i\sum_{i,\mu}\sigma_i(\partial t^i/\partial u^\mu)du^\mu = i\sum_i \sigma_i dt^i$. This yields $\mathrm{tr}\,(\mathcal{A}_0 \wedge \mathcal{A}_0 \wedge \mathcal{A}_0) = i^3\mathrm{tr}\,(\sigma_i\sigma_j\sigma_k)dt^i \wedge dt^j \wedge dt^k = i^3 3!\,\mathrm{tr}\,(\sigma_1\sigma_2\sigma_3)dt^1 \wedge dt^2 \wedge dt^3 = 3!\,2d\tau$, where $d\tau$ is the 3-volume element of the manifold $SU(2)$. The integration of the last expression of (8.47) is now replaced by an integration over the unit sphere $S^3 \sim SU(2)$ with the result $q = q_1 = -4\cdot 2\pi^2 = -8\pi^2$ where $2\pi^2$ is the volume of the unit sphere S^3 (see footnote on p. 53). Now, realize that $g_n = g_{n-1}g_1$. Since the gauge is a pure gauge on S^3, it can be gauged away on every trivial patch (chart) of S^3. Cover S^3 by U_N and U_S, the north and the south open hemisphere overlapping at the equator. Gauge away g_{n-1} from the north hemisphere, that is, deform $g_{n-1}(w^\mu)$ smoothly into $g'_{n-1}(w^\mu)$, where $g'_{n-1} = e$ is constant on the north hemisphere. In view of the bijection between $SU(2)$ and S^3 this amounts to a smooth coordinate transformation on S^3, which transforms the integral over $S^3 \setminus$ (north pole) into an integral over the south hemisphere. Likewise gauge away g_1 from the south hemisphere. Now, $g'_n = eg'_1 = g'_1$ on the north hemisphere and $g'_n = g'_{n-1}e = g'_{n-1}$ on the south hemisphere, and $-(1/3)\int_{S^3}\mathrm{tr}\,(\mathcal{A}'_n \wedge \mathcal{A}'_n \wedge \mathcal{A}'_n) = q_1 + q_{n-1}$. For negative n, the reversion of orientation of S^3 simply results in $q_{-n} = -q_n$. In summary,

$$q_n = -8\pi^2 n \quad \text{or}$$
$$n = \frac{1}{24\pi^2}\int_{S^3}\mathrm{tr}\,(g^{-1}dg \wedge g^{-1}dg \wedge g^{-1}dg) = \frac{1}{2}\int_{\mathbb{R}^4}\mathrm{tr}\left(\frac{i\mathcal{F}}{2\pi}\wedge\frac{i\mathcal{F}}{2\pi}\right), \tag{8.49}$$

that is, the Belavin–Polyakov instanton has a topologically quantized strength (topological charge, cf. Sect. 2.6). Note, that compared to Dirac's monopole there is not even a singularity string of the gauge potential in the present case; the gauge potential of the Belavin–Polyakov instanton is smooth everywhere in the base space $\mathbb{R}^4$. It is a soliton-like solution of the field equations, which per se also has no length scale: R was arbitrary in the choice of U_0 and U_∞, (8.48) does not depend on it. However, distinct from an ordinary soliton the instanton field strength $\mathcal{F}$ is non-zero in a vicinity of the origin of the four-space only: it is local and exists only an instant of time, hence instanton. Its presence imposes a gauge invariant non-trivial boundary condition for fields propagating in time from $-\infty$ to ∞.

Recall that in this case the quantization of (8.49) had its origin in the requirement that the gauge field vanishes at infinity, or the gauge potential is pure there. In fact, instead of $\mathbb{R}^4$ the compactified space $\overline{\mathbb{R}^4} \sim S^4$ was treated.

8.5 Characteristic Classes

The topological quantizations (8.37) and (8.49) have a common feature which reflects a very general algebraic structure admitting of a deep analysis. In both cases an integral of an exact r-form (coboundary) over an r-dimensional closed manifold (cycle) M is taken as a sum of integrals over charts, the items of which are transformed using Stokes' theorem into integrals over the boundaries of the charts. If the continuation of the preimage of the coboundary from one chart into the other is obstructed, then the total integral may be nonzero, but only depending on a cohomology class of $H^r_{dR}(M)$, called a characteristic class. (Compare also Sect. 8.2.) Recall, that a gauge potential is a local connection form on a principal fiber bundle and the gauge field is its local curvature form.

Let (P, M, π, G) be a principal fiber bundle with the Lie group G as its structure group and $\mathfrak{g}$ as its Lie algebra. In view of possible fiber bundles (E, M, π_E, F, G) associated with P, G and $\mathfrak{g}$ may be replaced in the following by any complex matrix representation in F. An $\mathrm{Ad}\, G$ **invariant** symmetric r-linear function p : $\mathfrak{g} \times \cdots \times \mathfrak{g}(r \text{ factors}) \to \mathbb{C}$ is a symmetric $(p(\ldots, X_i, \ldots, X_j, \ldots) = p(\ldots, X_j, \ldots, X_i, \ldots))$ r-linear function with the property

$$p(\mathrm{Ad}\, gX_1, \ldots, \mathrm{Ad}\, gX_r) = p(X_1, \ldots, X_r) \quad \text{for all } g \in G,\ X_i \in \mathfrak{g}. \tag{8.50}$$

In matrix representations,

$$p(gX_1g^{-1}, \ldots, gX_rg^{-1}) = p(X_1, \ldots, X_r). \tag{8.51}$$

Given a connection form ω on P, recall that the curvature form $\Omega = D\omega$ is a $\mathfrak{g}$-valued tensorial 2-form on P, that is, at every $q \in P$, $\langle \Omega, Z_1 \wedge Z_2 \rangle \in \mathfrak{g}$ for any pair of tangent vectors Z_1, Z_2 of $T_q(P)$. Define

$$p_\Omega(Z_1, \ldots, Z_{2r}) = \frac{1}{(2r)!} \sum_{\mathcal{P}} (-1)^{|\mathcal{P}|} p(\langle \Omega, Z_{\mathcal{P}(1)} \wedge Z_{\mathcal{P}(2)} \rangle, \ldots, \langle \Omega, Z_{\mathcal{P}(2r-1)} \wedge Z_{\mathcal{P}(2r)} \rangle), \tag{8.52}$$

where p is any $\mathrm{Ad}\, G$ invariant r-linear complex function, $\mathcal{P}$ means permutations of the numbers $(1, \ldots, 2r)$, and $Z_i \in T_q(P)$. Then, there holds the

Chern–Weil theorem *(a) There is a unique global $2r$-form on M equal to the local $2r$–forms $p_{\Omega_\alpha} = s^*_\alpha(p_\Omega)$ (p. 224) on trivializing neighborhoods U_α of the base space M of P, which is closed: $dp_{\Omega_\alpha} = 0$.*

(b) Let ω and ω' be two different connection forms on P. Then, the difference $p_{\Omega_\alpha} - p_{\Omega'_\alpha}$ is an exact $2r$-form on M, that is, the de Rham cohomology class associated with the glued together p_{Ω_α} in $H^{2r}_{dR}(M)$ is independent of ω.

Proof (a) As defined in Sect. 7.5, the local curvature form Ω_α is uniquely defined by Ω and is linearly depending on Ω for every coordinate neighborhood $U_\alpha \in M$: $\langle \Omega_{\alpha x}, X_{1x} \wedge X_{2x} \rangle = \langle s^*_\alpha(\Omega_{s_\alpha(x)}), X_{1x} \wedge X_{2x} \rangle = \langle \Omega_{s_\alpha(x)}, s^x_{\alpha *}(X_{1x}) \wedge s^x_{\alpha *}(X_{2x}) \rangle = \langle \Omega_{s_\alpha(x)},$

${}^h(s^x_{\alpha*}(X_{1x})) \wedge {}^h(s^x_{\alpha*}(X_{1x})))\rangle$. By r-linearity, $p_{\Omega_\alpha} = s^*_\alpha(p_\Omega)$ is uniquely defined by p_Ω through $Z_{is_\alpha} = s_{\alpha*}(X_i)$ for which inversely $\pi_*(Z_{is_\alpha}) = X_i$, where π is the bundle projection of the principal fiber bundle P. The tangent vector $Z_{is_\alpha(x)}$ may be pushed forward to any other point $s_\alpha(x)g$ on the fiber $\pi^{-1}(x)$ as $(R_g)_* Z_{is_\alpha(x)}$, and $\langle \Omega, (R_g)_* Z_1 \wedge (R_g)_* Z_2\rangle = \langle R^*_g \Omega, Z_1 \wedge Z_2\rangle = \mathrm{Ad}(g^{-1})\langle \Omega, Z_1 \wedge Z_2\rangle$, since Ω is a tensorial form of type $(\mathrm{Ad}, \mathfrak{g})$. Since $(R_g)_* Z_{is_\alpha(x)} - Z_{is_\alpha(x)}$ is vertical, also $\pi_*((R_g)_* Z_{is_\alpha(x)}) = X_i$, and in summary $\langle p_{\Omega_\alpha}, \pi_*(Z_1), \dots, \pi_*(Z_{2r})\rangle = \langle \pi^*(p_{\Omega_\alpha}), Z_1, \dots, Z_{2r}\rangle = \langle p_\Omega, Z_1, \dots, Z_{2r}\rangle$ or in short $\pi^*(p_{\Omega_\alpha}) = p_\Omega$, where $X_i = \pi_*(Z_i) = \pi_*({}^hZ_i)$ (since $\pi_*({}^vZ) = 0$). This also implies that $p_{\Omega_\alpha} = p_{\Omega_\beta}$ on $U_\alpha \cap U_\beta$, and hence the local forms p_{Ω_α} define a unique global $2r$-form on all M, which is pulled back to p_Ω on P by the bundle projection.

Next it is shown that for every n-form p on P which is equal to $\pi^*\tilde{p}$ for some n-form $\tilde{p}$ on M it holds that $dp = Dp$. Indeed, again with $X_i = \pi_*(Z_i) = \pi_*({}^hZ_i)$ and by linearity of π^* and d, $(dp)(Z_1, \dots, Z_{n+1}) = (d\pi^*\tilde{p})(Z_1, \dots, Z_{n+1}) = (\pi^* d\tilde{p})(Z_1, \dots, Z_{n+1}) = (d\tilde{p})(X_1, \dots, X_{n+1}) = (\pi^* d\tilde{p})({}^hZ_1, \dots, {}^hZ_{n+1}) = (dp)({}^hZ_1, \dots, {}^hZ_{n+1}) = (Dp)(Z_1, \dots, Z_{n+1})$.

Now, from Bianchi's identity, $0 = \sum p(\dots, \langle D\Omega, Z_i \wedge Z_j \wedge Z_k\rangle, \dots) = Dp(\dots, \langle \Omega, Z_j \wedge Z_k\rangle, \dots) = Dp_\Omega(Z_1, \dots, Z_{2r+1}) = dp_\Omega(Z_1, \dots, Z_{2r+1}) = dp_{\Omega_\alpha}(X_1, \dots, X_{2r+1})$, the last equality again by linearity of π^* and d. This completes the proof of (a).

(b) Let ω_0 and ω_1 be two connection forms on P, that is, two $\mathfrak{g}$-valued 1-forms with properties 1 and 2 given on p. 215. In view of the affine linearity of 1 and the linearity of 2 of these properties in $\omega, \omega_t = \omega_0 + t\theta, \theta = \omega_1 - \omega_0$, is another connection form for every $t \in [0, 1]$, and $\Omega_t = d\omega_t + [\omega_t, \omega_t] = d\omega_0 + [\omega_0, \omega_0] + t(d\theta + [\omega_0, \theta] + [\theta, \omega_0]) + t^2[\theta, \theta] = \Omega_0 + t(d\theta + [\omega_t, \theta] + [\theta, \omega_t]) - t^2[\theta, \theta] = \Omega_0 + tD_t\theta - t^2[\theta, \theta]$ is the corresponding curvature form. One has $d\Omega_t/dt = D_t\theta$. Moreover,

$$\begin{aligned} p_{\Omega_1} - p_{\Omega_0} &= \int_0^1 dt dp_{\Omega_t}/dt = r\int_0^1 dt p(\langle D_t\theta, \cdot \wedge \cdot\rangle, \langle \Omega_t, \cdot \wedge \cdot\rangle, \dots, \langle \Omega_t, \cdot \wedge \cdot\rangle) \\ &= r\int_0^1 dt D_t p(\langle \theta, \cdot\rangle, \langle \Omega_t, \cdot \wedge \cdot\rangle, \dots, \langle \Omega_t, \cdot \wedge \cdot\rangle) \\ &= r\int_0^1 dt dp(\langle \theta, \cdot\rangle, \langle \Omega_t, \cdot \wedge \cdot\rangle, \dots, \langle \Omega_t, \cdot \wedge \cdot\rangle) \\ &= d\int_0^1 dt r p(\langle \theta, \cdot\rangle, \langle \Omega_t, \cdot \wedge \cdot\rangle, \dots, \langle \Omega_t, \cdot \wedge \cdot\rangle) = d\Theta. \end{aligned} \tag{8.53}$$

The first equality is trivial, in the second the symmetry of p_{Ω_t} was used, in the third Bianchi's identity $D_t\Omega_t = 0$ was exploited, and in the fourth it was realized that D_t again applies to a pull back from M by π since θ like the connection forms ω_i is a pseudo-tensorial form of type $(\mathrm{Ad}, \mathfrak{g})$, pulled back from its local form on M by π. Now, since Θ, the integral of the last line, is a pseudo-tensorial form, in analogy to (a) a form $\Theta_\alpha = s_\alpha^*(\Theta)$ may be defined, so that $p_{\Omega_{1\alpha}} - p_{\Omega_{0\alpha}} = d\Theta_\alpha$ on M. According to (5.39), the de Rham cohomology classes, that is, the group elements of the de Rham cohomology group $H^{2r}_{dR}(M)$ are the sets of closed $2r$-forms which differ at most by an exact $2r$-form. Hence, $p_{\Omega_{1\alpha}}$ and $p_{\Omega_{0\alpha}}$ belong to the same element of $H^{2r}_{dR}(M)$. □

As in Sect. 8.2, the de Rham cohomology classes associated with p_{Ω_α} are called the **characteristic classes**. They depend on P and on the chosen $\mathrm{Ad}\,G$ invariant r-linear function p, but not on the chosen connection on P.

The set of formal sums of $\mathrm{Ad}\,G$ invariant symmetric r-linear functions (for all integer $r \geq 0$, complex numbers for $r = 0$) is made into a graded commutative algebra $I^*(G)$ by defining the product

$$pp'(X_1, \ldots, X_{r+s}) = \frac{1}{(r+s)!}\sum_{\mathcal{P}} p(X_{\mathcal{P}(1)}, \ldots, X_{\mathcal{P}(r)})p'(X_{\mathcal{P}(r+1)}, \ldots, X_{\mathcal{P}(r+s)}). \tag{8.54}$$

Note that r-linear functions by (8.52) give rise to forms of even degree $2r$.

Weil homomorphism *The mapping* $I^*(G) \to H^*_{dR}(M)$ *by* $p \mapsto \{p_\Omega\}$ *is a homomorphism of graded algebras.*

$\{p_\Omega\}$ means the de Rham cohomology class of p_Ω. This result is clear from the above, and by realizing that the image of the homomorphism consists of cohomology groups of even degree only and that in $H^*_{dR}(M)$ the $\wedge$-product of factors of even degree is commutative. Of course, the homomorphism depends on the topology of M. Hence, the whole mapping depends on the principal fiber bundle (P, M, π, G).

There is a one–one correspondence between symmetric r-linear functions and polynomials of degree r. Define the polynomial $p^{(r)}$ associated with p by

$$p^{(r)}(u) = p(u, \ldots, u), \quad r \text{ arguments}, \tag{8.55}$$

then $p(u_1, \ldots, u_n)$ is $(1/r!)$ times the coefficient of $t_1 \cdots t_r$ in $p^{(r)}(t_1u_1 + \cdots + t_ru_r)$; this is called the **polarization** of the polynomial $p^{(r)}$. It is clear that, iff $p(X_1, \ldots, X_r)$ is $\mathrm{Ad}\,G$ invariant, so is $p^{(r)}$, it is called an $\mathrm{Ad}\,G$ **invariant polynomial**. Now, $I^*(G)$ is isomorphic with the algebra of $\mathrm{Ad}\,G$ invariant polynomials.

An in a sense most general case is a complex vector bundle $(E, M, \pi_E, \mathbb{C}^k, Gl(k, \mathbb{C}))$ on a (real) m-dimensional base manifold M, associated with the principal fiber bundle $(P, M, \pi, Gl(k, \mathbb{C}))$. In this case there are k distinct $\mathrm{Ad}\,G$ invariant polynomials obtained from the characteristic polynomial of a general complex $(k \times k)$-matrix X as an element of $\mathfrak{gl}(k, \mathbb{C})$:

$$\det\left(\lambda\mathbf{1}_k - \frac{1}{2\pi i}X\right) = \sum_{r=0}^{k} p_{\mathrm{c}}^{(r)}(X)\lambda^{k-r}. \qquad (8.56)$$

The (in principle arbitrary) normalization convention of X ensures that the Chern numbers defined below will be real integers or fractions with small integer denominators. Since $\det(\lambda'\mathbf{1}_k - gXg^{-1}) = \det(g(\lambda'\mathbf{1}_k - X)g^{-1}) = \det g \det(\lambda'\mathbf{1}_k - X)\det(g^{-1}) = \det(\lambda'\mathbf{1}_k - X)\det(gg^{-1}) = \det(\lambda'\mathbf{1}_k - X)$, it is clear, that the polynomials $p_{\mathrm{c}}^{(r)}(X)$ are $\mathrm{Ad}\,G$ invariant. Let Ω be the curvature form of some connection form ω on P. The rth **Chern class** $C_r(E)$ of the complex vector bundle E is the de Rham cohomology class of the closed $2r$-form

$$c_r(Y_1,\ldots,Y_{2r}) = p_{c\Omega}(s_{\alpha*}(Y_1),\ldots,s_{\alpha*}(Y_{2r})), \quad Y_i \in \mathcal{X}(M), \qquad (8.57)$$

where $p_{c\Omega}(Z_1,\ldots,Z_{2r})$ is the polarization of $p_{\mathrm{c}}^{(r)}(Z)$. After introducing a base in $T_x(M)$, $x \in U_\alpha$, the 2-form Ω_α becomes a complex $(k\times k)$-matrix. A somewhat involved but straightforward calculation yields

$$c_r = \frac{(-1)^r}{(2\pi i)^r (r!)}\delta^{i_1\ldots i_r}_{j_1\ldots j_r}(\Omega_\alpha)^{j_1}_{i_1}\wedge\cdots\wedge(\Omega_\alpha)^{j_r}_{i_r}. \qquad (8.58)$$

Each matrix element $(\Omega_\alpha)^j_i$ is a 2-form on M. It can be shown that the Chern classes generate the whole image of $I^*(Gl(k,\mathbb{C}))$ in $H^*_{dR}(M)$. Their representation depends on E as seen from (8.58). The **total Chern class** corresponds to the direct sum over r of the c_r.

Some important other characteristic classes are:

Chern character Consider instead of (8.56) the expression

$$\mathrm{tr}\left(\exp\left(-\frac{1}{2\pi i}X\right)\right) = \mathrm{tr}\left(\sum_{r=0}^{\infty}\frac{1}{r!}\left(-\frac{1}{2\pi i}X\right)^r\right) = \sum_{r=0}^{\infty} p_{\mathrm{ch}}^{(r)}(X). \qquad (8.59)$$

It is easily seen that because of the trace the left expression is $\mathrm{Ad}\,G$ invariant. The rth Chern character $Ch_r(E)$ corresponds to $p_{\mathrm{ch}}^{(r)}$, in a base of $T_x(M)$,

$$ch_r = (1/r!)\mathrm{tr}\,(i\Omega_\alpha/(2\pi)\wedge\cdots\wedge i\Omega_\alpha/(2\pi)). \qquad (8.60)$$

(ch_r is related to $p_{ch\Omega}$, the polarization of $p_{ch}^{(r)}$, like (8.57).) The total Chern character is again the direct sum, which is finite, since the last expression vanishes, if $2r > \dim M$.

Todd classes Let E be the Whitney sum $E = L_1 \oplus \cdots \oplus L_k$, where each L_i is the complex line bundle over M. Let $x_i = c_1(L_i)$ be the first Chern class of L_i. The $2r$-form of the expansion of

$$td = \prod_{i=1}^{k}(\wedge)\frac{x_i}{1-\exp(-x_i)}, \qquad (8.61)$$

where the exponential is meant as the formal $\wedge$-expansion, is the rth Todd class $Td_r(E)$.

Pontrjagin classes They are the analogue of Chern classes for real vector bundles $(E, M, \pi_E, \mathbb{R}^k, Gl(k, \mathbb{R}))$. Instead of (8.56) one uses

$$\det\left(\lambda \mathbf{1}_k - \frac{1}{2\pi} X\right) = \sum_{r=0}^{k} p_{\mathrm{p}}^{(r)}(X) \lambda^{k-r}. \tag{8.62}$$

Replacing X in the determinant by the skew-symmetric matrix $\Omega = -\Omega^t$, one finds $\det(\mathbf{1}_k - \Omega) = \det(\mathbf{1}_k + \Omega^t) = \det(\mathbf{1}_k + \Omega)$ since $\det A = \det A^t$. From that it immediately follows that the odd Pontrjagin classes vanish. One finds

$$p_{2r} = \frac{1}{(2\pi)^{2r}(2r)!^2} \delta^{i_1 \dots i_{2r}}_{j_1 \dots j_{2r}} (\Omega_\alpha)^{j_1}_{i_1} \wedge \cdots \wedge (\Omega_\alpha)^{j_{2r}}_{i_{2r}}. \tag{8.63}$$

If one identifies the complex k-vector bundle E with the real $2k$-vector bundle E', then the rth Chern class becomes the $2r$th Pontrjagin class, $C_r(E) = P_{2r}(E')$.

Euler class Consider an orientable manifold M of even dimension $\dim M = 2m$ and let $(T(M), M, \pi_T, \mathbb{R}^{2m}, O(2m))$ be the tangent vector bundle on M associated with the reduced frame bundle $(L_O(M), M, \pi, O(2m))$ of orthonormal frames. The Euler class is given by

$$e = \frac{(-1)^m}{(4\pi)^m m!} \delta^{1 \dots 2m}_{i_1 \dots i_{2m}} (\Omega_\alpha)^{i_1}_{i_2} \wedge \cdots \wedge (\Omega_\alpha)^{i_{2m-1}}_{i_{2m}}, \tag{8.64}$$

in view of $\Omega_\alpha = -\Omega_\alpha^t$ implying $e^2 = (2\pi)^{-2m} \det(\Omega_\alpha)$ and hence being $\mathrm{Ad}\, O(2m)$ invariant.

Let $[z]$ be the homology class of a $2r$-dimensional cycle in M, and let $[p]$ be the cohomology class of a closed $2r$-form on M (cocycle). Equation 5.40 says that the integral $\int_z p$ depends only on the (co)homology classes of z and p, and hence is a topological invariant. Since characteristic classes are closed forms on M, they give rise to topological invariant numbers by integration over cycles in M. Best known is the

Gauss–Bonnet–Chern–Avez theorem *Let M be an orientable $2m$-dimensional compact manifold, let e be its Euler class and let $\chi(M)$ be its Euler characteristic (Euler–Poincaré characteristic)* (5.63). *Then,*

$$\chi(M) = \int_M e. \tag{8.65}$$

Particular cases of this general theorem are considered in [6]. For $2m = 2$ the theorem reduces to the well known Gauss–Bonnet theorem $\chi(M) = 1/(2\pi) \int_M K$ where $K = \Omega_\alpha$ is the curvature form of the surface M.

Integrals over a $2r$-cycle in M of an rth characteristic class are called Chern numbers. In field theory, for dim $M = 4$, the Chern numbers

$$\int_M ch_2 \quad \text{and} \quad \int_M ch_1 \wedge ch_1 \tag{8.66}$$

are of particular interest. In the real case with dim $M = 4$, $\int_M p_2$ is the Pontrjagin number.

Let (E, M, π_E, K^k, G) be a K-vector bundle associated with a principal fiber bundle (P, M, π, G), and let p_Ω be a $2r$-form (8.52) representing a characteristic class of P in a K-matrix representation. According to the Chern–Weil theorem, $p_{\Omega_\alpha} = s_\alpha^*(p_\Omega)$ for all trivializing neighborhoods $U_\alpha \in M$ defines a closed $2r$-form on all M. In view of Poincaré's lemma (end of Sect. 5.5) this implies that *locally*, but not in general globally, p_{Ω_α} is also exact, that is, there are local $(2r-1)$-forms q_α, so that

$$p_{\Omega_\alpha} = dq_\alpha. \tag{8.67}$$

Let $\omega = \omega_1$ be a connection form leading to the curvature form Ω. On a trivializing subbundle $U_\alpha \times G$ of P, the vertical 'unit' form ω_0 which is pulled back to $0 = \omega_{0\alpha} = s_\alpha^*(\omega_0)$ provides a flat connection on U_α. Let $\omega_{t\alpha} = t\omega_\alpha = ts_\alpha^*(\omega)$ on U_α. Then, after pulling back with s_α^* the chain of equations (8.53) in the proof of the second part of the Chern–Weil theorem yields

$$q_\alpha = r\int_0^1 dtp(\langle\omega_\alpha, \cdot\rangle, \langle\Omega_{t\alpha}, \cdot\wedge\cdot\rangle, \ldots, \langle\Omega_{t\alpha}, \cdot\wedge\cdot\rangle), \tag{8.68}$$

where $\Omega_{t\alpha} = td\omega_\alpha + t^2[\omega_\alpha, \omega_\alpha] = t\Omega_\alpha + (t^2 - t)[\omega_\alpha, \omega_\alpha]$. The $\mathfrak{g}$-valued (in the representation vector space K^k) local $(2r-1)$-form q_α on $U_\alpha \in M$ is called the **Chern–Simons form** of p_{Ω_α}.

Consider as an example the Chern–Simons $(2r-1)$-form of the rth Chern character Ch_r:

$$q_{\mathrm{ch}\alpha}^{(r)} = \frac{1}{(r-1)!}\left(\frac{i}{2\pi}\right)^r \int_0^1 dt\mathrm{tr}\,(\omega_\alpha \wedge \underbrace{\Omega_{t\alpha} \wedge \cdots \wedge \Omega_{t\alpha}}_{r-1\text{factors}}). \tag{8.69}$$

In particular

$$\begin{aligned} q_{\mathrm{ch}\alpha}^{(1)} &= \frac{i}{2\pi}\int_0^1 dt\mathrm{tr}\,\omega_\alpha = \frac{i}{2\pi}\mathrm{tr}\,\omega_\alpha, \\ q_{\mathrm{ch}\alpha}^{(2)} &= \left(\frac{i}{2\pi}\right)^2 \int_0^1 dt\mathrm{tr}\,(\omega_\alpha \wedge (td\omega_\alpha + t^2\omega_\alpha \wedge \omega_\alpha) \\ &= \frac{1}{2}\left(\frac{i}{2\pi}\right)^2 \mathrm{tr}\left(\omega_\alpha \wedge d\omega_\alpha + \frac{2}{3}\omega_\alpha \wedge \omega_\alpha \wedge \omega_\alpha\right), \\ &\ldots \end{aligned} \tag{8.70}$$

In local gauge field theory ω_α is commonly denoted $\mathcal{A}$ as the gauge potential and Ω_α is denoted $\mathcal{F}$ as the gauge field. In a one-dimensional vector bundle (line bundle) on M of two dimensions one has simply $\operatorname{tr}\mathcal{A} = \mathcal{A}$. The vector potential $A = (i/e)\mathcal{A}$ is just $(2\pi/e)q_{\text{ch}}^{(1)}$. Hence, (8.39) is a trivial case of a Chern character $eF/(2\pi)$ and a Chern–Simons form $eA/(2\pi)$. Likewise it is seen that $(-1/(8\pi^2))\operatorname{tr}(\mathcal{F}\wedge\mathcal{F})$ is a Chern character Ch_2, and that K of (8.46) is up to a factor of convention a Chern–Simons form: $K = -8\pi^2 q_{\text{ch}}^{(2)}$. Both relations (8.39, 8.46) are special cases of (8.67).

8.6 Geometric Phases in Quantum Physics

8.6.1 Berry–Simon Connection

Consider a quantum system under the influence of its surroundings. For the sake of simplicity non-relativistic quantum mechanics is considered, although more general cases could be treated similarly. The system is described by a Hamiltonian, and the influence of the surroundings is expressed by a set of in general time-dependent parameters the Hamiltonian depends on. Collect the parameters into a set R of real numbers which varies in some real m-dimensional manifold. Let the Hamiltonian and hence its eigenvalues, calculated for fixed R,

$$\hat{H}(R)|\Psi(R)\rangle = |\Psi(R)\rangle E(R), \quad \langle\Psi(R)|\Psi(R)\rangle = 1, \quad R \text{ fixed}, \tag{8.71}$$

continuously depend on R [7]. Let $E(R)$ be a non-degenerate and isolated eigenvalue of $\hat{H}(R)$ for some value R of the parameters. Then, a manifold M can always be found on which $E(R)$ varies continuously with R and remains isolated. Since $|\Psi(R)\rangle$ is defined up to a phase $e^{i\gamma}$, a Lie group $U(1)$ is attached to each point R of the manifold M, which makes it into a principal fiber bundle $(P, M, \pi, U(1))$. (The Lie group $U(1)$ is the symmetry group related to the conservation of $\langle\Psi|\Psi\rangle$ for *complex* $|\Psi\rangle$ which eventually is related to particle conservation).

Let R depend on time t through $\tau = t/T$ where T is a speed scaling factor for this dependence. The time-dependent Schrödinger equation reads ($\hbar = 1$)

$$i\frac{d|\Psi_T(R(\tau),t)\rangle}{dt} = \hat{H}(R(\tau))|\Psi_T(R(\tau),t)\rangle. \tag{8.72}$$

Let $|\Psi(R(\tau))\rangle$ be the state of (8.71). Then, the quantum adiabatic theorem says that

$$\lim_{T\to\infty} |\Psi_T(R(\tau),t)\rangle\langle\Psi_T(R(\tau),t)| = |\Psi(R(\tau))\rangle\langle\Psi(R(\tau))|, \tag{8.73}$$

where τ is kept constant in the limiting process. In order to determine the phase change of $|\Psi(R)\rangle$ on a path through M, put the ansatz

$$|\Psi_T(R(\tau),t)\rangle = |\Psi(R(\tau))\rangle \exp\left(i\gamma(t) - i\int_0^t dt' E(R(\tau'))\right) \tag{8.74}$$

into the equation (8.72) and find straightforwardly after multiplication from the left with $\langle\Psi_T(R(\tau),t)|$

$$\frac{d\gamma}{dt} = i\left\langle \Psi(R(\tau))\middle|\frac{d}{dt}\middle|\Psi(R(\tau))\right\rangle$$

or

$$\gamma(t) = i\int_0^t dt' \left\langle \Psi(R(\tau))\middle|\frac{d}{dt'}\middle|\Psi(R(\tau))\right\rangle = i\int_{R(0)}^{R(t)} dR\langle\Psi(R)|\frac{\partial}{\partial R}|\Psi(R)\rangle$$
$$= i\int_C \langle\Psi(R)|d|\Psi(R)\rangle, \tag{8.75}$$

where C is the considered path $R(\tau)$ through M. The phase $\gamma(t)$ is called **Berry's phase** (Berry, 1984).[1] It is in many instances a measurable quantity, and it took nearly 60 years since the foundation of the Hilbert space representation of quantum theory to realize that not every dynamical quantum observable is represented as a Hermitian operator.

B. Simon was the first to realize that the last integrand is a local connection form on $(P, M, \pi, U(1))$:

$$\mathcal{A} = \sum_i \mathcal{A}_i dR^i = \langle\Psi(R)|d|\Psi(R)\rangle = -(d\langle\Psi(R)|)|\Psi(R)\rangle. \tag{8.76}$$

The last relation is a direct consequence of the normalization of $|\Psi(R)\rangle$. It shows that $\mathcal{A}$ is anti-Hermitian, it is called the **Berry–Simon connection** . To see that it is a local connection form, consider two local sections $s_\alpha(R) = |\Psi(R)\rangle_\alpha$ and $s_\beta(R) = |\Psi(R)\rangle_\beta$ with the transition function $\psi_{\alpha\beta} = \exp(i\chi), |\Psi(R)\rangle_\beta = |\Psi(R)\rangle_\alpha \psi_{\alpha\beta}(R)$. Then,

$$\begin{aligned}\mathcal{A}_\beta(R) &= {}_\beta\langle\Psi(R)|d|\Psi(R)\rangle_\beta \\ &= \psi^{-1}_{\alpha\beta}\,{}_\alpha\langle\Psi(R)|d|\Psi(R)\rangle_\alpha \psi_{\alpha\beta} + \psi^{-1}_{\alpha\beta}\,{}_\alpha\langle\Psi(R)|\Psi(R)\rangle_\alpha d\psi_{\alpha\beta} \\ &= \psi^{-1}_{\alpha\beta}\mathcal{A}_\alpha(R)\psi_{\alpha\beta} + \psi^{-1}_{\alpha\beta} d\psi_{\alpha\beta} = \mathcal{A}_\alpha(R) + id\chi(R),\end{aligned}$$

which proves the required property. (In the first term on the second line d is meant to operate on $|\Psi(R)\rangle_\alpha$ only.) The corresponding curvature form $\mathcal{F} = D\mathcal{A} = d\mathcal{A}$ (the latter since $U(1)$ is Abelian) is called **Berry's curvature**, it is given by

[1] A collection of most of the relevant original papers on the subject is gathered in the volume [1].

$$\mathcal{F} = (d\langle\Psi(R)|) \wedge (d|\Psi(R)\rangle) = \frac{\partial\langle\Psi(R)|}{\partial R^i}\frac{\partial|\Psi(R)\rangle}{\partial R^j} dR^i \wedge dR^j. \tag{8.77}$$

A phase difference of two quantum states or two classical waves can be measured, if both waves are brought to interference. This happens, if parts of the wave propagate along different paths between the same start and end points or, equivalently, if a wave circuits along a closed loop C. In the latter case it interferes with itself according to the phase difference (8.75). Clearly, if $x \in M$ is a base point of loops, then all possible phase differences $\gamma(C)$ for all possible loops C based on x and running through M constitute the holonomy group H_x related to the connection on $(P, M, \pi, U(1))$ provided by the local connection form $\mathcal{A}$.

Let S be a two-dimensional surface in M bounded by $C = \partial S$. Then, Stokes' theorem yields

$$\gamma(C) = i \int_{C=\partial S} \mathcal{A} = i \int_S \mathcal{F}, \tag{8.78}$$

that is, Berry's phase equals i times the flux of Berry curvature through the surface S. This is suggestive of magnetism and of Aharonov–Bohm physics, but is much more general.

The expressions (8.76, 8.77) point out yet another important generalization of these considerations: As a connection in a physical parameter space, the Berry–Simon connection $\mathcal{A}$ and also Berry's phase (8.75) between *distinct* points of the parameter space is a **gauge potential** and hence gauge dependent and in general not measurable. On the contrary, Berry's curvature (8.77) is a **gauge field** and hence has physical relevance leading not only to a measurable quantity (8.78) *but is also measurable locally along any path through the parameter space, closed or not.*

8.6.2 Degenerate Case

Shortly after Berry's and Simon's papers, Wilczek and Zee pointed out that this concept has a relevant non-Abelian generalization. It happens that a quantum state has an isolated energy level which however is globally, that is on a whole parameter manifold M, N-fold degenerate. Think for instance of a **Kramers degenerate** doublet state of a molecule (see p. 337). Instead of (8.73), now

$$\sum_{a=1}^{N} |\Psi_a(R)\rangle\langle\Psi_a(R)|, \quad \langle\Psi_a(R)|\Psi_b(R)\rangle = \delta^a_b \tag{8.79}$$

is the adiabatic quantity, where *locally* the orthonormalized states $|\Psi_a(R)\rangle$ can always be chosen smoothly depending on the parameter set R ([7]).

With the ansatz

$$|\Psi(t)\rangle = \sum_{a=1}^{N} c^a(t)|\Psi_a(R(\tau))\rangle \tag{8.80}$$

one finds after projection on $\langle\Psi_b(R(\tau))|$

$$\frac{dc^b(t)}{dt} + \sum_{a=1}^{N}\left(\left\langle\Psi_b(R(\tau))\middle|\frac{d}{dt}\middle|\Psi_a(R(\tau))\right\rangle + iE(R(\tau))\delta^b_a\right)c^a(t) = 0 \tag{8.81}$$

with the formal solution

$$c^b(t) = \sum_{a=1}^{N}\left[\mathcal{T}\exp\int_0^t dt(-\mathcal{A}(R(\tau)) - iE(R(\tau))\mathbf{1}_N)\right]^b_a c^a(0), \tag{8.82}$$

where now the simple power series expressed by the exponentiation is to be replaced by a series which observes the order of factors with ascending time from right to left. This is formally expressed by the time-ordering operator $\mathcal{T}$ physicists are familiar with. In the adiabatic limit, the time integration can again be expressed as a path integration along the path parameter τ in the parameter space, leading to

$$\mathcal{A}^b_a = \sum_i \mathcal{A}^b_{ai}dR^i = \sum_i\left\langle\Psi_b(R)\middle|\frac{\partial}{\partial R^i}\middle|\Psi_a(R)\right\rangle dR^i = \langle\Psi_b(R)|d|\Psi_a(R)\rangle \tag{8.83}$$

for which the transition between local patches U_α and U_β of the m-dimensional parameter manifold M, $|\Psi_a(R)\rangle_\beta = \sum_b |\Psi_b(R)\rangle_\alpha \psi_{\alpha\beta}(R)^b_a$, in complete analogy to the case (8.76) yields

$$\mathcal{A}_\beta = \psi^{-1}_{\alpha\beta}\mathcal{A}_\alpha\psi_{\alpha\beta} + \psi^{-1}_{\alpha\beta}d\psi_{\alpha\beta}, \quad \mathcal{A}_\alpha = \left(\sum_i \mathcal{A}^b_{ai}dR^i\right)_\alpha, \tag{8.84}$$

that is, $\mathcal{A}$ is again a local connection form of a connection on (P,M,π,G): a 1-form on M, which is $\mathfrak{g}$-valued, where $\mathfrak{g}$ is the Lie algebra to the Lie group $G \ni \psi_{\alpha\beta}$ providing the degeneracy of quantum states. Note that due to (8.79) G is unitarily acting on the space $\mathbb{C}^N$ of wave functions $|\Psi(R)\rangle$ at given R. The geometric change of state along a closed loop is given by

$$\tilde{\psi}(C) = \mathcal{P}\exp\left(-\oint_C \mathcal{A}\right), \tag{8.85}$$

where $\mathcal{P}$ means path ordering from right to left surviving from the time ordering $\mathcal{T}$. For loops based on $x \in M$ it is again an element of the holonomy group H_x related to the connection on (P,M,π,G) provided by the local connection form

$\mathcal{A}$. In gauge field theory the corresponding quantity is called Wilson's loop integral.

The integral $\tilde{\psi}(C)$ is gauge covariant. Indeed, take any gauge transformation $\psi(g(R))$ on M, where $\psi(g)$ as on p. 261 is an element of the unitary representation of G in $\mathbb{C}^N$. It now corresponds to a smooth (on M) transition to new states $|\Psi'_a(R)\rangle = \sum_b |\Psi_b(R)\rangle\psi_a^b(g(R))$ alternative to (8.79). The corresponding change of the connection is $\mathcal{A}' = \psi^{-1}\mathcal{A}\psi + \psi^{-1}d\psi$. Exploit $\exp(B^{-1}AB) = B^{-1}(\exp A)B$, which holds for arbitrary matrices A, B. Consider the transformed loop integral

$$\begin{aligned}\tilde{\psi}(C) &= \prod_{dR}\exp\left(-\psi^{-1}\mathcal{A}\psi - \psi^{-1}dR\frac{\partial}{\partial R}\psi\right)\\ &= \prod_{dR}\psi^{-1}\exp(-\mathcal{A})\exp\left(-dR\frac{\partial}{\partial R}\right)\psi\\ &= \prod_{dR}\psi(g(R))^{-1}\exp(-\mathcal{A})\psi(g(R-dR)) = \psi(g(R_0))\left(\prod_{dR}\exp(-\mathcal{A})\right)\psi(g(R_0)),\end{aligned}$$

where the product is understood in path order which precisely leads to cancellation of the intermediate products $\psi(g(R-dR)\psi(g(R-dR))^{-1}$, and R_0 is the base point of the loop C. Hence $\tilde{\psi}'(C) = \psi(g(R_0))^{-1}\tilde{\psi}(C)\psi(g(R_0))$. This also means that $\tilde{\psi}(C)$ is gauge dependent and hence not directly measurable.

Recall from (8.70) that $i/(2\pi)\mathrm{tr}\,\mathcal{A}$ (with the trace taken in $\mathfrak{g}$) is the Chern–Simons form of the first Chern character $i/(2\pi)\mathcal{F}$ of the connection provided by $\mathcal{A}$; since $\mathcal{F} = \mathcal{D}\mathcal{A} = d\mathcal{A} + \mathcal{A}\wedge\mathcal{A}$, it follows that $\mathrm{tr}\,\mathcal{F} = d\,\mathrm{tr}\,\mathcal{A}$, because $\mathrm{tr}\,\mathcal{A}\wedge\mathcal{A} = 0$. Hence, if one takes the trace under the integrals of (8.78), one gets again a gauge invariant measurable Berry phase:

$$\gamma(C) = i\int_{C=\partial S}\mathrm{tr}\,\mathcal{A} = i\int_S \mathrm{tr}\,\mathcal{F}. \tag{8.86}$$

The above considerations show that

$$\mathrm{tr}\,\tilde{\psi}(C) = \mathrm{tr}\,\mathcal{P}\exp\oint_C \mathcal{A} \tag{8.87}$$

is another gauge-independent quantity which can be measured.

Finally, Aharonov and Anandan generalized the concept to general non-adiabatic situations. Although this seems not to lead to new measurable quantities, it provides a general classification of $U(N)$ principal fiber bundles and hence of all possible cases of geometric phases in quantum physics [2].

Nowadays there is a wealth of applications of this concept in solid state physics. The interested reader is referred to [1, 2] and citations therein. We only select a few typical examples.

8.6.3 Electrical Polarization

For details see the reviews [8, 9] and [10] and citations therein. This presentation follows closely [8]. Consider the bulk electric dipole density of a material, that is, the dipole density which is independent of the shape and the surfaces of a piece of material. This quantity is what is described by the thermodynamic limit, where the volume is let go to infinity with all average densities kept constant. To get rid of surface effects one uses periodic boundary conditions, that is, one replaces a volume L^3 by a 3-torus $x^1 \equiv x^1 + L$, $x^2 \equiv x^2 + L$, $x^3 \equiv x^3 + L$. Any charge density is forced to be periodic. For the sake of simplicity consider just one dimension. The electric charge density is $\rho(x) = \rho(x + L)$. Let it be represented by a generating function $R(x)$, $\rho(x) = dR(x)/dx$. For a neutral case, it must be $\int_a^{a+L} dx\rho(x) = \int_a^{a+L} dx(dR/dx) = R(a + L) - R(a) = 0$ for arbitrarily chosen a. Hence $R(x)$ is also periodic. Of course, an additive constant to R has no physical consequence and hence no physical meaning. Now calculate the 'average dipole density' with the help of integration by parts: $(1/L)\int_a^{a+L} dx\, x\rho(x) = -(1/L)\int_a^{a+L} dx(R(x) - R(a)) = -(1/L)\int_a^{a+L} dxR(x) + R(a)$. Due to periodicity of $R(x)$ the first term is independent of a. Hence, via the second term the result depends on the physically irrelevant reference position a. Although formally a 'bulk dipole density' seems to be defined, it can be given a quite arbitrary value, it is not at all related to the physics at hand. This flaw has entered many textbooks. In fact, the dipole density anticipated in physics, although a bulk property, is fixed by the surface of the sample which destroys periodicity. Opposite charges move in an applied electric field in the bulk in opposite directions and accumulate only at the surface, although the bulk determines how far charge is moving.

Consider a reference situation of an infinite crystal with zero electrical polarization for physical reasons, for instance since the crystal has a center of inversion. Let the system polarize by destroying this symmetry in an adiabatic process with keeping the periodicity fixed (that is, retaining some fixed periodicity without which the thermodynamic limit can hardly be dealt with), for instance by letting a ferroelectric slowly polarize by moving a (charged) sublattice of nuclei in some direction or by applying a spatially periodically oscillating electric field.

To treat these cases, the notion of lattices $\mathbb{L}_r \ni \boldsymbol{R}$ and $\mathbb{L}_k \ni \boldsymbol{G}$ inverse to each other is adopted and of the corresponding three-tori $\mathbb{T}_r^3$ and $\mathbb{T}_k^3$ as introduced in Sect. 5.9 on p. 160 ff to be the unit cells of those lattices. (Here, the notation $\boldsymbol{k} = \boldsymbol{p}/\hbar$ is used and $\sum_{\boldsymbol{R}} f(\boldsymbol{R})$ is written instead of $\sum_n f(\boldsymbol{R}_n)$, likewise for $\boldsymbol{G}$.) Recall that in infinite three-space

$$\delta(\boldsymbol{k}) = \frac{1}{(2\pi)^3}\int^{\infty} d^3r e^{-i\boldsymbol{k}\cdot\boldsymbol{r}} = \left(\frac{1}{(2\pi)^3}\int_{\mathbb{T}_r^3} d^3r e^{-i\boldsymbol{k}\cdot\boldsymbol{r}}\right)\left(\sum_{\boldsymbol{R}} e^{i\boldsymbol{k}\cdot\boldsymbol{R}}\right) \qquad (8.88)$$
$$= F(\boldsymbol{k})G(\boldsymbol{k}).$$

Here, F is clearly smooth and finite while the infinite sum G is a distribution like $\delta(\boldsymbol{k})$, with the obvious property $G(\boldsymbol{k}+\boldsymbol{G}) = G(\boldsymbol{k})$ due to $\boldsymbol{R}\cdot\boldsymbol{G} = 2\pi\cdot$ integer (5.102). Moreover, for $\boldsymbol{k}\to 0$ obviously $\delta(\boldsymbol{k}) = (|\mathbb{T}^3_{\boldsymbol{r}}|/(2\pi)^3)G(\boldsymbol{k})$ with the cell volume $|\mathbb{T}^3_{\boldsymbol{r}}|$, while $F(\boldsymbol{G})$ is easily found by direct calculation, together with a corresponding integral over the reciprocal cell,

$$\int_{\mathbb{T}^3_{\boldsymbol{r}}} d^3r e^{-i\boldsymbol{G}\cdot\boldsymbol{r}} = |\mathbb{T}^3_{\boldsymbol{r}}|\,\delta_{\boldsymbol{G}0}, \quad \int_{\mathbb{T}^3_{\boldsymbol{k}}} d^3k e^{-i\boldsymbol{k}\cdot\boldsymbol{R}} = |\mathbb{T}^3_{\boldsymbol{k}}|\delta_{\boldsymbol{R}0}, \tag{8.89}$$

with the Kronecker symbol $\delta_{\boldsymbol{G}0}$ on the lattice $\mathbb{L}_{\boldsymbol{k}}$ and $\delta_{\boldsymbol{R}0}$ on $\mathbb{L}_{\boldsymbol{r}}$, respectively. Altogether one has

$$\sum_{\boldsymbol{R}} e^{i\boldsymbol{k}\cdot\boldsymbol{R}} = |\mathbb{T}^3_{\boldsymbol{k}}|\sum_{\boldsymbol{G}}\delta(\boldsymbol{k}-\boldsymbol{G}), \quad \sum_{\boldsymbol{G}} e^{i\boldsymbol{G}\cdot\boldsymbol{r}} = |\mathbb{T}^3_{\boldsymbol{r}}|\sum_{\boldsymbol{R}}\delta(\boldsymbol{r}-\boldsymbol{R}), \tag{8.90}$$

where we also added the analogous relation for the reciprocal lattice, and $|\mathbb{T}^3_{\boldsymbol{k}}| = (2\pi)^3/|\mathbb{T}^3_{\boldsymbol{r}}|$. If one limits the variables $\boldsymbol{k}$ and $\boldsymbol{r}$ to the corresponding tori only (considering periodic functions), then only the single item with $\boldsymbol{G}=0$ and $\boldsymbol{R}=0$, respectively, survives on the right hand sides.

The electron charge density of a crystal may in principle rigorously be obtained from an effective one particle equation, the Kohn–Sham equation of density functional theory (e.g. [11]). The crystal orbitals themselves being eigenfunctions of the Kohn–Sham Hamiltonian, $H = -(\hbar^2/2m)\nabla^2 + U$ with $U(\boldsymbol{r}+\boldsymbol{R}) = U(\boldsymbol{r})$, are not lattice periodic; according to Bloch's theorem they carry a phase $e^{i\boldsymbol{k}\cdot\boldsymbol{r}}$ and are obtained from $H\psi_{n\boldsymbol{k}}(\boldsymbol{r}) = \psi_{n\boldsymbol{k}}(\boldsymbol{r})\varepsilon_{n\boldsymbol{k}}$ with the orthonormality condition $\int^{\infty} d^3r\psi^*_{n\boldsymbol{k}}\psi_{n'\boldsymbol{k}'} = \delta_{nn'}\delta(\boldsymbol{k}-\boldsymbol{k}')$. Comparison of this condition with the first equality (8.88) tells that in a constant potential U the state is $\psi_{\boldsymbol{k}} = (2\pi)^{-3/2}e^{i\boldsymbol{k}\cdot\boldsymbol{r}}$ which means $|\mathbb{T}^3_{\boldsymbol{r}}|/(2\pi)^3 = 1/|\mathbb{T}^3_{\boldsymbol{k}}|$ electrons per cell $\mathbb{T}^3_{\boldsymbol{r}}$. The states may, however, be represented as

$$\psi_{n\boldsymbol{k}}(\boldsymbol{r}) = \frac{e^{i\boldsymbol{k}\cdot\boldsymbol{r}}}{|\mathbb{T}^3_{\boldsymbol{k}}|^{1/2}}u_{n\boldsymbol{k}}(\boldsymbol{r}), \quad u_{n\boldsymbol{k}}(\boldsymbol{r}+\boldsymbol{R}) = u_{n\boldsymbol{k}}(\boldsymbol{r}), \quad u_{n,\boldsymbol{k}+\boldsymbol{G}}(\boldsymbol{r}) = u_{n\boldsymbol{k}}(\boldsymbol{r}), \tag{8.91}$$

where the periodic functions $u_{n\boldsymbol{k}}$ are obtained as eigenfunctions of the Hamiltonian $H_{\boldsymbol{k}} = e^{-i\boldsymbol{k}\cdot\boldsymbol{r}}He^{i\boldsymbol{k}\cdot\boldsymbol{r}} = (\hbar^2/2m)(-i\nabla+\boldsymbol{k})^2 + U$ which acts on functions on the torus $\mathbb{T}^3_{\boldsymbol{r}}$ and depends parametrically on $\boldsymbol{k}\in\mathbb{T}^3_{\boldsymbol{k}}$,

$$H_{\boldsymbol{k}}u_{n,\boldsymbol{k}}(\boldsymbol{r}) = u_{n,\boldsymbol{k}}(\boldsymbol{r})\varepsilon_{n\boldsymbol{k}}, \quad (u_{n\boldsymbol{k}}|u_{n'\boldsymbol{k}}) = \int_{\mathbb{T}^3_{\boldsymbol{r}}} d^3r u^*_{n\boldsymbol{k}}(\boldsymbol{r})u_{n'\boldsymbol{k}}(\boldsymbol{r}) = \delta_{nn'}. \tag{8.92}$$

The last orthonormality relation results from the orthonormality condition for the $\psi_{n\boldsymbol{k}}$ with (8.91) and the first equality (8.90). The functions $u_{n\boldsymbol{k}}$ still carry an arbitrary $\boldsymbol{k}$-dependent but now $\boldsymbol{r}$-independent phase as seen from the last eigenvalue problem; it is, however, essential for the following that this phase is chosen to be a periodic function on $\mathbb{T}^3_{\boldsymbol{k}}$ (and hence is the same for $\boldsymbol{k}$ and $\boldsymbol{k}+\boldsymbol{G}$). This assumption has already been made in the last relation (8.91).

In a semiconductor, there is an energy gap between all occupied energies $\varepsilon_{n\boldsymbol{k}}$ and all unoccupied energies. Then, for all occupied bands the n-non-ordered sets $\{\varepsilon_{n\boldsymbol{k}}\}$ and $\{u_{n\boldsymbol{k}}\}$ are smooth functions of $\boldsymbol{k}\in\mathbb{T}^3_{\boldsymbol{k}}$ (in an appropriate topology of a functional space of set-valued functions, [7]).

Instead of the periodically repeated functions $u_{n\boldsymbol{k}}$, multi-band Wannier functions

$$a_{n\boldsymbol{R}}(\boldsymbol{r})=\frac{1}{|\mathbb{T}^3_{\boldsymbol{k}}|}\int\limits_{\mathbb{T}^3_{\boldsymbol{k}}} d^3k e^{i\boldsymbol{k}\cdot(\boldsymbol{r}-\boldsymbol{R})}\sum_{n'}^{\text{occ.}} U_{nn'}(\boldsymbol{k})u_{n'\boldsymbol{k}}(\boldsymbol{r}),\quad U^\dagger(\boldsymbol{k})=U^{-1}(\boldsymbol{k}), \tag{8.93}$$

with a unitary-matrix function $U(\boldsymbol{k})$ may be introduced for the occupied bands. The matrix function $U(\boldsymbol{k})$ must again be periodic as function of $\boldsymbol{k}\in\mathbb{T}^3_{\boldsymbol{k}}$ but may otherwise be rather arbitrary. It is well known that, depending on its choice, for an energy-gap separated band group the Wannier functions can be exponentially localized in $\boldsymbol{r}$-space. With the relations above one easily verifies (exercise)

$$(a_{n\boldsymbol{R}}|a_{n'\boldsymbol{R}'})=\int^{\infty} d^3r a^*_{n\boldsymbol{R}}(\boldsymbol{r})a_{n'\boldsymbol{R}'}(\boldsymbol{r})=\delta_{n,n'}\delta_{\boldsymbol{R}\boldsymbol{R}'}. \tag{8.94}$$

It is another simple exercise to show that (n runs over the bands per spin and over the spin quantum number)

$$\sum_{\boldsymbol{R}}\sum_{n}^{\text{occ.}}|a_{n\boldsymbol{R}}(\boldsymbol{r})|^2=\sum_{n}^{\text{occ.}}\frac{1}{|\mathbb{T}^3_{\boldsymbol{k}}|}\int\limits_{\mathbb{T}^3_{\boldsymbol{k}}} d^3k|\psi_{n\boldsymbol{k}}(\boldsymbol{r})|^2=\rho(\boldsymbol{r}) \tag{8.95}$$

is the total electron density of the crystal, in the left expression written as a sum over the unit cells of the lattice. This lattice sum may be used to express a change of the average electron dipole density of the crystal as

$$\begin{aligned}\Delta\boldsymbol{P}_e&=-\frac{e}{|\mathbb{T}^3_{\boldsymbol{r}}|}\sum_{n}^{\text{occ.}}\int^{\infty} d^3r\boldsymbol{r}\Delta|a_{n0}(\boldsymbol{r})|^2\\&=\frac{e}{|\mathbb{T}^3_{\boldsymbol{r}}|}\Delta\left(\frac{-i}{|\mathbb{T}^3_{\boldsymbol{k}}|}\int\limits_{\mathbb{T}^3_{\boldsymbol{k}}} d^3k\sum_{n}^{\text{occ}}\int\limits_{\mathbb{T}^3_{\boldsymbol{r}}} d^3r(Uu)^\dagger_{n\boldsymbol{k}}\nabla_{\boldsymbol{k}}(Uu)_{n\boldsymbol{k}}\right)\end{aligned} \tag{8.96}$$

where e is the positive electric charge unit (proton charge). The last expression is obtained by inserting (8.93) into the previous one, using $\boldsymbol{r}e^{i\boldsymbol{k}\cdot\boldsymbol{r}}=-i\nabla_{\boldsymbol{k}}e^{i\boldsymbol{k}\cdot\boldsymbol{r}}$, integrating per parts, and again using $\int^\infty d^3re^{i\boldsymbol{k}\cdot\boldsymbol{r}}F(\boldsymbol{r})=\sum_{\boldsymbol{R}}e^{i\boldsymbol{k}\cdot\boldsymbol{R}}\int_{|\mathbb{T}^3_{\boldsymbol{r}}|}d^3re^{i\boldsymbol{k}\cdot\boldsymbol{r}}F(\boldsymbol{r})=|\mathbb{T}^3_{\boldsymbol{k}}|\delta(\boldsymbol{k})\int_{|\mathbb{T}^3_{\boldsymbol{r}}|}d^3re^{i\boldsymbol{k}\cdot\boldsymbol{r}}F(\boldsymbol{r})$ for a periodic function F (exercise).

By applying the Leibniz rule, the unitarity of the matrix $U(\boldsymbol{k})$ and (8.92), the last $\boldsymbol{r}$-integral in (8.96) is easily transformed,

$$\sum_{n}^{\text{occ.}}((Uu)_{n\boldsymbol{k}}|\nabla_{\boldsymbol{k}}(Uu)_{n\boldsymbol{k}}) = \sum_{n}^{\text{occ.}}(u_{n\boldsymbol{k}}|\nabla_{\boldsymbol{k}}u_{n\boldsymbol{k}}) + \operatorname{tr}\left(U^{-1}(\boldsymbol{k})\nabla_{\boldsymbol{k}}U(\boldsymbol{k})\right). \tag{8.97}$$

With (5.19) in the form $\det U = \det\exp\ln U = \exp\operatorname{tr}\ln U$ and the linearity of the trace the last term is cast into $\operatorname{tr}\nabla\ln U = \nabla\operatorname{tr}\ln U = \nabla\ln\det U = i\nabla\vartheta$; since U is unitary its determinant is $\det U = e^{i\vartheta}$. Now, recall that $U(\boldsymbol{k})$ was supposed periodic in $\boldsymbol{k}$, hence the same must hold for $e^{i\vartheta}$ which implies $\vartheta(\boldsymbol{k}) = \alpha(\boldsymbol{k}) + \sum'_{\boldsymbol{R}}\boldsymbol{k}\cdot\boldsymbol{R}$ with again a periodic function $\alpha(\boldsymbol{k})$ and some *finite selection* of lattice vectors $\boldsymbol{R}$; here, $\sum'_{\boldsymbol{R}}$ means a sum over finitely many items. We mention without proof (because this would go off to far from our subject) that for exponentially localized Wannier functions $\alpha(\boldsymbol{k})$ must be smooth. Finally, when put into (8.96), application of Stokes' theorem yields $\int_{\mathbb{T}^3_{\boldsymbol{k}}} d^3k\nabla_{\boldsymbol{k}}\alpha(\boldsymbol{k}) = \int_{\partial\mathbb{T}^3_{\boldsymbol{k}}} d^2k\alpha(\boldsymbol{k}) = 0$ since the torus $\mathbb{T}^3_{\boldsymbol{k}}$ has no boundary $\partial\mathbb{T}^3_{\boldsymbol{k}}$ (or equivalently, to each point on a face of a reciprocal cell there is an identical point on the opposite face with the same value of $\alpha(\boldsymbol{k})$ but opposite surface normal vector d^2k). There remains, however, a term

$$\sum_{\boldsymbol{R}}{}'\boldsymbol{P}_{e,\boldsymbol{R}} = \frac{e}{|\mathbb{T}^3_{\boldsymbol{r}}|}\sum_{\boldsymbol{R}}{}'\boldsymbol{R} \tag{8.98}$$

undetermined (per spin; if spin degeneracy holds as in normal ferroelectrics, then there always appears twice this term). Judged from (8.95, 8.96) this undetermined integer multiple of 'dipole quanta' $\boldsymbol{P}_{e,i} = (e/|\mathbb{T}^3_{\boldsymbol{k}}|)\boldsymbol{a}_i$, $i = 1,2,3$, where the $\boldsymbol{a}_i$ form a basis of the lattice $\mathbb{L}_{\boldsymbol{r}}$ of an infinite crystal, appears quite natural because the assignment of a Wannier function to a lattice position $\boldsymbol{R}$ has this arbitrariness. If a surface for a finite crystal is introduced, then a change of this assignment means a change of the surface contribution too which cancels the change of (8.98) rendering the total dipole moment unique. The remaining dipole density of the finite crystal is normally much smaller than the quanta $\boldsymbol{P}_{e,i}$, and the term (8.98) may be skipped when calculating $\Delta\boldsymbol{P}_e$.

In order to reveal the algebraic-topological structure of the obtained results we introduce lattice adapted (in general non-orthogonal) coordinates given by $\boldsymbol{r} = \sum_i r^i\boldsymbol{a}_i, \boldsymbol{k} = \sum_j k_j\boldsymbol{b}^j$ where the $\boldsymbol{a}_i$ and $\boldsymbol{b}^j$ form bases of the lattices $\mathbb{L}_{\boldsymbol{r}}$ and $\mathbb{L}_{\boldsymbol{k}}$, respectively; $\boldsymbol{a}_i\cdot\boldsymbol{b}^j = \boldsymbol{2\pi\delta}_i^j$, that is, $r^i = (2\pi)^{-1}\boldsymbol{b}^i\cdot\boldsymbol{r}$ and $k_j = (2\pi)^{-1}\boldsymbol{a}_j\cdot\boldsymbol{k}$. The cell volumes expressed in these bases are $|\mathbb{T}^3_{\boldsymbol{r}}| = (\boldsymbol{a}_1,\boldsymbol{a}_2,\boldsymbol{a}_3)$ and $|\mathbb{T}^3_{\boldsymbol{k}}| = (\boldsymbol{b}^1,\boldsymbol{b}^2,\boldsymbol{b}^3)$ with the triple scalar products $(\cdot,\cdot,\cdot)$ (4.46). On the tori the coordinates run from 0 to 1, so that the volume element in $\mathbb{T}^3_{\boldsymbol{k}}$ is $d^3k = |\mathbb{T}^3_{\boldsymbol{k}}|dk_1\wedge dk_2\wedge dk_3$ while $\nabla_{\boldsymbol{k}} = (2\pi)^{-1}\sum_i\boldsymbol{a}_i\partial/\partial k_i$. We have to cope with two dualities here, that between position and momentum and that between tangent vectors $\boldsymbol{dk}$ and forms on the torus of quasi-momenta $\boldsymbol{k}$. This is why here tangent vectors have lower indices and forms upper.

Consider now an adiabatic parameter λ changing from 0 to 1 with $\boldsymbol{P}_e(\lambda)$ changing from $\boldsymbol{P}_e(0) = 0$ to $\boldsymbol{P}_e(1) = \boldsymbol{P}_e$. Since it was important for the Wannier representation analysis to have energy-gap separated occupied bands in order that the unitary transformation matrices $U(\lambda, \boldsymbol{k})$ and the set of occupied states $\{u_{n\lambda\boldsymbol{k}}\}$ resulted in a smoothly $\boldsymbol{k}$-dependent phase $\alpha(\lambda, \boldsymbol{k})$, the crystal must remain semiconducting all the way along the λ-path. Combine $\boldsymbol{k}$ and λ in a four-dimensional manifold $M = [0,1] \times \mathbb{T}^3_{\boldsymbol{k}}$, with the volume form $|\mathbb{T}^3_{\boldsymbol{k}}|\, d\lambda \wedge dk_1 \wedge dk_2 \wedge dk_3$. The boundary of M is $\partial M = (1, \mathbb{T}^3_{\boldsymbol{k}}) - (0, \mathbb{T}^3_{\boldsymbol{k}})$ where the minus sign indicates that the surface normal at $\lambda = 0$ points into the negative λ-direction. ($\mathbb{T}^3_{\boldsymbol{k}}$ itself has no boundary.) Also, introduce the notation of a 1-form

$$du_{n\lambda\boldsymbol{k}} = \sum_j \frac{\partial u_{n\lambda\boldsymbol{k}}}{\partial k_j} dk_j, \quad (du_{n\lambda\boldsymbol{k}})^j = \frac{\partial u_{n\lambda\boldsymbol{k}}}{\partial k_j}. \tag{8.99}$$

Then, the result (8.96) may be cast into

$$\boldsymbol{P}_e = \frac{e}{|\mathbb{T}^3_{\boldsymbol{r}}|} \sum_j \frac{\boldsymbol{a}_j}{2\pi} \int_{\partial M} \mathcal{A}^j$$

$$\mathcal{A}^j = -i \sum_n^{\text{occ.}} (u_{n\lambda\boldsymbol{k}} | (du_{n\lambda\boldsymbol{k}})^j) dk_1 \wedge dk_2 \wedge dk_3 \tag{8.100}$$

$$= i \sum_n^{\text{occ.}} ((du_{n\lambda\boldsymbol{k}})^j | u_{n\lambda\boldsymbol{k}}) dk_1 \wedge dk_2 \wedge dk_3.$$

Though the expression for $\mathcal{A}^j$ is multiplied with the imaginary unit i, it is clear from the derivation that, like $\boldsymbol{P}_e$, the $\mathcal{A}^j$ are real. The last equality holds because of the constancy of normalization, $(u_{n\lambda\boldsymbol{k}} | u_{n\lambda\boldsymbol{k}}) = 1$. Comparison to (8.76) clearly shows that the polarization density component in $\boldsymbol{a}_i$-direction is given by $\mathcal{A}^i$, the wedge-product of a Berry–Simon connection form in $\boldsymbol{b}^i$-direction with the volume form of a two-dimensional section in $\mathbb{T}^3_{\boldsymbol{k}}$ perpendicular to the $\boldsymbol{a}_i$-direction (spanned by $\boldsymbol{b}^2$ and $\boldsymbol{b}^3$ in the case of $\boldsymbol{a}_1$). The integration domain ∂M contains the cycle from $k_j = 0$ to $k_j = 1$ while the λ-path in the adiabatic parameter space is not closed here. This particular case of a Berry phase in an adiabatic change of a band structure was first observed by Zak.[2]

---

[2] Phys. Rev. Lett. **62**, 2747–2750 (1989).

The expression with Berry's curvatures $\mathcal{F}^j = d\mathcal{A}^j$ corresponding to (8.100) is

$$\boldsymbol{P}_e = \frac{e}{|\mathbb{T}^3_{\boldsymbol{r}}|}\sum_j \frac{\boldsymbol{a}_j}{2\pi}\int_M \mathcal{F}^j,$$

$$\mathcal{F}^j = -2i\sum_n^{\text{occ.}}((du_{n\lambda\boldsymbol{k}})^0|(du_{n\lambda\boldsymbol{k}})^j)d\lambda\wedge dk_1\wedge dk_2\wedge dk_3, \quad (du_{n\lambda\boldsymbol{k}})^0 = \frac{\partial u_{n\lambda\boldsymbol{k}}}{\partial\lambda}. \tag{8.101}$$

(As usually $\partial^2 u_{n\lambda\boldsymbol{k}}/(\partial\lambda\partial k_j)\,d\lambda\wedge dk_j = 0$ since the second derivative is symmetric; the prefactor 2 is introduced since the matrix element in $\mathcal{F}^j$ is understood as element of an alternating tensor $(\cdot)^{0j} - (\cdot)^{j0}$, see below and (4.18)). While the connection $\mathcal{A}^i$ is only determined modulo an item (8.98), the curvature $\mathcal{F}^i$ does not have this ambiguity; since it must be smooth in M from $\lambda = 0$ to $\lambda = 1$ it cannot jump by discrete values (8.98).

Equation 8.101 allows for another physical interpretation. Let (with some adiabatic rescaling as in 8.72) $\lambda = t$ denote time. Then, (8.101) may be understood as $\boldsymbol{P}_e = \int_0^T dt \boldsymbol{J}_e(t)$ with the electronic charge current density $\boldsymbol{J}_e = (e/|\mathbb{T}^3_{\boldsymbol{r}}|)\sum_j(\boldsymbol{a}_j/2\pi)\int_{\mathbb{T}^3_{\boldsymbol{k}}}\mathcal{F}^j$ averaged over the unit cell of the crystal. This flowing charge against the lattice during the time T makes up the polarization density $\boldsymbol{P}_e$. This clearly shows that Berry's curvature is uniquely connected to a physical observable and hence not dependent on the gauge center $\boldsymbol{R}$ of the Wannier function.

To pursue the latter aspect further, a lattice periodic electric field $\boldsymbol{E}(\boldsymbol{r},t) = (\partial\boldsymbol{A}/c\partial t - \nabla_{\boldsymbol{r}}A_0)(\boldsymbol{r},t)$ is applied (cf. (5.80, 5.99)). Use a gauge with $A_0 = 0$, then $\boldsymbol{E} = -\partial\boldsymbol{A}/c\partial t$ and the Hamiltonian in (8.92) becomes (with $\lambda = t = k_0$, the electron charge $-e$ and the canonical momentum operator $\boldsymbol{p}$)

$$H_{t\boldsymbol{k}} = \frac{\hbar^2}{2m}\left(-i\nabla_{\boldsymbol{r}} + \boldsymbol{k} + \frac{e}{\hbar c}\boldsymbol{A}(\boldsymbol{r},t)\right)^2 + U(\boldsymbol{r}) = \frac{p^2}{2m} + U(\boldsymbol{r}) \tag{8.102}$$

implying the commutation relations

$$[\nabla_{\boldsymbol{k}}, H_{t\boldsymbol{k}}] = \hbar\frac{\boldsymbol{p}}{m}, \quad \left[\frac{\partial}{\partial k_0}, H_{t\boldsymbol{k}}\right] = -\frac{e}{m}\boldsymbol{E}\cdot\boldsymbol{p} \tag{8.103}$$

and, with $k_\mu = (k_0, k_i)$,

$$\left(u_{mk_\mu}\middle|\left[\frac{\partial}{\partial k_\nu}, H_{k_\mu}\right]\middle|u_{nk_\mu}\right) = (\varepsilon_{nk_\mu} - \varepsilon_{mk_\mu})\left(u_{mk_\mu}\middle|\frac{\partial}{\partial k_\nu}u_{nk_\mu}\right)$$
$$= (\varepsilon_{nk_\mu} - \varepsilon_{mk_\mu})(u_{mk_\mu}|(du_{nk_\mu})^\nu). \tag{8.104}$$

What enters the expression $\mathcal{F}^j$ in (8.101) is the alternating tensor

$$\sum_n^{\text{occ.}}\Big(((du_{nk_\mu})^0|(du_{nk_\mu})^j) - ((du_{nk_\mu})^j|(du_{nk_\mu})^0)\Big)$$
$$= \sum_n^{\text{occ.}}\sum_m\Big(((du_{nk_\mu})^0|u_{mk_\mu})(u_{mk_\mu}|(du_{nk_\mu})^j) - ((du_{nk_\mu})^j|u_{mk_\mu})(u_{mk_\mu}|(du_{nk_\mu})^0))\Big)$$
$$= \sum_n^{\text{occ.}}\sum_m\Big(((du_{nk_\mu})^0|u_{mk_\mu})(u_{mk_\mu}|(du_{nk_\mu})^j) - ((du_{mk_\mu})^0|u_{nk_\mu})(u_{nk_\mu}|(du_{mk_\mu})^j))\Big).$$

The last relation holds again because of $d(u_{mk_\mu}|u_{nk_\mu}) = 0$. The m-sum belongs to the inserted completeness relation and runs over all bands m, occupied and unoccupied. However, from the last line of the displayed chain of equations it is seen that the contributions from the occupied bands m cancel. Together with $(2\pi)^{-1}\sum_j \boldsymbol{a}_j\partial/\partial k_j = \nabla_{\boldsymbol{k}}$, Eq. 8.101 may now be rewritten as (exercise)

$$\boldsymbol{P}_e = \int_0^T dt \frac{i\hbar}{(2\pi)^3}\int_{\mathbb{T}^3_k} d^3k \sum_n^{\text{occ.}}\sum_m^{\text{unocc.}} \frac{(u_{mtk}|\boldsymbol{j}_e|u_{ntk})(u_{ntk}|\boldsymbol{j}_e\cdot\boldsymbol{E}|u_{mtk}) - \text{c.c.}}{(\varepsilon_{mtk} - \varepsilon_{ntk})^2}$$
$$= \int_0^T dt\, \sigma\boldsymbol{E}(t) = \int_0^T dt\boldsymbol{J}_e(t). \tag{8.105}$$

Here, $\boldsymbol{j}_e = -e\boldsymbol{p}/m$ is the current operator for one electron $(\delta H_{tk} = -(1/c)\boldsymbol{j}_e\cdot\boldsymbol{\delta A})$ and the whole expression which scalarly multiplies $\boldsymbol{E}$ is the longitudinal component of the conductivity tensor σ of the electrons expressed by Kubo's formula as a current–current correlation function (see Eq. 8.111). This proves that indeed (8.101) is the time integral over the current density $\sigma\boldsymbol{E}$ of the electrons, no matter whether $\boldsymbol{E}$ is a quasi-static applied electric ac field of some lattice-commensurable periodicity or the field produced by a shift of a sublattice of nuclei. Of course, to get the total polarization density of the solid the contribution of the nuclei has to be added.

In passing comparison of (8.105) with (8.101) shows that in the adiabatic limit the conductivity of the considered situation is a ground state property; the excited states do not figure in (8.101). Recall that a gaped band structure is considered corresponding to an insulating state. The current in the ground state has much in common with a supercurrent, it is not connected with a dissipative process.

The obtained results have many more facets [2]. For instance, consider the case where T is a full period of the time dependence of $\boldsymbol{E}$. Then, the domain M of (8.101) may be treated as a four-torus, that is, a cycle, $\partial M = \varnothing$. Consider the current component J^1_e in the direction of $\boldsymbol{a}_1$ through a section element $dk_2\wedge dk_3$ at fixed values k_i, $i = 2, 3$ which means considering the closed submanifold (two-torus) $M^1 = \mathbb{T}^2$ spanned by $k_0 = t/T$ and k_1, $0 \le k_0, k_1 \le 1$ with $k_\mu \equiv k_\mu + 1$, $\mu = 0, 1$, again with $\partial M^1 = \varnothing$. (The forms $\mathcal{A}$ and $\mathcal{F}$ considered below are scale invariant under rescaling of k_0, k_1, it makes no difference whether the circumferences of tori are taken to be 1 or 2π or any other value.) Ask for the amount of charge flowing through this section element in one period T of time and integrated from $k_1 = 0$ to $k_1 = 1$, that is, from a face of the $\boldsymbol{k}$-cell to the opposite face. It is proportional to $\int_{M^1} \mathcal{F}^1$. *Locally*, that is on any chart of M^1, $\mathcal{F}^1$ was obtained as the (covariant) derivation of $\mathcal{A}^1 = -i\sum_n^{\text{occ.}} \operatorname{tr}(U^{-1}(k_0, k_1)\, \partial^1 U(k_0, k_1))\, dk_1$, $\partial^\mu = \partial/\partial k_\mu$, which is proportional to the Chern–Simons form of the first Chern character on the principal fiber bundle $(P, \mathbb{T}^2, \pi, U(N))$, $\mathcal{A}^1 = -2\pi q_{ch}^{(1)}$, cf. (8.70). Here, the fiber $U(N)$ is the Lie group of global (in $\boldsymbol{r}$-space) unitary transformations of the N quantum states per $\boldsymbol{k}$-value (number of occupied bands, cf. (8.93, 8.96)). Would the relation $\mathcal{F}^1 = D\mathcal{A}^1 = d\mathcal{A}^1$ hold globally on M^1, then due to Stokes' theorem this amount of charge would be zero, $\int_{M^1} \mathcal{F}^1 = \int_{M^1} d\mathcal{A}^1 = \int_{\partial M^1} \mathcal{A}^1$, $\partial M^1 = \varnothing$. (The group $U(N)$ is non-Abelian, hence $\mathcal{F} = D\mathcal{A} = d\mathcal{A} + i[\mathcal{A}, \mathcal{A}]$, compare (8.20) where here the forms have an additional factor $(-i)$; however, because of the trace in $\mathcal{A}$ on has $[\mathcal{A}, \mathcal{A}] = 0$.) Although the curvature form $\mathcal{F}^1 = -2\pi ch_1$ (cf. (8.60)) is globally defined on M^1 on the basis of the Chern–Weil theorem and is related to an observable quantity J_e^1, the continuation of the *local* relation $\mathcal{F}^1 = d\mathcal{A}^1 = -2i\sum_n^{\text{occ.}} \operatorname{tr}(\partial_0(U^{-1}(k_0, k_1))\partial_1 U(k_0, k_1))dk_0 \wedge dk_1$ to all of M^j is obstructed by the topology of P; the quantum state may acquire a phase of a multiple of 2π around a cycle since M^1 is not simply connected. The amount of charge transported through the unit cell $\mathbb{T}_{\boldsymbol{r}}^3$ (after integration over the section of the k_i, $i = 2, 3$) may be non-zero, but is quantized. It is an integer multiple of $(e/|\mathbb{T}_{\boldsymbol{r}}^3|)\boldsymbol{a}_1$, compare (8.98). The charge quantum is proportional to the corresponding first Chern number C_1 (do not confuse it with the first Chern class $C_1(E)$ considered in the previous section),

$$C_1 = -\frac{1}{2\pi}\int_{M_1} \mathcal{F}^1$$
$$= \frac{i}{2\pi}\int dk_0 dk_1 \operatorname{tr}\left(\partial^0(U^{-1}(\lambda, k_1))\partial^1 U(\lambda, k_1) - (\partial^0\partial^1 \leftrightarrow \partial^1\partial^0)\right), \qquad (8.106)$$

of the first Chern character ch_1. It is integer for all values of N of the bundles $(P, \mathbb{T}^2, \pi, U(N))$ as was shown after (8.97). This fact was first mentioned by Thouless.[3]

If this charge quantum is not fixed to zero by independent physical reasons (e.g. mirror symmetry equivalence between $\boldsymbol{a}_1$ and $-\boldsymbol{a}_1$), then quantized charge per time period T can be pumped through the unit cell of the crystal, for instance, by vibrating nuclei driving a charge density wave. Also in agreement with the Chern–Weil theorem, the curvature form $-\mathcal{F}^1/(2\pi)$ is expressed by the current–current correlation function (8.105) and as such is independent of the gauge of the

[3] Phys. Rev. B **27**, 6083–6087 (1983).

electromagnetic potential and of the choice of the $\boldsymbol{k}$-dependence of the total phase of the ground state $\sum_n^{\text{occ.}} u_{nk}$ while the local connection form $-\mathcal{A}^1/(2\pi)$ depends on those choices.

8.6.4 Orbital Magnetism

The role of surface currents in producing the Landau diamagnetism of a non-interacting free-electron gas has been known as a paradox in physics over many decades [12]. On application of a homogeneous magnetic field the electrons start to move around the field lines on helical curves whose projections on the plane perpendicular to the field are circular cyclotron orbits with a homogeneous distribution of their centers over this plane so that all their currents mutually cancel. If the electron system is confined in some volume, then the cyclotron orbits crossing the boundary of this volume cannot be run through, the electrons are reflected at the surface and start a new orbit with a shifted center determined by the reflection conditions. It is easily seen that this shift of the cyclotron orbits produces a current tangent to the surface and circulating around the volume opposite to the circulation of a single complete cyclotron orbit. Classically the cancellation of all these currents still would persist; according to the well known Pauli–van Leeuwen theorem there cannot be orbital magnetic polarization in an equilibrium state in classical physics. Quantization of closed orbits is different for cyclotron orbits in the bulk and surface orbits around the bulk, it produces a slight difference in favor of the surface current. It creates a magnetization density in the volume which exactly equals the diamagnetic polarization of Landau's diamagnetism, including the superimposed de Haas–van Alphen oscillations at high magnetic fields. The free-electron model is of course an extreme idealization, the existence of such non-dissipative (as long as the applied static magnetic field is present) macroscopic currents on the surface of a sample was long denied for a realistic metal, and Landau obtained the magnetization from the thermodynamic potential instead. The non-dissipating edge currents were eventually observed in laboratory in two-dimensional quantum Hall samples. Meanwhile there is a large body of literature on the subject, see again [2].

Consider a two-dimensional crystalline sample, for the sake of simplicity with a square lattice with base vectors $\boldsymbol{a}_1 = a\boldsymbol{e}_x$, $\boldsymbol{a}_2 = a\boldsymbol{e}_y$, in a homogeneous magnetic field $\boldsymbol{B} = \nabla \times \boldsymbol{A}$ in z-direction and an electric field $\boldsymbol{E}$ in y direction. Use Landau gauge $A_x = 0$, $A_y = Bx - Ect$, $\boldsymbol{B} = B\boldsymbol{e}_z$, $\boldsymbol{E} = E\boldsymbol{e}_y$, both E and B independent of $\boldsymbol{r}$. The Hamiltonian becomes (recall that the electron charge is $-e$)

$$H = \frac{\hbar^2}{2m}\left(\left(-i\frac{\partial}{\partial x}\right)^2 + \left(-i\frac{\partial}{\partial y} + \frac{e}{\hbar c}Bx - \frac{e}{\hbar}Et\right)^2\right) + U(x, y) \qquad (8.107)$$

where $U(x, y)$ is the two-dimensional crystal potential. This Hamiltonian is not any more periodic in x-direction. However, if one introduces a magnetic translation operator

$$T(\boldsymbol{R}) = \exp\big(i\boldsymbol{R}\cdot\big(-i\nabla - (e/\hbar c)\boldsymbol{e}_y Bx\big)\big), \quad \boldsymbol{R} = n_1\boldsymbol{a}_1 + n_2\boldsymbol{a}_2, \tag{8.108}$$

with the commutation rules, easily verified by direct calculation,

$$T(\boldsymbol{R})T(\boldsymbol{R}') = \exp((ie/\hbar c)\boldsymbol{B}\cdot\boldsymbol{R}\times\boldsymbol{R}')\,T(\boldsymbol{R}')T(\boldsymbol{R}) \tag{8.109}$$

and defines a superlattice $\tilde{\boldsymbol{R}} = n_1\tilde{\boldsymbol{a}}_1 + n_2\tilde{\boldsymbol{a}}_2$ whose base vectors are integer multiples of $\boldsymbol{a}_1$ and $\boldsymbol{a}_2$, respectively, and such that $(e/\hbar c)\boldsymbol{B}\cdot\boldsymbol{a}_1\times\boldsymbol{a}_2 = 2\pi$, then there is a common eigenfunction system of the Hamiltonian and the magnetic translations $T(\tilde{\boldsymbol{R}}) = e^{i\tilde{\boldsymbol{R}}\cdot\hat{\boldsymbol{\kappa}}}$:

$$H_{\boldsymbol{\kappa}}u_{\boldsymbol{\kappa}} = u_{\boldsymbol{\kappa}}\varepsilon_{\boldsymbol{\kappa}}, \quad H_{\boldsymbol{\kappa}} = \frac{\hbar^2(\hat{\boldsymbol{\kappa}}+\boldsymbol{\kappa})^2}{2m} + U, \quad \hat{\boldsymbol{\kappa}} = -i\nabla + (e/\hbar c)\boldsymbol{e}_y(Bx - Ect). \tag{8.110}$$

Although the relation (8.109) is most easily directly calculated for a square lattice, it is written already in a form which holds for any two-dimensional lattice with base vectors $\boldsymbol{a}_1$, $\boldsymbol{a}_2$. The unit cell of the superlattice of vectors $\tilde{\boldsymbol{R}}$ is chosen such that it is penetrated by one quantum $\Phi_0 = \boldsymbol{B}\cdot\tilde{\boldsymbol{a}}_1\times\tilde{\boldsymbol{a}}_2 = 2\pi\hbar c/e$ of magnetic flux. The eigenvalues $\boldsymbol{\kappa}$ of the operator $\hat{\boldsymbol{\kappa}}$ run through the unit cell $\mathbb{T}^2_{\boldsymbol{\kappa}}$ of the lattice reciprocal to the superlattice.

Kubo's formula for the static conductivity tensor of the considered situation is

$$\sigma^{\alpha\beta} = \frac{i\hbar}{(2\pi)^2}\int d^2\kappa \sum_n^{\text{occ.}}\sum_m^{\text{unocc.}} \frac{(u_{m\boldsymbol{\kappa}}|j_e^\alpha|u_{n\boldsymbol{\kappa}})(u_{n\boldsymbol{\kappa}}|j_e^\beta|u_{m\boldsymbol{\kappa}}) - \text{c.c.}}{(\varepsilon_{m\boldsymbol{\kappa}} - \varepsilon_{n\boldsymbol{\kappa}})^2} \tag{8.111}$$

(Its derivation is beyond the scope of this text.[4]) Here, α and β are Cartesian indices, and we consider our two-dimensional case. This expression yields $\sigma^{xx} = 0$ (related to constant E, the Fermi level is in a gap between Landau levels) and $\sigma^{xy} = \sigma_{\rm H}$, the Hall conductivity. Here, as previously, $d^2\kappa = ((2\pi)^2/|\mathbb{T}^2_{\boldsymbol{r}}|)\,d\kappa_1\wedge d\kappa_2$ (the κ_i run from 0 through 1). Using again $\boldsymbol{j}_e = -(e/\hbar)[\nabla_{\boldsymbol{\kappa}}, H_{\boldsymbol{\kappa}}]$ and $\nabla_{\boldsymbol{\kappa}} = (2\pi)^{-1}\sum_i \tilde{\boldsymbol{a}}_i\,\partial/\partial\kappa_i$, $|\tilde{\boldsymbol{a}}_1\times\tilde{\boldsymbol{a}}_2| = |\mathbb{T}^2_{\boldsymbol{r}}|$, $\sigma_{\rm H}$ may be cast into

$$\sigma_{\rm H} = \frac{e^2}{2\pi\hbar}\left(\frac{1}{2\pi}\int_{|\mathbb{T}^2_{\boldsymbol{\kappa}}|}\sum_n^{\text{occ.}}(-2i)(\partial^1 u_{n\boldsymbol{\kappa}}|\partial^2 u_{n\boldsymbol{\kappa}})\,d\kappa_1\wedge d\kappa_2\right), \quad \partial^i = \frac{\partial}{\partial\kappa_i}. \tag{8.112}$$

The integrand is again the curvature form $\mathcal{F}$ of the local Chern–Simons connection form $\mathcal{A} = \sum_n^{\text{occ.}}(-i)(u_{n\boldsymbol{\kappa}}|\partial^j u_{n\boldsymbol{\kappa}})\,d\kappa_j$ on the bundle $(P, \mathbb{T}^2_{\boldsymbol{\kappa}}, \pi, U(N))$ over the closed manifold $\mathbb{T}^2_{\boldsymbol{\kappa}}$, and the integral expression in large parentheses is an integer, the first Chern number C_1, which due to the Chern–Weil theorem only depends on the bundle P and is independent of the actually chosen local connection form $\mathcal{A}$, that

[4] See for instance D. Cohen, Phys. Rev. B **68**, 155303-1–15 (2003).

is, of the gauge of the electromagnetic potential A_μ which determines the $\boldsymbol{r}$-dependent phases of the $u_{n\boldsymbol{\kappa}}$, and of the choice of (smoothly $\boldsymbol{\kappa}$-dependent) $\boldsymbol{r}$-independent unitary transformations of the $u_{n\boldsymbol{\kappa}}$. Hence, σ_{H} is quantized,

$$\sigma_{\mathrm{H}} = \frac{e^2}{2\pi\hbar}\, C_1, \quad C_1 \in \mathbb{Z}. \tag{8.113}$$

On this Chern number form of the Hall conductivity of a two-dimensional electron system the whole theory of the quantum Hall effect is essentially based. Note that demands of smoothness in the last expression again require an excitation gap of eigenenergies $\varepsilon_{n\boldsymbol{\kappa}}$ between occupied and unoccupied states. Here, this is the gap between adjacent Landau levels, which latter form a discrete sequence in two dimensions (but not in three). The chosen relation between the magnetic field $\boldsymbol{B}$ and the superlattice just ensures that the Fermi level lies in one such gap. The corresponding values of magnetic field strength would form a discrete sequence for an ideal two-dimensional crystal, and the corresponding values of the Hall voltage as function of the field strength would just lie on the classical straight line. However, potential perturbations present in a real sample may fill the gaps between Landau levels with disorder-localized states, still leaving a mobility gap. Since the role of the gap in the topological argument is essentially to localize the states, the Hall conductivity remains quantized (with an extremely high precision of less than one in 10^9) over some interval of magnetic field and the Hall voltage shows plateaus.

Sometimes the situation, although with N occupied bands, is interpreted with the $U(1)$-bundle $(P, \mathbb{T}^2_{\boldsymbol{\kappa}}, \pi, U(1))$. This is possible since only the trace of the $U(N)$ group element enters the first Chern number which is just a phase.

In the $(2+1)$-dimensional effective field theory of the situation, $x^\mu = (ct, x, y)$, $\mu = 0, 1, 2$, the transport equation is

$$j_e^\mu = \sigma_{\mathrm{H}}\, \delta_{012}^{\mu\nu\sigma}\, \partial_\nu A_\sigma = \frac{\delta S_{\mathrm{eff}}}{\delta A_\mu}, \quad S_{\mathrm{eff}} = \frac{\sigma_{\mathrm{H}}}{2!} \int d^3x\, \delta_{012}^{\mu\nu\sigma} A_\mu \partial_\nu A_\sigma, \quad \partial_\mu = \frac{\partial}{\partial x^\mu}, \tag{8.114}$$

where $j_e^0 = c\rho_e$ is given by the charge density ρ_e, A_σ is the potential of the electromagnetic field $F^\mu = \delta_{012}^{\mu\nu\sigma}\partial_\nu A_\sigma$ (U(1) gauge field), and the interaction part of the effective action is the Chern–Simons action S_{eff}. (Do not confuse A_σ and F^μ with the forms $\mathcal{A}$ and $\mathcal{F}$ above.) The settings before (8.107) yield $j_e^0 = c\rho_e = \sigma_{\mathrm{H}} B = \sigma_{\mathrm{H}} B^z$, $j_e^x = \sigma_{\mathrm{H}} E = \sigma_{\mathrm{H}} E^y$, $j_e^y = 0$. Under the action of an electric field in y-direction there is a Hall current in x-direction; the current does not gain energy out of the field, hence it does also not dissipate energy. The proportionality between the charge density ρ_e and the magnetic field strength perpendicular to the sample might seem strange; however, the time derivative of this relation together with the continuity relation for the charge density (charge conservation) yields $(1/c)\partial B^z/\partial t = \sigma_{\mathrm{H}}^{-1}\partial\rho_e/\partial t = -\sigma_{\mathrm{H}}^{-1}\nabla \boldsymbol{j}_e = -(\nabla \times \boldsymbol{E})^z$ which is Faraday's law of induction.

So far the analysis (8.108–8.113) was based on the consideration of an infinite periodic two-dimensional crystal which does not have edges. Consider as a more realistic situation a strip extending in y-direction and confined in x-direction. Assume it of length L in y-direction and closed to a loop (periodic boundary conditions in y-direction). Choose this direction to be that of $\tilde{\boldsymbol{a}}_2$ so that periodicity of the Hamiltonian in this direction is preserved and $\kappa_2 = \kappa$ remains a valid quantum number of stationary states. In x-direction the states $u_{n\kappa}$ are now localized across the strip with discrete quantum numbers included in n. The Hamiltonian (8.110) is changed into a one-dimensional Hamiltonian H_κ, $\kappa = \kappa(x) = -i\partial/\partial y + (e/\hbar c)(Bx - Ect)$. For simple tight-binding models this Hamiltonian can directly be solved for $E = 0$. There are still states sufficiently far from the edges of the strip grouped in Landau levels. Close to the edges there appear states whose energy as function of κ disperses across the gaps between Landau levels in a chiral manner; on the edge with larger x-coordinate their group velocity $\partial\varepsilon/\partial\kappa$ points in the negative y-direction only and vice versa. The corresponding edge currents in opposite directions on opposite edges do not lead to a net current along the strip nor do they dissipate energy. The current in x-direction cannot stationarily flow any more. Instead it polarizes the sample until the Hall voltage due to this polarization counterbalances the current driving force due to the voltage along the sample. Instead of applying an electric field E^y the magnetic field strength $B = B^z$ could adiabatically be changed in time which itself produces an electric field $E^y = -(x/c)\,\partial B/\partial t$ (in agreement with Faraday's law). Hence, $\partial\rho_e/\partial t = -\partial j^x_e/\partial x = -\sigma_{\mathrm{H}}\,\partial E/\partial x = (\sigma_{\mathrm{H}}/c)\,\partial B/\partial t$. The magnetic flux through the strip is $L_x L_y B$ where L_x and $L_y = L$ are the extensions of the strip. The total amount Q_e of charge which flows across the strip is given by $dQ_e/dt = L_x L_y \partial\rho_e/\partial t = (\sigma_{\mathrm{H}}/c) L_x L_y \partial B/\partial t$. Hence a change of flux by one flux unit Φ_0 transports precisely one charge unit across the strip polarizing it and contributing to the Hall voltage (Laughlin). Needless to say that this is only valid as long as the magnetic field varies within one quantum Hall plateau, because the whole approach is based on the position of the Fermi level in a mobility gap inside the strip.

This result also provides a connection with the situation of (8.101). Let the change in magnetic field ΔB be so weak that the variation of ΔBx through one unit cell can be neglected. (In semiconductor physics this is related to a switch from description with a microscopic wave function to one with envelope wave functions which are averaged over unit cells.) Small changes of E and B may then be treated as an adiabatic parameter $\lambda(x,t)$ on which the Hamiltonian $H_{\lambda(t,x)\kappa}$ and the quantum state $u_{ntx\kappa}$ depend, which, however, can be taken out of the matrix elements $(u_{mtx\kappa}\,|\,u_{ntx\kappa})$. One has $[(\partial/\partial t), H_{\lambda\kappa}] = -(x/c)(\partial B/\partial t)(e/m)\,p_y = (x/c)(\partial B/\partial t)\,j^y_e$. The polarization in x-direction is given like in (8.105) by $dP^x/dt = j^x$, hence $\partial j^x/\partial x = -\partial\rho/\partial t = \partial^2 P^x/(\partial t\partial x)$, and with a notation $(t,x,\kappa) = (q^0, q^2, q^2)$ one has

$$j^i = -\delta^{ij}_{01}\frac{\partial P^x}{\partial q^j}, \quad i,j = 0,1, \tag{8.115}$$

and with

$$\mathcal{A}_A = \sum_n^{\text{occ.}} (-i)(u_{nq}|\partial_A u_{nq}), \quad \mathcal{F} = d\mathcal{A}, \quad \partial_A = \frac{\partial}{\partial q^A}, \quad A = 0, 1, 2, \tag{8.116}$$

one finds $j^i = -1/(2\pi) \int d^2q \, \delta^{012}_{i2j} \, \mathcal{F}^{2j}$, in particular

$$\frac{dP^x_e}{dt} = j^x_e = \frac{e}{|\mathbb{T}^2_r|} \frac{|\boldsymbol{a}_2|}{2\pi} \int d^2q \, (-2i)(\partial_2 u_{nq})|\partial_0 u_{nq}) \, \frac{x}{c} \frac{\partial B}{\partial t} \, . \tag{8.117}$$

The t-integral of which compares precisely with a one-dimensional version of (8.101), if one identifies $(x/c)\,\partial B/\partial\lambda$ with the driving force polarizing the system. Cycling this force provides another mean to pump charge through the insulator in the ground state. The $(2+1)$-dimensional quantum field theory of the quantum Hall system reduces to a $(1+1)$-dimensional quasi-classical field theory for an adiabatic charge pumping process by replacing one wave-vector component by an adiabatically time-dependent parameter performing a closed loop in parameter space.

The whole story may be cast in yet another form relevant in most topical types of solids christened topological insulators, Chern insulators or topological semiconductors. Define phase space variables $q = (q^A) = (t, x^1, x^2, k_1, k_2)$, $A = 0, 1, 2, 3, 4$, $\partial_A = \partial/\partial q^A$, the corresponding electromagnetic potential $(A_A) = (A_0, A_1, A_2, 0, 0)$ and the Berry–Simon connection $((-i)\sum_n^{\text{occ.}} (u_{nq^3q^4}| \partial_A u_{nq^3q^4})) = (\mathcal{A}_A) = (0, 0, 0, \mathcal{A}_3, \mathcal{A}_4)$. Then, inserting (8.112) in the form $e^2/(\hbar(2\pi)^2) \int (\partial_3\mathcal{A}_4 - \partial_4\mathcal{A}_3)\, dq^3 dq^4$ for σ_{H}, the effective action (8.114) may be expressed as

$$S^{(2+1)}_{\text{eff}} = \frac{e^2}{\hbar} \frac{1}{2!(2\pi)^2} \int d^5q \, \delta^{ABCDE}_{01234} A_A \partial_B A_C \text{tr}\,(\partial_D \mathcal{A}_E) \tag{8.118}$$

Now, replace $k_1 + A_1$ by the adiabatic parameter $\lambda(t, x^1)$, replace ∂_3 by ∂_λ, $\mathcal{A}^{mn}_3 = (-i)(u_{mq^3q^4}|\partial_3 u_{nq^3q^4})$ by $\mathcal{A}^{mn}_\lambda = (-i)(u_{m\lambda q^4}|\partial_\lambda u_{m\lambda q^4})$, remove the integrations over $dq^1 = dx^1$ and over $dq^3/(2\pi) = dk_1/(2\pi)$ and let the only non-constant gauge potential A_2 depend on $t, x^1 = q^0, q^1$ only, so that $\delta^{ABC}_{012} A_A \partial_B A_C = 2(A_0\partial_1 A_2 - A_1\partial_0 A_2) = 2(A_0\partial_1\lambda - A_1\partial_0\lambda)$. This changes (8.118) into

$$\frac{e^2}{\hbar} \frac{1}{2\pi} \int dt dx^1 \, (A_0\partial_1\lambda - A_1\partial_0\lambda) \int dk_2 \, \text{tr}\,(\partial_{k_2}\mathcal{A}_\lambda - \partial_\lambda \mathcal{A}_{k_2}) \, .$$

Now, $u_{n\lambda q^4}$ depends on t and on x^1 through λ only, hence $(\partial_0\lambda)\partial_\lambda\mathcal{A}_{k_2} = \partial_0\mathcal{A}_{k_2}$ and $(\partial_1\lambda)\partial_\lambda\mathcal{A}_{k_2} = \partial_1\mathcal{A}_{k_2}$ as well as $(\partial_0\lambda)\partial_{k_2}\mathcal{A}_\lambda = \partial_{k_2}\mathcal{A}_0$ and $(\partial_1\lambda)\partial_{k_2}\mathcal{A}_\lambda = \partial_{k_2}\mathcal{A}_1$ which rewrites the displayed expression as

$$\frac{e^2}{\hbar} \frac{1}{2\pi} \int dt dx^1 dk_2 \, (A_0 \text{tr}\,(\partial_{k_2}\mathcal{A}_1 - \partial_1\mathcal{A}_{k_2}) - A_1 \text{tr}\,(\partial_{k_2}\mathcal{A}_0 - \partial_0\mathcal{A}_{k_2}))$$

or finally, renaming t, x^1, k_2 into q^A, q^B, q^C,

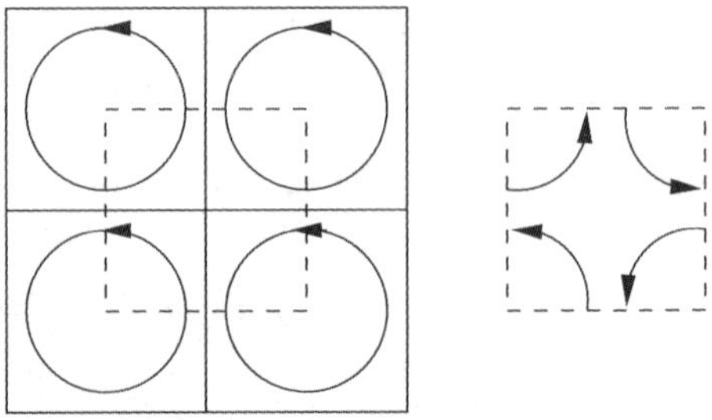

Fig. 8.5 Ring currents with respect to different periodicity volumes

$$S_{\text{eff}}^{(1+1)} = \frac{e^2}{\hbar}\frac{1}{2\pi}\int d^3q\,\delta_{012}^{ABC}\,A_A\text{tr}\,(\partial_B\mathcal{A}_C) \tag{8.119}$$

which is the phase space expression for the interaction part of the action in $(1+1)$ dimensions describing the quantum Hall strip, obtained by the *dimension reduction* procedure introduced by Qi et al. [13].

Only very recently activities were started to develop a Berry curvature theory of the orbital magnetic moment density in analogy to theory of the electric dipole density (see [14]). As mentioned above for the model of free electrons, it is produced by electric ring currents flowing in the bulk. This time, two dimensions are at least needed to consider ring currents (Fig. 8.5). While again the average physical moment density cannot be determined from the current density, it corresponds to a counterclockwise ring current density for a periodicity volume which is an integer multiple of the cells drawn in full lines on the left part of Fig. 8.5, and to a clockwise ring current for a periodicity volume which is an integer multiple of the dashed cell, the presence of a surface (not its shape) determines what are the physical ring currents adding up to a physically measurable edge current along the boundary of the sample volume.

8.6.5 Topological Insulators

This is a rapidly developing subject which was initiated in the first years of the present millennium. The aspects relevant in our context are best described in [13]. Consider the principal bundle $(P, \mathbb{T}^{2r}, \pi, U(N))$ over the $2r$-torus. The rth Chern number C_r of the rth Chern character is (cf. 8.61), $i\mathcal{F}^1$ replaced by $\mathcal{F}^1$ from above)

$$\begin{aligned} C_r &= \frac{1}{r!}\int \text{tr}\left(\frac{\mathcal{F}^1}{2\pi}\wedge\cdots\wedge\frac{\mathcal{F}^1}{2\pi}\right) = \text{tr}\left(\left(\frac{2}{2\pi}\int d^2k\,(\mathcal{F}^1)^{12}\right)^r\right) \\ &= \frac{1}{(2\pi)^r}\int \text{tr}\left((\mathcal{F}^1)^{12}(\mathcal{F}^1)^{34}\cdots(\mathcal{F}^1)^{2r-1,2r}\right) dk_1\wedge dk_2\wedge\cdots\wedge dk_{2r} \\ &= \frac{1}{r!2^r(2\pi)^r}\int d^{2r}k\,\delta_{i_1 i_2\cdots i_{2r}}^{12\cdots 2r}\,\text{tr}\left((\mathcal{F}^1)^{i_1 i_2}(\mathcal{F}^1)^{i_3 i_4}\cdots(\mathcal{F}^1)^{i_{2r-1}i_{2r}}\right) \end{aligned} \tag{8.120}$$

where $(\mathcal{F}^1)^{12}_{mn} = (-i)(\partial^1 u_{mk}|\partial^2 u_{nk}) + i[\mathcal{A},\mathcal{A}]^{12}_{mn}$ with $\mathcal{A}^1_{mn} = (-i)(U^{-1}(k)\partial^1 U(k))_{mn}$. Under the trace the matrices $\int d^2k\,\mathcal{F}^1_{mn}$ may be trigonalized without changing the trace; the skew-symmetric part $i[\mathcal{A},\mathcal{A}]_{mn}$ thereby remaining skew-symmetric and hence with zero diagonal. The trace of the rth power of triangular matrices equals the sum of the rth powers of the diagonal elements. Since we found earlier that $\mathrm{tr}\left((2/2\pi)\int d^2k(\mathcal{F}^1)^{12}\right)$ is an integer independent of the order of the matrix, all diagonal elements of this matrix are integers. Hence, C_r is also an integer.

Consider formally the $(2r+1)$-dimensional effective Chern–Simons gauge field theory with the interaction part of the action

$$\begin{aligned} S^{(2r+1)}_{\mathrm{eff}} &= \frac{e^2}{\hbar}\frac{C_r}{(2\pi)^r}\int A_0\partial_1 A_2\cdots\partial_{2r-1}A_{2r}\,dt\wedge dx^1\wedge\cdots\wedge dx^{2r} \\ &= \frac{e^2}{\hbar}\frac{C_r}{(r+1)!(2\pi)^r}\int d^{2r+1}x\,\delta^{\mu\nu\ldots\sigma\tau}_{01\ldots 2r}A_\mu\partial_\nu\cdots\partial_\sigma A_\tau, \\ j^\mu &= \frac{e^2}{\hbar}\frac{C_r}{r!(2\pi)^r}\delta^{\mu\nu\ldots\sigma\tau}_{01\ldots 2r}\partial_\nu A_\cdot\cdots\partial_\sigma A_\tau\,. \end{aligned} \tag{8.121}$$

Greek indices like μ run through $0,1,\ldots,2r$. In total as previously for $r=1$ the action integral may be written as

$$\begin{aligned} S^{(2r+1)}_{\mathrm{eff}} &= \frac{e^2}{\hbar}\frac{1}{r!(r+1)!(2\pi)^{2r}}\int d^{4r+1}q\,\delta^{A_0A_2\ldots A_{4r}}_{01\ldots 4r} \\ &\times A_{A_0}\partial_{A_1}\cdots\partial_{A_{2r-1}}A_{A_{2r}}\,\mathrm{tr}\left(D_{A_{2r+1}}\mathcal{A}_{A_{2r+2}}\cdots D_{A_{4r-1}}\mathcal{A}_{A_{4r}}\right) \end{aligned} \tag{8.122}$$

with $q=(q_A)=(t,x^1,\ldots,x^{2r},k_1,\ldots,k_{2r})$ and the subscripts A_i running from 0 through $4r$. Now, $(\mathcal{A}_A)_{mn} = (-i)\sum^{\mathrm{occ.}}(u_{mk}\,|\,\partial_A u_{nk})$ and $D\mathcal{A} = d\mathcal{A} + i[\mathcal{A},\mathcal{A}]$. This is called a $(2r+1)$-dimensional effective Chern–Simons field theory for a *fundamental* $2r$-dimensional lattice-periodic topological insulator with a Brillouin torus $\mathbb{T}^{2r}_{\boldsymbol{k}}$.

So far there is no experimental realization for $r>1$. However, a dimension reduction of the $(4+1)$-dimensional case yields a quasi-classical adiabatic $(3+1)$-dimensional theory of a three-dimensional topological insulator where in addition an $\mathbb{F}_2$ time-reversal symmetry plays a crucial role. Very recently, Bi_2Se_3, Bi_2Te_3, and Sb_2Te_3 have been found as realizations of this case. A further dimension reduction leads to a $(2+1)$-dimensional theory for a so called spin-Hall insulator, of which a HgTe quantum dot on CdTe is a realization. For more details see for instance [15].

The crux in all these cases is that crystal periodicity or periodic boundary conditions produce tori which on the one hand eliminate surface effects and preserve volume properties and on the other hand have non-trivial Chern numbers expressing those volume properties.

In order not to digress to far from our subject we used a Kohn–Sham approach in the whole presentation. In a more general many-body approach the same sort of theory may be formulated with quasi-particle Green's functions $G \sim |\cdot)(\cdot|/(\varepsilon - \Sigma)$ with a self-energy Σ, and where Kubo's formula takes on the form of an expression $\sim \mathrm{tr}\,(Gj_e Gj_e)$ and the (Berry) Chern–Simons form is $\sim \mathrm{tr}\,(G^{-1}dG)$. Otherwise all results remain the same.

8.7 Gauge Field Theory of Molecular Physics

In the Born–Oppenheimer adiabatic treatment of molecular motion and chemical reactions, the $3N$ coordinates $R = (\boldsymbol{R}^1, \ldots, \boldsymbol{R}^N)$ of the N involved nuclear positions form the adiabatic parameter manifold M for the electronic states $|\Psi_a(R)\rangle$. The adiabatic Hamiltonian is ($\hbar = |e| = m_e = 1$, with electron charge e and mass m_e, Z_n is the charge of nucleus n)

$$\hat{H}_{\mathrm{ad}}(R) = \sum_s \frac{\nabla_s^2}{2} + \sum_{ss'} \frac{1}{|\boldsymbol{r}^s - \boldsymbol{r}^{s'}|} - \sum_{sn} \frac{Z_n}{|\boldsymbol{R}^n - \boldsymbol{r}^s|} + \sum_{nn'} \frac{Z_n Z_{n'}}{|\boldsymbol{R}^n - \boldsymbol{R}^{n'}|}. \tag{8.123}$$

Via

$$\hat{H}_{\mathrm{ad}}(R)\langle r|\Psi_a(R)\rangle = \langle r|\Psi_a(R)\rangle E_a(R) \tag{8.124}$$

it yields the effective adiabatic potential $E_a(R)$ for the nuclear motion:

$$\hat{H}\langle r, R|\Psi_E\rangle = \langle r, R|\Psi_E\rangle E, \quad \langle r, R|\Psi_E\rangle = \sum_a \langle r|\Psi_a(R)\rangle\langle\Psi_a(R)|\langle R|\Psi_E\rangle. \tag{8.125}$$

$\hat{H} = \sum_n \nabla_n^2/(2M_n) + \hat{H}_{\mathrm{ad}}(R)$ is still the full Hamiltonian, $\langle r|\Psi_a(R)\rangle$ is an r-dependent wave function parametrically also depending on R, and $\langle\Psi_a(R)|\langle R|\Psi_E\rangle = \Psi_E^a(R)$ is a nuclear wave function depending only on R while the r-dependence is integrated out in the Hilbert scalar product. ($\langle r|$ and $\langle R|$ are position eigenstates.) So far (8.125) does not contain approximations.

In order to have a discrete spectrum of $\hat{H}$, the center of gravity of the nuclei is now eliminated, and further on R denotes relative nuclear coordinates rescaled in such a way that all corresponding relative masses μ are equal. This also implies neglecting the electron mass in the relative motion. Then, for the $(3N-3)$-dimensional Hilbert-space momentum vector operator $\check{P}$ corresponding to the relative coordinates and obeying $\langle r, R|\check{P}|\Psi_E\rangle = -i\nabla_R\langle r, R|\Psi_E\rangle$, where ∇_R means the covariant differential (7.28) in the R-manifold, one easily obtains

$$\langle\Psi_a(R)|\langle R|\check{P}|\Psi_E\rangle = \sum_b \left(-i\delta_b^a\nabla_R - i\mathcal{A}_b^a(R)\right)\Psi_E^b(R) \tag{8.126}$$

with

$$\mathcal{A}_b^a(R) = \langle\Psi_a(R)|\nabla_R|\Psi_b(R)\rangle. \tag{8.127}$$

It was already seen in (8.84) that $\mathcal{A}$ is a local connection form of a Berry–Simon connection and hence may be taken as a gauge potential. It is called a **Mead–Berry gauge potential**. Correspondingly, the exterior covariant derivative

$$D = \mathbf{1}\nabla_R + \mathcal{A}(R) \tag{8.128}$$

is introduced which casts (8.125) into

$$\left(-D^2/(2\mu) + E(R) - E\mathbf{1}\right)\Psi_E(R) = 0, \tag{8.129}$$

where $\Psi_E(R)$ means a column with elements $\Psi^a_E(R)$, $E(R)^a_b = \delta^a_b E_a(R)$, matrix multiplication is understood and (8.129) in principle contains an infinite column of equations, which in practical applications is cut off at a finite dimension for a small number of lowest eigenvalues $E_a(R)$. The ordinary Born–Oppenheimer approximation means taking only the lowest $E_a(R)$ and neglecting $\mathcal{A}$.

Equations 8.127–8.129 form the basis of the gauge theory of molecular physics. If the dimension of the problem (matrix dimension of $\mathcal{A}$) is not cut of, the gauge field vanishes and the potential $\mathcal{A}$ can *locally* be gauged away. To see this, consider the gauge field $\mathcal{F} = \mathcal{D}\mathcal{A} = d\mathcal{A} + \mathcal{A}\wedge\mathcal{A}$ in more detail:

$$\mathcal{F}^a_b = \sum_{ij}\left(\nabla_{R_i}\langle\Psi_a|\nabla_{R_j}|\Psi_b\rangle + \sum_c \langle\Psi_a|\nabla_{R_i}|\Psi_c\rangle\langle\Psi_c|\nabla_{R_j}|\Psi_b\rangle\right) dR^i \wedge dR^j,$$

where R_i and R_j are local coordinates in a coordinate patch of the R-manifold M. Because of orthonormality and completeness of the Ψ, the second term in parentheses is $-\sum_c\langle(\nabla_{R_i}\Psi_a)|\Psi_c\rangle\langle\Psi_c|\nabla_{R_j}|\Psi_b\rangle = -\langle(\nabla_{R_i}\Psi_a)|\nabla_{R_j}|\Psi_b\rangle$. This may be written as $-\nabla_{R_i}\langle\Psi_a|\nabla_{R_j}|\Psi_b\rangle + \langle\Psi_a|\nabla_{R_i}\nabla_{R_j}|\Psi_b\rangle$. Now, $\nabla_{R_i}\nabla_{R_j}|\Psi_b\rangle\, dR^i \wedge dR^j = d^2|\Psi_b\rangle = 0$. Hence, $\mathcal{F} = 0$ results. This does not mean that the full theory is trivial. In fact, (8.127) is not defined at points of term crossing, and hence all those points have to be excluded from the manifold M, which thus becomes homotopically highly non-trivial resulting in Aharonov–Bohm type situations. This is also the main case of application of this gauge field theory, if only a few lowest electronic terms are retained. If there is a degeneracy of lowest electronic levels on a whole manifold M, a case of non-Abelian gauge field theory is realized. A wealth of resulting phenomena is considered in [2].

References

1. Shapere, A., Wilczek, F. (eds.): Geometric Phases in Physics. World Scientific, Singapore (1989)
2. Bohm, A., et al.: The Geometric Phase in Quantum Physics. Springer, Berlin (2003)
3. Schwarz, A.S.: Topology for Physicists. Springer-Verlag, Berlin (1994)
4. t'Hooft, G. (ed.): 50 Years of Yang-Mills Theory. World Scientific, Singapore (2005)

5. Milton, K.A.: Theoretical and experimental status of magnetic monopoles. Rep. Prog. Phys. **69**, 1637–1711 (2006)
6. Nakahara, M.: Geometry, Topology and Physics. IOP Publishing, Bristol (1990)
7. Kato, T.: Perturbation Theory for Linear Operators. Springer, Berlin (1966)
8. Resta, R.: Macroscopic polarization in crystalline dielectrics: the geometric phase approach. Rev. Mod. Phys. **66**, 899–915 (1994)
9. Resta, R.: Manifestations of Berry's phase in molecules and condensed matter. J. Phys.: Condens. Matter **12**, R107–R143 (2000)
10. Resta, R., Vanderbilt, D.: In: Ahn, C.H., Rabe, K.M., Triscone, J.M. (eds.). Physics of Ferroelectrics: a Modern Perspective, pp. 31–68. Springer, Berlin (2007)
11. Eschrig, H.: The Fundamentals of Density Functional Theory. p. 226, Edition am Gutenbergplatz, Leipzig, (2003)
12. Peierls, R.: Surprises in Theoretical Physics. Princeton University Press, Princeton (1979)
13. Qi, X.-L., Hughes, T.L., Zhang, S.-C.: Topological field theory of timereversal invariant insulators. Phys. Rev. B **78**, 195424–1-43 (2008)
14. Ceresoli, D., Thonhauser, T., Vanderbilt, D., Resta, R.: Orbital magnetization in crystalline solids: Multi-band insulators, Chern insulators, and metals. Phys. Rev. **B 74**, 024408–1–13 (2006)
15. Hasan, M.Z., Kane, C.L.: Topological Insulators. Rev. Mod. Phys. **82**, 3045–3067 (2010)

Chapter 9
Riemannian Geometry

In Chaps. 3–8 the theory of manifolds was based on the smooth differentiable structure, a complete atlas compatible with a pseudogroup S of class C^∞ introduced at the beginning of Sect. 3.1. With the only exception of Hodge's star operator introduced at the end of Sect. 5.1, a metric was not needed on general manifolds and on bundles and was not introduced. Since the notion of manifold M was restricted to local homeomorphy with $\mathbb{R}^m$ in this text, by differentiation of real functions along paths in M the tangent space was defined in Sect. 3.3 as a local linearization of M. On this basis, tensor bundles, the tensor calculus and the exterior calculus as well as integration of exterior forms could be introduced and the whole theory up to here could be built without a metric on M. Now, by defining a metric of a norm on the tangent spaces, due to the locally linear relation between M and its tangent spaces, the manifold M itself is provided with a Riemannian metric. A connection compatible with this metric makes M into a Riemannian geometric space provided with a Riemannian geometry.

Despite the generality of Riemann's concepts, in his time and afterwards, the focus was on homogeneous manifolds having everywhere the same geometry, in particular the same curvature. Only the work of Einstein and Hilbert on general relativity provided an important application case of the concept of a general Riemannian space. In recent time, with R. Hamilton's concept of Ricci flow which is beyond the scope of this text, homogeneous manifolds came again into focus in classification problems of low-dimensional manifolds and in the theory of partial differential equations. So far these developments culminated into Perelman's proof of Poincaré's conjecture (see end of Sect. 2.5). Homogeneous manifolds are considered in Sect. 9.2.

The Riemannian geometry proper, with the unique linear connection for which everywhere $\nabla g = 0$, is treated in Sects. 9.3–9.5, while gravitation as the most important application of this case is shortly considered in Sect. 9.6.

Apart from a few occasions, this short excursion through topology and geometry was devoted to real manifolds. It is finally concluded with a brief outlook on some complex generalizations.

9.1 Riemannian Metric

Let M be a manifold of dimension m, and let g be a symmetric tensor field of type $(0,2)$, in a coordinate neighborhood

$$g = g_{ij}(\boldsymbol{x})dx^i \otimes dx^j, \quad g_{ij} = g_{ji}. \tag{9.1}$$

g is called a **non-degenerate rank 2 tensor** at point $x \in M$, if the linear equation system

$$g_{ij}(\boldsymbol{x})X^j = 0, \quad i = 1, \ldots, m \tag{9.2}$$

has $X^j = 0$ as its only solution, that is, det $g(x) \neq 0$. (For the sake of simplicity of notation, both the tensor g and the matrix $g = (g_{ij})$ in local coordinates are denoted by the same letter.) g is called a **positive definite rank 2 tensor** at point x, if the contraction

$$g(X,X) = C_{1,1}C_{2,2}(g \otimes X \otimes X) > 0 \quad \text{for all } X \neq 0,\ X \in T_x(M), \tag{9.3}$$

that is, the matrix $(g_{ij})(\boldsymbol{x})$ is positive definite in the sense of linear algebra.

A **generalized Riemannian manifold** is a manifold M provided with an everywhere on M non-degenerate symmetric tensor field g of type $(0,2)$, its **metric tensor** or **fundamental tensor**. If g is positive definite, then M is called a **Riemannian manifold.**[1]

The metric tensor of a Riemannian manifold M makes every tangent space $T_x(M)$ into a Euclidean space with **inner product** and **norm**

$$(X|Y) = g_{ij}(\boldsymbol{x})X^iY^j, \quad |X| = (X|X)^{1/2}. \tag{9.4}$$

For the first time in this text (besides mentioning it in passing in Sect. 5.1) this defines an angle

$$\angle(X,Y) = \arccos\left(\frac{(X|Y)}{|X||Y|}\right) \quad \text{for } |X| \neq 0 \neq |Y| \tag{9.5}$$

between tangent vectors at the same point x.

Being a smooth tensor field, (9.1) may be considered as a symmetric differential 2-form on M (to be distinguished from an exterior 2-form, which by definition is alternating), called the **metric form**, in the following sense: Let $C :]t_0, t_1[\to M$ be a smooth path in M and let $x = C(t)$ for some $t_0 < t < t_1$. In a coordinate neighborhood of x the path is given as $x^i = x^i(t)$, $i = 1, \ldots, m$. A tangent vector to the path is $X^i = dx^i/dt$. By definition of a tensor field (Sect. 4.1),

[1] Besides an indefinite metric there are many more generalizations of Riemannian manifold in the mathematical literature; the case of an indefinite metric is also called a pseudo-Riemannian manifold. In this text generalized Riemannian manifold just comprises Riemannian and pseudo-Riemannian manifold.

the expression (9.1) is independent of the choice of local coordinates. Hence, the square of the norm of Xdt,

$$ds^2 = g_{ij}\frac{dx^i}{dt}\frac{dx^j}{dt}dt^2 = g_{ij}dx^i dx^j, \tag{9.6}$$

is also independent of the choice of local coordinates, and the integral

$$s(C) = \int_{t_0}^{t_1} \sqrt{g_{ij}\frac{dx^i}{dt}\frac{dx^j}{dt}}\,dt = \int_C ds \tag{9.7}$$

too has a real value independent of the used local coordinates. This integral is a property of the path C alone and is called its **arc length**, $ds = \sqrt{ds^2}$ is called the element of the arc length.

Assigning in the case of a Riemannian manifold M this way an arc length to every piecewise smooth curve is said to define a **Riemannian metric** on every connected component of M. The corresponding distance function is

$$d(x,y) = \inf_{C(x,y)} \int_{C(x,y)} ds \quad \text{for } g > 0, \tag{9.8}$$

where $C(x,y)$ is any path from x to y in M, and the infimum is taken over all paths. As a distance function it must have the properties 1–3 given on p. 13. Let x be any point of M and take any coordinate neighborhood U_α of x. It is homeomorphic to a neighborhood $\boldsymbol{U}_\alpha = \boldsymbol{\varphi}_\alpha(\boldsymbol{U}_\alpha) \subset \mathbb{R}^m$ of $\boldsymbol{x} = \varphi_\alpha(x)$. As an open set of $\mathbb{R}^m$, for any point $y \neq x$, $\boldsymbol{U}_\alpha$ contains a closed ball $\overline{\boldsymbol{B}}_\delta(\boldsymbol{x})$ of some radius $\delta > 0$ not containing $\boldsymbol{y} = \boldsymbol{\varphi}_\alpha(y)$. $\overline{\boldsymbol{B}}_\delta(\boldsymbol{x})$ is compact, and hence the positive function of $\boldsymbol{x}$ equal to $\min_{\{X,|X|=1\}} g_{ij}(\boldsymbol{x})X^iX^j$ of $\boldsymbol{x}$ takes on a positive minimum value there. Let this value be ε^2. It is readily seen that $d(x,y) > \delta\varepsilon > 0$, no matter is $y \in U_\alpha$ or not: $C(x,y) \cap U_\alpha$ is part of any path $C(x,y)$ and the above minimum value ε yields an estimate of the integral (9.8) over this part from below. Hence $d(x,y) = 0$, iff $x = y$, and property 1 from p. 13 is fulfilled. Property 2 is obvious, and property 3 simply follows from concatenation of paths. Thus, (9.8) makes connected components of M into metric spaces.

In the case of a generalized Riemannian manifold with g not being positive definite, (9.6) is still called the element of the (sign carrying) arc length. Since in every coordinate neighborhood the matrix g_{ij} is symmetric, it can be diagonalized at a given point $x \in M$, yielding $ds^2 = g_{ii}(dx^i)^2$. Depending on the direction of dx^i in M, ds^2 may be positive, zero or negative. Hence, in this case g does not define a metric. Nevertheless, it is said to define an **indefinite Riemannian metric** on M. The expressions (9.4) are called an **indefinite inner product** and an **indefinite norm**; (9.5) defines an angle for $|X| \neq 0 \neq |Y|$, a non-zero vector X with $|X| = 0$ is said to be **isotropic**. Although (9.8) does not make sense in this case, large parts of the subsequent theory apply, and the indefinite Minkowski metric gives it relevance in physics.

There exists a Riemannian metric on any m-dimensional smooth (paracompact) manifold M.

Proof Recall that manifolds were supposed to be paracompact by definition. Hence, there exists a locally finite coordinate covering $\{(U_\alpha, \boldsymbol{\varphi}_\alpha)\}$ of M and a partition of unity $\{\varphi_\alpha \,|\, \mathrm{supp}\ \varphi_\alpha \subset U_\alpha,\ \varphi_\alpha \geq 0\}$, $\sum_\alpha \varphi_\alpha = 1$ on M. Then, $ds^2_\alpha = \sum_{i=1}^m (dx^i_\alpha)^2$ defines a positive definite symmetric 2-form on U_α ($g_{\alpha ij}$ is the unit matrix). Take a coordinate neighborhood U of $x \in M$ with $\boldsymbol{x} = (x^i) = \boldsymbol{\varphi}(x) \in \boldsymbol{U} \subset \mathbb{R}^m$ for which $\overline{U}$ is compact. It intersects with finitely many of the U_α only. Put

$$ds^2 = g_{ij} dx^i dx^j, \quad g_{ij} = \sum_\alpha \sum_{k=1}^m \varphi_\alpha \frac{\partial x^k_\alpha}{\partial x^i} \frac{\partial x^k_\alpha}{\partial x^j}.$$

The sum defining g_{ij} is finite, and hence the definition is correct. Since the Jacoby matrix $\partial x^k_\alpha / \partial x^i$ is regular and at least one of the $\varphi_\alpha(x)$, φ_β, say, is positive, $ds^2(x) \geq \varphi_\beta(x) \sum_i (dx^i_\beta)^2$ is positive definite and defines a Riemannian metric on M. □

This statement means that the bundle of symmetric covariant tensors of rank 2 on every manifold M has a positive definite section. This is remarkable. Recall from the end of Sect. 8.2, that the vector bundle over even a quite simple manifold, while always having a section, may not have a *non-zero* section. In general, there may also not exist an indefinite Riemannian metric on M.

Let $F : N \to M$ be an embedding of the manifold N into a Riemannian manifold M with metric form g. Then,

$$\bar{g}(X, Y) = g(F_* X, F_* Y) \quad \text{for all } X, Y \in T_x(N) \tag{9.9}$$

defines a Riemannian metric on N. Such a statement does again not hold for an indefinite metric. (Why?)

The following considerations apply to generalized Riemannian manifolds with both definite or indefinite metric.

The transformation law of the metric form between overlapping coordinate neighborhoods is

$$g_{\beta ij} = \frac{\partial x^k_\alpha}{\partial x^i_\beta} g_{\alpha kl} \frac{\partial x^l_\alpha}{\partial x^j_\beta}. \tag{9.10}$$

Denote the matrix inverse to g_{ij} by g^{ij}:

$$\sum_k g^{ik} g_{kj} = \sum_k g_{jk} g^{ki} = \delta^i_j. \tag{9.11}$$

Then, since the inverse of the Jacobi matrix $\partial x^k_\alpha / \partial x^j_\beta$ is $\partial x^k_\beta / \partial x^j_\alpha$ and since g_{kl} is symmetric, from (9.10) it follows that

$$g^{ij}_{\beta} = \frac{\partial x^i_\beta}{\partial x^k_\alpha} g^{kl}_\alpha \frac{\partial x^j_\beta}{\partial x^l_\alpha}, \tag{9.12}$$

that is, (9.11) defines a symmetric contravariant rank 2 tensor with components g^{ij} (tensor of type $(2, 0)$), uniquely attached with g.

Let $x \in M$ and $X \in T_x(M)$. Then, $\omega_X = g(X, .)$ is a 1-form: $\langle \omega_X, Y \rangle = g(X, Y)$ is a real number for every $Y \in T_x(M)$. Obviously, if X runs through the tangent vector space $T_x(M)$, ω_X runs through the cotangent vector space $T^*_x(M)$. Since g is non-degenerate, it provides a bijection between the tangent and cotangent vector spaces at every point $x \in M$, depending smoothly on x: If $X = \xi^i(\boldsymbol{x})(\partial/\partial x^i)$ is a tangent vector field on $U \subset M$, then $\omega_X = (g_{ij}(\boldsymbol{x})\xi^i(\boldsymbol{x}))dx^j$ is a cotangent vector field, since $(g_{ij}(\boldsymbol{x})\xi^i(\boldsymbol{x}))$ depends smoothly on $\boldsymbol{x}$ and transforms like a cotangent vector between overlapping coordinate neighborhoods, and $X = (g^{ik}\omega_{Xk})(\partial/\partial x^i) = (g^{ik}g_{kl}\xi^l)(\partial/\partial x^i) = (\delta^i_l\xi^l)(\partial/\partial x^i) = \xi^i(\partial/\partial x^i)$.

The metric tensor g establishes an isomorphism between tangent and cotangent spaces on M with its inverse mapping g^{-1} *locally given by* $g^{-1}(dx^i, dx^j) = g^{ij}$. *It likewise establishes an isomorphism between the spaces* $\mathcal{X}(M)$ *and* $\mathcal{D}^1(M)$ *of tangent vector fields and cotangent vector fields (1-forms). Together with the automorphism of the structure group which maps every group element to its inverse, it also establishes an isomorphism between the tangent bundle* $T(M)$ *and the cotangent bundle* $T^*(M)$.

The last of these statements is rather obvious. These isomorphisms extent by linearity to isomorphisms between tensors, tensor fields and tensor bundles of types (r, s) with $r + s$ fixed. In coordinate neighborhoods the corresponding mappings are obtained by **raising and lowering of tensor indices**, for example, with some n,

$$t^{i_1\ldots i_{r+1}}_{j_1\ldots j_{s-1}} = g^{i_{r+1}k} t^{i_1\ldots i_r}_{j_1\ldots j_{n-1}kj_n\ldots j_{s-1}}, \quad t^{i_1\ldots i_{r-1}}_{j_1\ldots j_{s+1}} = g_{j_1k} t^{i_1\ldots i_{n-1}ki_n\ldots i_{r-1}}_{j_2\ldots j_s}. \tag{9.13}$$

Recall the convention (4.4) that in a tensor of type (r, s) all contravariant indices precede all covariant indices. If the tensor t is not symmetric, the order of rising or lowering of indices must carefully be respected. Hence, in generalized Riemannian manifolds instead of types (r, s) of tensors only their **rank** $r + s$ matters. Moreover, the metric tensor g provides the following inner product in the tensor space of rank $r + s$:

$$(t \,|\, u) = t^{i_1\ldots i_r}_{j_1\ldots j_s} g^{j_1 l_1} \cdots g^{j_s l_s} g_{i_1 k_1} \cdots g_{i_r k_r} u^{k_1\ldots k_r}_{l_1\ldots l_s}. \tag{9.14}$$

In the case of an indefinite Riemannian metric it is an indefinite inner product.

9.2 Homogeneous Manifolds

Among the examples of principal fiber bundles at the end of Sect. 7.1, the notion of homogeneous manifold was introduced as the quotient space G/H of a Lie group G over its closed Lie subgroup H with the canonical projection $\pi : G \to G/H$.

The homogeneous manifold forms the base space of the principal fiber bundle $(G, G/H, \pi, H)$.

A subgroup H of a Lie group G is not automatically a Lie subgroup. According to the definition in Sect. 6.3, (H, Id) must also be an embedded submanifold of G, and that depends on the topology and differentiable structure introduced in H. However,

if a manifold structure of the subgroup H of the Lie group G exists which makes (H, Id) into a (second countable) submanifold of G, it is unique, and H is a Lie subgroup of G.

Proof Since H has a manifold structure, it has a tangent space $T_e(H)$ at the identity $e \in H \subset G$. By left translations in G, pushed forward to tangent vectors, a (dim H)-dimensional involutive distribution D on G is defined (Sect. 3.6). Clearly, at any $h \in H$ all tangent vectors of D_h are in the tangent space $T_h(H)$. The connected component H_e is an integral manifold of D on G through e. Indeed: let dim $H = k$, and let for some $h \in H$ the tangent space $T_h(H)$ be not contained in D_h. Then, there would be at least $k+1$ curves through h in H, which are smooth in G and have at least $k+1$ linearly independent tangent vectors in G. Left translating h back to e, the mapping $(x^1, \dots, x^{k+1}) \mapsto (F_1(x^1), \dots, F_{k+1}(x^{k+1}))$ could be completed by $m-k-1$, $m = \dim G$, further linearly independent curves to a diffeomorphism F of $\mathbb{R}^m$ to some neighborhood U of e in G. $F^{-1} \circ \mathrm{Id}_H$ would be a smooth immersion of $H \cap U$ into $\mathbb{R}^m$ containing some $\mathbb{R}^{k+1}$. This is not possible: H is second countable, and hence $H \cap U$ is an at most countable union of sets of dimension k. Now, with the left translation l_h, $h \in H$ in G, (H, l_h) is the uniquely defined (p. 78) smooth integral manifold of D through h and l_h is a (smooth) left translation of H. Moreover, $(h', h) \mapsto h'h^{-1}$ is smooth in G and hence smooth in (H, Id). Since Id is injective, H is an embedded submanifold and a uniquely defined Lie subgroup of G. □

Hence, a subgroup of a Lie group can only in a unique way (with a uniquely defined total atlas) be a Lie subgroup under the identity mapping (inclusion mapping). This answers uniqueness, it does not yet answer the question under which conditions a submanifold structure exists making (H, Id) into a Lie subgroup of G.

From the above proof it can be seen that, if H is an embedded submanifold of G in the relative topology from G, then it must be closed in G in this relative topology. Conversely, let G be a second countable Lie group, and let H be a subgroup of G closed in the relative topology from G. That implies that if U_n is a countable base of the topology of G, then $H \cap U_n$ is a countable base of the relative topology of H, and H is second countable. Let V be the closure of a coordinate neighborhood U of $e \in G$ being homeomorphic to some closed ball of $\mathbb{R}^m$. Then, $H \cup V$ is homeomorphic to a closed subset of this ball being complete in the metric of $\mathbb{R}^m$ and being a union of at most countably many closed subsets of H. Hence, $H \cup V$ is a Baire space, and at least one of its subsets has an open interior U_H. Translating U_H with all l_h, $h \in H$ yields an embedding of H in G as a submanifold with the relative topology. (A closed subset of $\mathbb{R}^2$ *in the relative topology* excludes the possibility of Example 5 on p. 76.)

(H, Id) *is a Lie subgroup of the (second countable) Lie group G, iff it is a subgroup closed in the relative topology from G.*

Hence, if G is a (second countable) Lie group and H is a closed subgroup of G (in the relative topology), then $(G, G/H, \pi, H)$ with the uniquely defined manifold structure of H as above is a principal fiber bundle, and G/H is a homogeneous manifold.

Let G be a Lie group acting smoothly from the left on a manifold M by $R: G \times M \to M$. Then, for a fixed $g \in G$, $R(g, \cdot) = R_g : M \to M$ is obviously a diffeomorphism of M into itself. Pick $m \in M$, then $G_m = \{g \in G \,|\, R_g(m) = m\}$ is a closed (as the preimage of the closed set $\{m\}$) subgroup of G, the **isotropy group** at m. (It may be trivial: $G_m = \{e\}$.) G_m acts also from the left on M by $R_m = R|_{G_m}$ and leaves m on place as a fixed point. By linearization, this action can be pushed forward to a mapping $R_{m*} : G_m \to \mathrm{Aut}(T_m(M))$ of the isotropy group into the automorphism group of the tangent space at m, yielding a Lie group representation of G_m in the vector space $T_m(M)$. (Exercise: Show that R_{m*} is a smooth homomorphism of Lie groups.) The image of the homomorphism R_{m*} is called the **linear isotropy group** at m.

Consider as an example the Lie group $O(n)$ of real orthogonal $(n \times n)$-matrices acting from the left on the unit sphere S^{n-1} of $\mathbb{R}^n$. The elements of $O(n)$ of the form

$$R = \begin{pmatrix} R' & 0 \\ 0 & 1 \end{pmatrix}, \tag{9.15}$$

where R' is an element of $O(n-1)$ and the zeros denote zero column and row, are precisely the elements of the isotropy group at the south pole $s = (0, \ldots, 0, 1)$ of S^{n-1}, and $O(n-1)$ is a closed subgroup of $O(n)$ (exercise). Since $T_s(S^{n-1}) \approx \mathbb{R}^{n-1}$, the linear isotropy group at s is again $O(n-1)$ in this case.

Now, let R be a transitive action of the Lie group G on M (p. 206), and let again $m \in M$. Then, the mapping

$$\tilde{R} : G/G_m \to M : gG_m \mapsto \tilde{R}(gG_m) = R_g(m) \tag{9.16}$$

is correctly defined, since for every $g' \in G_m$ one has $R_{gg'}(m) = R_g(R_{g'}(m)) = R_g(m)$. $\tilde{R}$ is onto and one–one: it is surjective since G acts transitively, and it is easily shown that $\tilde{R}(g_1 G_m) = \tilde{R}(g_2 G_m)$ implies $g_2^{-1} g_1 \in G_m$ and hence $\tilde{R}$ is injective. It can even be shown [1] that it is a diffeomorphism and hence M and G/G_m are equivalent homogeneous manifolds.

Returning to the above example $G = O(n)$, $M = S^{n-1}$, $G_s = O(n-1)$, the quotient space $O(n)/O(n-1)$ consists of the cosets $RO(n-1)$, $R \in O(n)$ of the subgroup $O(n-1)$ in $O(n)$. There is the diffeomorphism $O(n)/O(n-1) \to S^{n-1} : RO(n-1) \mapsto Rs$, $R \in O(n)$, and $O(n)/O(n-1)$ and S^{n-1} are equivalent homogeneous manifolds. As was shown on p. 194, $O(n)$ consists of two connected components for $\det R = \pm 1$, and $O(n)_e = SO(n)$. Therefore one has also an equivalence of the former two homogeneous manifolds with $SO(n)/SO(n-1)$, which was already considered in Sect. 2.6. Consider $SO(3)$, the group of all rotations of $\mathbb{R}^3$. If one fixes the south pole of the unit sphere S^2 in $\mathbb{R}^3$, then there remain the rotations with the rotation axis through the south pole fixed, which form the group $SO(2)$ of rotations of the equator of the sphere S^2 the elements of which are just given by the angle of rotation. Any coset $RSO(2)$, $R \in SO(3)$ consists of a rotation with the south pole fixed and a subsequent free rotation, which possibly moves the south pole to any point of the sphere S^2. This can also be achieved by first rotating the south pole to the new position and then making a rotation with that position fixed. (R and the element of $SO(2)$ related to the new axis are of course different in this case, since $SO(3)$ is non-Abelian.) Hence, all cosets of $SO(3)/SO(2)$ are in one–one and onto correspondence with all points of the sphere S^2. All those points can be obtained by an $SO(3)$-rotation Rs of the south pole $\boldsymbol{s}$.

Let G be a (second countable) Lie group and let H be its invariant closed subgroup. Then the homogeneous manifold G/H with its quotient group structure is a Lie group.

Since G/H is a manifold, it is only to check that the group operations are smooth. This is straightforward by use of local coordinates.

Now, let G be a (second countable) Lie group and let H be a compact subgroup (in the relative topology). Then, H is a Lie subgroup and G/H is a homogeneous manifold of cosets $x = g_xH$, $g_x \in G$ (g_x defines uniquely a coset x, but not vice versa) on which G acts transitively from the left by $G \times G/H \to G/H : (g, g_xH) \mapsto gg_xH$. Pick $x \in G/H$. The isotropy group at x is $G_x = \{g \in G \mid g(g_xH) = g_xH\}$ which implies $g_x^{-1}gg_xH = H$ and, since cosets are disjoint, $g_x^{-1}gg_x \in H$. It is not difficult to see (exercise) that together with H also G_x and the linear isotropy group R_{x*} of transformations of the tangent space $T_x(G/H)$ are compact. In a compact group, a finite invariant measure (Haar's measure) can be introduced. Take any positive scalar product in the vector space $T_x(G/H)$ and average it over the invariant measure with respect to the transformations of R_{x*}. The result is an invariant scalar product $g_x(X, Y) = g_{ij}(x)X^iY^j = g_{ij}(x)(g_*X)^i (g_*Y)^j = g_x(g_*X, g_*Y)$ for all $X, Y \in T_x(G/H)$ and all $g_* \in R_{x*}$. Since G acts transitively on G/H, for every $x' \in G/H$ there is $g_{xx'} \in G$ so that $x' = g_{xx'}x$. Then,

$$g_{x'}(X, Y) = g_x(g_{xx'*}^{-1}X, g_{xx'*}^{-1}Y) \tag{9.17}$$

is easily shown to be independent of the special choice of $g_{xx'}$. It defines an **invariant metric** on the homogeneous manifold G/H and makes this manifold into a **homogeneous Riemannian manifold**.

In the above example this is just the ordinary metric on the sphere S^{n-1} which *in orthogonal local coordinates* is given by the unit matrix g.

Let again G be a Lie group and consider the product manifold $G \times G$ (not the direct product of groups) with the group multiplication $(g_1, g_2)(g_1', g_2') = (g_1g_1', g_2g_2')$. It is easily seen that this makes $G \times G$ into another Lie group. Consider its action on G from the left as $(G \times G) \times G \to G : ((g_1, g_2), g) \mapsto$

$g_1 g g_2^{-1}$, and consider the corresponding isotropy group at $e \in G$. It is given as $\{(g_1, g_2) \,|\, g_1 e g_2^{-1} = e\}$ implying $g_1 = g_2$. Hence, the isotropy group $(G \times G)_e$ is the diagonal $D = \{(g_1, g_1) \,|\, g_1 \in G\} \approx G$. Moreover, $G \times G$ acts transitively on G, since for arbitrary $g, g' \in G$ there is $g' = g' g g^{-1} = (g', g) \cdot g$. Hence, $(G \times G)/D \to G : (g_1, g_2)D \mapsto g_1 e g_2^{-1}$ is a diffeomorphism of manifolds. By choosing $(g_2^{-1}, g_2^{-1}) \in D$ for a representative of the coset $(g_1, g_2)D$ one finds $(g_1 g_2^{-1}, e)$ which together with the coset is mapped to $g_1 g_2^{-1}$ by the above diffeomorphism, so that the diffeomorphism is also a homomorphism of groups and hence it is a Lie group isomorphism, $(G \times G)/D \approx G$. Consequently, if the Lie group G and hence also D is compact, $G \approx (G \times G)/D$ itself can be provided with an invariant metric and thus be made into a homogeneous Riemannian manifold.

Consider the Lie algebra $\mathfrak{g} \approx T_e(G)$ of the Lie group G and introduce on it the symmetric 2-form

$$\kappa(X, Y) = \mathrm{tr}(\mathrm{ad}(X) \circ \mathrm{ad}(Y)), \quad X, Y \in \mathfrak{g}. \tag{9.18}$$

It is called the **Killing form** or Killing–Cartan form of $\mathfrak{g}$. Since the elements of the Lie algebra $\mathfrak{g}$ are left invariant vector fields on G, (9.18) is clearly an invariant 2-form:

$$\kappa(X, Y) = \kappa(l^e_{g*} X, l^e_{g*} Y) \quad \text{for all } g \in G, \tag{9.19}$$

where l^e_{g*} is the left translation of tangent vectors from e to g in G like in Sect. 6.1. Recall from that section, that after choosing a base $\{X_1, \ldots, X_m\}$ in $\mathfrak{g}$, one gets $\mathrm{ad}(X_i)X_j = [X_i, X_j] = \sum_{k=1}^m c_{ij}^k X_k$ with the structure constants c_{ij}^k of the Lie group G. Thus, $\mathrm{ad}(X_i)_j^k = c_{ij}^k$ is cast into an $(m \times m)$-matrix acting on the m-dimensional vector space $\mathfrak{g} = \mathrm{span}_{\mathbb{R}}\{X_1, \ldots, X_m\}$, the composition of $\mathrm{ad}(Y)$ with $\mathrm{ad}(X)$ becomes the matrix multiplication and the trace becomes the matrix trace. From the theory of Lie algebras [2] and citations therein it is known that *the Killing form κ is non-degenerate, iff the Lie algebra $\mathfrak{g}$ is semi-simple; it is negative definite, iff $\mathfrak{g}$ is moreover the Lie algebra of a compact Lie group G*. Hence,

$$g(X, Y) = -\kappa(X, Y) \tag{9.20}$$

is an invariant metric making a compact semi-simple Lie group into a homogeneous Riemannian manifold, and it is an invariant indefinite metric making a non-compact semi-simple Lie group into a generalized homogeneous Riemannian manifold.

In the above considered simple case $G = O(n)$, the Lie algebra is $\mathfrak{g} \approx \mathbb{R}^{n-1}$, and after introducing a standard orthonormal base in $\mathbb{R}^{n-1}$, $g(X_i, X_j) = g_{ij}$ becomes the unit matrix, related to standard orthogonal local coordinates on the sphere S^{n-1}.

In physics, a closed finite piece of a Riemannian manifold with fixed time-independent distances between all of its points is called a **rigid body**. The peculiarity of a homogeneous manifold is that a piece of it as a rigid body can move through it without deformation. For that reason, after Riemann's habilitation talk where he introduced his revolutionary new concept of geometry, many scholars including Helmholtz thought that physical relevancc should be restricted to homogeneous manifolds, because the free motion of rigid bodies in space was still a dogma based on Kant's absolute space and time. At least after Einstein's theory of relativity it is known that rigid bodies are an abstraction from the real world, only possible in the limit of infinite velocity of light. Nevertheless, as this limit is often a very good approximation, homogeneous manifolds may continue to have relevance in kinematics and classical physics. Besides, Lagrangian Grassmannians are important homogeneous manifolds forming subspaces of the phase space of classical mechanics.

9.3 Riemannian Connection

Further on, M is a generalized Riemannian manifold with the metric form g. Like any tensor field, g can be considered as a section in a tensor bundle over M which is associated with the frame bundle $(L(M), M, \pi, Gl(m, \mathbb{R}))$ as its principal fiber bundle. In Sect. 7.7, linear connections on M were introduced as connections on $L(M)$, and covariant derivatives $\nabla_X t$ of tensor fields t in the direction of the tangent vector X as well as covariant differentials ∇t were associated with linear connections. A linear connection Γ on M is called a **metric connection**, if $\nabla g = 0$ on M, a metric connection is called a **Riemannian connection**, or **Levi–Civita connection**, if it is torsion free, $T = 0$ on M. This is subject of the **Fundamental Theorem of Riemannian Geometry**:

Every generalized Riemannian manifold allows for exactly one Riemannian connection. It is defined by the following expression for $\nabla_X Y$ valid for every $X, Y, Z \in \mathfrak{g}$:

$$\begin{aligned} 2g(\nabla_X Y, Z) &= Xg(Y, Z) + Yg(X, Z) - Zg(X, Y) \\ &\quad - g(X, [Y, Z]) - g(Y, [X, Z]) + g(Z, [X, Y]), \end{aligned} \tag{9.21}$$

where the first line on the right hand side is understood according to (7.26).

A Riemannian connection is said to define a **pseudo-Riemannian geometry** on a generalized Riemannian manifold with indefinite metric tensor, it is said to define a **Riemannian geometry** on a Riemannian manifold. A metric tensor defines uniquely not only a (possibly indefinite) metric on M, but also a (pseudo-) Riemannian geometry. A tensor bundle or more generally any vector bundle associated with $L(M)$ with a Riemannian connection is called a **Riemannian vector bundle** and g is called a **Riemannian structure** on it.

Put in local coordinates $X = \partial/\partial x^i$, $Y = \partial/\partial x^j$ and $Z = \partial/\partial x^k$ into (9.21). From (3.19) it is immediately seen that the second line of (9.21) vanishes in this case. Recall $g(\partial/\partial x^i, \partial/\partial x^j) = g_{ij}$ and, from (7.42), $\nabla_{\partial/\partial x^i}(\partial/\partial x^j) = \sum_l \Gamma^l_{ij}(\partial/\partial x^l)$. One readily obtains

$$\sum_l g_{lk}\Gamma^l_{ij} = \frac{1}{2}\left(\frac{\partial g_{jk}}{\partial x^i} + \frac{\partial g_{ik}}{\partial x^j} - \frac{\partial g_{ij}}{\partial x^k}\right) = \Gamma_{ikj}. \tag{9.22}$$

The last expression defines new **Christoffel symbols** for the Riemannian connection. For historical reasons they are also called Christoffel symbols of the first kind while the general Christoffel symbols introduced in Sect. 7.9 are called Christoffel symbols of the second kind. (Recall that Christoffel symbols are not tensors.) For reference, some properties of Christoffel symbols *valid for Riemannian connections* are given which are obvious from (9.22):

$$\Gamma^k_{ij} = \frac{1}{2}g^{kl}\left(\frac{\partial g_{il}}{\partial x^j} + \frac{\partial g_{jl}}{\partial x^i} - \frac{\partial g_{ij}}{\partial x^l}\right) = \Gamma^k_{ji}, \tag{9.23}$$

$$\Gamma_{ikj} = \Gamma_{jki}, \quad \frac{\partial g_{ij}}{\partial x^k} = \Gamma_{ijk} + \Gamma_{jik}. \tag{9.24}$$

Proof of the Fundamental Theorem Let a Riemannian connection be given and let ∇ be the corresponding covariant differential. Then, $\nabla g = 0$ and $\langle T, X \wedge Y\rangle = \nabla_X Y - \nabla_Y X - [X, Y] = 0$ (see (7.31)). From (7.27), $0 = (\nabla g)(Y, Z; X) = \nabla_X g(Y,Z) - g(\nabla_X Y, Z) - g(Y, \nabla_X Z) = Xg(Y,Z) - g(\nabla_X Y, Z) - g(Y, \nabla_Z X) - g(Y, [X,Z])$, since g is bilinear and $g(Y,Z)$ is a real function (see (7.26)). Likewise, $0 = \nabla_Y g(X,Z) - g(\nabla_Y X, Z) - g(X, \nabla_Y Z) = Yg(X,Z) - g(\nabla_X Y, Z) - g(Z, [Y,X]) - g(X, \nabla_Y Z)$ and $0 = \nabla_Z g(X,Y) - g(\nabla_Z X, Y) - g(X, \nabla_Z Y) = Zg(X,Y) - g(\nabla_Z X, Y) - g(X, \nabla_Y Z) - g(X, [Z,Y])$. Adding the first two relations and subtracting the last one yields (9.21).

Conversely, assume that (9.21) is valid. Since g is non-degenerate, (9.21) defines the action of ∇_X on tangent vector fields Y uniquely, its action on functions is understood according to (7.26) by definition. Hence, according to Sect. 7.7 the action of ∇_X on tensor fields is uniquely defined, if it is a derivative. Due to multi-linearity of the right hand side of (9.21), (7.18) is readily fulfilled, and the validity of (7.19) is easily checked. Hence, (9.21) defines a covariant derivative with Christoffel symbols (9.23) for which the demanded transformation properties (7.38) may straightforwardly calculated from the latter expression. This shows that (9.21) defines uniquely a linear connection on M. With (9.24) and (7.43), $\nabla g = 0$ is straightforwardly calculated, and hence Γ is a metric connection. It is torsion free which follows most easily from (7.47) and (9.23). □

The reader should perform the straightforward calculations of this proof as an exercise.

Useful relations are obtained from

$$g^{ij}\Gamma^k_{ij} = \frac{1}{2}g^{ij}g^{kl}\left(\frac{\partial g_{il}}{\partial x^j} + \frac{\partial g_{jl}}{\partial x^i} - \frac{\partial g_{ij}}{\partial x^l}\right) = g^{il}g^{kj}\frac{\partial g_{ij}}{\partial x^l} - \frac{1}{2}g^{kl}g^{ij}\frac{\partial g_{ij}}{\partial x^l}.$$

From $g^{kj}g_{ij} = \delta^k_i$ it follows that $g^{kj}\partial g_{ij}/\partial x^l = -g_{ij}\partial g^{kj}/\partial x^l$, and hence the first term of the displayed result is $-\partial g^{kl}/\partial x^l$. Since (g^{ij}) is the matrix inverse to (g_{ij}) the element g^{ij} is the minor of g_{ij} divided by det g, or det $g\, g^{ij} = \partial \det g/\partial g_{ij}$ and hence

$$\frac{\partial \det \mathrm{g}}{\partial x^l} = \frac{\partial \det \mathrm{g}}{\partial g_{ij}}\frac{\partial g_{ij}}{\partial x^l} = \det g\, g^{ij}\frac{\partial g_{ij}}{\partial x^l} = -\det g\, g_{ij}\frac{\partial g^{ij}}{\partial x^l}. \tag{9.25}$$

The last equality is just a special case of the preceding consideration. Now, one has

$$g^{ij}\Gamma^k_{ij} = -\frac{\partial g^{kl}}{\partial x^l} - \frac{1}{2}\frac{g^{kl}}{\det g}\frac{\partial \det \mathrm{g}}{\partial x^l} = -\frac{1}{\sqrt{|\det \mathrm{g}|}}\frac{\partial\left(\sqrt{|\det \mathrm{g}|}g^{kl}\right)}{\partial x^l}. \tag{9.26}$$

On the other hand, $\nabla g = 0$ implies $g^{kl}_{;l} = 0$, and with (7.44, 7.45) this means $-\partial g^{kl}/\partial x^l = g^{ml}\Gamma^k_{ml} + g^{km}\Gamma^l_{ml}$, which combines with (9.26) to

$$\Gamma^i_{ji} = \frac{1}{\sqrt{|\det \mathrm{g}|}}\frac{\partial\sqrt{|\det \mathrm{g}|}}{\partial x^j}. \tag{9.27}$$

The latter result yields for the **divergence** $C_{1,1}\nabla X$ of a vector field X

$$X^i_{;i} = \frac{\partial X^i}{\partial x^i} + \Gamma^i_{ji}X^j = \frac{1}{\sqrt{|\det g|}}\frac{\partial(\sqrt{|\det \mathrm{g}|}X^i)}{\partial x^i}. \tag{9.28}$$

A similar expression is obtained for the divergence $t^{ij}_{;j}$ of an alternating tensor t.

Let local coordinates be given in $L(M)$ as before (7.23), and let it be provided with a Riemannian connection with connection form ω. Then, the first structure equation (7.23) for vanishing torsion reads (as previously, the index α of a coordinate neighborhood is only occassionally used for the sake of clarity at local forms and is dropped here)

$$d\theta^i - \sum_j \theta^j \wedge \omega^i_j = 0. \tag{9.29}$$

In view of (7.43), the condition for ω to be a metric connection reads $0 = \nabla g = (\partial g_{ij}/\partial x^k - \Gamma^l_{ki}g_{lj} - \Gamma^l_{kj}g_{il})dx^i \otimes dx^j \otimes dx^k = (dg_{ij} - g_{lj}\omega^l_i - g_{il}\omega^l_j)\, dx^i \otimes dx^j$, where in the last equality the definition of the Christoffel symbols by the local connection form given on p. 241 was used. Hence, this condition reads

$$dg_{ij} - g_{lj}\omega^l_i - g_{il}\omega^l_j = 0, \tag{9.30}$$

where dg_{ij} is the ordinary differential of the component of the symmetric 2-form g. These last two relations yield an equivalent formulation of the **Fundamental Theorem of Riemannian Geometry**:

Given a generalized Riemannian manifold M with metric form g and given m linearly independent 1-forms θ^i, $i = 1, \ldots, m$ on every coordinate neighborhood U

of M, there exists a unique set of m^2 1-forms ω^i_j solving the equations (9.29, 9.30) *and forming local connection forms of a Riemannian connection. The metric form may then be expressed as $g_{kl}\theta^k \otimes \theta^l$.*

Use coordinate lines according to the integral curves of m standard horizontal vector fields X_i, $\langle\theta^k, X_i\rangle = \delta^k_i$ (p. 234). Then, $g_{ij} = g(X_i, X_j) = g_{kl}\langle\theta^k, X_i\rangle\langle\theta^l, X_j\rangle$. In the case of a Riemannian manifold (that is with a positive definite metric form) one may choose linearly independent standard horizontal vector fields for which $g_{ij} = \delta_{ij}$ and hence $ds^2 = \sum_i(\theta^i)^2$. Then, (9.30) yields a skew-symmetric local connection form: $\omega^i_j + \omega^j_i = 0$.

Since $g_{ij}(x)$ is a smooth function, the exterior derivative of its differential vanishes: $d \wedge dg_{ij} = 0$. The exterior derivative of (9.30) yields $0 = g_{kj}\omega^k_l \wedge \omega^l_i + g_{lk}\omega^k_j \wedge \omega^l_i + g_{lj}d\omega^l_i + g_{kl}\omega^k_i \wedge \omega^l_j + g_{ik}\omega^k_l \wedge \omega^l_j + g_{il}d\omega^l_j$. The second and the fourth terms cancel, and, after some renaming of summation indices, $(\omega^l_k \wedge \omega^k_i + d\omega^l_i)g_{lj} + g_{il}(\omega^l_k \wedge \omega^k_j + d\omega^l_j) = 0$ results. According to (7.23), the parentheses of the last relation contain the components of the curvature form $\Omega = d\omega + \omega \wedge \omega$, so that this relation reads

$$\Omega^l_i g_{lj} + g_{il}\Omega^l_j = 0. \tag{9.31}$$

$\Omega_{ij} = \Omega^l_i g_{lj}$ of a Riemannian connection (with possibly indefinite metric) is skew-symmetric.

Expressing (7.30) in local coordinates results in (exercise, see also the text of the next paragraph)

$$\Omega^i_j = \sum_{k<l} R^i_{jkl} dx^k \wedge dx^l, \tag{9.32}$$

where the curvature tensor R is given by (7.49). The **curvature tensor field** of a generalized Riemannian manifold,

$$R_{ijkl} = g_{im} R^m_{jkl}, \tag{9.33}$$

has the following properties:

$$\begin{aligned} R_{ijkl} &= -R_{jikl} = -R_{ijlk}, \\ R_{ijkl} &+ R_{iklj} + R_{iljk} = 0, \\ R_{ijkl} &= R_{klij}. \end{aligned} \tag{9.34}$$

The first line follows directly from (9.31, 9.32). To obtain the second line, first realize that (9.29) must hold locally for any linear independent system of 1-forms θ^i. Putting locally $\theta^i = dx^i$ yields $0 = \sum_j \omega^i_j \wedge dx^j$. take the exterior derivative of this relation and use it again to get $0 = \sum_j d\omega^i_j \wedge dx^j = \sum_j(\Omega^i_j - \sum_k \omega^i_k \wedge \omega^k_j) \wedge dx^j = \sum_j \Omega^i_j \wedge dx^j$. Insert (9.32) and obtain $\sum_{j<k<l} R^i_{jkl}\, dx^j \wedge dx^k \wedge dx^l = 0$ from which the second line follows. Interchanging in the second line i with j, subtracting

the result from the second line and observing the first line yields $2R_{ijkl} + R_{iklj} + R_{ljik} + R_{iljk} + R_{jkil} = 0$. Interchanging here ij with kl, subtracting and observing again the first line of (9.34) yields the third one.

Finally, exterior differentiation of $\Omega = d\omega + \omega \wedge \omega$ yields the Bianchi identities in the form $d\Omega = d\omega \wedge \omega - \omega \wedge d\omega = \Omega \wedge \omega - \omega \wedge \Omega$. This relation, which is still a relation of forms on the frame bundle, in the notation of (7.30) paired with tangent vectors X^* on $L(M)$, may be pulled back from the canonical section to the base manifold M as the same relation for the corresponding local forms on a trivializing coordinate neighborhood in M, which are paired with tangent vectors X on M as $\langle \Omega_\alpha, (\partial/\partial x^k) \wedge (\partial/\partial x^l)\rangle = R^i_{jkl} dx^j (\partial/\partial x^i)$ or $\Omega^i_{\alpha j} = \sum R^i_{jkl} dx^k \wedge dx^l$ (compare the text before (7.5) and before (7.14) as well as (7.46) and (9.32)). Insert this expression together with $\omega^i_{\alpha j}$ from p. 241 into the relation $d\Omega^i_{\alpha j} = -\omega^i_{\alpha n} \wedge \Omega^n_{\alpha j} + \Omega^i_{\alpha n} \wedge \omega^n_{\alpha j}$ and obtain

$$\sum \frac{\partial R^i_{jkl}}{\partial x^m} dx^m \wedge dx^k \wedge dx^l = \sum \left(-\Gamma^i_{mn} R^n_{jkl} + \Gamma^n_{mj} R^i_{nkl}\right) dx^m \wedge dx^k \wedge dx^l.$$

Complete the left hand side to a covariant derivative according to (7.43) and get

$$\sum R^i_{jkl;m}\, dx^m \wedge dx^k \wedge dx^l = -\sum \left(\Gamma^n_{mk} R^i_{jnl} + \Gamma^n_{ml} R^i_{jkn}\right) dx^m \wedge dx^k \wedge dx^l.$$

Now, this right hand side vanishes in the torsion free case, because according to (7.47) in this case the Christoffel symbols are symmetric in the lower indices. Hence, also the left hand side vanishes, which means that the corresponding alternating combination of $R^i_{jkl;m}$ in the last three subscripts must vanish. This is the sum with cyclic permutation minus the sum with anti-cyclic permutation of these subscripts. In view of the alternating dependence of R on its last two subscripts the six items can be combined into three. Thus,

$$R_{ijkl;m} + R_{ijlm;k} + R_{ijmk;l} = 0 \tag{9.35}$$

expresses the **Bianchi identities** for a Riemannian connection.

9.4 Geodesic Normal Coordinates

In this section, the linear connection on M is specialized step by step.

First a general manifold M with a linear connection is considered, that is, the frame bundle $L(M)$ with a linear connection Γ. In a coordinate neighborhood $U_\alpha \subset M$, the system of ordinary differential equations (7.51),

$$\frac{d^2x^i}{dt^2} + \sum_{jk} \Gamma^i_{jk} \frac{dx^j}{dt}\frac{dx^k}{dt} = 0, \quad i = 1, \ldots, m, \tag{9.36}$$

has the geodesics as solutions, which are the curves in M whose tangent vector is transported parallel to itself along the curve. As is well known from standard analysis, given initial conditions

$$x^i(0) = u^i, \quad \left.\frac{dx^i}{dt}\right|_0 = v^i, \tag{9.37}$$

there are a neighborhood $U \subset U_\alpha$ and two positive real numbers r, δ, so that for all $(u^i) \in \boldsymbol{U} = \boldsymbol{\varphi}_\alpha(U)$ and $(v^i) = \boldsymbol{v} \in \mathbb{R}^m$, $|\boldsymbol{v}| < r$ the system (7.51) has a unique solution

$$x^i = y^i(t, u^k, v^k) \quad \text{for } |t| < \delta, \tag{9.38}$$

where y^i depends smoothly on t, u^k, v^k. It may be thought of as a motion in time t through M passing at $t = 0$ through (u^k) with velocity vector (v^k). Rescaling the time by a factor c is equivalent to rescaling the velocity with $1/c$:

$$y^i(ct, u^k, v^k/c) = y^i(t, u^k, v^k) \quad \text{or} \quad y^i(ct, u^k, v^k) = y^i(t, u^k, cv^k) \tag{9.39}$$

In these relations, the left hand side is defined where the right hand side was defined in (9.38). Take $|c| < \delta$ and fix $(u^k) \in \boldsymbol{U}$. Then, with the first initial condition (9.37), the relation

$$x^i = y^i(1, u^k, v^k) \quad \text{with } u^i = y^i(0, u^k, v^k) = y^i(1, u^k, 0) = u^i \tag{9.40}$$

provides a mapping $(v^k) \mapsto (x^i)$ from a neighborhood of the origin of $T_u(M) \approx \mathbb{R}^m$ onto a neighborhood of $u \in M$ with the origin of $T_u(M)$ mapped to u. Indeed, now with the second initial condition (9.37),

$$\left.\frac{\partial y^i(1, u^k, v^k)}{\partial \boldsymbol{v}}\right|_{\boldsymbol{v}=0} \cdot \boldsymbol{v} = \left.\frac{\partial y^i(1, u^k, tv^k)}{\partial t}\right|_{t=0} = \left.\frac{\partial y^i(t, u^k, v^k)}{\partial t}\right|_{t=0} = v^i$$

and hence $\left(\partial y^i(1, u^k, v^k)/\partial v^j\right)_{\boldsymbol{v}=0} = \delta^i_j$ so that the mapping is regular at $\boldsymbol{v} = 0$ and provides a bijection of some neighborhood of $\boldsymbol{v} = 0$ onto a neighborhood of $u \in M$. This latter neighborhood is a coordinate neighborhood of M with coordinates v^i whose transition to the original coordinates x^i is defined by the first relation (9.40). The coordinates v^i are called **geodesic normal coordinates** at u or in short normal coordinates. Of course, they are determined up to the choice of a base in $T_u(M)$, that is up to a non-degenerate linear transformation. (So far, M is a linear connection space, a metric in M and angles in $T_u(M)$ were not yet introduced.) For every $u \in M$ and every fixing of a base in $T_u(M)$ there is a neighborhood $U(u)$ and a local coordinate system in $U(u)$ of geodesic normal coordinates at u; it is called a **normal coordinate neighborhood** of u.

Choose a coordinate neighborhood $U_\alpha \subset M$ (with respect to coordinates x^i), a point $u \in U_\alpha$ and a base $\{X_{iu} \mid i = 1, \ldots, m\}$ in $T_u(M)$, and consider a tangent vector $X_u = \sum_i \bar{v}^i X_{iu}$. The points with geodesic normal coordinates $(v^i = t\bar{v}^i)$ form a

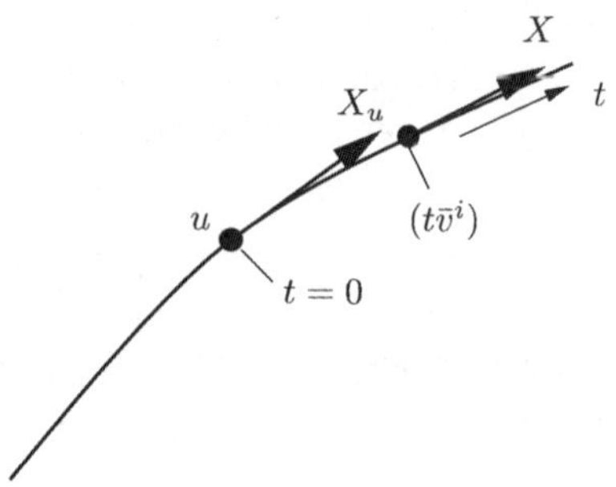

Fig. 9.1 A geodesic through the point u with the corresponding geodesic normal coordinates

geodesic through the point $u \in M$ (at $t = 0$) which has X_u as its tangent vector at u. If this tangent vector is parallel transported along the geodesic, it remains tangent to it in all points of the geodesic (Fig. 9.1). From the above analysis it is clear that given a point $u' \in U(u)$ with normal coordinates v^i (with respect to u), the tangent vector X_u of a geodesic from u to u' is uniquely defined up to a scaling factor with an inverse scaling of the curve parameter t of the geodesic. Such a rescaling does of course not change the geodesic curve itself, it changes only its parametrization.

In a normal coordinate neighborhood $U(u)$ of $u \in M$, there is for every point $u' \in U(u)$ a uniquely defined geodesic curve connecting u and u'.

For small enough $\rho > 0$, there is a **ρ-ball neighborhood**

$$U_\rho(u) = \left\{ x \,\middle|\, \sum_{i=1}^{m} (v^i(x))^2 < \rho^2 \right\}, \quad \boldsymbol{U}_\rho(u) = \left\{ \boldsymbol{x} \,\middle|\, \sum_{i=1}^{m} (v^i(\boldsymbol{x}))^2 < \rho^2 \right\} \tag{9.41}$$

of u contained in the normal coordinate neighborhood $U(u)$ $(\boldsymbol{U}_\rho(u) = \boldsymbol{\varphi}_\alpha(U_\rho(u)))$. Moreover, if $B_\delta(0)$ is the open ball of radius δ centered at the origin of the Euclidean space $\mathbb{R}^m$, then there is $\delta > 0$ so that through every point $u' \in U_\rho(u)$ and for every $\boldsymbol{v} \in B_\delta(0)$ there is a unique geodesic through u' with tangent vector $\boldsymbol{v}$ at u' and given by $x^i = y^i(t, u'^k, v^k)$, $|t| < 2$. In summary, there is a mapping

$$\psi : \boldsymbol{U}_\rho(u) \times B_\delta(0) \to \boldsymbol{U}_\rho(u) \times \boldsymbol{U}_\alpha : (u'^k, \boldsymbol{v}) \mapsto \psi(u'^k, \boldsymbol{v}) = (u'^k, y^k(1, u'^k, v^k)). \tag{9.42}$$

For every $u \in U_\alpha$ there are such positive numbers ρ and δ (depending on u). Of basic importance for the following is the statement visualized in Fig. 9.2:

For any point u in a linear connection manifold M there exists a neighborhood W of u such that every point $u' \in W$ has a normal coordinate neighborhood $U(u')$ that contains W.

Proof By the analysis after (9.39), the Jacobian of the mapping ψ is $[\partial(u'^k, y^k)/\partial(u'^k, v^k)]_{(u,0)} = 1$. Hence, the differential $\psi_*^{(u,0)}$ is a linear isomorphism, and from the inverse function theorem (Sect. 3.5, case 3) it follows that there is a neighborhood V of $(u, 0) \in \boldsymbol{U}_\rho(u) \times B_\delta(0)$ and a positive number $\epsilon < \delta$ so that the restriction

Fig. 9.2 See statement on the previous page

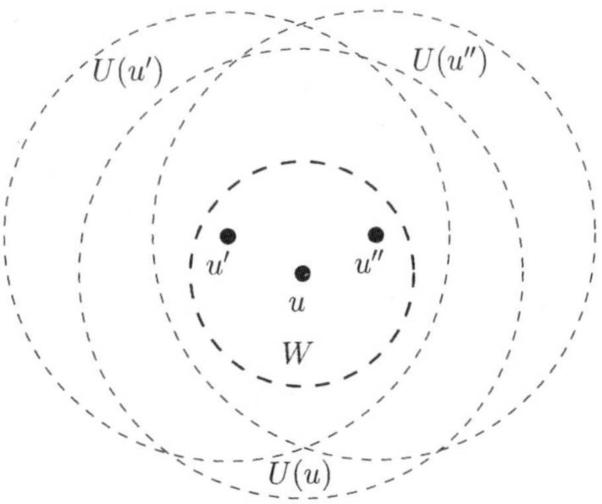

$\psi|_V : V \to U_\epsilon(u) \times U_\epsilon(u)$ is a diffeomorphism. For every point $u' \in U_\epsilon(u)$ take the set of vectors $B_{u'} = \{v \in B_\epsilon(0) | (u', v) \in V\}$. Then, the set of points $x \in U_\alpha$ with α-coordinates $x^i = y^i(1, u'^k, v^k)$, $v \in B_{u'}$ is contained in a normal coordinate neighborhood $U(u')$ of u' and is diffeomorphic to $U_\epsilon(u) = W$. □

An immediate consequence is that every pair of points of W can be connected by a geodesic.

Now, let the linear connection space W additionally be *torsion-free*. The normal coordinates along a geodesic through a given point u may be chosen $v^i = t\bar{v}^i$, hence $d^2v^i/dt^2 = 0$, with $t = 0$ at u, and for instance unit vectors $\bar{v}$, and the system (9.36) at u, expressed with the normal coordinates v^i instead of the x^i, reads $\sum_{jk} \Gamma^i_{jk}(0)\bar{v}^j\bar{v}^k = 0$, where now the Christoffel symbols $\Gamma^i_{jk}(v^l)$ refer to the normal coordinates at u. Since in the torsion-free case these symbols are symmetric in the lower indices in any coordinate system, it follows that $\Gamma^i_{jk}(0) = 0$ for all $i, j, k = 1, \ldots, m$.

At any point u of a torsion-free linear connection manifold M, by choosing geodesic normal coordinates at u the Christoffel symbols $\Gamma^i_{jk}(0)$ at u are made vanish.

This is again an indication that the Christoffel symbols do not form a tensor, for a tensor vanishing in some coordinate system would vanish in any coordinate system.

Another remarkable property of torsion-free linear connection manifolds is that *the local connection form of a torsion-free linear connection manifold M is uniquely defined by the curvature tensor.*

Proof Let $u \in M$ be any point and let $U(u)$ be a normal coordinate neighborhood of u. Fix a frame $(u, X_1, \ldots, X_m)$ at u and choose corresponding geodesic normal coordinates in the following way: $\sum_i \bar{v}^i X_i$, $\sum_i (\bar{v}^i)^2 = 1$, is a tangent vector at u in any direction given by the components $\bar{v}^i$ with respect to the fixed frame; $(v^i) = (t\bar{v}^i)$ are geodesic normal coordinates in $U(u)$ on geodesics having the tangent vector $\sum_i \bar{v}^i X_i$ at u. Build a frame field in $U(u)$ by transporting the chosen frame at u parallel (horizontally in $L(M)$) along geodesics. Tangent vectors to geodesics, expressed in the polar coordinates $\bar{v}^i, t$, $\sum_i (v^i)^2 = 1$ are $\partial/\partial t$ and have constant components $\bar{v}^i$ with respect to this frame field along geodesics through u. Since the frame field is spanned by horizontal vectors, m linearly independent vector fields X_j with constant components $\bar{v}^i_j$ extend to standard horizontal vector fields $B(\bar{\boldsymbol{v}}_j)$ on $L(M)$ (compare (7.7)). Hence, according to (7.24), the canonical 1-form θ and the connection 1-form ω are given as $\theta^i = \bar{v}^i dt + \bar{\theta}^i$, $\omega^i_j = \bar{\omega}^i_j$, where $\bar{\theta}^i$ and $\bar{\omega}^i_j$ depend on the $\bar{v}^i$ and on t but do not contain dt (annihilate tangent vectors to geodesics $\bar{\boldsymbol{v}} = \text{const.}$). Moreover, since at $t = 0$ a tangent vector in any direction is proportional to dt (away from $t = 0$ this holds only for vectors tangent to the constructed geodesics through u), $\bar{\omega}^i_j|_{t=0} = 0$ and $\bar{\theta}^i|_{t=0} = 0$.

In view of the vanishing torsion, the structure equations (7.23) read

$$d\theta^i + \sum_j \omega^i_j \wedge \theta^j = 0, \quad d\omega^i_j + \sum_k \omega^i_k \wedge \omega^k_j = \Omega^i_j. \tag{9.43}$$

Since each component Ω^i_j of the tensorial curvature form Ω is a horizontal 2-form, it can be expressed as $\Omega^i_j = \sum_{k<l} \bar{R}^i_{jkl}\theta^k \wedge \theta^l$. In fact, $\bar{R}^i_{jkl}$ is the (bijective) u-transformation of the curvature tensor $R^i_{jkl} = u^i_p \bar{R}^p_{qrs}(u^{-1})^q_j(u^{-1})^r_k(u^{-1})^s_l$ in the meaning of (7.30) and (7.6); the u-transformation is not to be confused with the point u above. The components of both tensors R and $\bar{R}$ are functions of the $\bar{v}^i$ and of t. Comparing in the equations (9.43) of forms only the terms proportional to dt yields

$$\left(d\bar{v}^i - \frac{\partial\bar{\theta}^i}{\partial t} + \sum_j \bar{\omega}^i_j \bar{v}^j\right) \wedge dt = 0, \quad \left(\frac{\partial\bar{\omega}^i_j}{\partial t} - \sum_{kl} \bar{R}^i_{jkl}\bar{v}^k\bar{\theta}^l\right) \wedge dt = 0. \tag{9.44}$$

(Note that $d\bar{\theta}^i = dt \wedge (\partial\bar{\theta}^i/\partial t) + \cdots = -(\partial\bar{\theta}^i/\partial t) \wedge dt + \cdots$.) Hence, the expressions in parentheses must vanish. A further differentiation of the first expression with respect to t (t and $\bar{v}^i$ are independent) and then insertion of the second results in

$$\frac{\partial^2\bar{\theta}^i}{\partial t^2} - \sum_{jkl} \bar{R}^i_{jkl}\bar{v}^j\bar{v}^k\bar{\theta}^l = 0. \tag{9.45}$$

Given R^i_{jkl} and hence $\bar{R}^i_{jkl}$, this last equation is in fact an ordinary differential equation for $\bar{\theta}^i$, which with the initial conditions $\theta^i|_{t=0} = 0$ and $(\partial\bar{\theta}^i/\partial t)_{t=0} = d\bar{v}^i + \sum_j \bar{v}^j\bar{\omega}^i_j|_{t=0} = d\bar{v}^i$ has a unique solution $\bar{\theta}^i(\bar{v}^k, t) = \sum_j \bar{\bar{\theta}}^i_j(\bar{v}^k, t)d\bar{v}^j + \bar{\bar{\theta}}^i_t(\bar{v}^k, t)dt$. (Recall that the components $\bar{\theta}^i$ are 1-forms, (9.45) is an equation of forms and contains an equation for every $\bar{\bar{\theta}}^i_j$ and every $\bar{\bar{\theta}}^i_t$.)

Finally, take any point in $U(u)$ with coordinates $(v^i) = (t\bar{v}^i)$. For every fixed $l = 1, \ldots, m$ there is a (non-unique) regular linear transformation w with $\bar{v}^j = \sum_k w^j_k \delta^k_l$. Inserting this into the first equation (9.44) yields

$$\sum_j \omega^i_j w^j_l = \frac{\partial \bar{\theta}^i}{\partial t} - d\bar{v}^i, \quad l = 1, \ldots, m,$$

which after matrix multiplication with w^{-1} from the right uniquely determines $\omega^i_j(\bar{v}^k, t)$ on $U(u)$. □

Now, let M be a Riemannian manifold and consider the *Riemannian geometry* provided by the Riemannian connection on M. Let $u \in M$ be any point. Modify the choice of the fixed frame at u from the above proof into an orthonormal frame with respect to the Riemannian metric tensor: $(X_{iu} \mid X_{ju}) = \delta_{ij}$. (Compare (9.4, 9.5).) Then, t acquires the meaning of the arc length of geodesics measured from the point u, and the frame field becomes a local section in the orthonormal frame bundle $L_O(M)$. With respect to the corresponding normal coordinates $v^i = t\bar{v}^i$, a tangent vector on M at u, that is, at $v = 0$ is $X_u = \partial/\partial t = \sum_j (\partial v^j/\partial t)(\partial/\partial v^j) = \sum \bar{v}^j(\partial/\partial v^j)$. By the above choice, $g_{ij}(u) = g(X_{iu}, X_{ju}) = \delta_{ij}$, or, with $\langle \theta^k, X_i \rangle = \delta^k_i$, $g = \sum_{i=1}^m \theta^i \otimes \theta^i$. Since, by definition of a Riemannian connection, g does not change upon parallel transport, the last expression holds on the whole normal coordinate neighborhood $U(u)$. (Caution: $\theta^i \neq dv^i$ in general on $U(u)$ and g is not $\sum_i dv^i \otimes dv^i$.) Moreover, for the local connection forms themselves in these normal coordinates $\omega^i_j + \omega^j_i = 0$ (see text after the second formulation of the Fundamental Theorem of Riemannian Geometry in the previous section). There are m standard horizontal vector fields $X_i, i = 1, \ldots, m$ whose values at point u are X_{iu} and which on $U(u)$ hence make up the above frame field, and dual to them there are m components θ^i of the canonical one form so that $\langle \theta^i, X_j \rangle = \delta^i_j$ (p. 234) yielding

$$ds^2 = g_{ij}\theta^i\theta^j = \sum_i (\theta^i)^2 \tag{9.46}$$

for the element of the arc length. The symmetric 2-form (9.46) is called the **first fundamental form** of a Riemannian geometry. It is obviously covariant under $O(m)$-transformations of the structure group of the orthonormal frame bundle $L_O(M)$. The corresponding natural definition of a **volume form** of a Riemannian geometry is

$$\tau = \theta^1 \wedge \cdots \wedge \theta^m = \sqrt{|\det g|}\, dx^1 \wedge \cdots \wedge dx^m, \tag{9.47}$$

where the first expression is invariant under $SO(m)$-transformations and only changes sign under $O(m)$-transformations with determinant -1 of the transformation matrix (and hence switching orientation of the frame), while the second expression holds in any local coordinate system as was shown for (5.84).

As in the previous proof on p. 316, $\theta^i = \bar{v}^i dt + \bar{\theta}^i$, where $\sum_i (\bar{v}^i)^2 = 1$, $\bar{\theta}^i|_{t=0} = 0$, $(\partial\bar{\theta}^i/\partial t)_{t=0} = d\bar{v}^i$ and $\bar{\theta}^i$ does not contain dt. The element of the arc length in $U(u)$ is obtained as

$$ds^2 = dt^2 + 2dt \sum_i \bar{v}^i \bar{\theta}^i + \sum_i (\bar{\theta}^i)^2.$$

However, $\sum_i \bar{v}^i \bar{\theta}^i|_{t=0} = 0$, and by the first equality (9.44), $(\partial/\partial t)(\sum_i \bar{v}^i \bar{\theta}^i) = \sum_i \bar{v}^i d\bar{v}^i + \sum_{ij} \bar{v}^i \bar{\omega}^i_j \bar{v}^j = 0$; the first sum vanishes because of the normalization of $(\bar{v}^i)$ and the second because of the skew-symmetry of ω^i_j. Hence, $\sum_i \bar{v}^i \bar{\theta}^i = 0$ on $U(u)$, and

$$ds^2 = dt^2 + \sum_i (\bar{\theta}^i)^2, \quad \bar{\theta}^i = t d\bar{v}^i + \sum_j \bar{\bar{\theta}}^i_j(\bar{v}^k, t) d\bar{v}^j,$$
$$\bar{\bar{\theta}}^i_j\Big|_{t=0} = 0, \quad \frac{\partial \bar{\bar{\theta}}^i_j}{\partial t}\Big|_{t=0} = 0. \tag{9.48}$$

In particular, (9.48) implies $(dt|\bar{\theta}^i) = 0$, $i = 1, \ldots, m$ on $U(u)$. Observe that $m-1$ linearly independent linear combinations of the m (linearly dependent because of the linear dependence of the $d\bar{v}^i$) cotangent vectors $\bar{\theta}^i$ span locally (infinitesimally) the hypersurface $t^2 = \sum_i (v^i)^2 = \text{const.}$, while dt is a cotangent vector on the geodesic through u. (Since with respect to normal coordinates $g_{ij} = \delta^i_j$, tangent vectors X with components X^i and cotangent vectors σ with components $\sigma_i = g_{ij} X^j = X^i$ are equivalent.)

In a Riemannian geometry, in every normal coordinate neighborhood of a point u, geodesics through u are orthogonal to hypersurfaces (hyperspheres) $\sum_i (v^i)^2 = \text{const.}$

It is easy now to prove that

in a Riemannian geometry M there exists at every point $u \in M$ a normal coordinate neighborhood W such that (i) *every point $u' \in W$ has a normal coordinate neighborhood that contains W and* (ii) *the geodesic curve that connects u and $u' \in W$ is the unique shortest curve in W connecting these two points.*

Proof (i) was already proved to hold true for any linear connection manifold. Hence, it holds in particular for the Riemannian connection of a Riemannian manifold.

To prove (ii), let $W = U_\epsilon(u)$ with sufficiently small ϵ be the neighborhood (9.41) of u with respect to normal coordinates v^i at u. Pick $u' \in W$, and let $C_0 : v^i = \bar{v}^i t$, $\bar{v}^i = \text{const.}$, $0 \le t \le t_0$ be the geodesic connecting u with u', t_0 being its arc length, that is, u' has normal coordinates $\bar{v}^i t_0$. Let $C : v^i = v^i(t)$ be any curve in W from u to u' with parameter t, without loss of generality chosen to be the geodesic arc length from u to the point $u'' \in C$ with normal coordinates $v^i(t)$. Then, the arc length of C is

$$s = \int_0^{t_0} ds = \int_0^{t_0} \Big(dt^2 + \sum_i (\bar{\theta}^i)^2\Big)^{1/2} \ge \int_0^{t_0} dt = t_0.$$

Hence, s cannot be shorter then t_0, and $s = t_0$ implies $\bar{\theta}^i = 0$ for $i = 1, \ldots, m$. In view of (9.48), this means that for $0 \leq t \leq t_0$

$$d\bar{v}^i + \sum_j \frac{\bar{\bar{\theta}}^i_j(\bar{v}^k, t)}{t} d\bar{v}^j = 0,$$

while $d\bar{v}^i$ is independent of t, and $\lim_{t\to 0}(\bar{\bar{\theta}}^i_j(\bar{v}^k, t)/t) = 0$ due to the initial conditions of (9.48). Hence, it follows $d\bar{v}^i = 0$ and $\bar{v}^i = $ const. for $0 \leq t \leq t_0$, which means that C is the unique geodesic C_0 connecting u and u'. □

In the normal coordinate neighborhood W of this theorem the distance (9.8) from u to u' equals the arc length of the unique geodesic connecting the two points. This is in general not true globally; for instance, on a sphere geodesics are the great circles, and for two points on a great circle not being antipodes there are two arcs of different length connecting them.

Sufficiently small ϵ-ball normal coordinate neighborhoods have important properties in relation to geodesics which are analogues of corresponding properties of balls of any radius in Euclidean space in relation to straight lines. This section is concluded with their consideration.

Let $U(u)$ be a normal coordinate neighborhood of u in a Riemannian geometry and let

$$S_\rho = \left\{ x \,\middle|\, \sum_{i=1}^{m} (v^i(x))^2 = \rho \right\} \tag{9.49}$$

be a hypersphere of radius ρ in $U(u)$. Then, (i) *there exists a positive number ϵ so that for any $\rho, 0 < \rho < \epsilon$, any geodesic curve $v^i(s)$ tangent to S_ρ at $s = 0$ is outside S_ρ for small values of the curve parameter s and has $s = 0$ as its only one point in common with S_ρ in a neighborhood of $s = 0$;* (ii) *there exists a positive ρ such that for any two points in the ρ-ball neighborhood $U_\rho(u)$ of u there is a unique geodesic in $U_\rho(u)$ connecting the two points.*

A neighborhood of u having the two properties of $U_\rho(u)$ of this theorem is called a **geodesic convex neighborhood**; the theorem states the existence of a geodesic convex neighborhood of every point in a Riemannian geometry.

Proof Consider the real function $F(v^1, \ldots, v^m) = (1/2)(\sum_i (v^i)^2 - \rho^2)$. Then, S_ρ is determined by $F(v^i(x)) = 0$. Let $C : v^i(s)$ be a geodesic through $u' \in S_\rho$ and being tangent to S_ρ at this point, with curve parameter s passing through zero at u'. Since it was already shown that a geodesic $C_0 : v^i = tv^i(s = 0)$ from u to u' is orthogonal to S_ρ at u', the latter condition amounts to saying that C is orthogonal to C_0 at u'. Denote tangent vectors to these geodesics at u' by X_C and X_{C_0}, respectively. Their orthogonality spells

$$0 = (X_{C_0} \,|\, X_C) = \sum_{i=1}^{m} v^i(s = 0) \frac{dv^i}{ds}\bigg|_{s=0}.$$

This implies

$$\left.\frac{dF(v^k(s))}{ds}\right|_{s=0} = \sum_i v^i(0)\left.\frac{dv^i}{ds}\right|_{s=0} = 0,$$

$$\left.\frac{d^2F(v^k(s))}{ds^2}\right|_{s=0} = \sum_i \frac{dv^i}{ds}\frac{dv^i}{ds} - \sum_{ijk} v^k(0)\Gamma^k_{ij}(v^l(0))\frac{dv^i}{ds}\frac{dv^j}{ds},$$

where all derivatives are taken for $s = 0$ and (9.36) was used in the second derivative. Therefore,

$$F(v^k(s)) = \frac{s^2}{2}\left(\sum_i\left(\frac{dv^i}{ds}\right)^2 - \sum_{ijk} v^k(0)\Gamma^k_{ij}(v^l(0))\frac{dv^i}{ds}\frac{dv^j}{ds}\right) + o(s^2).$$

Since with respect to normal coordinates $\Gamma^i_{jk}(0) = 0$, its value at u', that is, for $v^l(0)$ can be made arbitrarily small by taking a small value of ρ. Therefore, for small enough ρ, $F(v^k(s))$ is zero for $s = 0$ and strictly positive for small non-zero values of s. This proves (i).

Next, chose $\rho \le \epsilon/4$ where ϵ is the value of statement (i). For the distance of two points $u_1, u_2 \in U_\rho(u)$ the triangle inequality yields $d(u_1,u_2) \le d(u,u_1) + d(u,u_2) < 2\rho \le \epsilon/2$. Hence, $u_2 \in U_{\epsilon/2}(u_1)$, and for any $x \in U_{\epsilon/2}(u_1)$ it holds that $d(u,x) \le d(u,u_1) + d(u_1,x) < 3\epsilon/4$. According to the theorem visualized in Fig. 9.2, $U_\epsilon(u)$ may be chosen such that every point $u' \in U_\epsilon(u)$ has a normal coordinate neighborhood $U(u')$ that contains $U_\epsilon(u)$. Then, $U_{\epsilon/2}(u_1) \subset U_\epsilon(u) \subset U(u_1)$, and normal coordinates at u_1 can be used in $U_{\epsilon/2}(u_1)$. By the previous theorem above there is a unique geodesic C connecting u_1 with u_2 whose arc length is $d(u_1,u_2)$, and for any intermediate point x on this geodesic $d(u_1,x) \le d(u_1,u_2)$. $C \subset U_{\epsilon/2}(u_1) \subset U_\epsilon(u)$, and hence $d(u,x)$ is bounded by ϵ on C, while $d(u,u_1)$ and $d(u,u_2)$ are less than $\rho \le \epsilon/4$. Suppose that the maximum of $d(u,x)$ on C is at an intermediate point x_0 and is equal to $d_0 < \epsilon$. Then, C is tangent to S_{d_0} and hence by virtue of (i) d_0 is a local minimum on C. This contradiction proves $d(u,x) < \rho$, that is, C lies in $U_\rho(u)$. □

Again, in that last theorem no statement is made on the global behavior of geodesics. For large enough values of the curve parameter s, a geodesic tangent to S_ρ at u' may return arbitrarily close to u' and cross S_ρ, and likewise two points in $U_\rho(u)$ may in addition to the claimed unique geodesic *in* $U_\rho(u)$ be connected by a geodesic (maybe the same) that intermediately leaves $U_\rho(u)$; think for instance again of a great circle on a spherical manifold.

Modifying the last proof in such a way that one works only with Euclidean distances in coordinate neighborhoods $\boldsymbol{U}$, the existence of geodesic convex neighborhoods can be proved for any linear connection manifold independent of the presence of a Riemannian metric [3].

From the above analysis it is also clear that a curve $x^i = x^i(t)$ in a linear connection space M is a geodesic, iff it is the projection on M of an integral curve of a standard horizontal vector field on the frame bundle $L(M)$. Recall, that every integral curve of a vector field is contained in a maximal integral curve corresponding to some open interval $a<t<b$ of the parameter t of the local 1-parameter group. A linear connection is called a **complete linear connection**, if every geodesic with parameter t may be continued to a geodesic for $-\infty<t<\infty$.

9.5 Sectional Curvature

For the case of a Riemannian geometry a more detailed geometric meaning of the curvature tensor can now be found.

The Riemannian curvature tensor (9.33), as any tensor of type $(0,4)$ may be considered as a quadrilinear function $R : T_x(M) \times T_x(M) \times T_x(M) \times T_x(M) \to \mathbb{R}$:

$$R(W,X,Y,Z) = \langle R, W \otimes X \otimes Y \otimes Z \rangle = R_{ijkl} W^i X^j Y^k Z^l. \tag{9.50}$$

In particular, in arbitrary local coordinates,

$$R_{ijkl} = R\left(\frac{\partial}{\partial x^i}, \frac{\partial}{\partial x^j}, \frac{\partial}{\partial x^k}, \frac{\partial}{\partial x^l}\right). \tag{9.51}$$

The properties (9.34) transfer to corresponding properties of the quadrilinear function:

$$\begin{aligned} &R(W,X,Y,Z) = -R(X,W,Y,Z) = -R(W,X,Z,Y), \\ &R(W,X,Y,Z) + R(W,Y,Z,X) + R(W,Z,X,Y) = 0, \\ &R(W,X,Y,Z) = R(Y,Z,W,X). \end{aligned} \tag{9.52}$$

On p. 311 f it was shown that the third line of (9.34) follows from the first two lines. Hence, the same is true for (9.52). Moreover, let R and R' be two quadrilinear functions on some vector space V, having the properties of the first two lines of (9.52). If

$$R(X,Y,X,Y) = R'(X,Y,X,Y) \quad \text{for all } X,Y \in V, \tag{9.53}$$

then $R = R'$. Indeed, consider $R'' = R' - R$ and suppose $R''(X,Y,X,Y) = 0$ for all $X,Y \in V$. Then, for all $Z \in V$, $0 = R''(X,Y+Z,X,Y+Z) = R''(X,Y,X,Z) + R''(X,Z,X,Y) = 2R''(X,Y,X,Z)$ due to the third line of (9.52) which follows from the first two lines. Hence, $R''(X,Y,X,Z) = 0$ for all $X,Y,Z \in V$ and therefore $0 = R''(W+Y,X,W+Y,Z) = R''(W,X,Y,Z) + R''(Y,X,W,Z) = R''(W,X,Y,Z) + R''(W,Z,Y,X) = R''(W,X,Y,Z) - R''(W,Z,X,Y)$, where in the third equality the third line of (9.52) was used and in the last equality the first line. Thus, also $R''(W,X,Y,Z) = R''(W,Z,X,Y)$ and, by simply renaming X,Y,Z into Y,Z,X, $R''(W,Y,Z,X) = R''(W,X,Y,Z)$ for all $W,X,Y,Z \in V$. From these last two relations, $3R''(W,X,Y,Z) = R''(W,X,Y,Z) + R''(W,Y,Z,X) + R''(W,Z,X,Y) = 0$ for all $W,X,Y,Z \in V$ and hence $R = R'$.

A quadrilinear function determined solely by the metric tensor (9.3), which has the same properties (9.52), is

$$G(W,X,Y,Z) = G(W,Y)G(X,Z) - G(W,Z)G(X,Y). \tag{9.54}$$

For $X, Y \in T_x(M)$,

$$G(X,Y,X,Y) = |X|^2|Y|^2 - (X|Y)^2 = (|X||Y|\sin\angle(X,Y))^2. \tag{9.55}$$

This is the square of the area of a parallelepiped spanned by the vectors X and Y. Let X', Y' span the same two-dimensional subspace E of $T_x(M)$ as X, Y, that is, $(X',Y')^t = A(X,Y)^t$, where A is a regular (2×2)-matrix, $\det A \neq 0$. From the properties (9.52) it is easily found (exercise) that

$$R(X',Y',X',Y') = (\det A)^2 R(X,Y,X,Y),$$
$$G(X',Y',X',Y') = (\det A)^2 G(X,Y,X,Y).$$

Hence, the quotient of R and G is an invariant of E. Its negative is called the **sectional curvature** of M at (x,E):

$$K(x,E) = -\frac{R(X,Y,X,Y)}{G(X,Y,X,Y)}. \tag{9.56}$$

Since G is uniquely defined by the metric tensor, it follows from the above that *the curvature tensor of a Riemannian manifold M at point x is uniquely determined by the sectional curvatures of all the two-dimensional subspaces of the tangent space $T_x(M)$.*

Gauss' theory of surfaces uses a parameter representation

$$\boldsymbol{r} = \boldsymbol{r}(x^1,x^2), \quad \boldsymbol{r} = (r^1,r^2,r^3) \tag{9.57}$$

with parameters x^1, x^2 for a two-dimensional smooth surface $\mathcal{E}$ embedded in the three-dimensional Euclidean space $\mathbb{R}^3$, with Cartesian coordinates r^i (with respect to an orthonormalized base $\{\boldsymbol{a}_1,\boldsymbol{a}_2,\boldsymbol{a}_3\}$, $\boldsymbol{r} = \sum_i r^i \boldsymbol{a}_i$). The frame (Fig. 9.3)

$$\boldsymbol{e}_1 = \frac{\partial \boldsymbol{r}}{\partial x^1}, \quad \boldsymbol{e}_2 = \frac{\partial \boldsymbol{r}}{\partial x^2}, \quad \boldsymbol{n} = \frac{\boldsymbol{e}_1\times\boldsymbol{e}_2}{|\boldsymbol{e}_1\times\boldsymbol{e}_2|}, \quad \boldsymbol{e}_1\times\boldsymbol{e}_2 \neq 0, \tag{9.58}$$

at point $\boldsymbol{r}(x_0^1,x_0^2) \in \mathbb{R}^3$ describes the tangent plane on $\mathcal{E}$ at point $(x_0^1,x_0^2) \in \mathcal{E}$ as the plane spanned by the vectors $\boldsymbol{e}_1$ and $\boldsymbol{e}_2$ and having the normal $\boldsymbol{n}$ in $\mathbb{R}^3$. In the metric inherited from $\mathbb{R}^3$, the element of arc length on $\mathcal{E}$ is given by

$$d\boldsymbol{s} = \boldsymbol{e}_1 dx^1 + \boldsymbol{e}_2 dx^2 = \frac{\partial \boldsymbol{r}}{\partial x^1}dx^1 + \frac{\partial \boldsymbol{r}}{\partial x^2}dx^2,$$
$$ds^2 = \sum_{i=1}^{3}\sum_{j,k=1}^{2}\frac{\partial r^i}{\partial x^j}dx^j\frac{\partial r^i}{\partial x^k}dx^k = E(dx^1)^2 + 2F dx^1dx^2 + G(dx^2)^2 = g_{jk}dx^j dx^k, \tag{9.59}$$
$$E = \sum_{i=1}^{3}\left(\frac{\partial r^i}{\partial x^1}\right)^2, \quad F = \sum_{i=1}^{3}\frac{\partial r^i}{\partial x^1}\frac{\partial r^i}{\partial x^2}, \quad G = \sum_{i=1}^{3}\left(\frac{\partial r^i}{\partial x^2}\right)^2, \quad (g_{jk}) = \begin{pmatrix} E & F \\ F & G\end{pmatrix}.$$

The second line is Gauss' first fundamental form of the surface $\mathcal{E}$. The volume form of $\mathcal{E}$ (surface area element, without a definition of sign of orientation) is

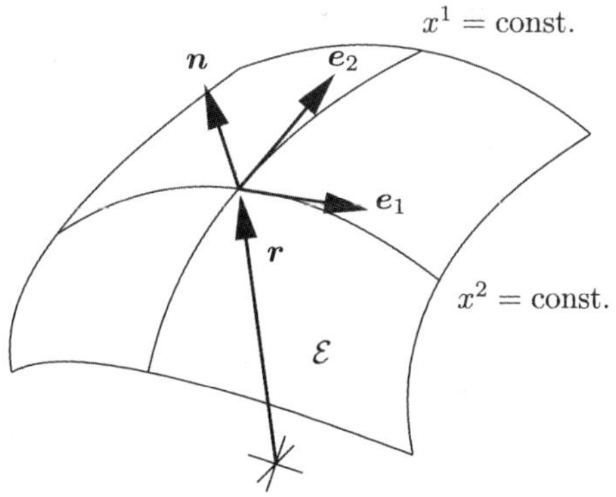

Fig. 9.3 A smooth surface E embedded in the Euclidean space $\mathbb{R}^3$

$$
\begin{aligned}
|\tau| = |\boldsymbol{e}_1 dx^1 \times \boldsymbol{e}_2 dx^2| = (|\boldsymbol{e}_1||\boldsymbol{e}_2| - \boldsymbol{e}_1\cdot\boldsymbol{e}_2)^{1/2} dx^1 dx^2 = \sqrt{EG-F^2}\,dx^1 dx^2 \\
- \sqrt{\det g}\,dx^1 dx^2.
\end{aligned}
\tag{9.60}
$$

From $d\boldsymbol{n} = |\boldsymbol{e}_1 \times \boldsymbol{e}_2|^{-1}(d\boldsymbol{e}_1 \times \boldsymbol{e}_2 + \boldsymbol{e}_1 \times d\boldsymbol{e}_2) + \boldsymbol{n}\cdots$ and $\boldsymbol{n}\cdot d\boldsymbol{s} = 0$, for Gauss' second fundamental form one easily finds

$$
\begin{aligned}
&-d\boldsymbol{n}\cdot d\boldsymbol{s} = L(dx^1)^2 + 2M dx^1 dx^2 + N(dx^2)^2 = b_{jk} dx^j dx^k, \\
&L - (EG-F^2)^{-1/2}\det\begin{pmatrix} r^1_{11} & r^2_{11} & r^3_{11} \\ r^1_1 & r^2_1 & r^3_1 \\ r^1_2 & r^2_2 & r^3_2 \end{pmatrix}, \\
&M = (EG-F^2)^{-1/2}\det\begin{pmatrix} r^1_{12} & r^2_{12} & r^3_{12} \\ r^1_1 & r^2_1 & r^3_1 \\ r^1_2 & r^2_2 & r^3_2 \end{pmatrix}, \qquad (b_{jk}) = \begin{pmatrix} L & M \\ M & N \end{pmatrix}, \\
&N = (EG-F^2)^{-1/2}\det\begin{pmatrix} r^1_{22} & r^2_{22} & r^3_{22} \\ r^1_1 & r^2_1 & r^3_1 \\ r^1_2 & r^2_2 & r^3_2 \end{pmatrix}
\end{aligned}
\tag{9.61}
$$

with the abbreviations $r^i_{jk} = \partial^2 r^i/\partial x^j \partial x^k$, $r^i_j = \partial r^i/\partial x^j$. While the first fundamental form depends only on the metric on $\mathcal{E}$, the second fundamental form depends also on the embedding of $\mathcal{E}$ in $\mathbb{R}^3$ (on the differential $d\boldsymbol{n}$ of the normal vector $\boldsymbol{n}$ on $\mathcal{E}$ in $\mathbb{R}^3$).

Any point $\boldsymbol{r} \in \mathcal{E}$ may be chosen as origin of $\mathbb{R}^3$, and the Cartesian base $\{\boldsymbol{a}_1, \boldsymbol{a}_2, \boldsymbol{a}_3\}$ may in particular be chosen such that $\boldsymbol{a}_1, \boldsymbol{a}_2$ span the tangent plane on $\mathcal{E}$ and $\boldsymbol{a}_3 = \boldsymbol{n}$. Then, in a neighborhood of this point, with the just introduced notation of derivatives the surface $\mathcal{E}$ is described as $r^3(x^1,x^2) = (r^3_{11}(x^1)^2 + 2r^3_{12}x^1x^2 + r^3_{22}(x^2)^2)/2 + \cdots$, $x^1 = r^1$, $x^2 = r^2$, where the derivatives are taken at the origin. By a rotation of the (x^1,x^2)-plane around $\boldsymbol{n}$ this expression may be brought to the form $r^3(x^1,x^2) = (k_1(x^1)^2 + k_2(x^2)^2)/2 + \cdots$ from which it is immediately seen that k_j are the two **principal curvatures** in the two (orthogonal to each other) principal curvature directions $\boldsymbol{e}_1 = \boldsymbol{a}_1$ and $\boldsymbol{e}_2 = \boldsymbol{a}_2$, that is, the inverse values of the curvature radii of the corresponding parabolic intersection lines of the planes spanned by $\boldsymbol{e}_j$ and $\boldsymbol{n}$ with $\mathcal{E}$. The **total**

curvature or **Gaussian curvature** is defined as the product $K(x^1, x^2) = k_1 k_2$. The signs of k_1 and k_2 depend on the orientation of $\mathcal{E}$ (or $\boldsymbol{n}$, determined by the choice of subscripts of $\boldsymbol{e}_1$ and $\boldsymbol{e}_2$). Their product K, however, is a true scalar. The special choice of coordinates implies $r^i_j = \delta^i_j, r^i_{jk} = 0$ for $i = 1, 2$ and $r^3_{jk} = k_j \delta_{jk}$. Hence, $LN - M^2 = K$, $EG - F^2 = 1$, and the canonical fundamental forms are (compare (9.46))

$$ds^2 = (dx^1)^2 + (dx^2)^2, \quad -d\boldsymbol{n} \cdot d\boldsymbol{s} = k_1 (dx^1)^2 + k_2 (dx^2)^2. \tag{9.62}$$

Now, K has a geometric meaning of the embedding of $\mathcal{E}$ in $\mathbb{R}^3$ which is independent of used coordinates. On the other hand, g_{jk} and b_{jk} are tensors of type $(0, 2)$ in two dimensions, and hence the quotient det b/det g is also independent of used coordinates. Hence,

$$K = \frac{\det b}{\det g} = \frac{LN - M^2}{EG - F^2} \tag{9.63}$$

holds independently of chosen coordinates.

Let the frame $\{\boldsymbol{e}_1, \boldsymbol{e}_2, \boldsymbol{n}\}$ move on $\mathcal{E}$. The corresponding derivatives of $\boldsymbol{e}_j$ at (x^1_0, x^2_0) may be re-expanded into the frame there:

$$\frac{\partial \boldsymbol{e}_j}{\partial x^k} = \sum_{l=1}^{2} \Gamma^l_{jk} \boldsymbol{e}_l + b_{jk} \boldsymbol{n}. \tag{9.64}$$

The first term is an intrinsic relation on $\mathcal{E}$ and does not make use of the embedding in $\mathbb{R}^3$. Identifying in view of the first line of (9.59) $\boldsymbol{e}_j$ with $\partial/\partial x^j$, this term becomes a case of (7.41) and recovers Γ^l_{jk} as the Christoffel symbols of the two-dimensional Riemannian manifold $\mathcal{E}$, which according to (9.23) are expressed in terms of derivatives of the metric tensor g. As regards the second term, the general property $\boldsymbol{e}_j \cdot \boldsymbol{n} = 0$ of the considered frame implies $d\boldsymbol{e}_j \cdot \boldsymbol{n} = -\boldsymbol{e}_j \cdot d\boldsymbol{n}$, and scalar multiplication of (9.64) with $\boldsymbol{n}\, dx^j dx^k$, dx^j and dx^k *arbitrary*, and summation over the two values of j and k yields $b_{jk} dx^j dx^k = (\partial \boldsymbol{e}_j/\partial x^k) \cdot \boldsymbol{n}\, dx^j dx^k = -\boldsymbol{e}_j \cdot (\partial \boldsymbol{n}/\partial x^k) dx^j dx^k = -d\boldsymbol{s} \cdot d\boldsymbol{n}$, which agrees with (9.61). Equation 9.64 is Gauss' equation for the moving frame. For the change of $\boldsymbol{n}$, Weingarten's equation

$$\frac{\partial \boldsymbol{n}}{\partial x^k} = -\sum_{j=1}^{2} g^{jl} b_{lk} \boldsymbol{e}_j, \quad g^{jl} g_{lk} = \delta^j_k, \tag{9.65}$$

is obtained, where summation over $l = 1, 2$ as tensor multiplication is understood and as usual (g^{jk}) is the inverse of (g_{jk}). First of all, $\boldsymbol{n}^2 = 1$ implies $\boldsymbol{n} \cdot d\boldsymbol{n} = 0$, and hence a term proportional to $\boldsymbol{n}$ is missing on the right hand side. Scalar multiplication of (9.65) with $\boldsymbol{e}_l dx^l dx^k$, the latter two again arbitrary, and summation yields $(\partial \boldsymbol{n}/\partial x^k) \cdot \boldsymbol{e}_l dx^l dx^k = d\boldsymbol{n} \cdot d\boldsymbol{s} = -b_{lk} dx^l dx^k$ and hence $(\partial \boldsymbol{n}/\partial x^k) \cdot \boldsymbol{e}_l = -b_{lk}$. The final result follows since $g_{jk} = \boldsymbol{e}_j \cdot \boldsymbol{e}_k$ implies $\sum_{lj} \boldsymbol{e}_l g^{lj} \boldsymbol{e}_j = 1$.

The relations (9.64, 9.65) comprise 18 differential equations for the 9 functions $\boldsymbol{e}_1, \boldsymbol{e}_2, \boldsymbol{n}$ of x_1 and x_2, and hence for their solubility by smooth functions integrability conditions must be imposed on their right hand sides. These are the 9 conditions

$$\frac{\partial}{\partial x^i}\Big(\sum \Gamma^l_{jk} \boldsymbol{e}_l + b_{jk} \boldsymbol{n}\Big) = \frac{\partial^2 \boldsymbol{e}_j}{\partial x^i \partial x^k} = \frac{\partial}{\partial x^k}\Big(\sum \Gamma^l_{ji} \boldsymbol{e}_l + b_{ji} \boldsymbol{n}\Big), \tag{9.66}$$

In the present context, most important of the implications of a straightforward but lengthy analysis (preferably in the above particular coordinates of $\mathbb{R}^3$) of these integrability conditions is Gauss' **theorema egregium** (exquisite theorem)

$$K(x_0) = \frac{R_{1212}}{\det g}, \quad R_{1212} = \frac{\partial \Gamma_{122}}{\partial x^1} - \frac{\partial \Gamma_{121}}{\partial x^2} + \sum_{j=1}^{2} (\Gamma^j_{11}\Gamma_{2j2} - \Gamma^j_{12}\Gamma_{2j1}). \tag{9.67}$$

Gauss found it amazing since the two principal curvatures k_1 and k_2 clearly depend on (b_{jk}) and hence on the second fundamental form, that is, on the embedding of $\mathcal{E}$ in $\mathbb{R}^3$ while their product does not. It is uniquely defined by the metric (g_{jk}) of $\mathcal{E}$ and hence can be determined by measurements on $\mathcal{E}$ alone without reference to the embedding in $\mathbb{R}^3$. For instance, on a cylinder one of the principal curvatures is zero and the other is non-zero. On a plane both values are zero. The total curvature K is zero in both cases, and in fact both manifolds per se are isometric and locally essentially equivalent, their embedding in $\mathbb{R}^3$, however, is locally different.

Comparison of (9.67) with (7.48) shows, that R_{1212} is the component of the curvature tensor, which due to the general properties (9.34) in the case of a two-dimensional Riemannian geometry is up to permutations of subscripts the only non-zero tensor component of the curvature tensor R_{ijkl}: for $i = j$ or $k = l$ it is zero, and the remaining components are equal up to a sign.

Returning to the general case, let M again be a Riemannian geometry of dimension m, let $u \in M$, and let X and Y span the two-dimensional subspace $E \subset T_u(M)$. Choose an orthogonal frame $\{\boldsymbol{e}_1, \ldots, \boldsymbol{e}_m\}$ at u for which $\boldsymbol{e}_1, \boldsymbol{e}_2$ span E. Consider the submanifold $\mathcal{E}$ of M formed by the points of all geodesics through u and tangent to E. Then, $\mathcal{E}$ is given by the conditions

$$\mathcal{E} : v^r = 0, \quad r = 3, \ldots, m, \tag{9.68}$$

for the normal coordinates v^i at u. The submanifold $\mathcal{E}$ is a two-dimensional Riemannian geometry with metric tensor (cf. (9.9))

$$\bar{g}_{ij}(v^1, v^2) = g_{ij}(v^1, v^2, 0, \ldots, 0), \quad 1 \le i, j \le 2, \tag{9.69}$$

inherited from the metric tensor g_{ij} of M, and v^1, v^2 are normal coordinates at u in $\mathcal{E}$ since the shortest path in M between two points of $\mathcal{E}$, which is completely in $\mathcal{E}$, is a fortiori the shortest path in $\mathcal{E}$ between the points. Clearly, the Christoffel symbols $\overline{\Gamma}_{ijk}$ of $\mathcal{E}$ are

$$\frac{1}{2}\left(\frac{\partial \bar{g}_{jk}}{\partial v^i} + \frac{\partial \bar{g}_{ij}}{\partial v^k} - \frac{\partial \bar{g}_{ik}}{\partial v^j}\right) = \overline{\Gamma}_{ijk}(v^1, v^2) \tag{9.70}$$

again for $1 \le i, j, k \le 2$. In particular, since v^i are normal coordinates at u in both manifolds, $\overline{\Gamma}_{ijk}(0) = \Gamma_{ijk}(0) = 0$ at u. Hence,

$$R_{1212}(0) = \left(\frac{\partial \Gamma_{122}}{\partial v^1} - \frac{\partial \Gamma_{121}}{\partial v^2}\right) = \left(\frac{\partial \overline{\Gamma}_{122}}{\partial v^1} - \frac{\partial \overline{\Gamma}_{121}}{\partial v^2}\right) = \bar{R}_{1212}(0). \tag{9.71}$$

At u one has $\det \bar{g}(0) = g_{11}g_{22} - g_{12}^2 = G(\boldsymbol{e}_1, \boldsymbol{e}_2, \boldsymbol{e}_1, \boldsymbol{e}_2)$, so that finally the important result

$$K(u, E) = -\frac{R(\boldsymbol{e}_1, \boldsymbol{e}_2, \boldsymbol{e}_1, \boldsymbol{e}_2)}{G(\boldsymbol{e}_1, \boldsymbol{e}_2, \boldsymbol{e}_1, \boldsymbol{e}_2)} = -\frac{\bar{R}_{1212}}{\det \bar{g}} = \bar{K}(u) \tag{9.72}$$

follows. (Recall, that the value of $K(u,E)$ is independent of the tangent vectors $X,Y \in T_u(M)$ used which span E.)

The sectional curvature $K(u,E)$ in a Riemannian geometry M is equal to the total curvature of the surface $\mathcal{E}$ formed by all geodesics through u which are tangent to E, with the inherited metric.

A Riemannian geometry M is called **wandering** at u, if $K(u,E) = K(u)$ is independent of $E \subset T_u(M)$. Since $R(W,X,Y,Z)$ is uniquely determined by all $R(X,Y,X,Y)$ for all linearly independent $X,Y \in T_u(M)$, and likewise for G,

$$R_{ijkl}(u) = -K(u)\big(g_{ik}(u)g_{jl}(u) - g_{il}(u)g_{jk}(u)\big), \quad \text{iff } M \text{ is wandering at } u. \tag{9.73}$$

M is called a **constant curvature space**, if it is wandering at every point and $K(u) = K$ is independent of u.

In three dimensions, for every real value of K there is up to positioning exactly one two-dimensional connected constant curvature space; for $K > 0$ it is a sphere (of radius $1/K$), for $K = 0$ it is a plane, and for $K<0$ it is called a pseudo-sphere (obtained by rotating a tractrix around its asymptote; it is singular on its largest circumference).

9.6 Gravitation

(For further studies [4] is recommended.)

Identical physical entities as atoms or molecules from which one can say that they are positioned at fixed mutual distance emit characteristic light with identical frequency spectra. This allows for the definition of an absolute time scale (unit of time). Phenomenologically there is no velocity of propagation of information observed exceeding the velocity of light in vacuum, which is found to be the same in all directions. Therefore the speed of light is assumed to be a universal constant. Distances on the other hand of remote events can only be measured reliably by propagating light signals between them. It is natural to define distances by the time interval a light signal needs to propagate forth and back between the events. This is why Minkowski's metric

$$ds^2 = g_{\mu\nu}\, dx^\mu dx^\nu = (dx^0)^2 - (dx^1)^2 - (dx^2)^2 - (dx^3)^2, \quad dx^0 = cdt, \tag{9.74}$$

where t is time, c is the vacuum speed of light, and x^1, x^2, x^3 are taken to be Cartesian spatial coordinates, is assumed to have physical reality.

Think of a clock fixed with an isolated particle (for instance realized by the phase of a vibration mode of a molecule). In a frame attached to the particle, $\sum_{i=1}^{3}(dx'^i)^2 = 0$ and hence $ds^2 = (cdt')^2$. An observer, who sees the particle move

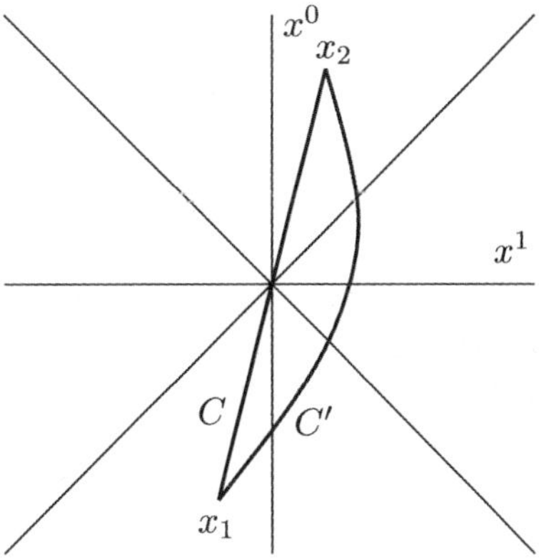

Fig. 9.4 World lines of observer, C, and a particle, C'

with velocity v relative to himself, observes in the same interval ds (which is an invariant and hence the same in all reference systems) the particle run a distance $\sum_i (dx^i)^2 = (vdt)^2$. Hence, $(cdt')^2 = ds^2 = (cdt)^2 - (vdt)^2$ or

$$dt' = dt\sqrt{1 - \left(\frac{v}{c}\right)^2}. \tag{9.75}$$

The time interval dt' is the **proper time** interval experienced by the particle. Consider an observer moving with constant speed on a straight world line C (Fig. 9.4) and observing a particle departing from his own position x_1 and returning to him at position x_2. Since the particle moves relative to the observer, its proper time t' stays all the way behind that of the observer, t. (In Fig. 9.4 an inertial reference system is used which moves relative to both observer and particle. Of course both must have velocities not exceeding c, indicated by the two lines $x^1 = \pm ct$ in the figure.) Hence,

$$t_2 - t_1 = \int_C ds > \int_{C'} ds = t_2' - t_1',$$

and clearly the straight world line has the maximum length of all world lines between two given world points. (The situation is asymmetric between C and C' because in vacuum a set of reference systems which move relative to each other at most with constant speed is distinguished as inertial systems.)

A massive particle propagates between world points (events) x_1^μ and x_2^μ on a world line C which is determined by the principle of least action. If nothing is present besides the particle, the only invariant action that can be formed is

$$S = -mc\int_{x_1}^{x_2} ds = \int_{t_1}^{t_2} Ldt, \tag{9.76}$$

where the prefactor is convention and sets a mass or energy scale while the minus sign is needed because in Minkowski's metric, as just shown, the integral has a maximum but no minimum. It yields the Lagrange function $L = -mc^2 (1 - (v/c)^2)^{1/2}$, where v is the velocity of the particle. In general geometric terms this action means that the world line of the particle is a geodesic in the Minkowski space, which is a straight line.

However, if a force acts on the particle, its motion is accelerated, and it turns out that, if the force is that of a gravitational field, all accelerations of all particles passing a region where the field has a certain value (measured statically by means of a spring-balance) are the same independent of the particle mass. Exactly the same way force-free particle motions are seen by an accelerated observer, that is, if the particle coordinates are described with respect to an accelerated frame. Because these two situations are indistinguishable and (9.74) is invariant under mere changes of the coordinates, the same action (9.76) must describe the motion of a particle under the action of a gravitational field, while the metric tensor $g_{\mu\nu}$ now may be much more general than in the case of use of mere general coordinates in the above discused case of absence of gravitation. Hence, the world line of a particle in a gravitational field is still a geodesic, but now in a more general metric. If one takes the geodesic arc length s as curve parameter, the equation of motion is

$$\frac{d^2x^\lambda}{ds^2} + \sum_{\mu\nu=0}^{3} \Gamma^\lambda_{\mu\nu} \frac{dx^\mu}{ds}\frac{dx^\nu}{ds} = 0, \tag{9.77}$$

with the Christoffel symbols $\Gamma^\lambda_{\mu\nu}$ of the Riemannian connection of the metric tensor g now describing the action of the **gravitational field**. The Christoffel symbols are obtained from derivatives of the metric tensor with respect to the coordinates, hence the metric tensor itself may be viewed as forming **gravitational potentials**.

Since coordinates are now arbitrary and cannot have any more an immediate physical meaning, their relation to time intervals and distances in ordinary space must be analyzed. In this section, systematically Latin tensor indices run from 1 to 3, Greece indices run from 0 to 3, and t means the physical time, that is the proper time of a material entity. Although a consequent covariant formulation of physical laws as used in the theory of gravitation would tolerate the use of completely arbitrary (smooth) coordinate systems, every point of physical space time is the apex of a cone of future and of a cone of past, and the convention is reasonable to use only coordinate systems where the coordinate line for x^0 runs from the past to the future while the coordinate lines of x^i are outside of the cones. Such coordinates may be realized by material constructions (sample holders and clocks). If in such a case $dx^i = 0$, then $ds^2 = g_{00}(dx^0)^2 = (cdt)^2$, and

$$dt = \frac{1}{c}\sqrt{g_{00}(x)}dx^0, \quad t = \frac{1}{c}\int \sqrt{g_{00}(x)}dx^0, \quad g_{00}(x) > 0, \tag{9.78}$$

describes the **proper time** of a world line $x^\mu(s)$. To find an expression for spatial distances, consider two neighboring world lines C and C' with their proper times t and t'. Send a light signal from C to C' which arrives at C' at world point x^μ and is reflected back to C. It departs from C at world point $x^\mu + dx_1^\mu$ and returns to C at $x^\mu + dx_2^\mu$. For the propagation forth and back $(a = 1, 2)$ one has

$$0 = g_{\mu\nu}dx_a^\mu dx_a^\nu = g_{00}(dx_a^0)^2 + 2g_{0i}dx_a^0 dx_a^i + g_{ij}dx_a^i dx_a^j.$$

Assume the two world lines kept at constant spatial distance, $dx_1^i = dx_2^i = dx^i$. Then,

$$dx_a^0 = -\frac{1}{g_{00}}\left(g_{0i}dx^i \pm \sqrt{(g_{0i}dx^i)^2 - g_{00}g_{ij}dx^i dx^j}\right)$$

with the upper sign for $a = 1$ and the lower sign for $a = 2$ yield the proper time interval on world line C from departure to return of the light signal as

$$dt = \frac{1}{c}\sqrt{g_{00}}(dx_2^0 - dx_1^0) = \frac{2}{c\sqrt{g_{00}}}\sqrt{(g_{0i}g_{0j} - g_{00}g_{ij})dx^i dx^j}.$$

This time is apparently two times the spatial distance between the world lines divided by c, hence the square of the distance is

$$dl^2 = \bar{g}_{ij}dx^i dx^j, \quad \bar{g}_{ij} = \frac{g_{0i}g_{0j}}{g_{00}} - g_{ij}, \quad \bar{g}^{ij} = -g^{ij}, \tag{9.79}$$

with the metric tensor $\bar{g}^{ij}$ of the spatial submanifold in a coordinate neighborhood of materializable coordinates x^μ of space–time given by $x^0 = \text{const}$. The last relation, in which as usual $(g^{\mu\nu})$ is the inverse of $(g_{\mu\nu})$ and likewise for $(\bar{g}^{ij})$, is a simple exercise.

For physical reasons, since it describes spatial distances, $\bar{g}_{ij}$ must be a positive definite metric. This implies for the submatrices

$$\bar{g}_{11} > 0, \quad \det\begin{pmatrix} \bar{g}_{11} & \bar{g}_{12} \\ \bar{g}_{21} & \bar{g}_{22} \end{pmatrix} > 0, \quad \det \bar{g} > 0. \tag{9.80}$$

The last equations (9.78) and (9.79) then require

$$g_{00} > 0, \quad \det\begin{pmatrix} g_{00} & g_{01} \\ g_{10} & g_{11} \end{pmatrix} < 0, \quad \det\begin{pmatrix} g_{00} & g_{01} & g_{02} \\ g_{10} & g_{11} & g_{12} \\ g_{20} & g_{21} & g_{22} \end{pmatrix} > 0, \quad \det g < 0. \tag{9.81}$$

(Caution: Some authors of physics textbooks, including [4], use a convention $ds^2 = -g_{\mu\nu}dx^\mu dx^\nu$, then only $<$ signs appear in (9.81), and the second and third relations (9.79) have reversed signs.) These relations must be fulfilled in any coordinate system realized by matter constructions.

As a simple example, consider an observer at rest and use cylinder coordinates (x'^0, r', φ', z'),

$$ds^2 = (dx'^0)^2 - (dr')^2 - r'^2(d\varphi')^2 - (dz')^2.$$

Let the observer rotate around the z'-axis with angular velocity ω. He now makes observations with respect to coordinates $x^0 = x'^0, r = r', \varphi = \varphi' - (\omega/c)x^0$, $z = z'$, and

$$ds^2 = \left(1 - \frac{\omega^2 r^2}{c^2}\right)(dx^0)^2 - 2\frac{\omega r^2}{c} d\varphi dx^0 - (dr)^2 - r^2(d\varphi)^2 - (dz)^2.$$

For $r > c/\omega$ one would have $g_{00} < 0$ and hence outside of this radius the considered rotating coordinate systems would not be materializable. Matter at radius $r < c/\omega$, seen at rest by the rotating observer, 'in truth' moves with velocity $v = \omega r$ in the primed coordinates. The proper time t of the rotating matter is related to the time t' of an observer at rest as

$$dt = \sqrt{1 - \frac{\omega^2 r^2}{c^2}} dt'. \tag{9.82}$$

Compared to a time interval $\Delta t' = \Delta x'^0/c = \Delta x^0/c$, its proper time interval is $\Delta t(r) = \sqrt{1 - \omega^2 r^2/c^2}\, \Delta t' = \sqrt{1 - \omega^2 r^2/c^2}\, \Delta t(0)$. Light emitted from this matter at r by atomic vibrations is observed at $r = 0$ by the observer who is measuring with his atomic clocks at rest ticking with frequency $\nu(r = 0) = \sqrt{1 - \omega^2 r^2/c^2}\, \nu(r)$. Since the clocks of the observer tick at lower frequency, he registers the incident light coming from r blue shifted and approaching infinite frequency for $r \to c/\omega$ from below.

Consider the observed circumference of a circle of radius r:

$$L = \oint_r dl = \int_0^{2\pi} \sqrt{g_{\varphi\varphi}}\, d\varphi = 2\pi\sqrt{\frac{g_{0\varphi}^2}{g_{00}} - g_{\varphi\varphi}} = \frac{2\pi r}{\sqrt{1 - \omega^2 r^2/c^2}}. \tag{9.83}$$

The observed radius would be unchanged: $R = \int_0^r \sqrt{g_{rr}}\, dr = \int_0^r dr = r$. The observed ratio L/R approaches the familiar value 2π only for $r \to 0$ and is otherwise larger than 2π, diverging for $r = c/\omega$. The reason is the following. Assume, in the primed system at rest an observer calibrates a measuring rod to have length $l' = r'/n = r/n$, $n \gg 1$, and lays it down tangent to the circle with radius r centered at $r' = r = 0$. Its ends seen from the origin form an angle $\delta\varphi' = l'/r = 1/n$. The rotating observer sees the rod move with velocity $v = \omega r$ (against the rotation direction), and hence sees it Lorentz contracted to length $l = l'\sqrt{1 - \omega^2 r^2/c^2}$. Its ends seen from him at the origin form an angle $\delta\varphi = \delta\varphi'\sqrt{1 - \omega^2 r^2/c^2} = \sqrt{1 - \omega^2 r^2/c^2}/n$. This comes about by synchronization of clocks at both ends of the rod in the rotating system. A by a factor

$1/\sqrt{1-\omega^2 r^2/c^2}$ larger number of rods is needed to cover the full angle of 2π, or, alternatively expressed, the seen angle summed up of rods filling one circumference is smaller than 2π. At $r = c/\omega$ the rods would be seen shrunk to zero. The observed 2-space perpendicular to z is not Euclidean any more, it has got a negative curvature. (On a sphere having positive curvature, obviously $L/R < 2\pi$, if the radius is measured from a pole along a great circle.)

The proper time t of rotating matter (including electromagnetic matter as light) so that it is at rest in the rotating system is related to the time t' of the observer at rest at $r = 0$ as (9.82). Compared to his it runs slower and slower at increasing distances from the observer, and, since the vacuum speed of light is the same constant if measured with proper time (9.78) and is defining lengths (9.79), if he sends out a light signal radially, it will propagate less and less distance, if less and less of time is passing. At $r = c/\omega$ it seems to come to a halt. In fact it does not propagate radially in the rotating coordinates. Instead its ray bends against the direction of rotation and finally it nestles to the sphere $r = c/\omega$. Similarly, a light ray emitted from rotating matter inward is bent in the opposite direction. The observer is unable to trace any *process* outside the radius c/ω since any two signals coming close in time from the same direction are originating from a-causal events (separated by $\Delta s < 0$). Nevertheless, the particular role of $r = c/\omega$ is solely caused by the chosen rotating coordinate system. The physics is not special there, and for instance $\det g$, which is invariant under linear coordinate transformations, in the considered case is regular in both systems, $\det g' = \det g = r^2$ as directly calculated from the above two expressions for ds^2.

The observer at $r = 0$ would see the same physics regardless whether he is rotating or a gravitational field is in effect which acts on all masses with a force equal to the fictitious forces of the rotating system *in the domain* $r < c/\omega$. As soon as he is able to be convinced to see real physical processes outside this domain simply by changing his rotation frequency with respect to his reference system, he can put an upper limit to such a gravitational field and finally exclude its existence by extending his observations to unlimited distances. So much about the example.

Since the space-time manifold M (generalization of the Minkowski space) has got a metric, the Hodge operator (4.89) may be defined and applied to the external calculus on M. Since $\det g < 0$, according to (4.87) and (9.47) the invariant volume form is

$$\tau = \sqrt{-\det g}\, dx^0 \wedge dx^1 \wedge dx^2 \wedge dx^3, \tag{9.84}$$

and the **Levi–Civita pseudo-tensor** (5.85) is

$$E_{\kappa\lambda\mu\nu} = \sqrt{-\det g}\, \delta^{0123}_{\kappa\lambda\mu\nu}, \quad E^{\kappa\lambda\mu\nu} = \frac{-1}{\sqrt{-\det g}} \delta^{0123}_{\kappa\lambda\mu\nu}. \tag{9.85}$$

For various alternating tensors ω, (5.87) now reads

$$
\begin{aligned}
& & (*1)_{\kappa\lambda\mu\nu} &= E_{\kappa\lambda\mu\nu}, \\
(*\omega)^{\nu} &= E^{\kappa\lambda\mu\nu}\omega_{\kappa\lambda\mu}, & (*\omega)_{\lambda\mu\nu} &= \tfrac{1}{6}E_{\kappa\lambda\mu\nu}\omega^{\kappa}, \\
(*\omega)^{\mu\nu} &= \tfrac{1}{2}E^{\kappa\lambda\mu\nu}\omega_{\kappa\lambda}, & (*\omega)_{\mu\nu} &= \tfrac{1}{2}E_{\kappa\lambda\mu\nu}\omega^{\kappa\lambda}, \\
(*\omega)^{\lambda\mu\nu} &= \tfrac{1}{6}E^{\kappa\lambda\mu\nu}\omega_{\kappa}, & (*\omega)_{\nu} &= E_{\kappa\lambda\mu\nu}\omega^{\kappa\lambda\mu}, \\
(*1)^{\kappa\lambda\mu\nu} &= E^{\kappa\lambda\mu\nu}. & &
\end{aligned}
\tag{9.86}
$$

These rules provide a duality between alternating tensors ω of rank n and alternating tensors $*\omega$ of rank $4-n$. The dual δ to the external differentiation d is (5.90) for $m = 4$:

$$\delta\omega = *(d(*\omega)), \tag{9.87}$$

and for $\omega, \sigma \in \mathcal{D}_c^n(M)$, $0 \leq n \leq 4$

$$[\omega|\sigma] = \int_M \omega \wedge *\sigma \tag{9.88}$$

is an invariant integral to which Stokes' theorem can be applied. For contravariant alternating tensors ω the forms $*\omega$ of the right column of (9.86) are often called tensor densities, which name, however, has no deeper meaning; the understanding of tensor and tensor density may be interchanged in (9.88). Among many other things these relations allow to apply readily the equations (5.96) to (5.101) and their consequences to the case of Maxwell fields in the presence of a gravitational field.

As is seen from (9.77), gravitation is related to deviations of space–time from being flat. In a flat space-time, $\Gamma^{\lambda}_{\mu\nu} \equiv 0$ for chosen coordinates, all free particles move along straight lines of an inertial system of coordinates which means that no gravitational field is present (or it is exactly compensated in an accelerated reference system of the observer, as for instance in the case of an observer in a freely falling lift; these two possibilities can principally not be distinguished). In order to construct an action integral for the gravitational field, one needs a scalar formed from curvature of space–time. The Christoffel symbols cannot directly be used since they are not covariant at all. Scalars can systematically be formed by contraction of tensors. The Riemannian curvature tensor (9.33) allows only for one non-zero contraction

$$R_{\mu\nu} = g^{\kappa\lambda}R_{\kappa\mu\lambda\nu}, \quad R_{\mu\nu} = R_{\nu\mu}, \tag{9.89}$$

since contraction of the first or last pair of indices of $R_{\kappa\lambda\mu\nu}$ yields a vanishing result due to the alternating behavior (first line of (9.34)) and the remaining contractions are equal up to a sign. The obtained **Ricci tensor** is symmetric due to the last line of (9.34) and hence may have again a non-zero contraction, the **curvature scalar**

$$R = g^{\mu\nu}R_{\mu\nu}. \tag{9.90}$$

Einstein's action is

$$\mathcal{S}[g] = \frac{c}{\kappa} \int R\tau, \tag{9.91}$$

where κ is Einstein's gravitational constant, but the prefactor is convention as long as no interaction with matter is considered. However, the action integral would not have a minimum unless this prefactor is positive. The volume form τ was given in (9.84). True, this action contains second derivatives of the potentials $g_{\mu\nu}$, however, they can be eliminated by an integration by parts. The variation of this action yields

$$\begin{aligned} \delta\left(R\sqrt{-\det g}\right) &= \delta\left(g^{\mu\nu}R_{\mu\nu}\sqrt{-\det g}\right) \\ &= R_{\mu\nu}\sqrt{-\det g}\,\delta g^{\mu\nu} + R\delta\sqrt{-\det g} + g^{\mu\nu}\sqrt{-\det g}\,\delta R_{\mu\nu}. \end{aligned}$$

With (9.25),

$$\delta\sqrt{-\det g} = -\frac{\delta g}{2\sqrt{-\det g}} = -\frac{1}{2}\sqrt{-\det g}\, g_{\mu\nu}\delta g^{\mu\nu},$$

and the above variation may be cast into

$$\delta\left(R\sqrt{-\det g}\right) = \left(R_{\mu\nu} - \frac{1}{2}Rg_{\mu\nu}\right)\sqrt{-\det g}\,\delta g^{\mu\nu} + \sqrt{-\det g}\, g^{\mu\nu}\delta R_{\mu\nu}.$$

The last expression is a total derivative, which can be seen from calculating it in normal coordinates at x^μ, where then the $\Gamma^\lambda_{\ \mu\nu}$ and the derivatives of the $g_{\mu\nu}$ vanish at x^μ. (See p. 315; $\nabla g = 0$, and for vanishing $\Gamma^\lambda_{\ \mu\nu}$ one has from (7.43) $\nabla_{\partial/\partial x^\mu} g = \partial g/\partial x^\mu$.) From (7.48), in normal coordinates,

$$g^{\mu\nu}\delta R_{\mu\nu} = g^{\mu\nu}\left(\frac{\partial}{\partial v^\kappa}\delta\Gamma^\kappa_{\ \mu\nu} - \frac{\partial}{\partial v^\nu}\delta\Gamma^\kappa_{\ \mu\kappa}\right) = \frac{\partial}{\partial v^\kappa}\left(g^{\mu\nu}\delta\Gamma^\kappa_{\ \mu\nu} - g^{\mu\kappa}\delta\Gamma^\nu_{\ \mu\nu}\right) = \frac{\partial w^\kappa}{\partial v^\kappa}.$$

In these expressions, $\delta\Gamma^\kappa_{\ \mu\nu}$ is the variation of $\Gamma^\kappa_{\ \mu\nu}$ under a variation of the metric $g_{\mu\nu}$. Such a variation does not affect the second term on the right hand side of (7.38) which only depends on the transition function between two local coordinate systems. Hence, it drops out of the variation of (7.38), and from what remains it is seen that unlike $\Gamma^\kappa_{\ \mu\nu}$ itself its variation is a tensor. (It measures the difference of two parallel transported tensors along the same path element in different metrics and hence takes the difference of two tensors in the same point of the manifold.) Thus, w^κ is a vector, and since the left hand side of the above chain of equations is a scalar, the right hand side must also be a scalar. This allows to write down this relation in arbitrary coordinates as $g^{\mu\nu}\delta R_{\mu\nu} = w^\kappa_{;\kappa}$. Now, from (9.91), $\sqrt{-\det g}g^{\mu\nu}\delta R_{\mu\nu} = \partial(\sqrt{-\det g}w^\kappa)/\partial x^\kappa$, and this derivative can be omitted in the variation of the action (9.91). For a free gravitational field the final equation of motion is $R_{\mu\nu} - (1/2)Rg_{\mu\nu} = 0$ or

$$R^\mu_\nu - \frac{1}{2} R \delta^\mu_\nu = 0. \tag{9.92}$$

Linearization of this equation at $g_{0\mu\nu}$, equal to the Minkowski metric (9.74), that is, linearization in $h_{\mu\nu}$ from $g_{\mu\nu} = g_{0\mu\nu} + h_{\mu\nu}$, yields the wave equation $\Box h^\nu_\mu = \left((\partial/\partial ct)^2 - \sum_i (\partial/\partial x^i)^2\right) h^\nu_\mu = 0$ for gravitational waves in vacuum. With (9.85–9.87) it is a simple exercise to get the relation $(\delta\omega)^\kappa = (1/2) g^{\kappa\lambda} (\partial/\partial x^\mu) \omega_\lambda^{\ \mu}$ for an alternating tensor field ω and *a homogeneous metric*, $g_{\mu\nu} = \text{const}$. Hence, $(d\delta\omega)_\lambda^{\ \kappa} = (1/2) g^{\kappa\nu} \partial^2 / (\partial x^\lambda \partial x^\mu) \omega_\nu^{\ \mu}$, from which $(d\delta)_{\lambda\mu}^{\kappa\nu} = (1/2) g^{\kappa\nu} \partial^2 / (\partial x^\lambda \partial x^\mu)$ results. Similarly, $(\delta d)_{\lambda\mu}^{\kappa\nu}$ yields the same result. That means that the d'Alembert operator $\Box$ for a homogeneous metric is to be replaced by the Laplace–Beltrami operator $(\delta d + d\delta)$ in covariant relations for an arbitrary metric. Therefore, $(\delta d + d\delta) h^\nu_\mu = 0$ is the wave equation for the propagation of gravitational waves on top of any gravitational field.

If in addition to a gravitational field matter fields are present, then similar to (9.91) an action integral

$$S_m = \frac{1}{2c} \int T^{\mu\nu} g_{\mu\nu} \tau \tag{9.93}$$

may additionally be introduced with again a prefactor of convention and of setting an energy scale, where $T^{\mu\nu}$ is the energy–momentum tensor of matter. Its expression for various forms of matter can be found in theoretical physics textbooks. (T^{00}/c is an energy density in 3-space, and its multiplication with the 4-volume gives an action.) The mere demand of covariance couples matter to the gravitational potential $g_{\mu\nu}$. (By using anti-commutators in the symmetry group, gravity as supergravity becomes a Yang–Mills theory and hence a gauge field theory [5, Chapter 18].) Varying $g_{\mu\nu}$ in the sum of (9.91) and (9.93) yields Einstein's final field equations for the gravitational field as

$$R^\mu_\nu - \frac{1}{2} R \delta^\mu_\nu = \frac{\kappa}{c^2} T^\mu_\nu, \quad R = -\frac{\kappa}{c^2} T, \quad R^\mu_\nu = \frac{\kappa}{c^2} \left(T^\mu_\nu - \frac{1}{2} T \delta^\mu_\nu \right) \tag{9.94}$$

in which the energy-momentum tensor appears as the source term for gravitation as phenomenologically expected. The second and hence the third equation simply follow from tensor contraction of the first. Note that this equation, on which nowadays the construction of highest resolution GPS is based, was established prior to all experimental observations of its consequences.

Let

$$g_{00} = 1 + \frac{2\varphi}{c^2}, \quad g_{0i} = 0, \quad g_{ij} = -\delta_{ij} \left(1 - \frac{2\varphi}{c^2} \right), \quad \frac{2\varphi}{c^2} \ll 1, \tag{9.95}$$

and let $T_0^0 = mc^2\delta(\boldsymbol{r}) = T$, $\boldsymbol{r} = (x^1, x^2, x^3)$ be the only non-zero component of the energy-momentum tensor for a point particle of mass m at rest at the coordinate origin. Then, from the last equation (9.94),

$$R_0^0 = \frac{1}{2}\kappa m\delta(\boldsymbol{r}), \quad R_i^0 = R_0^i = 0, \quad R_k^i = -\delta_k^i \frac{1}{2}\kappa m\delta(\boldsymbol{r}).$$

From (7.48) and (9.23), in lowest order of $1/c$,

$$R_0^0 \approx R_{00} \approx \frac{\partial \Gamma^i_{00}}{\partial x^i} \approx \frac{1}{2} g^{ii} \frac{\partial^2 g_{00}}{\partial x^{i2}} = \frac{1}{c^2}\Delta\varphi \approx -R_i^i$$

with the Laplacian Δ. Hence,

$$\Delta\varphi = \frac{c^2\kappa m}{2}\delta(\boldsymbol{r}) = 4\pi k m\delta(\boldsymbol{r}), \quad \kappa = \frac{8\pi k}{c^2}, \tag{9.96}$$

which is Newton's law, if k is Newtons phenomenological gravitational constant. This finally determines Einstein's constant in (9.91).

A solution of (9.96) is $\varphi(\boldsymbol{r}) = -km/r$. It has a singularity at $r = 0$, and for an observer on a world line $x^\mu(s)$ proper time and length element (9.78, 9.79) are

$$dt = \frac{1}{c}\sqrt{1 - \frac{2km}{c^2 r}}dx^0, \quad dl^2 = \left(1 + \frac{2km}{c^2 r}\right)\sum_{i-1}^{3}(dx^i)^2.$$

Although (9.96) is the right equation for the gravitational field of a point charge only in lowest order in $1/c$ and hence only for large enough distances from the center $r = 0$ (in truth, in modified coordinates (r, θ, ϕ) in which $2\pi r$ means the length of a circle centered at the mass center so that due to the curvature r is not any more the proper radius, $dl^2 = dr^2/(1 - 2km/(c^2 r)) + r^2(\sin\theta\, d\phi^2 + d\theta^2))$, a qualitative feature is retained in the solution of the exact equation: there is a horizon, the Schwarzschild radius $r_0 = 2km/c^2$ at which dt vanishes for any world line $x^\mu(s)$. A clock on this world line becomes so fast that any motion with velocity $v \le c$ measured with this clock seems to come to a halt. No passing of the horizon by matter or light seems to be possible. The latter is indeed what happens with matter or light from smaller radii. However, in the above mentioned exact Schwarzschild solution of the problem, g_{rr} and hence dl^2 diverges at the same radius r_0, and all world lines of matter are bent inward for $r < r_0$, which radius thus can be passed only in one direction. Although g_{00} and g_{rr} are singular *in the chosen coordinates*, det g for the exact solution has again no singularity there. For an observer from outside, in regularized coordinates the surface of the r_0-sphere shrinks to a point in which all information approaching it disappears. Inside r_0 no materializable rest frame exists. All matter moves eventually only further inward; there is no cone of future opening outward from r_0. If for a real spherical mass distribution of high enough mass density ρ there is a solution of $r = r_0(m(r))$, where $m(r)$ is the total mass inside r, then this situation of a **black hole** is realized. Even at large enough values of r for which (9.96) is justified, clocks, for instance atomic vibrations, slow down with increasing r. For an observer at a fixed value r_1, $r_0 < r_1 < r$ light from atomic spectra at r arrives at r_1 red shifted and its rays are bent. This gravitational red shift and bending from the earth's mass has to be taken into account in the high resolution GPS.

9.7 Complex, Hermitian and Kählerian Manifolds

Throughout this text real manifolds were considered with differentiable structures (complete atlases) for simplicity assumed smooth although many statements on manifolds generalize to less restrictive cases of C^m-manifolds for sufficiently large m (between 0 and 3 for most statements). If M has a complete atlas of coordinate neighborhoods U_α which are homeomorphic to open sets $\boldsymbol{U}_\alpha \in \mathbb{C}^m$ in such a way that all transition functions $\boldsymbol{\varphi}_\beta \circ \boldsymbol{\varphi}_\alpha^{-1} : \boldsymbol{\varphi}_\alpha(U_\alpha \cap U_\beta) = \boldsymbol{U}_\alpha \cap \boldsymbol{U}_\beta \to \boldsymbol{U}_\alpha \cap \boldsymbol{U}_\beta = \boldsymbol{\varphi}_\beta(U_\alpha \cap U_\beta)$, are analytic ($C^\omega$), then M is called an m-dimensional **complex manifold**.

Recall from complex function theory, that a complex function $F(z^1, \dots, z^m) = \operatorname{Re} F(x^1, y^1, \dots, x^m, y^m) + i \operatorname{Im} F(x^1, y^1, \dots, x^m, y^m)$, $z^j = x^j + iy^j$ is analytic (holomorphic) at point (z^j), iff it obeys the **Cauchy–Riemann conditions**

$$\frac{\partial \operatorname{Re} F}{\partial x^j} = \frac{\partial \operatorname{Im} F}{\partial y^j}, \quad \frac{\partial \operatorname{Re} F}{\partial y^j} = -\frac{\partial \operatorname{Im} F}{\partial x^j}, \quad j = 1, \dots, m, \tag{9.97}$$

or iff it can be Taylor expanded in a neighborhood of (z^j). If a mapping is analytic and a homeomorphism, then the inverse mapping exists and is also analytic. Recall also that a complex function F analytic on some domain cannot assume an extremal absolute value $|F|$ at an inner point of that domain unless it is a constant. Hence, any analytic mapping of a compact, connected complex manifold without boundary into the $\mathbb{C}^m$ must be a mapping to one single point of $\mathbb{C}^m$. This gives some flavor of how restrictive the condition of analyticity on complex manifolds is.

As the simplest example, consider a two-dimensional oriented real manifold M with a Riemannian metric $ds^2 = (\omega^1)^2 + (\omega^2)^2$, where ω^j are two 1-forms on M, in a coordinate neighborhood given by $\omega^j = \omega^j_x(x,y)dx + \omega^j_y(x,y)dy$. A complex-valued 1-form ω in two real dimensions is called an **analytic 1-form**, if it can be written as $\omega = f(z)dz$, $z = x + iy$, with an analytic complex function f. If in the above metric $\omega^1 + i\omega^2$ is analytic, then $ds^2 = (\omega^1 + i\omega^2)(\omega^1 - i\omega^2)$ is called an **analytic metric**. This metric can be written as

$$ds^2 = |f(z)|^2 dz d\bar{z} = |f(z)|^2 (dx^2 + dy^2). \tag{9.98}$$

Thus, locally an analytic Riemannian metric on a two-dimensional real manifold can always be brought into the form (9.98). Coordinates x, y in this form are called **isothermal coordinates**. If two metrics g^1 and g^2 are related as $g^2 = fg^1$ where f is a positive function, then it is readily seen from (9.4, 9.5) that all angles have the same values in both metrics. The Riemannian geometries with such two metrics are called **conformal**, and mappings $F: M_1 \to M_2$ between geometries M_1 and M_2 which preserve all angles are called **conformal mappings**. Hence, an analytic Riemannian metric on a two-dimensional manifold is always locally conformal to the Euclidean metric. If there are coordinates $w = u + iv$ in which ds^2 may be written as $ds^2 = |\phi(w)|^2 dw d\bar{w}$, then either dw is proportional to dz (orientation preserving) or to $d\bar{z}$ (orientation reversing). In the first case w is an analytic function of z. If $F(z)$ is an analytic complex function with $dF/dz = f(z)$, then locally $ds^2 = dF d\bar{F}$ and $\mathrm{Re}\, F$ and $\mathrm{Im}\, F$ are Cartesian local coordinates in the above considered manifold M.

The metric (9.98) defines a one-dimensional complex Riemannian manifold which is also called a **Riemannian surface**. Note that $(\omega^1 + i\omega^2) = f\, dz$ means $\omega^1_x = \omega^2_y = \mathrm{Re} f, \omega^1_y = -\omega^2_x = \mathrm{Im} f$. It is seen that every two-dimensional orientable real manifold allows for a complex manifold structure with a locally Euclidean metric that makes it into a Riemannian surface.

Let M be a complex manifold of dimension m and consider the tangent space $T_z(M)$ on M at point $z = (z^1, \ldots, z^m)$. The linear mapping $J: T_z(M) \to T_z(M)$: $JX = iX$ or $JX = -iX$ has the obvious property $J^2 = J \circ J = -\mathrm{Id}_{T_z(M)}$. Conversely, if V is a vector space, then a linear transformation $J: V \to V$, $J^2 = -\mathrm{Id}_V$ is called a **complex structure** on V. Treat V as a real vector space, then two vectors X and JX cannot be proportional to each other, $JX = \lambda X$ with real λ, because then $J^2 X = \lambda^2 X$, $\lambda^2 \geq 0$, against the assumption. Hence, X and JX span a two-dimensional subspace E_X of V which is clearly invariant under J, $JE_X = E_X$, and every invariant subspace of J in V is two-dimensional. A complex structure can only exist on V, if as a real vector space it is even-dimensional. If V^* is the dual space of V, then a complex structure J on V induces a complex structure (also called J) on V^* by the definition

$$\langle \omega, JX \rangle = \langle J\omega, X \rangle \quad \text{for all } \omega \in V^*,\ X \in V. \tag{9.99}$$

From this definition, $J^2 = -\mathrm{Id}_{V^*}$ readily follows.

An example from physics is the time reversal operator $\widehat{T}$ in quantum mechanics. As an operator in the space of spinor quantum states Ψ it has the property $\widehat{T}^2 = -\mathrm{Id}$ and hence is a complex structure. If a Hamiltonian $\widehat{H}$ commutes with $\widehat{T}$, then $\widehat{H}\Psi = \Psi E$ implies $\widehat{H}(\widehat{T}\Psi) = \widehat{T}\widehat{H}\Psi = (\widehat{T}\Psi)E$, and all eigenvalues E of $\widehat{H}$ are twofold degenerate. This is called **Kramers degeneracy**. Much more general implications of a complex structure on quantum physics are discussed in [6].

Treat further on V as a real even-dimensional vector space provided with a complex structure J. Consider the set of complex valued linear functions on V as the complexification $V^* \otimes \mathbb{C}$ of the dual space V^*. Since V^* is a real vector space of dimension m, $V^* \otimes \mathbb{C}$ is a complex vector space of complex dimension m. Any element $\Omega \in V^* \otimes \mathbb{C}$ can be written as $\Omega = \omega^1 + i\omega^2$, $\omega^j \in V^*$. Extend J from V^* to $V^* \otimes \mathbb{C}$ by complex linearity. The linear operator J has eigenvectors as elements of $V^* \otimes \mathbb{C}$ with eigenvalues $\pm i$, which are called elements of type $(1,0)$ (upper sign) and $(0,1)$ (lower sign), respectively. Denote the subspace of $V^* \otimes \mathbb{C}$ of all elements of type $(1,0)$ by $V^*_{\mathbb{C}}$ and the subspace of all elements of type $(0,1)$ by $\bar{V}^*_{\mathbb{C}}$. Let Ω be any vector in $V^* \otimes \mathbb{C}$ and let $\Omega^1 = (1 - iJ)\Omega/2$, $\Omega^2 = (1 + iJ)\Omega/2$. Then, $\Omega = \Omega^1 + \Omega^2$ and $J\Omega^1 = (J + i)\Omega/2 = i\Omega^1$, $J\Omega^2 = (J - i)\Omega/2 = -i\Omega^2$. Hence, $\Omega^1 \in V^*_{\mathbb{C}}$, $\Omega^2 \in \bar{V}^*_{\mathbb{C}}$ and $V^* \otimes \mathbb{C} = V^*_{\mathbb{C}} \oplus \bar{V}^*_{\mathbb{C}}$. Moreover, J maps $V^*_{\mathbb{C}}$ onto $\bar{V}^*_{\mathbb{C}}$ under complex conjugation. Indeed, let again $\Omega = \omega^1 + i\omega^2$ and let it be in $V^*_{\mathbb{C}}$. Then, $J\Omega = i(\omega^1 + i\omega^2) = -\omega^2 + i\omega^1$. Hence, $J\omega^1 = -\omega^2$, $J\omega^2 = \omega^1$, and $J\bar{\Omega} = J(\omega^1 - i\omega^2) = -\omega^2 - i\omega^1 = -i\bar{\Omega}$, which means $\bar{\Omega} \in \bar{V}^*_{\mathbb{C}}$ for $\Omega \in V^*_{\mathbb{C}}$. Both spaces $V^*_{\mathbb{C}}$ and $\bar{V}^*_{\mathbb{C}}$ are complex vector spaces of dimension $m/2$ each. If $\{\Omega^j \,|\, j = 1,\dots,m/2\}$ is any base in $V^*_{\mathbb{C}}$, then $\{\Omega^j, \bar{\Omega}^j \,|\, j = 1,\dots,m/2\}$ is a base in $V^* \otimes \mathbb{C}$. Let

$$\Omega^j = \omega^j + i\omega^{m/2+j} \in V^*_{\mathbb{C}}, \quad \bar{\Omega}^j = \omega^j - i\omega^{m/2+j} \in \bar{V}^*_{\mathbb{C}},\ j = 1,\dots,m/2 \tag{9.100}$$

with $\omega^k \in V^*$, $k = 1,\dots,m$. Since $\{\omega^j = (\Omega^j + \bar{\Omega}^j)/2, \omega^{m/2+j} = (\Omega^j - \bar{\Omega}^j)/(2i) | j = 1,\dots,m/2\}$ is just another base in $V^* \otimes \mathbb{C}$ with real base vectors ω^j, $j = 1,\dots,m$, it is also a base in the real space V^*. Because $J\Omega^j = i\Omega^j, J\bar{\Omega}^j = -i\bar{\Omega}^j$,

$$J\omega^j = -\omega^{m/2+j}, \quad J\omega^{m/2+j} = \omega^j,\ j = 1,\dots,m/2. \tag{9.101}$$

The base $\{\omega^j \,|\, j = 1,\dots,m\}$ in V^* is called a ***J*-adapted base**. For the dual base $\{X_k \,|\, k = 1,\dots,m\}$ in V, $\langle \omega^j, X_k \rangle = \delta^j_k$, this immediately implies

$$JX_k = X_{m/2+k}, \quad JX_{m/2+k} = -X_k,\ k = 1,\dots,m/2. \tag{9.102}$$

It is likewise a J-adapted base. With these notations one has $\omega^j \wedge J\omega^j = -\omega^j \wedge \omega^{m/2+j} = -(i/2)\Omega^j \wedge \bar{\Omega}^j$, whence the volume element is (with the obvious notation $\wedge_{j=1}^m \omega^j = \omega^1 \wedge \cdots \wedge \omega^m$)

$$\tau = \bigwedge_{j=1}^{m/2} (\omega^j \wedge \omega^{m/2+j}) = \left(\frac{i}{2}\right)^{m/2} \bigwedge_{j=1}^{m/2} (\Omega^j \wedge \bar{\Omega}^j). \tag{9.103}$$

Any linear base transformation $\Pi^j = \sum_{k=1}^{m/2} \Omega^k A^j_k$ with a complex $((m/2) \times (m/2))$-matrix A yields $\wedge_j (\Pi^j \wedge \bar{\Pi}^j) = |\det A|^2 \wedge_j (\Omega^j \wedge \bar{\Omega}^j)$ and hence it corresponds to a linear base transformation in V which preserves the orientation of V.

Summarizing: *If J is a complex structure on a real vector space V, then V is of even dimension m, and there exists a base $\{X_k, JX_k \,|\, k = 1,\ldots,m/2\}$ in V, and any two such bases yield the same orientation of V. Conversely, if the dimension m of V is even and there exists a direct sum decomposition $V^* \otimes \mathbb{C} = V^*_{\mathbb{C}} \oplus \bar{V}^*_{\mathbb{C}}$ of the complexified dual of V such that complex conjugation maps bijectively $V^*_{\mathbb{C}}$ onto $\bar{V}^*_{\mathbb{C}}$, then there is a complex structure on V for which the elements of $V^*_{\mathbb{C}}$ are of type $(1,0)$ and the elements of $\bar{V}^*_{\mathbb{C}}$ are of type $(0,1)$.*

For the last statement one simply puts $J\Omega = i\Omega$ in $V^*_{\mathbb{C}}$ and $J\Omega = -i\Omega$ in $\bar{V}^*_{\mathbb{C}}$.

A $(p+q)$-linear alternating complex function on V which may be written as

$$\sum_{j_1<\cdots<j_p,\bar{j}_1<\cdots<\bar{j}_q} C_{j_1\ldots j_p,\bar{j}_1\ldots\bar{j}_q}\Omega^{j_1}\wedge\cdots\wedge\Omega^{j_p}\wedge\bar{\Omega}^{\bar{j}_1}\wedge\cdots\wedge\bar{\Omega}^{\bar{j}_q} \tag{9.104}$$

is called an **exterior (p,q)-form**. Obviously, if Ω is an exterior (p,q)-form, then $\bar{\Omega}$ is an exterior (q,p)-from, and if Π is an exterior (r,s)-form, then $\Omega\wedge\Pi$ is an exterior $(p+r,q+s)$-from.

Like the J-adapted base $\{\omega^i \,|\, j = 1,\ldots,m\}$ could be used as a base in both the real space V^* and the complex space $V^*\otimes\mathbb{C}$, the J adapted base $\{X_k \,|\, k = 1,\ldots,m\}$ in V can also be used in the complex m-dimensional space $V\otimes\mathbb{C}$. Then, introduce in the latter space the alternative base

$$Z_k = \frac{1}{2}\big(X_k - iX_{m/2+k}\big),\quad \bar{Z}_k = \frac{1}{2}\big(X_k + iX_{m/2+k}\big),\ k = 1,\ldots,m/2 \tag{9.105}$$

and find $\langle\Omega^j, Z_k\rangle = \langle(\omega^j + i\omega^{m/2+j}), (X_k - iX_{m/2+k})\rangle/2 = \delta^j_k = \langle\bar{\Omega}^j, \bar{Z}_k\rangle$, $j,k = 1,\ldots m/2$ since ω^j and X_k were dual to each other for $j,k = 1,\ldots,m$. Likewise one finds $\langle\Omega^j, \bar{Z}_k\rangle = 0 = \langle\bar{\Omega}^j, Z_k\rangle$. The bases $\{\Omega^j, \bar{\Omega}^j\}$ and $\{Z_k, \bar{Z}_k\}$ are dual to each other in the complex spaces $V^*\otimes\mathbb{C}$ and $V\otimes\mathbb{C}$.

Let V again be a real vector space with a complex structure J, and let $H: V\times V\to\mathbb{C}$ be a complex function of two vector variables with the properties (physics convention, mathematics convention would interchange the first with the second argument, compare p. 19)

1. $H(X, \lambda_1X_1 + \lambda_2X_2) = \lambda_1H(X,X_1) + \lambda_2H(X,X_2),\quad X, X_j\in V,\ \lambda_j\in\mathbb{R},$
2. $H(Y,X) = \overline{H(X,Y)},\quad X,Y\in V,$
3. $H(X,JY) = iH(X,Y).$

Then, H is called a **Hermitian structure**. The first two conditions mean that $\operatorname{Re}H$ is a symmetric bilinear function, and $\operatorname{Im}H$ is an alternating bilinear function while it follows from the second and third properties that

$$\begin{aligned}&\operatorname{Re}H(X,JY) = -\operatorname{Im}H(X,Y),\quad \operatorname{Im}H(X,JY) = \operatorname{Re}H(X,Y),\\ &H(JX,JY) = H(X,Y).\end{aligned} \tag{9.106}$$

If besides the complex structure J a symmetric bilinear function $F(X,Y)$ on V is given (or an alternating bilinear function $G(X,Y)$), then it defines with the relations (9.106) a Hermitian structure $\operatorname{Re} H = F$ ($\operatorname{Im} H = G$). The Hermitian structure H is called positive definite, if $\operatorname{Re} H = F$ is a positive definite function. It is easily seen that a positive definite Hermitian structure defines a (real) inner product in V,

$$(X|Y) = \operatorname{Re} H(X,Y) = \frac{1}{2}(H(X,Y) + H(Y,X)), \tag{9.107}$$

that is invariant under the mapping J, $(JX|JY) = (X|Y)$.

Expanded in the J-adapted base (9.102), general vectors $X, Y \in V$ are

$$X = \sum_{k=1}^{m/2}\left(\xi^k X_k + \xi^{m/2+k} JX_k\right), \quad Y = \sum_{k=1}^{m/2}\left(\eta^k X_k + \eta^{m/2+k} JX_k\right), \quad \xi^k, \eta^k \in \mathbb{R}$$

and hence,

$$H(X,Y) = \sum_{j,k=1}^{m/2}\left(\xi^j - i\xi^{m/2+j}\right)\left(\eta^k + i\eta^{m/2+k}\right)H(X_j,X_k).$$

Now, $\xi^j = \langle \omega^j, X\rangle$, and hence, with (9.100), $\xi^j - i\xi^{m/2+j} = \langle \bar{\Omega}^j, X\rangle$, $\eta^k + i\eta^{m/2+k} = \langle \Omega^k, Y\rangle$. Therefore,

$$H(X,Y) = \sum_{j,k=1}^{m/2} H_{\bar{j}k}\langle \bar{\Omega}^j, X\rangle\langle \Omega^k, Y\rangle, \quad H_{\bar{j}k} = H(X_j,X_k) = \bar{H}_{k\bar{j}}, \tag{9.108}$$

that is, the Hermitian structure is given by a Hermitian 2-form

$$H = \sum_{j,k=1}^{m/2} H_{\bar{j}k}\bar{\Omega}^j \otimes \Omega^k \tag{9.109}$$

expressed in a J-adapted base in $V^* \otimes \mathbb{C}$.

The alternating 2-form (exterior 2-from) G with $\langle G, X\wedge Y\rangle = \operatorname{Im} H(X,Y) = \left(H(X,Y) - \overline{H(X,Y)}\right)/(2i) = \left(\sum_{j,k=1}^{m/2} H_{\bar{j}k}\langle\bar{\Omega}^j,X\rangle\langle\Omega^k,Y\rangle - \langle\bar{\Omega}^j,Y\rangle\langle\Omega^k,X\rangle\right)/(2i) = \left\langle(-i/2)\sum_{j,k=1}^{m/2} H_{\bar{j}k}\bar{\Omega}^j\wedge\Omega^k, X\wedge Y\right\rangle$, that is,

$$iG = \frac{1}{2}\sum_{j,k=1}^{m/2} H_{\bar{j}k}\bar{\Omega}^j \wedge \Omega^k, \tag{9.110}$$

is called the **Kählerian form** of the Hermitian structure H. Hence, $H = F + iG$, where $F = \operatorname{Re} H$ is a symmetric 2-form.

If V is a complex vector space of dimension $m/2$, then a **Hermitian structure** on the complex vector space is a function $H : V \times V \to \mathbb{C}$ with the properties

1. $H(Z, \lambda_1 Z_1 + \lambda_2 Z_2) = \lambda_1 H(Z, Z_1) + \lambda_2 H(Z, Z_2), \quad Z, Z_j \in V, \ \lambda_j \in \mathbb{C},$
2. $H(Z_2, Z_1) = \overline{H(Z_1, Z_2)}, \quad Z_j \in V.$

It is said to be positive definite, if $H(Z,Z) > 0$. Thus, if V is an m-dimensional real vector space with a complex structure J, then by defining $iX = JX$ for all $X \in V$ it extends by complex linearity to a complex vector space of dimension m; if it has a Hermitian structure H, then this is both a Hermitian structure of the real and complex vector spaces V.

A real manifold M of dimension m with a given smooth real tensor field J of type $(1,1)$ such that at every point $x \in M$ the tensor J_x provides the tangent space $T_x(M)$ with a complex structure is called an **almost complex manifold** and J is called an **almost complex structure** on M. Clearly, M must be orientable and of even dimension. However, it can be shown that not every orientable manifold of even dimension can have an almost complex structure. Suppose M is an almost complex manifold and x^j are local coordinates in M. Then,

$$J_x\left(\frac{\partial}{\partial x^j}\right) = \sum_{k=1}^{m} J_j^k(x) \frac{\partial}{\partial x^k}, \quad \sum_{l=1}^{m} J_j^l(x) J_l^k(x) = -\delta_j^k, \tag{9.111}$$

with smooth functions $J_j^k(x)$.

As will be seen now, every complex manifold is an almost complex manifold, the opposite, however, is in general only true for $m = 2$. Let M be an $m/2$-dimensional complex manifold. It can be viewed as an m-dimensional real manifold. Consider a coordinate neighborhood with local complex coordinates $(z^1, \ldots, z^{m/2})$, $z^j = x^j + iy^j$, $x^j, y^j \in \mathbb{R}$. Define the tensor field J by

$$J_z\left(\frac{\partial}{\partial x^j}\right) = \frac{\partial}{\partial y^j}, J_z\left(\frac{\partial}{\partial y^j}\right) = -\frac{\partial}{\partial x^j}. \tag{9.112}$$

Obviously, $J_z^2 = -\mathrm{Id}_{T_z(M)}$. Of course, the definition of J must have a coordinate independent meaning. Let $z^j = z^j(w^k)$ be $m/2$ analytic complex functions in a neighborhood of the point $z \in M$, so that $w^k = u^k + iv^k$, $u^k, v^k \in \mathbb{R}$ are alternative local coordinates there. Due to the Cauchy–Riemann conditions (9.97) for the functions z^j, one gets from (9.112)

$$J_z\left(\frac{\partial}{\partial u^k}\right) = J_z\left(\sum_j \frac{\partial x^j}{\partial u^k}\frac{\partial}{\partial x^j} + \sum_j \frac{\partial y^j}{\partial u^k}\frac{\partial}{\partial y^j}\right) = \frac{\partial}{\partial v^k},$$

$$J_z\left(\frac{\partial}{\partial v^k}\right) = J_z\left(\sum_j \frac{\partial x^j}{\partial v^k}\frac{\partial}{\partial x^j} + \sum_j \frac{\partial y^j}{\partial v^k}\frac{\partial}{\partial y^j}\right) = -\frac{\partial}{\partial u^k}.$$

Thus, J is correctly defined. This almost complex structure defined by (9.112) is called the **canonical almost complex structure** of the complex manifold M. From (9.99) it follows that

$$J_z(dx^j) = -dy^j, \quad J_z(dy^j) = dx^j, \tag{9.113}$$

and hence $dz^j = dx^j + idy^j$ is a complex differential $(1,0)$-form, $J_z(dz^j) = idz^j$, and $d\bar{z}^j = dx^j - idy^j$ is a complex differential $(0,1)$-form, $J_z(d\bar{z}^j) = -id\bar{z}^j$, both as elements of the complex vector space $T_z^*(M) \otimes \mathbb{C} = T_z^*(M)_\mathbb{C} \oplus \overline{T_z^*(M)}_\mathbb{C}$. Accordingly, in the complexified tangent space $T_z(M) \otimes \mathbb{C}$,

$$\frac{\partial}{\partial z^j} = \frac{1}{2}\left(\frac{\partial}{\partial x^j} - i\frac{\partial}{\partial y^j}\right), \quad \frac{\partial}{\partial \bar{z}^j} = \frac{1}{2}\left(\frac{\partial}{\partial x^j} + i\frac{\partial}{\partial y^j}\right), \; j = 1, \ldots, m/2, \tag{9.114}$$

are tangent vectors of type $(1,0)$, $J_z(\partial/\partial z^j) = i(\partial/\partial z^j)$, and $(0,1)$, $J_z(\partial/\partial \bar{z}^j) = -i(\partial/\partial \bar{z}^j)$, respectively, which together span the complex tangent space $T_z(M) \otimes \mathbb{C} = T_z(M)_\mathbb{C} \oplus \overline{T_z(M)}_\mathbb{C}$. The real cotangent space of M taken as an m-dimensional real manifold is spanned by the combinations $dx^j = (dz^j + d\bar{z}^j)/2$, $dy^j = (dz^j - d\bar{z}^j)/(2i)$ and likewise for the real tangent space. As is seen, any complex manifold is indeed an almost complex manifold.

Conversely, let M be an almost complex manifold of (real) dimension m with an almost complex structure given by the smooth tensor field J of type $(1,1)$. Consider this almost complex structure locally given by $m/2$ linearly independent differential 1-forms Ω^j, $j = 1, \ldots, m/2$ of type $(1,0)$ and $m/2$ corresponding differential 1-forms $\bar{\Omega}$ of type $(0,1)$. Let $dx^j = (\Omega^j + \bar{\Omega}^j)/2$ and $dy^j = (\Omega^j - \bar{\Omega}^j)/(2i)$. If these differentials integrate locally to real analytic coordinates (that is, to coordinates belonging to an analytic complete atlas with only analytic transition functions of M), then M is a complex manifold. By virtue of Frobenius' theorem this is the case, iff

$$d\Omega^j \equiv 0 \mod \left(\Omega^k,\, 1 \le k \le m/2\right), \tag{9.115}$$

because this also implies $d\bar{\Omega}^j \equiv 0 \bmod (\bar{\Omega}^k,\, 1 \le k \le m/2)$, and in that case $\bar{\Omega}^j = 0$, $j = 1, \ldots, m/2$ defines an analytic submanifold of M, the complexified tangent spaces to which are $T_{(x,y)}(M)_\mathbb{C}$ and in which hence $m/2$ analytic complex coordinates z^j may be chosen. Since $\{\Omega^j, \bar{\Omega}^j\}$ is a base in $T^*_{(x,y)}(M) \otimes \mathbb{C}$, the exterior differential 2-form $d\Omega^j$ may be written as

$$d\Omega^j = \sum_{k<l} A^j_{kl}\Omega^k \wedge \Omega^l + \sum_{kl} B^j_{kl}\Omega^j \wedge \bar{\Omega}^l + \sum_{k<l} C^j_{kl}\bar{\Omega}^k \wedge \bar{\Omega}^l,$$

and (9.115) means $C^j_{kl} \equiv 0$. Since these expressions for $d\Omega^j$ are tensor relations, C^j_{kl} is a tensor and its vanishing and hence the condition (9.115) is invariant under linear transformations of the Ω^j as it must be. If an almost complex manifold fulfills the condition (9.115), it is called **integrable**.

It has been proved by quite technical means that real analyticity need not be presupposed, so that *every (smooth) integrable almost complex manifold M derives its almost complex structure from a canonical complex structure of a complex manifold structure of M.*

Since for $m = 2$ the condition (9.115) imposes no restriction ($\Omega = dx + idy$ in that case and hence $d\Omega = 0$), a two-dimensional almost complex manifold always derives from a one-dimensional complex manifold, that is, from a Riemannian surface. For all even $m > 2$ there exist almost complex manifolds which cannot be given the structure of a complex manifold. (This holds for instance for the real sphere S^4.)

Let M be a real manifold of even dimension m and let an almost complex structure J on M be given. Let $(x^1, \ldots, x^m)$ be arbitrary local coordinates in M. In view of (9.111), $\langle J_x(dx^j), \partial/\partial x^k\rangle = \langle dx^j, J_x(\partial/\partial x^k)\rangle = \sum_l^m J_k^l(x)\langle dx^j, \partial/\partial x^l\rangle = J_k^j(x)$. Hence, $J_x(dx^j) = \sum_k^m J_k^j(x)dx^k$. Now, $\sum_k^m (J_k^j(x) + i\delta_k^j)dx^k = (J_x + i)dx^j$ and $J_x(J_x + i)dx^j = i(i + J_x)dx^j$, $J_x(J_x - i)dx^j = -i(-i + J_x)dx^j$. This proves that

$$\sum_k^m (J_k^j(x) + i\delta_k^j)dx^k \quad \text{is an exterior differential (1,0)-form, and}$$

$$\sum_k^m (J_k^j(x) - i\delta_k^j)dx^k \quad \text{is an exterior differential (0,1)-form.}$$

Suppose the integrability condition (9.115) is valid, It follows that

$$d\left(\sum_k^m (J_k^j(x) + i\delta_k^j)dx^k\right) = \sum_{l<k}^m J_{kl}^j dx^l \wedge dx^k$$
$$\equiv 0 \mod \left(\sum_p^m \left(J_p^n(x) + i\delta_p^n\right)dx^p,\ 1 \le n \le m\right),$$

where

$$J_{kl}^j = \frac{\partial J_k^j}{\partial x^l} - \frac{\partial J_l^j}{\partial x^k}. \tag{9.116}$$

The 1-forms $\sum_p^m (J_p^n(x) + i\delta_p^n)dx^p$, $1 \le p \le m$, of type $(1,0)$ annihilate the subspace of tangent vectors $\sum_n^m (J_p^n(x) - i\delta_p^n)\partial/\partial x^n$, $1 \le p \le m$, of type $(0,1)$. Hence, the conditions above (9.116) can be expressed as

$$\sum_{kl}^m J_{kl}^j\left(J_p^k - i\delta_p^k\right)\left(J_q^l - i\delta_q^l\right) = 0, \quad \text{for all } 1 \le j,p,q \le m,$$

or, with (9.116) and the second relation (9.111),

$$T_{kl}^j = 0, \quad T_{kl}^j = \sum_p^m \left(J_{kp}^j J_l^p - J_{lp}^j J_k^p\right). \tag{9.117}$$

The tensor field T^j_{kl} of type $(1,2)$ is called the **torsion tensor field** of the almost complex structure J (not to be confused with the torsion tensor of a linear connection).

The almost complex structure J on M is integrable, iff its torsion tensor field is zero.

A very systematic complex external calculus can be developed on an integrable almost complex manifold (and hence on a complex manifold). If the components $C_{j_1\ldots j_p,\bar{j}_1\ldots\bar{j}_q}$ of (9.104) are smooth functions on M, then (9.104) is called an **exterior differential** (p,q)**-form**. Like in the real case, the algebra of all complex exterior forms over the ring of smooth complex functions on M may be considered, of which the exterior (p,q)-forms are homogeneous elements. Clearly, like in the case of the 1-form Ω^j, if Ω is an exterior (p,q)-form, then $d\Omega$ is a sum of a $(p+2,q-1)$-form (in general non-zero, if $q>0$), a $(p+1,q)$-form, a $(p,q+1)$-form and a $(p-1,q+2)$-form (in general non-zero, if $p>0$). Let $\partial\Omega$ be the part of $d\Omega$ which is a $(p+1,q)$-form and let $\bar{\partial}\Omega$ be the part of $d\Omega$ which is a $(p,q+1)$-form. Take the coordinate representation (9.104) of Ω and consider the integrability condition (9.115). It is easily seen that *either of the conditions*

$$d=\partial+\bar{\partial},\quad \bar{\partial}^2=0 \tag{9.118}$$

is equivalent to the integrability condition (9.115).

Proof Indeed, in either case (9.115) and (9.118), left condition, the $(p+2,q-1)$-part and the $(p-1,q+2)$-part of $d\Omega$ vanish, and the other two parts are obtained under both conditions. Moreover, $d^2=0$ which holds generally implies $\partial^2+\partial\circ\bar{\partial}+\bar{\partial}\circ\partial+\bar{\partial}^2=0$ because of the left condition of (9.118), and, since forms of different type are linearly independent, $\partial^2=0$, $\partial\circ\bar{\partial}+\bar{\partial}\circ\partial=0$, $\bar{\partial}^2=0$. Conversely, for a smooth function F one obtains $\partial F=\sum_j F_j\Omega^j$ with smooth functions F_j. Now, $\bar{\partial}^2F$ is the $(0,2)$-part of $d(\bar{\partial}F)=d(d-\partial)F=-d(\partial F)=-\sum_j(dF_j)\wedge\Omega^j-\sum_j F_j d\Omega^j$. The first term has no $(0,2)$-part, and that of the second term is $-(1/2)\sum_{jkl}F_jC^j_{kl}\bar{\Omega}^k\wedge\bar{\Omega}^l$. Hence, $\bar{\partial}^2=0$ implies $C^j_{kl}=0$ which is equivalent to (9.115). □

With $dz^j=dx^j+idy^j$ and (9.114), the differential of a smooth complex function F on M may be written as

$$dF=\sum_{j=1}^{m/2}\left(\frac{\partial F}{\partial x^j}dx^j+\frac{\partial F}{\partial y^j}dy^j\right)=\sum_{j=1}^{m/2}\left(\frac{\partial F}{\partial z^j}dz^j+\frac{\partial F}{\partial \bar{z}^j}d\bar{z}^j\right),$$

hence

$$\partial F=\sum_{j=1}^{m/2}\frac{\partial F}{\partial z^j}dz^j,\quad \bar{\partial}F=\sum_{j=1}^{m/2}\frac{\partial F}{\partial \bar{z}^j}d\bar{z}^j. \tag{9.119}$$

For a general exterior differential (p,q)-form

$$\Omega = \sum_{k_1<\cdots<k_p,\bar{l}_1<\cdots<\bar{l}_q} \Omega_{k_1\ldots k_p,\bar{l}_1\ldots\bar{l}_q} dz^{k_1}\wedge\cdots\wedge dz^{k_p}\wedge d\bar{z}^{l_1}\wedge\cdots\wedge d\bar{z}^{l_q} \tag{9.120}$$

one has (cf. (3.40))

$$\partial\Omega = \sum_{r=1}^{p+1}(-1)^{r+1} \sum_{k_1<\cdots<k_{p+1},\bar{l}_1<\cdots<\bar{l}_q} \frac{\partial\Omega_{k_1\ldots k_{r-1}k_{r+1}\ldots k_{p+1},\bar{l}_1\ldots\bar{l}_q}}{\partial z^{k_r}} dz^{k_1}\wedge\ldots\wedge dz^{k_p}\wedge d\bar{z}^{l_1}\wedge\ldots\wedge d\bar{z}^{l_q} \tag{9.121}$$

and

$$\bar{\partial}\Omega = \sum_{s=1}^{q+1}(-1)^{p+s+1} \sum_{k_1<\cdots<k_p,\bar{l}_1<\cdots<\bar{l}_{q+1}} \frac{\partial\Omega_{k_1\ldots k_p,\bar{l}_1\ldots\bar{l}_{s-1}\bar{l}_{s+1}\ldots\bar{l}_{q+1}}}{\partial\bar{z}^{l_s}} dz^{k_1}\wedge\cdots\wedge dz^{k_p}\wedge d\bar{z}^{l_1}\wedge\cdots\wedge d\bar{z}^{l_{q+1}}. \tag{9.122}$$

Splitting F in (9.119) into its real and imaginary parts as in (9.97) it is readily seen that

F is analytic, iff $\bar{\partial}F = 0$.

Likewise, if Ω is an exterior differential $(p,0)$-form, it is analytic, iff $\bar{\partial}\Omega = 0$, and then it holds that $d\Omega = \partial\Omega$.

If M is a complex manifold of (complex) dimension n and $(z^1,\ldots,z^n)$ is a local coordinate system, then the tangent vectors $\partial/\partial z^j$ span the (complex) tangent space $T_z(M)$. If a Hermitian structure is given on $T_z(M)$ for every $z\in M$ so that in every coordinate neighborhood the functions $H_{\bar{j}k}(z) = H((\partial/\partial z^j),(\partial/\partial z^k))$ are smooth functions of z, then M is called a **Hermitian manifold**. If moreover, the Kählerian form G of H, in local coordinates

$$iG = \frac{1}{2}\sum_{j,k=1}^{n} H_{\bar{j}k}dz^j\wedge dz^k, \tag{9.123}$$

where G is a real-valued differential $(1,1)$-form, is a closed differential form, that is, $dG=0$, then M is called a **Kählerian manifold**.

The definition of bundles and connections transfers readily to complex manifolds M. If the typical fiber of a fiber bundle is a complex vector space V of dimension n, a complex vector bundle on M is obtained. Complex tangent and cotangent bundles, as well as tensor and exterior bundles on the basis of the former, are the constructions in which the so far considered tangent, cotangent and tensor fields and complex exterior differential forms live. In particular, the consideration of complex frame fields leads to linear connections, for which the notion of torsion and curvature forms can be generalized. Complex bundles were occassionally already discussed as complex Lie groups and their Lie algebras or in connection with characteristic classes. If a positive Hermitian structure is smoothly assigned to the fibers of a complex vector bundle, then it is called a Hermitian vector bundle. It turns out that a Hermitian manifold is Kählerian, iff its Hermitian tangent bundle is torsion-free. For further reading see for instance [7] or [3].

References

1. Warner, F.W.: Foundations of Differentiable Manifolds and Lie Groups. Springer, New York (1983)
2. Fuchs, J.: Affine Lie Algebras and Quantum Groups. Cambridge University Press, Cambridge (1992)
3. Kobayashi, S., Nomizu, K.: Foundations of Differential Geometry, Interscience, vol. I and II. New York (1963, 1969)
4. Landau, L.D., Lifshits, E.M.: Classical Theory of Fields. Pergamon Press, London (1989)
5. t'Hooft, G. (eds.): 50 Years of Yang–Mills Theory. World Scientific, Singapore (2005)
6. Scolarici, G., Solombrino, L.: On the pseudo-Hermitian nondiagonalizable Hamiltonians. J Math Phys **44**, 4450–4459 (2003)
7. Chern, S.S., Chen, W.H., Lam, K.S.: Lectures on Differential Geometry. World Scientific, Singapore (2000)

Compendium

C.1 Basic Algebraic Structures

See for instance [1].

Monoid $(M, \cdot)$ or shortly M:

- $g_1 \cdot g_2 \in M$ for all $g_1, g_2 \in M$,
- $(g_1 \cdot g_2) \cdot g_3 = g_1 \cdot (g_2 \cdot g_3)$,
- $e \in M$ with $e \cdot g = g \cdot e = g$ for all $g \in M$, (frequent notation $e =: 1$)

Group $(G, \cdot)$ or G:

- $(G, \cdot)$ is a monoid,
- $g^{-1} \in G$ with $g \cdot g^{-1} = g^{-1} \cdot g = e$ for all $g \in G$.

Abelian (commutative) group $(G, +)$:

- $g_1 + g_2 = g_2 + g_1, \quad e =: 0$.

Ring $(K, +, \cdot)$ or K (with unity):

- $(K, +)$ is an Abelian group with $e =: 0$,
- $(K, \cdot)$ is a monoid with $e =: 1$,
- $k_1 \cdot (k_2 + k_3) = k_1 \cdot k_2 + k_1 \cdot k_3$, $(k_1 + k_2) \cdot k_3 = k_1 \cdot k_3 + k_2 \cdot k_3$, (distributivity; if $1 = 0$ then $K = \{0\}$).

Division ring:

- $1 \neq 0$,
- $(K \setminus \{0\}, \cdot)$ is a group.

Field

- K is a division ring with commutative multiplication.

(Some authors consider rings without unity of multiplication and call a division ring also a field.)

Module $(V, K, +)$ or V over the ring K (K-module):

- $(V, +)$ is an Abelian group with $e =: 0$,
- for all $\alpha, \beta \in K$ and for all $a, b \in V$, $\alpha a \in V$ holds with
  - $\alpha(a + b) = \alpha a + \alpha b$,
  - $(\alpha + \beta)a = \alpha a + \beta a$,
  - $\alpha(\beta a) = (\alpha \cdot \beta)a$.

  K itself is a special (one-dimensional) module over K.

Vector space over K (K-vector space):

- K is a field. (Often $K = \mathbb{R}$ or $K = \mathbb{C}$.)

K itself is a special (one-dimensional) vector space over K.

Algebra $(A, K, +, \cdot)$ or A (K-algebra):

- K is a commutative ring (with unity),
- $(A, K, +)$ is a module,
- $(A, \cdot)$ is a monoid, and for all $a, b, c \in A$ and for all $\alpha \in K$
  - $a \cdot (b + c) = a \cdot b + a \cdot c, \quad (a + b) \cdot c = a \cdot c + b \cdot c$,
  - $\alpha(a \cdot b) = (\alpha a) \cdot b = a \cdot (\alpha b)$.

K itself is a special (one-dimensional) algebra over K.

Associative algebra:

- $(a \cdot b) \cdot c = a \cdot (b \cdot c)$.

Algebra with unity:

- $e \in A$ with $a \cdot e = e \cdot a = a$ for all $a \in A$.

In an associative algebra with unity $(A, \cdot)$ is a ring.

Commutative algebra:

- $a \cdot b = b \cdot a$.

Algebraic homomorphism $f : \mathrm{AS} \to \mathrm{AS}'$, where f commutes with all algebraic operations, for instance

$$f(g_1 \cdot g_2) = f(g_1) \cdot f(g_2) \text{ etc.}$$

In cases of modules or vector spaces and algebras, usually only $K = K'$ is of interest. Homomorphisms of modules or of vector spaces are called **linear mappings** or **linear functions** or **linear operators**.

Kernel of the homomorphism: $\mathrm{Ker}\, f = f^{-1}(e')$.

$f : \mathrm{AS} \to \mathrm{AS}'$ is called an **algebraic isomorphism**, if f and f^{-1} are algebraic homomorphisms. AS and AS$'$ are called **algebraically isomorphic** in this case: $\mathrm{AS} \approx \mathrm{AS}'$.

A homomorphism $f : \mathrm{AS} \to \mathrm{AS}$ is called **Endomorphism** ($f \in \mathrm{End}\,(AS)$).

An isomorphism $f : \mathrm{AS} \to \mathrm{AS}$ is called **Automorphism** ($f \subset \mathrm{Aut}\,(AS)$).

AS is called a **sub-AS** of AS$'$, if it is isomorphic to a part of AS$'$. For every homomorphism $f : \mathrm{AS} \to \mathrm{AS}'$, $f(\mathrm{AS})$ is a sub-AS of AS$'$.

Let S be part of an AS. The intersection of all sub-AS containing S is called the sub-AS **generated** by S. A part S generating a module is said to be **linearly independent** if every finite subset of S is linearly independent; a linearly independent set S generating a module is called its **algebraic base**. The cardinality of an algebraic base of a vector space is its **algebraic dimension**.

Invariant subgroup (normal subgroup) H of G:

H is a subgroup of G and

$$gHg^{-1} \subset H \qquad \text{for all } g \in G.$$

In this case there exists a homomorphism $f : G \to G'$ with $\mathrm{Ker}\, f = H$ and $f(G) = G'$.

G' is the **quotient group** or factor group G/H.

For any $g' \in G/H$, $f^{-1}(g') \subset G$ is called **coset** of H in G.

If $G' = G/H$ is also an invariant subgroup of G, then $G = H \times G'$ is a **direct product** of groups, that is,

$$G = \{g = g_1 \cdot g_2 = g_2 \cdot g_1 \,|\, g_1 \in H,\ g_2 \in G',\ g \cdot g' = (g_1 \cdot g_1') \cdot (g_2 \cdot g_2')\}.$$

Ideal ((twosided) invariant subring) I of a ring K:
I is a subring of K and

$$aI \subset I,\ Ia \subset I \quad \text{for all } a \in K.$$

In this case there exists a homomorphism $f : K \to K'' \supset K'$ with $\operatorname{Ker} f = I$ and $f(K) = K'$.

K' is the **quotient ring** (factor ring) K/I.

If $K' = K/I$ is also an ideal of K, then $K = I \oplus K'$ is a **direct sum** of rings, that is,

$$K = \{a = a_1 + a_2 \,|\, a_1 \in I,\, a_2 \in K',\, a \cdot a' = a_1 \cdot a_1' + a_2 \cdot a_2'\}.$$

Simple group a group $G \neq \{e\}$ that has no non-trivial invariant subgroups, that is, no invariant subgroups besides G and $\{e\}$.

Representation of a group G is a homomorphism $D : G \to \operatorname{Aut}(S)$ into the group of transformations (permutations) of a non-empty set S.

$$D(g_1 \cdot g_2) = D(g_1) \cdot D(g_2) \quad (\Rightarrow D(e) = \operatorname{Id}_S).$$

The **adjoint representation** $\operatorname{Ad} : G \to \operatorname{Aut}(G)$ represents $g \in G$ as the transformation $g' \mapsto gg'g^{-1}$ of G,

$$\operatorname{Ad}(g) : G \to G : g' \mapsto \operatorname{Ad}(g)g' = gg'g^{-1}.$$

In this case, a common notation is

$$\operatorname{Ad}(g^{-1})g' = g^{-1}g'g = (g')^g, \quad ((g')^g)^h = (g')^{gh}, \quad (g')^e = g'.$$

Another most important special case of group representation is that S is a complex vector space V ($\mathbb{C}$-vector space).

Representation of a K-algebra A is a homomorphism $D : A \to \operatorname{End}(V)$ into the K-algebra of endomorphisms of a K-module V, the **representation module**, which commutes with the action of K into both algebras:

- $D(e) = \operatorname{Id}_V$ if A is an algebra with unity,
- $D(\alpha a + \beta b) = \alpha D(a) + \beta D(b)$, $\alpha, \beta \in K$, $a, b \in A$,
- $D(a \cdot b) = D(a) \cdot D(b)$.

For both groups and algebras there exists always the **trivial representation** (unit representation, null representation)

- $D(g) = \operatorname{Id}_S$ for all $g \in G$ and
- $D(a) = 0$ for all $a \in A$.

If $G(A)$ itself is a group (algebra) of automorphisms (endomorphisms), then the identical isomorphism is called the **identical representation**.

A representation is called **faithful**, if the representing homomorphism is injective.

If a group G or an algebra A is represented in a finite-dimensional vector space V and V contains an invariant subspace V_1, that is, $D(g)V_1 \subset V_1$ for all $g \in G$ or

A, then the representation D is called **algebraically reducible**, otherwise it is **algebraically irreducible**. If D is reducible, then there exists a suitable basis in V so that

$$D(g) = \begin{pmatrix} D_1(g) & \bigstar \\ 0 & D_2(g) \end{pmatrix} \quad \text{for all } g \in G \text{ or } g \in A.$$

If $V_2 = V/V_1$ is also invariant, then $D = D_1 \oplus D_2$ is a **direct sum of representations**, that is,

$$D(g) = \begin{pmatrix} D_1(g) & 0 \\ 0 & D_2(g) \end{pmatrix} \quad \text{for all } g \in G \text{ or } g \in A.$$

Two representations D and D' in V and V' are called **equivalent**, $D \approx D'$, if there is an isomorphism $L : V \to V'$ so that

$$D(g) = L^{-1}D'(g)L \quad \text{for all } g \in G \text{ or } g \in A.$$

Schur's lemma If D is a finite-dimensional irreducible representation in V, then each linear operator L in V which commutes with all $D(g)$, $g \in G$ or A is proportional to the unit operator, $L = \lambda\, \mathrm{Id}_V$.

Derivation $\boldsymbol{\delta}$ Linear mapping $\delta : A \to A$ of an algebra A to itself obeying the **Leibniz rule** $\delta(g_1 g_2) = \delta(g_1)g_2 + g_1\delta(g_2)$ for all $g_1, g_2 \in A$.

Category $\mathcal{A}$: Consisting of

- a class $\mathrm{Ob}(\mathcal{A}) = \{A, B, C, \ldots\}$ of objects,
- a class $\mathrm{Ar}(\mathcal{A})$ of **morphisms** (arrows) f, g,... with the properties
  - for each pair $(A, B) \in \mathrm{Ob}(\mathcal{A}) \times \mathrm{Ob}(\mathcal{A})$ there is a set $\mathrm{Mor}(A, B) \in \mathrm{Ar}(\mathcal{A})$ so that the composition rule

    $$\mathrm{Mor}(B, C) \times \mathrm{Mor}(A, B) \to \mathrm{Mor}(A, C) : (g, f) \mapsto g \circ f$$

    holds,
  - $\mathrm{Ar}(\mathcal{A}) = \cup_{(A,B)} \mathrm{Mor}(A, B)$ is a disjoint union,
  - for each $A \in \mathrm{Ob}(\mathcal{A})$ there is $\mathrm{Id}_A \in \mathrm{Mor}(A, A) = \mathrm{End}(A)$ with

    $$\mathrm{Id}_B \circ f = f = f \circ \mathrm{Id}_A \text{ for every } f \in \mathrm{Mor}(A, B),$$

  - when it is defined, the composition of morphisms is associative:

    $$(h \circ g) \circ f = h \circ (g \circ f).$$

Isomorphism and automorphism have their usual meaning; End (A) is obviously a monoid with respect to composition, Aut (A) is a group (its morphisms are called permutations or transformations). Hence there are group homomorphisms from any group G into Aut (A) of *any* category.

Examples of categories are sets with mappings, various AS with algebraic homomorphisms, topological spaces with continuous functions, topological AS with homomorphisms, topological, differentiable, smooth, analytic manifolds with corresponding mappings, fiber spaces with homomorphisms and so on.

Diagrams For $f \in$ Mor (A, B) the diagram

$$A \xrightarrow{f} B$$

is used. If $\mathcal{A}$ is a category then its arrows may be taken as objects of a new category $\mathcal{M}$: Ar $(\mathcal{A}) =$ Ob $(\mathcal{M})$. A morphism $f \to g, f, g \in$ Ar $(\mathcal{A})$ of $\mathcal{M}$ is a pair (φ, ψ) so that the following diagram is commutative:

$$\begin{array}{ccc} A & \xrightarrow{\varphi} & C \\ {\scriptstyle f}\downarrow & & \downarrow{\scriptstyle g} \\ B & \xrightarrow[\psi]{} & D \end{array}$$

The **commutative diagram** means $\psi \circ f = g \circ \varphi$. If $B = D$ and $\psi =$ Id $_B$ are fixed, then a category $\mathcal{A}_B$ is obtained with morphisms φ.

Covariant functor F from category $\mathcal{A}$ into category $\mathcal{B}$:

- Ob $(\mathcal{A}) \ni A \to F(A) \in$ Ob $(\mathcal{B})$ with $F(\mathrm{Id}_A) = \mathrm{Id}_{F(A)}$,
- Mor $(A, B) \ni f \to F(f) \in$ Mor $(F(A), F(B))$ with $F(g \circ f) = F(g) \circ F(f)$. Instead of $F(f)$ the notation f_* is often used: **push forward**.

A cleaning covariant functor maps a category into a simpler category.

Let $\mathcal{S}$ be the category of sets, and let some category $\mathcal{A}$ and an object $A \in$ Ob $(\mathcal{A})$ be fixed. The functor $M_A : \mathcal{A} \to \mathcal{S}$ given by

- $M_A(X) =$ Mor (A, X) (taken as a set of mappings) for every $X \in$ Ob $(\mathcal{A})$,
- $M_A(\varphi)$: Mor $(A, X) \to$ Mor $(A, X') : \psi \mapsto \varphi \circ \psi$ for every $\varphi \in$ Mor $(X, X') \subset$ Ar $(\mathcal{A})$

is called a representing covariant functor.

Contravariant functor F from category $\mathcal{A}$ into category $\mathcal{B}$:

- Ob $(\mathcal{A}) \ni A \to F(A) \in$ Ob $(\mathcal{B})$ with $F(\mathrm{Id}_A) = \mathrm{Id}_{F(A)}$,
- Mor $(A, B) \ni f \to F(f) \in$ Mor $(F(B), F(A))$ with $F(g \circ f) = F(f) \circ F(g)$. Instead of $F(f)$ the notation f^* is often used: **pull back**.

Let again some category $\mathcal{A}$ and an object $A \in$ Ob $(\mathcal{A})$ be fixed. The functor $M^A : \mathcal{A} \to \mathcal{S}$ given by

- $M^A(X) = \text{Mor}\,(X, A)$ (taken as a set of mappings) for every $X \in \text{Ob}\,(\mathcal{A})$,
- $M^A(\varphi)$: $\text{Mor}\,(X,A) \to \text{Mor}\,(X',A) : \psi \mapsto \psi \circ \varphi$ for every $\varphi \in \text{Mor}\,(X', X) \subset \text{Ar}(\mathcal{A})$

is called a representing contravariant functor.

A functor puts isomorphisms into isomorphisms.

Representing functors are used to transfer certain structures on sets to arbitrary categories. (If e.g. A has a group structure, then Mor (X, A) has also a group structure by $(fg)(x) = f(x)g(x)$, and M^A is a functor from the category $\mathcal{A}$ into the category of groups. Conversely, if Mor (X, A) has some group structure, then, by the same relation, A has a group structure.)

Complexes of K-modules

- $C = \oplus_{r\in\mathbb{Z}} C_r$ with $C_r \cdot C_{r'} \subset C_{r+r'}$ is a **graded** (by r) **module** (vector space, algebra) over a ring (field) K,
- $d = \{d_r \mid r \in \mathbb{Z}\} : C \to C'$ is a **graded morphism** of degree s, $d_r : C_r \to C'_{r+s}$ from the graded module C into the graded module C',
- a **complex** (C, d) consists of a graded K-module C and a graded endomorphism $d : C \to C$ of degree 1,
- a **morphism** $f : (C, d) \to (C', d')$ **of complexes** of degree s is a graded morphism for which the diagram

$$\begin{array}{ccc} C_{r-1} & \xrightarrow{f_{r-1}} & C'_{r-1+s} \\ {\scriptstyle d_{r-1}}\downarrow & & \downarrow{\scriptstyle d'_{r-1+s}} \\ C_r & \xrightarrow[f_r]{} & C'_{r+s} \end{array}$$

  commutes.

Homology and cohomology of a complex (C, d):

- $Z_r = \text{Ker}\, d_r$ is the module of r-**cocycles**,
- $B_r = \text{Im}\, d_{r-1}$ is the module of r-**coboundaries**,
- $H_r = Z_r/B_r$ is the rth **cohomology module** of the cohomology $H(C)$ of the complex (C, d).
- if $f : (C, d) \to (C', d')$ is a graded morphism of degree s, then it is pushed forward to a canonical homomorphism $f_* : H(C) \to H(C')$ of degree s of their cohomologies,
- **homology** is the same for a complex (C, d) with d of degree -1; this is included into the above scheme by the mapping $r \to -r$.

Exact sequences

- a sequence of morphisms of AS,

$$\cdots \to A \to B \to \cdots$$

is called exact, if the image of each morphism in the sequence is the kernel of the next,

- $0 \to G \xrightarrow{f} H$ means that f is **injective**,
- $G \xrightarrow{f} H \to 0$ means that f is **surjective**,
- $0 \to G \xrightarrow{f} H \to 0$ means that f is an **isomorphism**,
- for Abelian groups or modules, the sequence

$$0 \to H \to G \to G/H \to 0$$

  is called a **short exact sequences**,
- let

$$0 \to (C, d_C) \xrightarrow{f} (D, d_D) \xrightarrow{g} (E, d_E) \to 0$$

  be a short exact sequence of graded morphisms, without loss of generality of degree 0, of complexes; then there exists canonically a graded morphism

$$H(E) \xrightarrow{\delta} H(C)$$

  of degree 1 of their cohomologies, so that the long sequence

$$\begin{aligned} \cdots \xrightarrow{\delta} H_r(C) \xrightarrow{f_*} H_r(D) \xrightarrow{g_*} H_r(E) \\ \xrightarrow{\delta} H_{r+1}(C) \xrightarrow{f_*} H_{r+1}(D) \xrightarrow{g_*} H_{r+1}(E) \xrightarrow{\delta} \cdots \end{aligned} \tag{3.124}$$

  is exact (**'snake lemma'**).

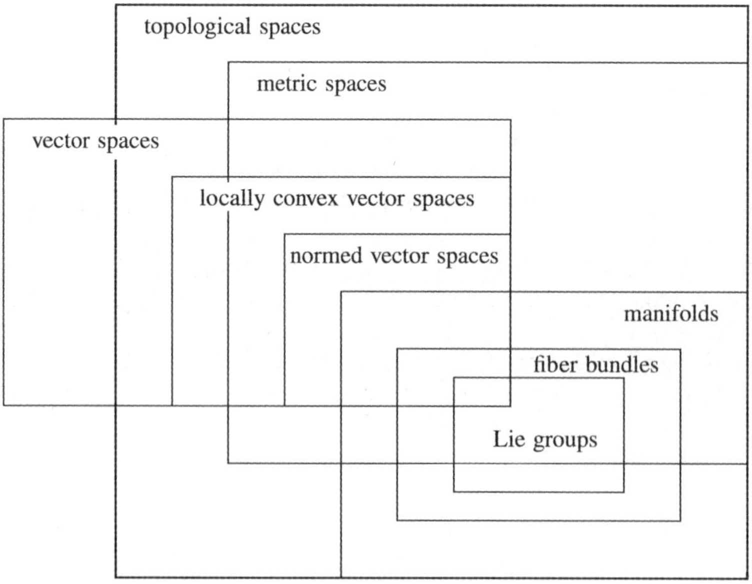

Schematic interrelation of various topological spaces

C.2 Basic Topological (Analytic) Structures

See for instance [2]; for homotopy groups, [3].

Topological space $(X, \mathcal{T})$ or shortly X:

- $\mathcal{T}$ is the set of **open subsets** of $X; \emptyset \in \mathcal{T}, X \in \mathcal{T}$,
- Every union of sets $T \in \mathcal{T}$ belongs to $\mathcal{T}$,
- Every intersection of finitely many sets $T \in \mathcal{T}$ belongs to $\mathcal{T}$,

$\mathcal{T}$ is called the **topology** on X.

Coarser (finer) topology $\mathcal{T}'$ on $X : \mathcal{T}' \subset \mathcal{T}, (\mathcal{T}' \supset \mathcal{T})$. **Discrete topology**: $\mathcal{T} = \mathcal{P}(X)(\mathcal{P}(X)$: set of all subsets of X). **Trivial topology**: $\mathcal{T} = \{\emptyset, X\}$.

Closed sets $C : C = X \backslash U$ for some $U \in \mathcal{T}$. In the discrete topology every set is open-closed. The **closure** $\overline{A}$ of a set $A \subset X$ is the smallest closed set containing A; the **interior** $\overset{\circ}{A}$ is the largest open set contained in A. The **boundary** ∂A of A is $\overline{A} \setminus \overset{\circ}{A}$. A is **dense** in X, if $X = \overline{A}$; A is **nowhere dense** in X, if $(\overline{A})^\circ = \emptyset$. X is **separable**, if $X = \overline{A}$ for some countable A.

Neighborhood of a point $x \in X : U(x) \in \mathcal{T}, x \in U(x)$. A neighborhood of a set is a neighborhood of every point of the set. **Inner point** x of $A \subset X$: $U(x) \subset A$ for some neighborhood $U(x)$, $x \in \overset{\circ}{A}$. **Point of closure** x of $A \subset X$: $A \cap U(x) \neq \emptyset$ for any neighborhood $U(x)$, $x \in \overline{A}$. **Cluster point** x of $A \subset X : (A \setminus \{x\}) \cap U(x) \neq \emptyset$ for any neighborhood $U(x)$.

Relative topology $\mathcal{T}_A$ on $A \subset X$ related to $(X, \mathcal{T}) : \mathcal{T}_A = \{A \cap T \mid T \in \mathcal{T}\}$.

Hausdorff topology Every pair of points $x \neq y \in X$ has a pair of disjoint neighborhoods, $U(x) \cap U(y) = \emptyset$. Then, every one point set $\{x\}$ is closed. The limes of a sequence of points is unique if it exists. **Regular topology**: Every non-empty open set contains the closure of another non-empty open set. **Normal topology**: Every one point set is closed and every pair of disjoint closed sets has a pair of disjoint neighborhoods.

Continuous function F: $X \to Y$ from a topological space $(X, \mathcal{T})$ into a topological space $(Y, \mathcal{U})$: For every $U \in \mathcal{U}$, $F^{-1}(U) \in \mathcal{T}$.

Homeomorphism F: $X \to Y$: a bicontinuous bijection F: F^{-1} exists and F and F^{-1} are continuous. X and Y are **homeomorphic**, $X \sim Y$, if a homeomorphism F: $X \to Y$ exists. A **topological invariant** is a property preserved under homeomorphisms.

Base of the topology $\mathcal{T}$: Family $\mathcal{B} \subset \mathcal{T}$, so that every $T \in \mathcal{T}$ is a union of sets $B \in \mathcal{B}$. **Neighborhood base** $\mathcal{B}(x)$ at $x \in X$: Every $B \in \mathcal{B}(x)$ is a neighborhood of x, and for every neighborhood $U(x)$ there is a $B \in \mathcal{B}(x)$ with $B \subset U(x)$. X is **first countable**, if every $x \in X$ has a countable neighborhood base, X is **second countable** if it has a countable base of topology.

Product topology $(X, \mathcal{T}) = (\prod_{\alpha \in A} X_\alpha, \{\prod_{\alpha \in A} U_\alpha\})$ where $U_\alpha \in \mathcal{T}_\alpha$ and $U_\alpha = X_\alpha$ for all but finitely many U_α (Tychonoff's product; the set $\prod_{\alpha \in A} X_\alpha$ is the set of all functions $F : \alpha \mapsto x_\alpha \in X_\alpha$ on A.)

Quotient topology The finest topology on X/R (R is an equivalence relation in X) in which the canonical projection π: $X \to X/R$ is continuous. ($U \in X/R$ is open, iff $\pi^{-1}(U) \in X$ is open.)

Metric topology with base $\mathcal{B} = \{B_{1/n}(x) | x \in X, 0 < n \in \mathbb{Z}\}$ of **open balls** $B_r(x) = \{x' | d(x, x') < r, r > 0\}; d : X \times X \to \mathbb{R}_+$ is the **distance function**:

- $d(x, y) = 0$ iff $x = y$,
- $d(x, y) = d(y, x)$,
- $d(x, z) \leq d(x, y) + d(y, z)$ (triangle inequality).

Metrizable space A topological space with topology which has a base of open balls in some metric d.

Cauchy sequence in a metric space X: $\{x_n\}$ with $\lim_{m,n} d(x_m, x_n) = 0$.

Complete metric space X Every Cauchy sequence converges in X. Every contracting sequence $C_1 \supset C_2 \supset \cdots$ of closed balls has a non-empty intersection.

Topological vector space X over the field $K : K \times X \to X : (\lambda, x) \mapsto \lambda x, X \times X \to X : (x, x') \mapsto x + x', \lambda \in K, x, x' \in X$ are continuous. Sum of subsets: $A + B = \{x + x' \; x \mid \in A, x' \in B$. If $\{B_\alpha(0)\}$ is a neighborhood base at $x = 0$, then $\{x + B_\alpha(0)\}$ is a neighborhood base at x. ($x + A =: \{x\} + A$.)

Functional F: $X \to K$.

Linear independence of a set $E \subset X$ If $\sum_{n=1}^{N} \lambda_n x_n$ holds for any set of $N < \infty$ distinct $x_n \in E$, then all $\lambda_n = 0$. With arbitrary $\lambda_n \in K$, all possible such linear combinations with all possible $N < \infty$ form the **span** $\mathrm{span}_K E$ of E. If $\overline{\mathrm{span}_K E} = X$, then the cardinality of E is the **dimension** of X and E forms a **base** in the vector space X. X is separable, if it admits a countable base. If E is a base of X and F is a base of Y, then $E \cup F$ taken as a linear independent disjunct sum is a base of $X + Y$, the **direct sum** of the topological vector spaces X and Y with the product topology; $X \cap Y = \{0\}$.

Seminorm in a vector space: a function $p : X \to \mathbb{R}_+$ with the properties

- $p(x + x') \leq p(x) + p(x')$ (subadditivity),
- $p(\lambda x) = |\lambda| p(x)$
- If moreover $p(x) = 0$, iff $x = 0$, then $p(x) = \|x\|$ is called a **norm**.

A family of seminorms $\{p_\alpha \mid \alpha \in A\}$ in the vector space X is **separating**, if for every $x \in X$, $x \neq 0$ there is some α with $p_\alpha(x) > 0$.

Locally convex space Vector space topologized with a separating family of seminorms and with the neighborhood base at 0 of open sets $B_{\{(\alpha_i,\varepsilon_i)\},n}(0) = \{x \in X \mid p_{\alpha_i} < \varepsilon_i (i = 1, \ldots, n), \varepsilon_i > 0\}$, $0 < n < \infty$.

A locally convex space with a countable neighborhood base at 0 can be metrized with the distance function $d(x, y) = \sum_{i=1}^{\infty} 2^{-i} p_{\alpha_i}(x - y)(1 + p_{\alpha_i}(x - y))^{-1}$.

Normed vector space A neighborhood base at 0 and a metric are defined by a single norm: $d(x, y) = \|x - y\|$.

Fréchet space Complete metrizable locally convex space.

Banach space Complete normed vector space.

Banach algebra Banach space X with a norm continuous multiplication $X \times X \to X : (x, y) \mapsto xy$ which makes X into an algebra with unity e, and

- $\|e\| = 1$,
- $\|xy\| \leq \|x\| \|y\|$.

Linear function (linear operator) L: $X \to Y$, X, Y normed vector spaces: $L(\lambda x + \lambda' x') = \lambda L(x) + \lambda' L(x')$.

L is **bounded** if $\|L\| = \sup_{0 \neq x \in X} \|L(x)\|_Y / \|x\|_X < \infty$. A bounded linear function is continuous.

$\mathcal{L}(X, Y) = \{L : X \to Y \mid \|L\| < \infty\}$ is a normed vector space by $(\lambda L + \lambda' L')(x) = \lambda L(x) + \lambda' L'(x)$, λ, λ' scalars in Y. It is Banach, if Y is Banach.

Topological dual $X^* = \mathcal{L}(X, K)$ of a topological vector space X Set of all continuous linear functionals $f : X \to K : x \mapsto \langle f, x \rangle \in K$, $\langle f, \lambda x + \lambda' x' \rangle = \lambda \langle f, x \rangle + \lambda' \langle f, x' \rangle$, made into a vector space by $\langle \lambda f + \lambda' f', x \rangle = \lambda \langle f, x \rangle + \lambda' \langle f', x \rangle$ (observe the difference to an inner product). X^* is always Banach (since K is Banach).

There is an embedding $J : X \to X^{**} = (X^*)^* : x \mapsto \tilde{x}$, were $\tilde{x}$ is defined by $\langle \tilde{x}, f \rangle = \langle f, x \rangle$ for all $f \in X^*$. If J is surjective, $X = X^{**}$ is called **reflexive**.

Weak topology on X Given by the neighborhood base at 0 of open sets $\{x \in X \mid |\langle f, x \rangle| < 1/n\}$, $n = 1, 2, \ldots$ for all $f \in E^*$, a base of the vector space X^*. In general, X is only locally convex in the weak topology; in a finite dimensional X weak and norm topologies are equal.

Weak* topology on X^* Given by the neighborhood base at 0 of open sets $\{f \in X^* | |\langle f,x\rangle| < 1/n\}$, $n = 1,2,\ldots$ for all $x \in E$, a base of $X \subset X^{**}$.

Inner product space A sesquilinear form (inner product) $X \times X \to \mathbb{C}$ (or $\mathbb{R}$) : $(x,y) \mapsto (x|y)$ is defined with the properties

- $(x|y) = \overline{(y|x)}$, $\bar{z}$: complex conjugate of z, ($\bar{z} = z$ for $z \in \mathbb{R}$),
- $(x|y_1 + y_2) = (x|y_1) + (x|y_2)$,
- $(x|\lambda y) = \lambda(x|y)$(physics convention),
- $(x|x) > 0$ for $x \neq 0$.

$\|x\| = (x|x)^{1/2}$ is a norm for which the **Schwarz inequality** $|(x|y)| \leq \|x\|\|y\|$ holds. Two vectors x, $y \in X$ are called **orthogonal** to each other, if $(x|y) = 0$.

Hilbert space Complete inner product space. In a **direct sum** $X \oplus Y$ of Hilbert spaces, X and $Y = X^\perp$ are orthogonal by definition. The **tensor product** $X \otimes Y$ of Hilbert spaces is the product space with the inner product $(x \oplus y|x' \oplus y') = (x|x')(y|y')$.

Unitary space Finite dimensional complex Hilbert space.

Euclidean space Finite dimensional real Hilbert space. Angles are defined by $\cos \measuredangle(x,y) = (x|y)/(\|x\|\|y\|)$.

Directional derivative of a function F: $\Omega \to Y$ from an open set Ω of a normed vector space X into a topological vector space Y:

$$D_xF(x_0) = \frac{d}{dt}F(x_0 + tx)\Big|_{t=0} = \lim_{\substack{t\neq 0,\, t\to 0\\ x_0+tx\in\Omega}} \frac{F(x_0 + tx) - F(x_0)}{t}, \qquad \|x\| = 1.$$

Gâteaux derivative, if $D_xF(x_0)$ is a continuous linear function of $x \in X$.

Total derivative (Fréchet derivative) of a function F: $\Omega \to Y$ from an open set Ω of a normed vector space X into a normed vector space Y : $DF(x_0) \in \mathcal{L}(X,Y)$ so that

$$F(x_0 + x) - F(x_0) = DF(x_0)x + R(x)\|x\|, \quad \lim_{x\to 0} R(x) = 0,$$

If $D_xF(x_0)$ is a continuous function of $x_0 \in \Omega$, then $D_xF(x_0) = DF(x_0)x$.

Functional derivative of a functional F: $\Omega \to K$ from an open set Ω of a normed vector space X into its scalar field K.

If $X = L^p(K^n, d^nz) \ni f(z)$ then $DF(f_0) = g(z) \in L^q(K^n, d^nz)$, $1/p + 1/q = 1$.

Taylor expansion

$$F(x_0+x)=F(x_0)+DF(x_0)x+\frac{1}{2!}D^2F(x_0)xx+\cdots+\frac{1}{k!}D^kF(x_0)\underbrace{xx\cdots x}_{(k\text{ factors})}+\cdots,$$

provided x_0 and $x_0 + x$ belong to a convex domain $\Omega \subset X$ on which F is defined and has total derivatives to all orders, which are continuous functions of x_0 in Ω and provided this Taylor series converges in the norm topology of Y. Here, $D^kF(x_0) \in \mathcal{L}(X, \mathcal{L}(X, \ldots, \mathcal{L}(X, Y)\ldots))$ is a k-linear function from $X \times X \times \cdots \times X$ (k factors) into Y.

$$\|L_k\|_{\underbrace{\mathcal{L}(X,\mathcal{L}(X,\ldots,\mathcal{L}(X,Y)\ldots))}_{\text{depth k}}}=\sup_{x^{(1)}\cdots x^{(k)}\in X}\frac{\|L_kx^{(1)}\cdots x_Y^{(k)}\|_Y}{\|x^{(1)}\|_X\cdots\|x^{(k)}\|_X}.$$

Chain rule F: $X \supset \Omega \to Y$, $F(\Omega) \subset \Omega'$, G: $Y \supset \Omega' \to Z$ and $H = G \circ F$: $X \supset \Omega \to Z$. Then,

$$DH(x_0)=DG(F(x_0))\circ DF(x_0)$$

if the right hand side derivatives exist. In this case, $DF(x_0) \in \mathcal{L}(X,Y)$ and $DG(F(x_0)) \in \mathcal{L}(Y,Z)$ and hence $DH(x_0) \in \mathcal{L}(X,Z)$. Moreover, if $DF : \Omega \to \mathcal{L}(X,Y)$ is continuous at $x_0 \in \Omega$ and $DG : \Omega' \to \mathcal{L}(Y,Z)$ is continuous at $F(x_0) \in \Omega'$, then $DH : \Omega \to \mathcal{L}(X,Z)$ is continuous at $x_0 \in \Omega$.

Class $C^n(\Omega, Y)$ function $n = 0, 1, \ldots, \infty, \omega$: F: $(\Omega \subset X) \to Y$ which for $n = 0$ is continuous, has continuous derivatives up to order n, $n = 1, 2,\ldots,\infty$, and has a converging Taylor expansion for $n = \omega$. For $n = \infty$ it is called **smooth**, for $n = \omega$ it is called **analytic**.

C^n **diffeomorphism** a bijective mapping from $\Omega \subset X$ onto $\Omega' \subset Y$ which, along with its inverse, is C^n, $n > 0$, (C^∞, C^ω).

Derivative of a product in a C^n-algebra, $n \geq 1$:

$$D(FG)=(DF)G+F(DG).$$

Implicit function theorem Let X, Z be normed vector spaces and let Y be a Banach space. Let $F \in C^1((\Omega \subset X \times Y), Z)$ and consider the equation

$$F(x,y)=c,\quad c\in Z\text{ fixed}.$$

Assume that $F(a, b) = c$ and that $Q = D_yF(a, b)$ is a linear bijection from Y onto Z, so that $Q^{-1} \in \mathcal{L}(Z,Y)$. Then, there are open sets $A \ni a$ and $B \ni b$ in X and Y, so that for every $x \in A$ the above equation has a unique solution $y \in B$ which defines a continuous function $G : A \to Y : x \mapsto y = G(x)$ implicitly by the above equation. The function G has a continuous total derivative at $x = a$, and

$$DG(a) = -(D_yF(a,b))^{-1} \circ D_xF(a,b),\ b = G(a).$$

Open cover of a set Ω in a topological space X: Family of open sets $U_\alpha, \alpha \in A$ so that $\cup_\alpha U_\alpha \supset \Omega$. A subfamily which is also a cover is called a **subcover**.

Compact set C Every open cover of C contains a finite subcover, $\cup_{i=1}^n U_i \supset C$. A Hausdorff compact set is a **compactum**. A set is **relatively compact**, if its closure is compact.

A compactum is closed. A closed subset of a compact set is compact.

Every infinite subset of a compact set has a cluster point.

Every sequence in a compact set has a convergent subsequence.

Locally compact topological space X Every point $x \in X$ has a relatively compact neighborhood.

Compact function (operator) It is continuous and maps bounded sets into relatively compact sets.

Lower (upper) semicontinuous function $F : X \to \overline{\mathbb{R}}$: For every $x \in X$ and every $\varepsilon > 0$ there is a neighborhood $U_\varepsilon(x)$ in which $F > F(x) - \varepsilon (F < F(x) + \varepsilon)$.

F is **finite from below**, if $F(x) > -\infty$ for all x.

Extremum problems A continuous real-valued function on a compact set takes on its maximum and minimum values.

If F is a finite from below and lower semicontinuous function from a non-empty compactum A into $\mathbb{R}$, then F is even bounded below and the minimum problem $\min_{x \in A} F(x) = \alpha$ has a solution $x_0 \in A$, $F(x_0) = \alpha$.

Fixed point theorems Banach: A strict contraction $F : X \to X$, $d(Fx, Fx') \leq \lambda d(x, x'), \lambda < 1$, on a complete metric space X has a unique fixed point.

Tychonoff: A continuous mapping F: $C \to C$ in a compact convex set C of a locally convex space has a fixed point.

Schauder: A compact mapping F: $C \to C$ in a closed bounded convex set C of a Banach space has a fixed point.

Banach-Alaoglu theorem The unit ball of the dual X^* of a Banach space X is compact in the weak* topology.

The unit ball of a reflexive Banach space is compact in the weak topology.

A Banach space is in general not first countable in the weak and weak* topologies; this is why instead of sequences nets are needed.

Net Set of points $x_a \in X$ indexed with a directed set $I \ni a$ (every pair $(a, b) \in I \times I$ has an upper bound $c \in I$, $a \leq c$, $b \leq c$).

Every net in a compact set has a convergent subnet.

Function of compact support F: $X \to Y$, X locally compact space, Y normed vector space, supp F is contained in some compactum; supp F is the smallest closed set outside of which $F(x) = 0$. A **class** $C_0^n(\Omega, Y)$ **function** is a class $C^n(\Omega, Y)$ function with compact support ($n = 0, \ldots, \infty$).

Partition of unity A family $\{\varphi_\alpha \,|\, \alpha \in A\}$ of $C_0^\infty(X, \mathbb{R})$-functions such that

- there is a **locally finite open cover**, $X \subset \cup_{\beta\in B} U_\beta$ (every $x \in X$ has a neighborhood W_x which intersects only with finitely many U_β),
- the support of each φ_α is in some U_β,
- $0 \leq \varphi_\alpha(x) \leq 1$ on X for every α,
- $\sum_{\alpha\in A} \varphi_\alpha(x) = 1$ on X.

The partition of unity is called **subordinate** to the cover $\cup_{\beta\in B} U_\beta$

Paracompact space A Hausdorff topological space for which every open cover, $X \subset \cup_{\alpha\in A} U_\alpha$, has a **locally finite refinement**, that is a locally finite open cover $\cup_{\beta\in B} V_\beta$ for which every V_β is a subset of some U_α.

It is a space which permits a partition of unity.

Connectedness A topological space is **connected**, if

- it is not a union of two disjoint non-empty open sets, or equivalently,
- it is not a union of two disjoint non-empty closed sets, or equivalently,
- the only open-closed sets are the empty set and the space itself.

Otherwise it is **disconnected**.

The **connected component** of a point x of a topological space X is the largest connected set in X containing x.

X is **totally disconnected**, if its connected components are all its one point sets $\{x\}$. (A discrete space is totally disconnected; the rational line $\mathbb{Q}$ in the relative topology as a subset of $\mathbb{R}$ is totally disconnected, but not discrete.)

X is **locally connected**, if every point has a neighborhood base of connected neighborhoods.

The image $F(A)$ of a connected set A in a continuous mapping is a connected set.

Homotopy Continuous function H: $[0, 1] \times X \to Y$ translating the continuous function $F_1 : X \to Y$ into the continuous function $F_2 : X \to Y$: $H(0, \cdot) = F_1$ and $H(1, \cdot) = F_2$. F_1 and F_2 are called **homotopic**, $F_1 \cong F_2$. By concatenating two homotopies, H_1 translating F_1 into F_2 and H_2 translating F_2 into F_3, their product H_2H_1 is defined as a homotopy translating F_1 into F_3. Homotopic, $\cong$, is an equivalence relation dividing $C^0(X, Y)$ into **homotopy classes** $[F]$. The homotopy class of a constant function mapping X into a single point of Y is called the **null-homotopy class**.

Homotopy equivalent Two topological spaces X and Y are homotopy equivalent, if their exists continuous functions $F: X \to Y$ and $G: Y \to X$ so that $F \circ G \cong \mathrm{Id}_Y$ and $G \circ F \cong \mathrm{Id}_X$. X is called **contractible**, if it is homotopy equivalent to a one point space.

Pathwise connected (also called arcwise connected) A topological space X is pathwise connected, if for every pair (x, x') of points of X there is a continuous function $H: [0, 1] \to X$, $H(0) = x$, $H(1) = x'$. A general space X consists of the set $\pi_0(X)$ of its **pathwise connected components**. If X is a topological group, then $\pi_0(X)$ is its zeroth homotopy group.

Locally pathwise connected A space X is locally pathwise connected, if every point has a neighborhood base of pathwise connected sets.

***n*th homotopy group** $\pi_n(X)$ of a pathwise connected topological space X: The homotopy classes of functions from the n-dimensional sphere S_n into X, mapping the north pole of the sphere into a fixed point of X. By an intermediate homeomorphism from the n-sphere to the one-point compactified n-cube, two mappings may be concatenated along the x^1-axis of the cube. Concatenation as group operation yields a group structure in $\pi_n(X)$. If X is a (not necessarily pathwise connected) topological group, then the group multiplication yields an isomorphic group structure on $\pi_n(X^e)$ for the pathwise connected component X^e of X, and $\pi_n(X) = \pi_0(X) \times \pi_n(X^e)$. $\pi_1(X)$ is called the **fundamental group** of X.

***n*-connected** A topological space is called n-connected (also n-simple), if every continuous image in X of the n-dimensional sphere S^n is contractible. A topological group X is n-connected, if $\pi_n(X) \approx \pi_0(X)$. A 0-connected space is pathwise connected, a 1-connected space is called **simply connected**.

C.3 Smooth Manifolds

See for instance [4].

Finite-dimensional smooth manifold M

- M is a paracompact topological space, or slightly more special, M is locally compact, Hausdorff and second countable,
- every point $x \in M$ has a neighborhood U_α which is homeomorphic to an open subset $\boldsymbol{U}_\alpha$ of the Euclidean space $\mathbb{R}^m$ by a homeomorphism $\boldsymbol{\varphi}_\alpha : U_\alpha \to \boldsymbol{U}_\alpha$, m is the **dimension** of M,

- the **charts** $(U_\alpha, \boldsymbol{\varphi}_\alpha)$ form an **atlas** of M, that is, the U_α form a cover of M, and there is a complete family of C^∞-diffeomorphisms $\boldsymbol{\psi}_{\beta\alpha}$ from $\boldsymbol{U}_\alpha$ to $\boldsymbol{U}_\beta$, compatible with the $\boldsymbol{\varphi}_\alpha$ with respect to composite mappings.
- There exists always a **complete atlas** $\mathcal{A}_M$ which is not a proper subset of any other atlas of M; it is also called the **differentiable structure** of M.

Coordinate neighborhood of $x_0 \in M$:

- A basc $\{\boldsymbol{e}_1, \ldots, \boldsymbol{e}_m\}$ with $\boldsymbol{\varphi}_\alpha(x_0)$ as origin in $\mathbb{R}^m$ so that $\boldsymbol{x} = \sum x^i \boldsymbol{e}_i \in \boldsymbol{U}_\alpha$ and $\boldsymbol{\varphi}_\alpha(x) = (\varphi^1_\alpha(x), \ldots, \varphi^m_\alpha(x))$ with $\varphi^i_\alpha = \pi^i \circ \boldsymbol{\varphi}_\alpha$ and $\pi^i(\boldsymbol{x}) = x^i$,
- **transition functions** $\boldsymbol{\psi}_{\beta\alpha} = (\psi^1_{\beta\alpha}(\boldsymbol{x}), \ldots, \psi^m_{\beta\alpha}(\boldsymbol{x})$ with a regular Jacobian.
- The set $\{\varphi^i_\alpha\}$ is called a **local coordinate system** on $U_\alpha \in M$; $\{\varphi^i_\alpha(x)\}$ are local coordinates of $x \in M$.

Orientable manifold A smooth manifold M which permits a complete atlas the transition functions of which all have only positive Jacobians.

Product manifold of two manifolds $(M, \mathcal{A}_M)$ and $(N, \mathcal{A}_N)$ is the manifold $(M \times N, \mathcal{A}_M \times \mathcal{A}_N)$ with the product topology and the complete atlas containing the natural product of the complete atlases $\mathcal{A}_M$ and $\mathcal{A}_N$; its dimension is dim M + dim N.

Smooth parametrized curve in $M : \mathbb{R} \supset]a, b[\ni t \mapsto x(t) \in M$, so that every $\boldsymbol{x}_\alpha(t) = \boldsymbol{\varphi}_\alpha \circ x(t)$ for a restriction of $x(t)$ to the corresponding open interval of t is a smooth vector function with values in $\boldsymbol{U}_\alpha$.

Smooth real function on $M : M \ni x \mapsto F(x) \in \mathbb{R}$. It defines in every coordinate neighborhood U_α a smooth real function $F_\alpha(x^1_\alpha, \ldots, x^m_\alpha) = (F \circ \boldsymbol{\varphi}^{-1}_\alpha)(\boldsymbol{x}_\alpha)$.

Tangent vector X_{X_0} on M at x_0: For every smooth real function F, X_{x_0} : $F \mapsto X_{x_0} F \in \mathbb{R}$: in local coordinates

$$X_{x_0} F = \sum_i \xi^i_\alpha \frac{\partial F_\alpha}{\partial x^i_\alpha}, \quad \xi^i_\beta = \sum_j \xi^j_\alpha (\psi_{\beta\alpha})^i_j, \quad (\psi_{\beta\alpha})^i_j = \pi^i \circ \boldsymbol{\psi}_{\beta\alpha} \circ \boldsymbol{\pi}_j,$$

where $\boldsymbol{\pi}_j(x^j) = (0, \ldots, 0, x^j, 0, \ldots, 0) \in \mathbb{R}^m$; the tangent vectors at x_0 span an m-dimensional real vector space, the **tangent space** $T_{x_0}(M)$.

Cotangent vector ω_{x_0} on M at x_0: For every $X_{x_0}, \omega_{x_0} : X_{x_0} \mapsto \langle \omega_{x_0}, X_{x_0} \rangle \in \mathbb{R}$: in local coordinates

$$\langle \omega_{x_0}, X_{x_0} \rangle = \sum_i \omega^\alpha_i \xi^i_\alpha, \quad \omega^\beta_i = \sum_j (\psi^{-1}_{\beta\alpha})^j_i \omega^\alpha_j, \quad (\psi^{-1}_{\beta\alpha})^j_i = (\psi_{\alpha\beta})^j_i,$$

the cotangent vectors span the m-dimensional real **cotangent space** $T^*_{x_0}(M)$; in particular

$$dF_{x_0} = \sum_i \frac{\partial F_\alpha}{\partial x^i_\alpha}\bigg|_0 dx^i_\alpha, \quad \langle dF_{x_0}, X_{x_0}\rangle = X_{x_0}F.$$

Local bases

$$X_{x_0} = \sum_i \xi^i_\alpha \partial/\partial x^i_\alpha, \quad \omega_{x_0} = \sum_i \omega^\alpha_i dx^i_\alpha,$$

$$\langle dx^i_\alpha, \partial/\partial x^j_\alpha\rangle = \delta^i_j, \quad dx^i_\beta = \sum_j dx^j_\alpha (\psi_{\beta\alpha})^i_j, \quad \partial/\partial x^i_\beta = \sum_j (\psi^{-1}_{\beta\alpha})^j_i \partial/\partial x^j_\alpha.$$

Tangent and cotangent vector fields Functions $X : M \ni x \mapsto X_x \in T_x(M)$ and $\omega : M \ni x \mapsto \omega^x \in T^*_x(M)$, so that $XF : x \mapsto X_xF$ and $\omega(X) : x \mapsto \langle \omega^x, X_x\rangle$ are smooth real functions on M for every smooth F and every smooth X, respectively. In any local coordinate system the components ξ^i of X and ω_i of ω are smooth real functions of the local coordinates.

Lie product of two tangent vector fields: It is a tangent vector field

$$[X, Y] = XY - YX, \quad [X, Y]F = \sum_{ij}\left(\xi^j \frac{\partial \eta^i}{\partial x^j} - \eta^j \frac{\partial \xi^i}{\partial x^j}\right)\frac{\partial}{\partial x^i}F;$$

the tangent vector fields on M form a **Lie algebra**.

Smooth mapping of manifolds F: $M \to N$ induces at every point $x \in M$ as a push forward a linear mapping F^x_*: $T_x(M) \to T_{F(x)}(N)$ and at every point $F(x) \in N$ as a pull back a linear mapping $F^*_{F(x)}$: $T^*_{F(x)}(N) \to T^*_x(M)$ with

$$\langle F^*_{F(x)}(\omega_{F(x)})), X_x\rangle = \langle \omega_{F(x)}, F^x_*(X_x)\rangle;$$

the composite of F: $M \to N$ and G: $N \to P$ yields $(G \circ F)^x_* = G^{F(x)}_* \circ F^x_*$ and $(G \circ F)^*_{G(F(x))} = F^*_{F(x)} \circ G^*_{G(F(x))}$, that is, F_* composes covariantly and F^* composes contravariantly; if F^x_* is injective at every point $x \in M$, then the mapping F is an **immersion** of M into N; if in addition F itself is injective, then it is an **embedding** of M into N and M is an **embedded submanifold** of N.

Tensor fields on M: Functions t mapping $x \in M$ into a tensor product of tangent and cotangent spaces on M at x, in local coordinates

$$t(x) = t^{i_1\ldots i_r}_{j_1\ldots j_s}(x)\frac{\partial}{\partial x^{i_1}} \otimes \cdots \otimes \frac{\partial}{\partial x^{i_r}} \otimes dx^{j_1} \otimes \cdots \otimes dx^{j_s},$$

with **Einstein summation** over pairs of equal upper and lower indices understood, so that the component functions $t^{i_1\ldots i_r}_{j_1\ldots j_s}(x)$ are smooth in every coordinate neighborhood and transform for each index like tangent and cotangent vector components, respectively; t is of **type** (r, s).

Tensor product $t \otimes t'$, in coordinate neighborhoods

$$(t \otimes t')^{i_1\ldots i_{r+r'}}_{j_1\ldots j_{s+s'}} = t^{i_1\ldots i_r}_{j_1\ldots j_s} t'^{i_{r+1}\ldots i_{r+r'}}_{j_{s+1}\ldots j_{s+s'}}.$$

Tensor contraction $C_{p,q}(t)$, in coordinate neighborhoods

$$C_{p,q}(t)^{i_1\dots i_{r-1}}_{j_s\dots j_{s-1}} = t^{i_1\dots i_{p-1}ki_p\dots i_{r-1}}_{j_1\dots j_{q-1}kj_q\dots j_{s-1}}.$$

Lie derivative L_X with respect to a tangent vector field X, in coordinate neighborhoods

$$(L_X u)^{i_1\dots i_r}_{j_1\dots j_s} = \xi^i \frac{\partial u^{i_1\dots i_r}_{j_1\dots j_s}}{\partial x^i} - \sum_{p=1}^{r} \frac{\partial \xi^{i_p}}{\partial x^i} u^{i_1\dots i_{p-1}ii_{p+1}\dots i_r}_{j_1\dots j_s} + \sum_{q+1}^{s} \frac{\partial \xi^j}{\partial x^{j_q}} u^{i_1\dots i_r}_{j_1\dots j_{q-1}jj_{q+1}\dots j_s},$$

it maps tensors of type (r, s) to tensors of the same type and is the derivative of u along integral curves ('field lines') of X; $L_{[X,Y]} = [L_X, L_Y]$.

Exterior r-form is an alternating tensor of type $(0, r)$.

Exterior product For r-, s-, and t-forms ω, σ, and τ and functions F and G,

- $\omega \wedge \sigma = (-1)^{rs}\sigma \wedge \tau$,
- $\omega \wedge (F\sigma + G\tau) = F\omega \wedge \sigma + G\omega \wedge \tau$,
- $(\omega \wedge \sigma) \wedge \tau = \omega \wedge (\sigma \wedge \tau)$;

the general r-form in a coordinate neighborhood is

$$\omega = \sum_{i_1<\dots<i_r} \omega_{i_1\dots i_r}(\boldsymbol{x})dx^{i_1} \wedge \cdots \wedge dx^{i_r}, \quad \omega = 0 \quad \text{if } r > m,$$

$$\omega = t_{i_1\dots i_r}\, dx^{i_1} \otimes \cdots \otimes dx^{i_r}, \quad \omega_{i_1\dots i_r} = r!\, t_{i_1\dots i_r}, \quad t \text{ alternating.}$$

$$[\omega, \sigma] = \sum_{i_1<\cdots<i_{r+s}} \left(\omega_{i_1\dots i_r}\sigma_{i_{r+1}\dots i_{r+s}} - \sigma_{i_{r+1}\dots i_{r+s}}\omega_{i_1\dots i_r}\right)dx^{i_1} \wedge \cdots \wedge dx^{i_{r+s}},$$

1-forms: $[\omega, \sigma] = \omega \wedge \sigma$ (in general $\neq -\sigma \wedge \omega$ for non-commutative quantities ω_i and σ_j).

Exterior derivative d in coordinate neighborhoods

$$d\omega = \sum_{s=1}^{r+1}(-1)^{s+1} \sum_{i_1<\cdots<i_{r+1}} \frac{\partial \omega_{i_1\dots i_{s-1}i_{s+1}\dots i_{r+1}}}{\partial x^{i_s}} dx^{i_1} \wedge \cdots \wedge dx^{i_{r+1}}.$$

Interior multiplication For a tangent vector field X and an r-form ω,

$$\iota_X(\omega) = C_{1,1}(X \otimes \omega) = r \sum_{i,i_1<\cdots<i_r} \xi^i \omega_{ii_1\dots i_{r-1}} dx^{i_1} \wedge \cdots \wedge dx^{i_{r-1}}.$$

Exterior calculus Basic relations are

$$L_X = d \circ \iota_X + \iota_X \circ d, \quad [d, L_X] = 0, \quad [\iota_Y, L_X] = \iota_{[Y,X]}, \quad d^2 = 0, \quad (\iota_X)^2 = 0.$$

Stokes' theorem

$$\int_{\Omega} d\omega = \int_{\partial\Omega} \omega$$

where Ω is a domain in M with boundary $\partial\Omega$ and ω is an m-form, $m = \dim M$.

C.4 Topological Groups

See for instance [5, 4].

Topological group G

- G is a group (the abstract group of G),
- G is a topological space,
- $G \times G \to G : (g, h) \mapsto gh^{-1}$ is continuous.

If $\mathcal{B}^e = \{B^e_\alpha\}$ is a neighborhood base of the unity $e \in G$, then $\mathcal{B}^g = \{gB^e_\alpha\}$ is a neighborhood base of g for every $g \in G$.

Subgroup H

- H is an abstract subgroup of G,
- H is a *closed* subset of G.

If H is an abstract subgroup of G, then $\overline{H}$ is a subgroup of the topological group G.

Homomorphism $f : G \to G'$:

- f is an algebraic homomorphism,
- f is continuous.

An **isomorphism** is an algebraic isomorphism and a homeomorphism.

The kernel $N = \ker f$ of a homomorphism $f : G \to G'$ is an invariant subgroup of G, and the quotient group G/N with the quotient topology is a topological group.

The direct product of groups of two topological groups is a topological group in the product topology.

If U is any neighborhood of the unity $e \in G$, then $\cup_{n=1}^{\infty} U^n = G^e$ is the connected component of e of G. G^e is an invariant subgroup of G, and $G^* = G/G^e$ is totally disconnected. (If G^e is pathwise connected, then $G^* \doteq \pi_0(G)$, the zeroth homotopy group of G.)

Lie group G:

- G is a topological group,

- G is a smooth (real or complex) manifold,
- $G \times G \to G : (g, h) \mapsto gh^{-1}$ is smooth.

Since G as a smooth manifold is second countable and locally Euclidean, it is locally compact, locally pathwise connected, semi-locally one-connected (every point $x \in M$ has a neighborhood U such that every loop in U with base point x is contractible in M into x), and its connected components are pathwise connected.

For every Lie group G, the differentiable manifold structure which makes the topological group G into a Lie group is uniquely defined (up to Lie group isomorphisms, see below, this will always be understood under uniqueness). It has even a uniquely defined substructure of an analytic manifold for which the mappings $(g, h) \mapsto gh^{-1}$ are analytic.

Left and right translations Let G be a Lie group, let g be a fixed element and let h be a running element of G.

$$l_g : h \mapsto gh, \quad r_g : h \mapsto hg$$

are $C^\infty(C^\omega)$-diffeomorphisms *of manifolds* of G onto G.

Left (right) invariant vector field $X \in \mathcal{X}(G)$ with

$$l_{g*}X_h = X_{gh} \quad (r_{g*}X_h = X_{hg}),$$

that is, X is pushed forward by a translation from its value at h to its own value at gh (hg) and hence is uniquely defined by its value X_e at $e \in G$ (called infinitesimal generators of G^e in physics). Invariant vector fields are automatically smooth (analytic), they form a dim G-dimensional subspace of $\mathcal{X}(G)$ which is isomorphic to the tangent space $T_e(G)$ on G at e.

The Lie product $[X, Y]$ of left (right) invariant vector fields is again an invariant vector field; left (right) invariant vector fields form two isomorphic realizations of a Lie algebra, the **Lie algebra** $\mathfrak{g}$ **of the Lie group** G. If right invariant vector fields are distinguished by a tilde, then the isomorphism is $X \mapsto \tilde{X}$, $[X, Y] \mapsto [\tilde{Y}, \tilde{X}]$.

Left (right) invariant r**-form:** $\omega \in \mathcal{D}(G)$ with

$$l_g^*\omega_{gh} = \omega_g \quad (r_g^*\omega_{hg} = \omega_g),$$

that is, ω is pulled back by a translation from its value at gh (hg) to its own value at h. Left (right) invariant 1-forms form a dim G-dimensional subspace of $\mathcal{D}^1(G)$ which is dual to $\mathfrak{g}$. They are called the **Maurer-Cartan forms** of G.

Lie group homomorphisms and representations Mappings $f : G \to G'$ with

- f is an algebraic group homomorphism,
- f is a smooth (analytic) mapping of manifolds.

f is a **Lie group isomorphism**, if it is an algebraic isomorphism and a diffeomorphism of manifolds.
A Lie group homomorphism $f : G \to G'$ is pushed forward to a Lie algebra homomorphism $f_* : \mathfrak{g} \to \mathfrak{g}'$.

If V is some K'-vector space, then a homomorphism r: $G \to \mathrm{Aut}\ (V)$ is a **representation** of G; if it is finite-dimensional, $Gl(\dim V, K') \approx \mathrm{Aut}\ (V)$ is used. $K' = \mathbb{R}$ or $\mathbb{C}$, but it is not in general directly related to the local Euclidean structure $K^{\dim G}$ of G (it may contain K as a subfield, see end of this section).

Lie subgroup H of the Lie group G:

- $\tilde{H}$ is an abstract subgroup of G,
- $(\tilde{H},\ \mathrm{Id}_{\tilde{H}})$ is an embedded submanifold of G,
- There is a Lie group H which is algebraically isomorphic to $\tilde{H}$.

H is a regular embedding into G, that is, with the relative topology as a submanifold of G, iff $\tilde{H}$ is closed in the topology of G.

If $f : G \to G'$ is a Lie group homomorphism, then Ker f is a closed Lie subgroup of G; the quotient group $G/\mathrm{Ker} f$ is also a Lie group.

Covering space $\tilde{M}$ of a topological space M:

- A continuous surjective mapping $\pi : \tilde{M} \to M$,
- every $x \in M$ has a neighborhood U which is **evenly covered**, that is, $\pi^{-1}(U)$ is a (possibly infinite) union of sets V_α each of which is homeomorphic to U.

Universal covering group $\tilde{G}$ of a connected Lie group G;

- $\tilde{G}$ is a connected, simply connected covering space of G,
- $\pi : \tilde{G} \to G$ is a Lie group homomorphism.

The kernel of π is a discrete subgroup of $\tilde{G}$.

Every connected Lie group has (up to isomorphisms) a uniquely defined connected, simply connected covering group.

There is a one-one correspondence between connected, simply connected Lie groups and Lie algebras.

The connection between a Lie algebra $\mathfrak{g}$ and its connected, simply connected Lie group $\tilde{G}$ is obtained by the **exponential mapping** $\exp : \mathfrak{g} \ni X \mapsto \exp(X) \in \tilde{G}$. Any other Lie group G with the same Lie algebra $\mathfrak{g}$ is a Lie subgroup of $\tilde{G}$ where the homomorphism from $\tilde{G}$ to G has a discrete kernel. *This largely reduces the study of Lie groups to the study of Lie algebras.*

Lie algebra $\mathfrak{g}$ over $K = \mathbb{R}$ or $\mathbb{C}$ of *finite* dimension:

- $\mathfrak{g}$ is a K-vector space and a multiplicative monoid with respect to the Lie product $(X, Y) \mapsto [X, Y]$, $X, Y \in \mathfrak{g}$,
- $[X, Y] + [Y, X] = 0$ (anti-commutativity),
- $[X, [Y, Z]] + [Y, [Z, X]] + [Z, [X, Y]] = 0$ (Jacobi identity).

Infinite-dimensional Lie groups and Lie algebras are not considered here.

With respect to a given base $\{X_1, \ldots, X_n\}$, $n =$ dim $\mathfrak{g}$ in the vector space $\mathfrak{g}$,

$$[X_i, X_j] = \sum_{k=1}^{n} c_{ij}^k X_k, \quad c_{ij}^k + c_{ji}^k = 0, \quad \sum_{k=1}^{n} c_{ij}^k c_{kl}^m + c_{jl}^k c_{ki}^m + c_{li}^k c_{kj}^m = 0;$$

the **structure constants** c_{ij}^k depend on the chosen base, nevertheless, they determine the Lie algebra uniquely.

With respect to the same base, the structure constants of the Lie algebra of right invariant vector fields of a Lie group differ from those of left invariant vector fields by a sign.

With respect to the dual base $\{\omega^1, \ldots, \omega^n\}$ in $\mathfrak{g}^*$ for Maurer-Cartan forms, $\langle \omega^i, X_j \rangle = \delta^i_j$, the **Maurer-Cartan equations** or **structure equations**

$$d\omega^k = -\sum_{1 \le i < j \le n} c_{ij}^k \omega^i \wedge \omega^j, \quad c_{ij}^k = \omega^k([X_i, X_j])$$

hold.

Lie subalgebra $\mathfrak{h}$ of $\mathfrak{g}$:

- $\mathfrak{h}$ is a linear subspace of the vector space $\mathfrak{g}$,
- $\mathfrak{h}$ is itself a Lie algebra.

Ideal $\mathfrak{h}$ **of a Lie algebra** $\mathfrak{g}$:

- $\mathfrak{h}$ is a Lie subalgebra of $\mathfrak{g}$,
- $[\mathfrak{g}, \mathfrak{h}] \subset \mathfrak{h}$ with $[\mathfrak{g}, \mathfrak{h}] = \mathrm{span}_K\{[X, Y] \,|\, X \in \mathfrak{g},\ Y \in \mathfrak{h}\}$

$\mathfrak{g}$ is a **simple Lie algebra**, if it contains no ideals except $\mathfrak{g}$ itself and $\{0\}$; it is a **semi-simple Lie algebra**, if it is a direct sum of simple Lie algebras.

The ideal $[\mathfrak{g}, \mathfrak{g}]$ of $\mathfrak{g}$ is the **derived algebra** of $\mathfrak{g}$. The series $\mathfrak{g}^{(0)} = \mathfrak{g}, \ldots, \mathfrak{g}^{(k)} = [\mathfrak{g}^{(k-1)}, \mathfrak{g}^{(k-1)}], \ldots$ is the **derived series** of $\mathfrak{g}$. A Lie algebra is called **solvable**, if the derived series ends up with the trivial ideal $\{0\}$ after a finite number of items. The **radical** $\mathfrak{g}_{\mathrm{rad}}$ is the maximal solvable ideal of $\mathfrak{g}$; $\mathfrak{g}$ is solvable, if $\mathfrak{g}_{\mathrm{rad}} = \mathfrak{g}$. The radical of a semi-simple Lie algebra is zero, $\mathfrak{g}^{(1)} = \mathfrak{g}^{(k)} = \mathfrak{g}$; thus semi-simplicity is a strong opposite of solvability.

The series of ideals $\mathfrak{g}_{(1)} = \mathfrak{g}, \ldots, \mathfrak{g}_{(k)} = [\mathfrak{g}, \mathfrak{g}_{(k-1)}], \ldots$ is the **lower central series** of $\mathfrak{g}$. A Lie algebra is called **nilpotent**, if the lower central series ends up with the trivial ideal $\{0\}$ after a finite number of items. A maximal nilpotent ideal $\mathfrak{g}_0$ of $\mathfrak{g}$ is a **Cartan subalgebra**.

Center $\mathfrak{z}(\mathfrak{g})$ of a Lie algebra $\mathfrak{g} : \mathfrak{z}(\mathfrak{g}) = \{Y \mid [Y, X] = 0 \text{ for all } X \in \mathfrak{g}\}$.

Lie algebra homomorphisms and representations $F : \mathfrak{g} \to \mathfrak{g}'$:

- F is a linear mapping of vector spaces,
- $F([X, Y]) = [F(X), F(Y)], X, Y \in \mathfrak{g}$.

If $\mathfrak{g}' \subset \text{End}\ (V) = \mathfrak{gl}(\dim\ \mathfrak{g}, K')$ for some K'-vector space V, the **representation space**, then the homomorphism $R : \mathfrak{g} \to \text{End}\ (V)$ with

$$R([X, Y]) = R(X) \circ R(Y) - R(Y) \circ R(X),$$

where $\circ$ means the composition of endomorphisms of V, is a representation of $\mathfrak{g}$. After introducing a base in $VR(X)$ is given by a dim $V \times$ dim V-K'-matrix, and $\circ$ is the matrix multiplication.

A representation is **irreducible**, if V does not contain proper subspaces invariant under R. Every irreducible representation of a solvable Lie algebra is one-dimensional.

If $R : \mathfrak{g} \to \text{End}\ (V) = \mathfrak{gl}(\dim\ \mathfrak{g}, K')$ is a representation of the Lie algebra $\mathfrak{g}$, then $\exp(R) : G \to \text{Aut}\ (V) = Gl(\dim\ \mathfrak{g}, K')$ is a representation of the Lie group G, $\exp(R)$ is irreducible, iff R is irreducible.

If $V = \mathfrak{g}$ ($\mathfrak{g}$ taken as vector space) then $\text{ad} : \mathfrak{g} \to \text{End}\ (\mathfrak{g}) : X \mapsto \text{ad}\ (X)$ with ad $(X)Y = [X, Y]$ for all $Y \in \mathfrak{g}$ is the **adjoint representation** of $\mathfrak{g}$; its dimension is dim $\mathfrak{g}$; also $\exp(\text{ad}) = \text{Ad}$, the adjoint representation of G (see p. 350).

$\kappa(X, Y) = \text{tr}\,(\text{ad}(X) \circ \text{ad}(Y))$ is the **Killing form** of $\mathfrak{g}$. It is a bilinear form on the vector space $\mathfrak{g}$, which is non-degenerate, iff $\mathfrak{g}$ is semi-simple.

Universal enveloping algebra $\mathfrak{U}(\mathfrak{g})$ **of a Lie algebra** $\mathfrak{g}$:
$\mathfrak{U}(\mathfrak{g})$ is the quotient algebra of the tensor algebra, $\mathbb{T}(\mathfrak{g})/\mathbb{J}(\mathfrak{g})$, with $\mathfrak{g}$ taken as the vector space, where the ideal $J(\mathfrak{g})$ is generated by all tensors $X \otimes Y - Y \otimes X - [X, Y]$. $\mathfrak{U}(\mathfrak{g})$ is a graded algebra with $\mathfrak{U}(\mathfrak{g})_0 = K, \mathfrak{U}(\mathfrak{g})_1 = \mathfrak{g}$, and all $\mathfrak{U}(\mathfrak{g})_k$ consisting of symmetric tensors of type $(k, 0)$ for $k \geq 2$.

Casimir element $C_{\mathfrak{h}}$ of the algebra $\mathfrak{U}(\mathfrak{g})$ corresponding to a simple ideal $\mathfrak{h}$ of the Lie algebra $\mathfrak{g}$:

- Let $\{X_i\}$ be any base of the vector space $\mathfrak{h}$,
- let κ be the Killing form of $\mathfrak{h}$,
- $C_{\mathfrak{h}} = \sum_{i,j} \kappa(X_i, X_j) X_i \otimes X_j$; it belongs to the center of $\mathfrak{U}(\mathfrak{g})$ and hence is a constant for every irreducible representation of $\mathfrak{g}$.

Classification of all finite-dimensional complex simple Lie algebras $\mathfrak{g}$:

- Let r be the (unique) dimension of a Cartan subalgebra $\mathfrak{g}_0$, and choose a base $\{H_i \mid i = 1, \ldots, r\}$ of $\mathfrak{g}_0$, compatible with the relations below (Chevalley basis),
- $\mathfrak{g}$ is generated by $3r$ generators $\{E_i^{\pm}, H_i \mid i = 1, \ldots, r\}$ with the Lie products

$$[H_i, H_j] = 0, \quad [H_i, E_j^{\pm}] = \mp n^j_i E_j^{\pm}, \quad [E_i^{+}, E_j^{-}] = \delta^j_i H_j,$$

  the Jacobi identities and the Serre relations $(\mathrm{ad}(E_i^{\pm}))^{1+n^j_i} E^{\pm_j} = 0$ for $i \neq j$,
- the (by convention negative of the) **Cartan matrix** n^i_j has the diagonal $n^i_j = -2$, and the only possibilities for the off-diagonal elements are $n^i_j = n^j_i = 0$ or $n^i_j = 1$, $n^j_i = 1$, 2, or 3, while $\det n < 0$; this also fixes the normalization of the generators (mathematics convention, in physics one uses half of the values n^j_i so that the ladder elements $E_i^{\pm}$ shift the eigenvalues of H_i by $\pm$ 1 instead of by ± 2),
- the **Dynkin diagram** for $\mathfrak{g}$ consists of r dots connected by $v^i_j = \max(n^i_j, n^j_i)$ lines and an arrow from i to j if $n^j_i > n^i_j$; the Dynkin diagram must be connected for a simple Lie algebra,
- a purely combinatorial task yields the following complete set of solutions:

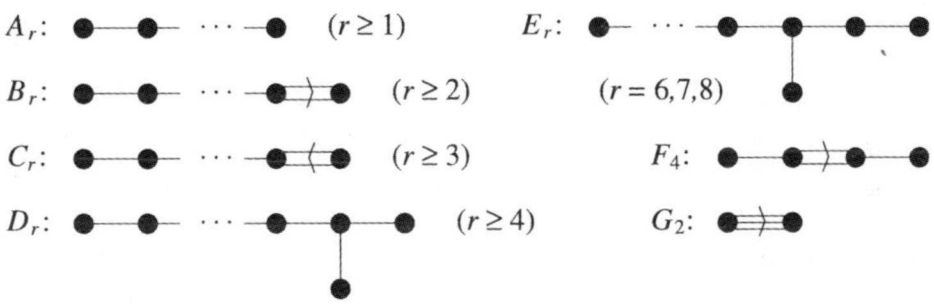

  The restrictions given for r are made since otherwise one would have $C_2 \approx B_2, D_3 \approx A_3, E_4 \approx A_4, E_5 \approx D_5$.

The four infinite series are called classical Lie algebras, the other five are called exceptional.

In the notation of Sect. 6.6, the classical Lie algebras are:

$A_r \approx \mathfrak{sl}(r+1, \mathbb{C}), B_r \approx \mathfrak{so}(2r+1, \mathbb{C}), C_r \approx \mathfrak{sp}(r, \mathbb{C}),$ and $D_r \approx \mathfrak{so}(2r, \mathbb{C})$.

There is (up to isomorphisms) a unique way of **complexification** of a real Lie algebra by linear extension of the field K. However, several non-isomorphic real Lie algebras may result in the same complex one. For instance, the real Lie algebras $\mathfrak{sl}(r, \mathbb{R}), \mathfrak{su}(r, \mathbb{C})$ and several others yield the same complex Lie algebra $\mathfrak{sl}(r, \mathbb{C})$. From this it is also clear that a real Lie algebra may have relevant representations in a complex (or even in a quaternionic) representation space.

Among the real simple Lie algebras $\mathfrak{g}$ whose complexification yields the same complex simple Lie algebra there is always only one for which the simple Lie group $\exp(\mathfrak{g})$ is compact as a manifold. These so called compact real Lie algebras for A_r, B_r, C_r, and D_r are in turn

$$\mathfrak{su}(r+1,\mathbb{C}), \mathfrak{so}(2r+1,\mathbb{R}), \mathfrak{sp}(2r), \text{ and } \mathfrak{so}(2r,\mathbb{R});$$

$\mathfrak{sp}(2r) \approx \mathfrak{u}(r,\mathbb{H})$ where $\mathbb{H}$ means the quaternion field).

An example of infinite-dimensional Lie algebras large parts of which can be completely classified are the **Kac-Moody algebras** of smooth mappings of the circle S^1 (as a manifold) to finite-dimensional Lie algebras. They are related to the so called quantum groups in physics [6].

C.5 Fiber Bundles

See for instance [7].

Principal fiber bundle (P, M, π, G), in short P:

- P is a manifold,
- the Lie group G acts freely on P from the right: $R_g : P \times G \to P : (p, g) \mapsto pg = R_g p$, $R_{gh^{-1}} = R_{h^{-1}} R_g$, $R_g p \neq p$ for $g \neq e$,
- $M = P/G$ and the natural projection π: $P \to M$ is smooth,
- P is locally trivial: $M = \cup_\alpha U_\alpha$, $\pi^{-1}(U_\alpha) \sim U_\alpha \times G$: diffeomorphism $\psi_\alpha(p) = (\pi(p), \phi_\alpha(p))$ for all $p \in \pi^{-1}(U_\alpha)$: $\phi_\alpha(pg) = \phi_\alpha(p)g$ for all $g \in G$.

For every $x \in M$: $\pi^{-1}(x) \sim G$ is the **fiber** over x, the **structure group** G is the **typical fiber**, M is the **base space**, P is the **total space**, and π is the **bundle projection.**

Set of **transition functions** $\psi_{\beta\alpha}(\pi(p)) = \phi_\beta(p)\phi_\alpha^{-1}(p)$ with $\psi_{\gamma\alpha}(x) = \psi_{\gamma\beta}(x)\psi_{\beta\alpha}(x)$ for all $x \in U_\alpha \cap U_\beta \cap U_\gamma$.

Bundle homomorphism$(F, \bar{F}, \bar{\bar{A}})$:

$$\begin{array}{ccc} P' & \xrightarrow{F} & P \\ {\scriptstyle \pi'}\downarrow & & \downarrow{\scriptstyle \pi} \\ M' & \xrightarrow[\bar{F}]{} & M \end{array} \qquad \bar{\bar{F}} : G' \to G.$$

P' is a **subbundle** of P, if $M' \to M$ is an embedding; if $M' = M$ and $\bar{F} = \mathrm{Id}_M$, then F is called a **reduction of the structure group** G to G'.

Frame bundle $(L(M), M, \pi, Gl(m, K))$ is an important case of a principal fiber bundle, with $p = (x, X_1, \ldots, X_m) \in L(M)$, where any ordered base $(X_1, \ldots, X_m)$ in the tangent space $T_x(M)$ is a frame, and frames in $T_x(M)$ are transformed into one another by $g \in Gl(m, K)$. The structure group $Gl(m, K)$ can be reduced to $U(m)$ for $K = \mathbb{C}$ and to $O(m)$ for $K = \mathbb{R}$. Local coordinates: $\psi_\alpha(p) = (x^k(p), u_i^k(p), i = 1, \ldots, m), \pi(p) = x, u(p) : K^m \to T_x(M) : e_i \mapsto u(p)e_i$.

Section $s : M \to P$ with $\pi \circ s = \mathrm{Id}_M$; a **local section**$s$: $M \supset U \to P$ always exists, **canonical local section**: $s_\alpha : U_\alpha \to \pi^{-1}(U_\alpha) : x \mapsto s_\alpha(x) = \psi_\alpha^{-1}(x, e)$. P has a (global) section, iff it is trivial, that is, $P = M \times G$.

Fundamental vector field X^* on P: Let $\mathfrak{g} = \{X\}$ be the Lie algebra of G, then, $R_*^* : \mathfrak{g} \to \mathcal{X}(\pi^{-1}(x)) : X \mapsto X^* = R_*^*(X)$ is an isomorphism between left invariant vector fields on G (elements of $\mathfrak{g}$) and left invariant (with respect of the action of G) vector fields on each fiber $\pi^{-1}(x)$ of P.

Connection Γ on a principal fiber bundle P:

- $T_p(P) = G_p \oplus Q_p = [\text{vertical space } G_p = T_p(\pi^{-1}(x))] \oplus [\text{horizontal space}]$,
- $Q_{pg} = (R_g)_* Q_p$ for every $p \in P$and every $g \in G$,
- Q_p depends smoothly on $p \in P$.

Connection form ω ($\mathfrak{g}$-valued 1-form on P in one-one correspondence with Γ):

- $\omega(R_*^*(X)) = X$ for every $X \in \mathfrak{g}$,
- $((R_g)^*\omega)(X^*) = (\mathrm{Ad}\,(g^{-1})\omega)(X^*)$ for every $g \in G$ and every $X^* \in \mathcal{X}(P)$.

$Q_p = \{X^* \in T_p(P) \mid \langle \omega, X^* \rangle = 0\}$, for every $p, Q_p \approx T_{\pi(p)}(M)$.

Unique decomposition of any tangent vector at p, $X_p = {}^vX_p + {}^hX_p$, into its **vertical** and **horizontal components**.

Local connection forms $\omega_\alpha = \overset{*}{s_\alpha}(\omega)$ are $\mathfrak{g}$-valued 1-forms on U_α:

- $\omega_\beta(X) = \psi_{\alpha\beta}^{-1}\, \omega_\alpha(X)\, \psi_{\alpha\beta} + \psi_{\alpha\beta}^{-1}\, \psi_{\alpha\beta\,*}(X)$,
- every set of local $\mathfrak{g}$-valued 1-forms with this transition property defines a connection on P.

Holonomy

- **lift** F^***of a path** F: $[0, 1] \to M$: F^*: $[0, 1] \to P$ with $\pi \circ F^* = F$ and with a horizontal tangent vector in every point p,
- **parallel transport** of the fiber over x_0 to the fiber over x_1 along the path F: isomorphism $\tilde{F} : \pi^{-1}(x_0) \to \pi^{-1}(x_1)$ provided by all lifts of F,
- **holonomy group** H_x is the automorphism group of $\pi^{-1}(x)$ due to all closed loops F with base point x,

- **restricted holonomy group** H_x^0: group of automorphisms due to null-homotopic loops,
- **holonomy group** H_p **with reference point** $p : H_p = \{g|p$ may be parallel transported into $pg\} \subset G$, likewise H_p^0.

Pseudo-tensorial ***r*****-form of type** $(\mathrm{Ad}, \mathfrak{g}) : (R_g)^*\sigma = \mathrm{Ad}\,(g^{-1})\,\sigma$ for every $g \in G$.

Tensorial ***r*****-form** ${}^h\sigma$ of σ: $\langle {}^h\sigma, X_1 \wedge \ldots \wedge X_r\rangle = \langle \sigma, {}^hX_1 \wedge \ldots \wedge {}^hX_r\rangle$.

Exterior covariant derivative: $D\sigma = {}^h(d\sigma)$.

Curvature form Ω of the connection form $\omega : \Omega = D\omega, \quad d\omega = -[\omega, \omega] + \Omega$.

Bianchi identities: $D\Omega = 0$.

Fiber bundle (E, M, π_E, F, G), in short E:

- E is associated with a principal fiber bundle (P, M, π, G),
- G acts on F from the left, that is, $G \times F \to F$: $(g, f) = gf$, $g \in G, f \in F$,
- $E = P \times {}_GF$, that is, $(p, f) = (pg, g^{-1}f)$ is an equivalence relation R in $P \times F$, and $E = (P \times F)/R$, the elements of E are denoted $p(f)$,
- $\pi_E : E \to M : p(f) \mapsto \pi(p)$,
- every local diffeomorphism $\pi^{-1}(U) \sim U \times G$, $U \subset M$, induces a local diffeomorphism $\pi_E^{-1}(U) \sim U \times F$.

Now F is the **typical fiber,** $\pi_E^{-1}(x)$ is the **fiber** over x, G is the **structure group,** E is the **bundle space**, and π_E is the **bundle projection**. **Sections** and local sections in E are defined in analogy to those in P.

Vector bundle $(E, M, \pi_E, V = K^n, G)$, $G \subset Gl(n, K)$.

Whitney sum of vector bundles: $(E \oplus E', M, \pi_E \oplus \pi_{E'}, V \oplus V', G)$.

Tensor product of vector bundles: $(E \otimes E', M, \pi_E \otimes \pi_{E'}, V \otimes V', G)$, analogously **exterior product of vector bundles.**

Tangent bundle: $T(M) = (T(M), M, \pi_T, K^m, Gl(m, K))$, $m = \dim M$, its dual is the **cotangent bundle** $T^*(M) = (T^*(M), M, \pi_{T^*}, K^m, Gl(m, K))$, both associated with the frame bundle $L(M)$.

Tensor bundle $T_{r,s}(M)$ of type (r, s) over M: tensor product of tangent and cotangent bundles, **exterior** ***r*** **bundle** $\Lambda_r^*(M)$ over M.

Vertical and horizontal spaces $T_\epsilon(E) = F_\epsilon(E) \oplus Q_\epsilon(E), \epsilon = p(f), \pi_E(\epsilon) = x$:

- Vertical space $F_\epsilon(E) = T_\epsilon(\pi_E^{-1}(x)) \approx T_f(F)$,
- Horizontal space $Q_\epsilon(E) = \pi_{f*}(Q_p) \approx Q_p \approx T_x(M), \pi_f : P \to E : p \mapsto p(f) = \{(pg, g^{-1}f) | g \in G\}$ for fixed f.

Vector bundles

Vector field: local or global section $s\colon M \supset U \to E$, $\pi_E \circ s = \mathrm{Id}_U$.

covariant derivative

$$\nabla_X F = XF, \quad \nabla_X s(x_t) = \lim_{\delta\to 0} \frac{\Phi^*_{(t+\delta,t)}(s(x_{t+\delta})) - s(x_t)}{\delta},$$

$\Phi^*_{(t+\delta,t)}$ is the parallel transport from $x_{t+\delta}$ to x_t along the path Φ in M, X is tangent to Φ at $x_t = \Phi(t)$.

Tensor bundles $(X^*_p \in T_p(L(M)),\ X_x = \pi_*(X^*_p) \in T_x(M))$

Connections on $L(M)$ are called **linear connections.**

Canonical form θ: $\mathbb{R}^m$-valued 1-form on $L(M)$, defined by

$$\theta_p(X^*_p) = u^{-1}(\pi_*(X^*_p)), \quad X^*_p \in T_p(L(M)).$$

Torsion form $\Theta = D\theta$, it depends on the linear connection form via D.

Structure equations: $d\theta = -\omega\wedge\theta + \Theta, \quad d\omega = -\omega\wedge\omega + \Omega.$

Bianchi identities: $D\Theta = \Omega\wedge\theta, \quad D\Omega = 0.$

Torsion tensor $T: \langle T, X\wedge Y\rangle = u\langle\Theta, X^*\wedge Y^*\rangle = \nabla_X Y - \nabla_Y X - [X,Y].$

Curvature tensor R: $C(\langle R, X\wedge Y\rangle\otimes Z) = u(\langle\Omega, X^*\wedge Y^*\rangle(u^{-1}Z)), \langle R, X\wedge Y\rangle = [\nabla_X, \nabla_Y] - \nabla_{[X,Y]}$.

Expressions in local coordinates $x = (x^1,\dots,x^m)$, $X_i = \sum_k X_i^k(\partial/\partial x^k)$:
Canonical form: $\theta^i = \sum_j \tilde{X}^i_j dx^j, \quad \sum_j X^k_j\tilde{X}^j_i = \sum_j \tilde{X}^k_j X^j_i = \delta^k_i,$
Connection form: $\omega = \sum_{ik}\omega^i_k E^k_i$ with base E^k_i of $\mathrm{gl}(m,\mathbb{R})$, $\omega^i_{\alpha k} = \sum_j \Gamma^i_{jk}dx^j$,

$$\Gamma^i_{\beta jk} = \sum_{lmn}\Gamma^l_{\alpha mn}(\psi_{\beta\alpha})^i_l(\psi^{-1}_{\beta\alpha})^m_j(\psi^{-1}_{\beta\alpha})^n_k + \sum_l(\psi_{\beta\alpha})^i_l(d\psi^{-1}_{\beta\alpha})^l_{jk},$$

Every set of symbols Γ^i_{jk} with this transition property defines a connection form.
Covariant derivative of a tensor field t:

$$(\nabla_X t(x))^{i_1\dots i_r}_{j_1\dots j_s} = X^k\left(\frac{\partial t^{i_1\dots i_r}_{j_1\dots j_s}}{\partial x^k} + \sum_{\mu=1}^r \Gamma^{i_\mu}_{kl} t^{i_1\dots i_{\mu-1} l i_{\mu+1}\dots i_r}_{j_1\dots j_s} - \sum_{\mu=1}^s \Gamma^l_{kj_\mu} t^{i_1\dots i_r}_{j_1\dots j_{\mu-1} l j_{\mu+1}\dots j_s}\right),$$

convention: $\left(\nabla_{\partial/\partial x^k} t(x)\right)^{i_1\dots i_r}_{j_1\dots j_s k} = t^{i_1\dots i_r}_{j_1\dots j_s;k}(x), \quad (\nabla^n t))^{i_1\dots i_r}_{j_1\dots j_s k_1\dots k_n} = t^{i_1\dots i_r}_{j_1\dots j_s;k_1;\dots;k_n},$

Torsion and curvature tensors:

$$T^i_{jk} = \Gamma^i_{jk} - \Gamma^i_{kj}, \quad R^i_{jkl} = \frac{\partial \Gamma^i_{lj}}{\partial x_k} - \frac{\partial \Gamma^i_{kj}}{\partial x^l} + \Gamma^m_{lj}\Gamma^i_{km} - \Gamma^m_{kj}\Gamma^i_{lm},$$

Geodesic: curve on which the tangent vector to it is transported parallel to itself,

$$\frac{d^2x^i}{dt^2} + \sum_{jk} \Gamma^i_{jk} \frac{dx^j}{dt}\frac{dx^k}{dt} = 0, \quad i = 1, \ldots, m.$$

Exact homotopy sequence for fiber bundles

$$\cdots \to \pi_n(F, f_0) \to \pi_n(E, \epsilon_0) \to \pi_n(M, x_0) \to \pi_{n-1}(F, f_0) \to \cdots$$

Local gauge field theories Fiber bundle with the vector space of the matter field vector as typical fiber, associated with a principal fiber bundle on the physical space-time as base space and the inner symmetry group of the gauge fields as structure group.

gauge potential	$\leftrightarrow$	local connection form
gauge covariant derivative	$\leftrightarrow$	exterior covariant derivative
gauge field	$\leftrightarrow$	local curvature form
homogeneous field equations	$\leftrightarrow$	Bianchi identities
pure gauge	$\leftrightarrow$	flat connection.

C.6 Basic Geometric Structures

See for instance [7, 8].

Metric tensor (fundamental tensor) g of type (0, 2) on a manifold M: In local coordinates

$$g = g_{ij}(\boldsymbol{x})dx^i \otimes dx^j, \quad g_{ij} = g_{ji}, \quad \det g \neq 0,$$

$$g^{ik}g_{kj} = g_{jk}g^{ki} = \delta^i_j,$$

Raising and lowering of tensor indices

$$t^{i_1 \ldots i_{r+1}}_{j_1 \ldots j_{s-1}} = g^{i_1 k} t^{i_2 \ldots i_{r+1}}_{j_1 \ldots j_{n-1} k j_n \ldots j_{s-1}}, \quad t^{i_1 \ldots i_{r-1}}_{j_1 \ldots j_{s+1}} = g_{j_1 k} t^{i_1 \ldots i_{n-1} k i_n \ldots i_{r-1}}_{j_2 \ldots j_s},$$

Inner product in tangent spaces and homogeneous tensor spaces on M:

$$(X|Y) = g_{ij}X^iY^j, \quad (t|u) = t^{i_1 \ldots i_r}_{j_1 \ldots j_s} g^{j_1 l_1} \cdots g^{j_s l_s} g_{i_1 k_1} \cdots g_{i_r k_r} u^{k_1 \ldots k_r}_{l_1 \ldots l_s}.$$

Generalized Riemannian manifold M: manifold with a metric tensor g. If g is not positive definite (has negative eigenvalues), then it is said to define an indefinite metric.

Element of the arc length: $ds = \sqrt{ds^2}, ds^2 = g_{ij}dx^i dx^j$,

Length of a path C in M: $s(C) = \int_C ds$.

Riemannian manifold M: manifold with a positive definite metric tensor g.

Euclidean geometry in the tangent space:

$$|X| = (X|X)^{1/2}, \quad \angle(X,Y) = \arccos\left(\frac{(X|Y)}{|X|\,|Y|}\right) \quad \text{for} \quad |X| \neq 0 \neq |Y|,$$

Riemannian metric on M: To every picewise smooth curve C in M a positive arc length $s(C) = \int_C ds$ is assigned.

Example of a homogeneous Riemannian manifold: Let G be a Lie group considered as a manifold, and let $\mathfrak{g}$ be its Lie algebra.

$$g_{ij}X^iY^j = g(X,Y) = -\kappa(X,Y) = -\text{ tr }(\text{ad}(X)\circ\text{ad}(Y)), \quad X,Y \in \mathfrak{g}$$

with the **Killing form** κ is an invariant metric on G.

Metric connection Γ: A linear connection on M for which $\nabla g = 0$. If it is torsion-free, it is called a **Riemannian connection** or a **Levi-Civita connection.** It is uniquely determined by g and defines a **pseudo-Riemannian geometry** for an indefinite metric and a **Riemannian geometry** for a positive definite metric.

Christoffel symbols of the Riemannian connection for g:

$$\sum_l g_{lk}\Gamma^l_{ij} = \frac{1}{2}\left(\frac{\partial g_{jk}}{\partial x^i} + \frac{\partial g_{ik}}{\partial x^j} - \frac{\partial g_{ij}}{\partial x^k}\right) = \Gamma_{ikj}, \quad \Gamma_{ikj} = \Gamma_{jki},$$

$$\frac{\partial g_{ij}}{\partial x^k} = \Gamma_{ijk} + \Gamma_{jik},$$

$$\Gamma^i_{ji} = \frac{1}{\sqrt{|\det g|}}\frac{\partial\sqrt{|\det g|}}{\partial x^j}, \quad X^i_{;i} = \frac{\partial X^i}{\partial x^i} + \Gamma^i_{ji}X^j = \frac{1}{\sqrt{|\det g|}}\frac{\partial(\sqrt{|\det g|}X^i)}{\partial x^i}.$$

Curvature tensor field R (see p. 375): $R_{ijkl} = g_{im}R^m{}_{jkl}$

$$R_{ijkl} = -R_{jikl} = -R_{ijlk}, \quad R_{ijkl} + R_{iklj} + R_{iljk} = 0, \quad R_{ijkl} = R_{klij},$$

$$R_{ijkl;m} + R_{ijlm;k} + R_{ijmk;l} = 0, \quad \text{(Bianchi identities)}.$$

Sectional curvature: Gaussian curvature of $\mathcal{E} \subset M$ formed by geodesics through x and tangent to E,

$$K(x,E) = -\frac{R(X,Y,X,Y)}{|X|^2|Y|^2 - (X|Y)^2}, \quad E = \text{span}\{X,Y\}.$$

Ricci tensor: The only non-zero contraction of the Riemann curvature tensor,

$$R_{\mu\nu} = g^{\kappa\lambda} R_{\kappa\mu\lambda\nu}, \quad R_{\mu\nu} = R_{\nu\mu}.$$

Curvature scalar: $R = g^{\mu\nu} R_{\mu\nu}$.

References

1. Lang, S.: Algebra. Addison-Wesley, Reading (1965)
2. Reed, M., Simon, B.: Methods of Modern Mathematical Physics. Vol. I: Functional Analysis. Academic Press, New York (1973)
3. Steenrod, N.E.: The Topology of Fiber Bundles, 6th ed. Princeton University Press, Princeton (1967)
4. Warner, F.W.: Foundations of Differentiable Manifolds and Lie Groups. Springer, New York (1983)
5. Pontrjagin, L.S.: Topological Groups, 2nd ed. Gordon and Breach, New York (1966)
6. Fuchs, J.: Affine Lie Algebras and Quantum Groups. Cambridge University Press, Cambridge (1992)
7. Kobayashi, S., Nomizu, K.: Foundations of Differential Geometry (Interscience, New York, 1963 and 1969), Vol. I and II.
8. Chern, S.S., Chen, W.H., Lam, K.S.: Lectures on Differential Geometry. World Scientific, Singapore (2000)

List of Symbols

Index